W9-BNY-818

ALSO BY JOYCE E. CHAPLIN

Benjamin Franklin's Political Arithmetic:
A Materialist View of Humanity

The First Scientific American:
Benjamin Franklin and the Pursuit of Genius

Subject Matter: Technology, the Body, and Science
on the Anglo-American Frontier, 1500–1676

An Anxious Pursuit: Agricultural Innovation and
Modernity in the Lower South, 1730–1815

Round About

Circumnavigation

Joyce E. Chaplin

the Earth

from Magellan to Orbit

Simon & Schuster

New York   London   Toronto   Sydney   New Delhi

NEW HANOVER COUNTY
PUBLIC LIBRARY
201 CHESTNUT STREET
WILMINGTON, NC 28401

 Simon & Schuster
1230 Avenue of the Americas
New York, NY 10020

Copyright © 2012 by Joyce E. Chaplin

All rights reserved, including the right to reproduce this book or
portions thereof in any form whatsoever. For information address
Simon & Schuster Subsidiary Rights Department,
1230 Avenue of the Americas, New York, NY 10020.

First Simon & Schuster hardcover edition October 2012

SIMON & SCHUSTER and colophon are registered trademarks
of Simon & Schuster, Inc.

For information about special discounts for bulk purchases,
please contact Simon & Schuster Special Sales at
1-866-506-1949 or business@simonandschuster.com.

The Simon & Schuster Speakers Bureau can bring authors
to your live event. For more information or to book an event,
contact the Simon & Schuster Speakers Bureau at
1-866-248-3049 or visit our website at www.simonspeakers.com.

Maps by Jeffrey L. Ward

Manufactured in the United States of America

10 9 8 7 6 5 4 3 2 1

Library of Congress Cataloging-in-Publication Data
Chaplin, Joyce E.
  Round about the earth : circumnavigation from Magellan to orbit / Joyce E. Chaplin.
    p. cm.
  Includes bibliographical references and index.
  1. Voyages around the world—History.   I. Title.
  G439.C44 2012
    910.4'1—dc23   2012016459
ISBN 978-1-4165-9619-6
ISBN 978-1-4391-0006-6 (ebook)

NEW HANOVER COUNTY
PUBLIC LIBRARY
201 CHESTNUT STREET
WILMINGTON, NC 28401

This book is dedicated to
two circumnavigators

Enrique de Malacca
and
Laika

who reluctantly made history

Contents

List of Maps

Prologue

I'll put a girdle round about the Earth
In forty minutes . . .
    —*Puck,* A Midsummer Night's Dream

Imagine that you have purchased, to accompany this book, a model of a theater in which you can stage your very own around-the-world voyage. The whole setup will require special delivery because, for some reason, the theater, which is round, turns out to be about 8.25 meters in circumference (just under 9 feet across). That is only the first of several Alice-in-Wonderland surprises, as you discover when you enter the theater, which seems empty. Luckily, there is an envelope pinned to the theater door which contains a map, with "X" marking the spot where you can find the most important piece of scenery, the small ship that will make a staged circumnavigation. Once you locate the X, you must remove a magnifying glass (helpfully provided in the envelope) to see the ship. It is the size of the head of a pin and scattered with what look like bits of dust. Nearby is a microscope, and only with its assistance can you make out that the dust particles are actually tiny dolls that represent the ship's sailors.

The theater kit has these odd components and dimensions because, correctly proportioned to each other, the theater is the Earth, the pin-head ship is the *Victoria,* the first ship to go around the world, and the near-invisible dolls within her represent sailors, the angels that dance on the head of this particular pin. Given the high mortality rates during most of the history of around-the-world travel, sailors were indeed angels in the making. Their tininess, and that of their ship, make a circumnavigation of the enormous globe seem impossible. And yet it happened, over and over, hundreds of thousands of times, and even more circumnavigations are taking place right now, as you read.

With those repeated embraces of the globe, from 1519 onward, human beings have established what is now a nearly five-hundred-year history of going around the world. It is the longest tradition of a human activity done on a planetary scale. Around-the-world travelers make a grand gesture, as big as the physical world itself, even though they are individually so small that the huge global stage on which they act makes them hard to find.

I found this book in the middle of the Atlantic Ocean. Six years ago, in St. George's, Bermuda, I embarked on a 140-foot sailing ship, the Sea Education Association's SSV *Corwith Cramer*. I would be at sea for three weeks, away from telephone, Internet, and physical libraries. Yet I was in the middle of a research project on Benjamin Franklin that required me to read material in French. I decided to use my time at sea to revive my French by reading a novel in that language. The book I chose was a small paperback edition of Jules Verne's *Le Tour du monde en quatre-vingts jours,* or *Around the World in Eighty Days,* first published as a newspaper serial in 1872. When I wasn't on watch or otherwise busy, I slowly made my way through the book.

My French was good enough—to my surprise—that I actually enjoyed the story and, as a historian, I appreciated its period detail, especially the nature of the bet that sends Verne's protagonist, the Englishman Phileas Fogg, racing around the world. At his London club, Fogg remarks that scheduled travel services could take a person around the globe in a period of eighty days. Prove it, the club men challenge him, and he's off. That eighty-day measure was only conceivable by the late nineteenth century. In the age of sail, getting around the world had taken months or years. (The speed of my sailing ship would have lost Fogg his wager.) It was the invention of steampower, but also the creation of regimented European empires around the globe, the opening of the Suez Canal, and the emergence of commercial travel services that together made it just possible, by the 1870s, to do the global circuit in eighty days.

The second thing that impressed me about Verne's story was how the material developments that sped up global travel required a dramatically increased use of natural resources. When Fogg leaves London, he takes his new valet (and invaluable comic foil) Passepartout. The two men board a night train which has scarcely departed London when Passepartout lets out "a real cry of despair":

> "... in the rush ... my state of confusion ... I forgot ... to switch off the gas lamp in my bedroom."

"Well, my dear fellow," Phileas Fogg replied coldly, "you'll be paying the bill." The gas lamp is the novel's running joke. True, it is a small part of the journey's cost, but we present-day readers of Verne quickly realize that the joke is on us. We are, notoriously, the first generation that has realized what the planetary bill for centuries of burning fossil fuel is going to be.[1]

In Verne's era, coal was a costly but essential part of modern progress. Yet Fogg's steam-powered exploits, set at the height of European imperialism, represent a phase of the past that truly is history, over and done with. Airplanes have replaced the coal-burning engines and ships that hurtled Fogg around the world; the empires that protected some people at the expense of others have been replaced with other political regimes. It's now difficult to cross the surface of the world in eighty days, though easy to fly around it in hours, if you can afford the ticket.

Back on land, I looked for a history of around-the-world travel. There was none. Two books have given narrative histories of circumnavigations, but only in the age of sail; a third book examines some nineteenth-century examples, among other long-distance voyages; a fourth looks at small-boat circumnavigations in the early twentieth century. That's not much. For that reason, there is no separate Library of Congress heading for histories of around-the-world travel—the histories are simply lumped with first-person accounts of the travelers. There are many more studies of individual around-the-world travels, especially the really famous attempts by Magellan, Cook, Earhart, Gagarin, Chichester, and others. There are analyses of fictional circumnavigators, including Fogg, and even Shakespeare's Puck. And there is Raymond John Howgego's incomparable *Encyclopedia of Exploration*, which includes historical circumnavigations among its thousands of entries. Finally, there are other encyclopedias that chronicle circumnavigations or orbital travels or, especially in electronic form, just list the journeys themselves.[2]

All of these works take for granted that an around-the-world voyage represents a distinct human activity, something unlike other voyages or expeditions. But none of them explains *why* they are distinctive. Why do they matter?

The answer is not obvious. To find out what is different about around-the-world voyages requires setting aside everything they might have in common with other kinds of travel and exploration. Circumnavigators made geographic discoveries; but so did other explorers. They changed the world map, but other expeditions did too. They encountered lands and people new to them; so did many others. They expanded and defended empires—many actors did. Even if circumnavigators did more of these activities—more geographic discoveries, more cultural encounters—that is a difference in quantity, not quality.

The temptation is to find an encyclopedic solution, to inventory every around-the-world journey that took place, with details of what happened on each. But a work that simply listed all of these journeys, let alone details of what happened on them, would take a great many volumes, or a vast electronic database, yet still not establish why they are special.

They are special because any voyager who goes all the way around the world thinks of himself or herself on a planetary scale, as an actor on a stage the size of the world. This is unique. No other form of travel, and, really, hardly any other human experience, is truly planet-encompassing. In fact, an around-the-world voyage is distinctively planetary in three ways, as measured by time, space, and death.

Around-the-world travel is *time* travel. Verne's *Around the World in Eighty Days* has as its plot twist Fogg's eastward progress around the globe, through which he gains a day that wins him his bet. Because of what is now called the International Date Line, the day gained or lost during an around-the-world voyage is a fundamental and unique characteristic of such a journey. Any traveler changes his or her place on the globe, but only circumnavigators change their date on the calendar. That experience reorients them in relation to the Sun, timekeeper of the hours and days on Earth and throughout the solar system. Circumnavigators were the first people to discover, by mind-blowing experience, that time is not universal but relative.

The *space* over which a circumnavigator must travel is also distinctive because it represents the whole world, which is painfully larger than any of the individual humans who might contemplate getting around it. Shakespeare's Puck may have been able to zip round about the Earth with ease, but humans had to go more slowly, and they would always need a protected space around their small selves, whether physical, imagined, or both.

It is conventional in many cultures to consider the human body as a microcosm, a smaller, metaphorical version of the world or cosmos. So too have ships been considered microcosms, little wooden worlds that ride the waves. To put a body around the world on a ship (or in another vehicle) puts these three worlds into a dynamic relationship to each other. The mismatch in size makes the encounter both comic and cosmic, as with the ill-matched theater kit that represents world, ship, and sailors in proportion to each other, with the sailors reduced to specks of dust.

Because of the size of the planet, and the inadequacy of early sailing ships to cover global distances while protecting those within them, sailors quickly learned that a circumnavigation meant *death*. For the first 250 years, more people died than survived circumnavigations. The planet simply shrugged them off. Many forms of travel and exploration were badly dangerous, but around-the-world travel was the most consistently bad. Only by the end of the eighteenth century did circumnavigations begin to have lower mortality rates, even as their breathtaking risks have lingered as part of their mystique.

And yet mere mortals did manage to get their tiny bodies around the vast planet in vessels that, even if bigger than the *Victoria*, were (and are) strikingly small. To commemorate their achievement, the survivors established traditions for their experience. It did not happen at once—the first around-the-world voyage was too surprising, and only repeated successes teased out internationally observed conventions about what such journeys meant. Circumnavigators were represented in certain kinds of portraits, they were granted characteristic coats of arms, and their journeys were represented on maps and globes in particular ways.

Above all, the stories they told about their voyages fell into certain patterns, especially by stressing the elements of time travel, planetary distance, and frequent fatalities. Circumnavigators knew they were generating a tradition. They read each others' accounts as models for their own voyages and narratives: William Dampier read Sir Francis Drake; George Anson read Dampier; Louis-Antoine de Bougainville read Anson; Adam Johann von Krusenstern read Bougainville; Charles Darwin read Krusenstern—and so on, even to the present day, when around-the-world sailors still read Joshua Slocum (who read Darwin). Even people who would eventually go around the world by other means—steamships, railroads, bicycles, airplanes, spaceships—would refer back to the longer maritime tradition on which they thought they were building.

As a result, circumnavigators have generated nearly half a millennium's worth of evidence of humanity's direct, tangible, and conscious connection to something usually perceived in the abstract, the whole Earth. Of course, humans have had a relationship to the planet for as long as they have existed. And yet consciousness of that relationship is thought to be rather recent. Most people in the past are not supposed to have been aware of themselves on a planetary scale, except metaphorically. They are otherwise supposed to have thought small, to have been tucked into cosy patches of a larger landscape, unable even to give consistent labels to continents and oceans, and incapable of conceiving of their actions in relation to the natural world on a long time scale. It was allegedly only by the nineteenth century that mastery of the oceans and penetration into continents gave western Europeans the world's first version of a planetary consciousness, augmented by developments in science and geography, and complete with an understanding of enormous spaces and the swift yet long passage of geological time. And it was only supposedly with the famous Apollo 8 photographs of the Earth, taken in 1968, that humans first truly grasped their place on the Earth as a whole.[3]

Today, we are hyper-aware of our connection to the planet, given our ongoing environmental crisis, most of which can be attributed to human actions in generating matter that overheats the atmosphere, or clogs land, air, and water. We use our planetary consciousness both to praise and criticize ourselves, to emphasize our greater scientific wisdom about the Earth, for example, but also to stress our unprecented transformation of it. Plus there is our sense of globalization. Surely we must have a greater awareness of the planet, given the globalized culture in which we live, the global connections we know we now have with people all over the world?

But it may just be the case that no one has looked hard enough to find a longer history of planetary consciousness. Globalization has hogged all the attention. Global is social—it implies the social relations that extend over the globe. In contrast, planetary is physical, implying the physical planet itself. Far more studies have focused on the former than the latter. That is because human-to-human interactions have been historians' major focus. Only recently have human relations with the non-human parts of nature been put into dialogue with those human relationships; only recently have scholars begun to reread historical documents to discover our past sense of our place within nature. Circumnavigators' long and self-aware tradition of engagement with the planet questions our sense of uniqueness and may teach us something worth knowing about why we think of the Earth the way we do.[4]

Toward that end, this book defines around-the-world travel as a *geodrama*, from the Greek for "Earth" (*Ge* or *Gaia*) and for "action" (*drama*). Within the European countries that sponsored the first circumnavigations, there was an established tradition of considering the world as a theater: *theatrum mundi*. It was an ancient Greek idea, sustained through Roman antiquity and the Renaissance, and exemplified in Shakespeare's claim that "all the world's a stage." The *theatrum mundi* was a metaphor, but around-the-world travelers made it a reality by presenting themselves as actors on a stage of planetary dimensions. Over time, circumnavigations would be presented as dramatic entertainments, first in print, then on stage, and later in film.

Geodrama is different from geography, meaning depictions of the Earth made by writing (*graphos*). While geography engages the human

eye, hand, and mind, geodrama requires all of a human being, the entire body and its range of physical experiences, in relation to Gaia, the
Earth. That whole-body experience of the whole Earth is well documented in accounts of circumnavigation, which describe what it felt
like, from agonizing to exhilarating. Most people never go around the
world but, by now, almost everyone has some idea of the big statement
that such a journey makes.

For that reason, published, first-person accounts of circumnavigators are this book's principal sources. Again, the point is not to include
every available account, as in an encyclopedia, but to examine some of
the most visible in order to show how to read any one of the others.
And, frankly, the more famous and influential the account the better.
The famous narratives disseminated a sense of what it was like to go
around the world, even for people who would never make the journey
themselves. Together, the accounts constitute the longest and most
sustained way in which people have been able to consider themselves
as actors within a geodrama, even as the drama has changed over time.

The changes can be understood as three acts in the drama, three
phases in human beings' comprehension of themselves as actors on
the physical planet. In the first act, which lasted from Magellan's departure on the first circumnavigation to James Cook's death in Hawai'i—
that is, from 1519 to 1779—mariners who went around the world did
so in *fear*. It was reasonable for them to be fearful, given the dangers of
such a voyage in the age of sail. In this initial phase, the longest in the
history of circumnavigation, death prevailed. Humans might take on
the planet, but the planet fought them every inch of the way.

From the 1780s until the 1920s, however, travelers who made their
way around the world did so with a striking *confidence* that they could
survive the experience. That was because Western societies had generated technology and political networks that seemed to have conquered
the globe. It was not only possible to go around the world, but it had
become a popular pastime. Representations of doing a circumnavigation became playful, enticing, joyous. There were costs, not all of them
hidden, but they seemed outweighed by the glories of making an easy
swing around the planet. One result of the new optimism was the generation of a new meaning for "around the world": to blanket the Earth,

rather than simply girdle it, even though confidence about the blanketing depended on prior success with the girdling.

Over the twentieth century, and now into the twenty-first, the confidence has given way to *doubt*. Technologically newer forms of travel, especially airplanes and rocket-propelled space capsules, revived the sense of extreme danger that had faded during the relatively safer nineteenth century. Equally, it is now clear that imperialism had smoothed the way for most earlier circumnavigators, under political and social conditions that would be unwise and unjust to perpetuate, let alone re-create. Above all, there is a growing sense that the planet is again beginning to bite back, now that the environmental costs of planetary domination have begun to haunt us.

We live with all three legacies of around-the-world travel: a re-emerging fear that the planet could simply shrug us off; continuing confidence that we might be able to generate technologies and political alliances to dominate the planet; but doubt that it is always wise to do so. It is especially apparent that the characteristic confidence of the long nineteenth century was the shortest of planetary experiences, yet has been the most difficult to relinquish. Our current doubts seem to be taking us back to the fears of the early modern period, a circular return that matches the swings around the planet that themselves went through the three acts of geodrama.

But there were always more hopeful elements to the story. Those bright moments matter too, and they make clear that the human past is as complicated and contradictory as its present condition, whether seen on a small scale or a large one, even the largest of all, a geodrama in three acts.

Act One

Fear

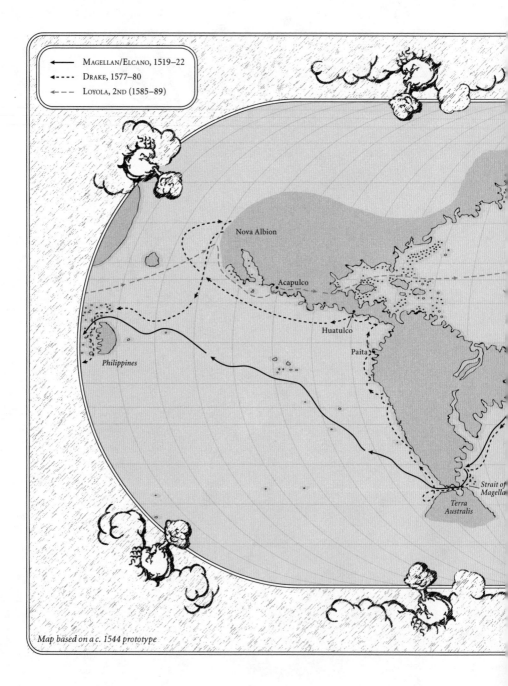

Map based on a c. 1544 prototype

MAGELLAN/ELCANO, 1519–22
DRAKE, 1577–80
LOYOLA, 2ND (1585–89)

Nova Albion

Acapulco

Huatulco

Philippines

Paita

Strait of
Magellan

Terra
Australis

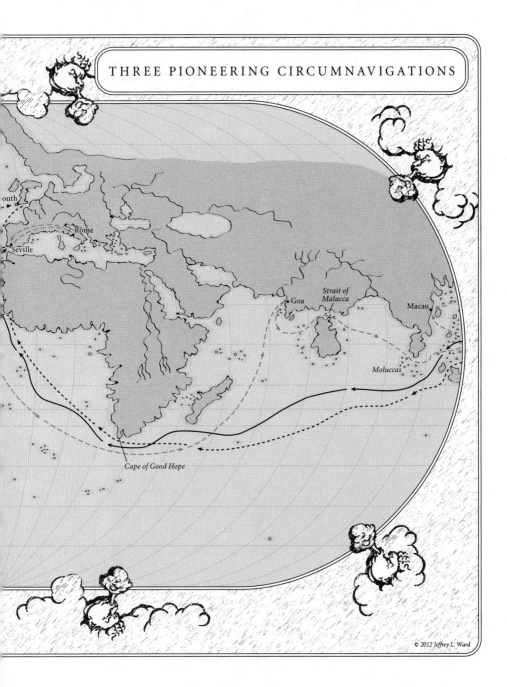

THREE PIONEERING CIRCUMNAVIGATIONS

outh

Rome

Seville

Goa

Strait of
Malacca

Macau

Moluccas

Cape of Good Hope

© 2012 Jeffrey L. Ward

Chapter 1

Magellan agonistes

*M*any ancient legends feature bold voyagers: Ka-ulu in the Pacific, Odysseus in the Mediterranean, Sinbad in the Seven Seas, Sun Wukong in China and Japan. And several early travelers undertook voyages that became legendary, including the Italian Marco Polo and the Arab Ibn Battuta. But only one person before 1519 was said to have tried to go around a spherical world. It's just as well he's a mythical character, given what happened.[1]

In Greek legend, the gods cannot resist human females and one result is Phaeton, son of Helios, the Sun, and a mortal woman. When he is grown, Phaeton visits his father, who sits in state, surrounded by the intervals of time—"Day, Month, Year . . . and the Hours"—that he alone determines. Phaeton asks to guide Helios' solar chariot around the world for a day. He gets his wish, but cannot control the chariot's horses. They pull the burning car far off course, over the icy poles and close to the Earth. Forests ignite, populations perish, dormant Arctic monsters thaw and shudder into life. When Gaia begins to crack open from the heat, she begs Zeus, king of the gods, to stop Phaeton. Zeus flings a thunderbolt and knocks the youth from the chariot. Hair afire, Phaeton screams like a meteor into the Eridanus River and drowns. Nymphs bury him beneath a stone engraved: "Here Phaethon lies: his father's car he tried—/Though proved too weak, he greatly daring died."[2]

The dramatic story made good reading and retelling through the

Middle Ages and into the Renaissance, especially the version in Ovid's *Metamorphoses*. To warn all mortals that the lesson applies to them, Ovid twice compares the Sun's chariot to a ship riding the waves. Death awaits anyone who dares to imitate the celestial bodies that circle the Earth, which was presumed to be the center of the universe.

Little wonder that the first around-the-world voyage was unintended—and deadly. Fernão de Magalhães e Sousa, better known as Ferdinand Magellan, is famous for something he never did. Ask anyone, "Who was the first person to go around the world?" and they will probably say, "Ferdinand Magellan," though he never did, and never meant to. His agonizing and unplanned circumnavigation killed him and most of his men. He made going around the world seem possible, although extremely difficult, and therefore heroic. Ironically, he died while trying to reach the lands of the wondrous spices that Europeans assumed would bless them with health and long life.

Cloves, nutmeg, mace, cinnamon, ginger, cardamom: in Magellan's day, a well-stocked European kitchen smelled like Christmas all year round. Disregard the old slurs about medieval cookery. Europeans did not eat spices to mask the flavor of rotting meat—they ate spices because they adored them. They consumed all the spices now associated with holiday fare, plus many others. Medieval and early modern recipes recommended spicing everything, from fish pies to love potions. The amounts were stupendous; one cook specified several hundred pounds of spices for a grand banquet. Today, gingerbread and nutmeg-dusted eggnog faintly hint at the early modern palate, as do deviled ham, mulled wine, and steak sauce tangy with cloves. Clearly, it was an acquired taste. But acquired from where? [3]

Europeans thought that spices tasted of paradise. It was no restaurant reviewer's metaphor. Paradise meant just that, the Eden from which humanity had been expelled. That primordial event always influenced Europeans' view of themselves and of the physical world around them, a world that they saw as both a real, everyday place and a mystical entity embedded with cosmic truths.

To interpret the world, they combined Christian beliefs with Greek philosophy. Like us, they assumed the world was a globe—the idea that they thought the Earth was flat is a myth invented in the eighteenth century. Their major difference from us was their belief in a geocentric cosmos, with an unmoving Earth at the center of all other heavenly bodies. Whereas the stars and planets were perfect and eternal, revolving in crystalline spheres around the Earth, the globe at the center was imperfect. Its four basic elements—earth, air, fire, and water—were in constant flux and decay. Only Eden escaped decay, and by leaving it, Adam and Eve had forfeited immortality. Their progeny were destined to sicken and die because their bodies were microcosms of the imperfect world, composed of its base elements.[4]

Humans differed from each other because of the globe's own physical differences. Ancient geographers had laid a grid over the spherical globe, to divide it into north-south degrees of latitude and east-west degrees of longitude. The lines dividing the world from east to west lack any natural foundation, in contrast to latitudes, which are natural divisions—the Sun gives north-south latitudes their differing temperatures, with the hottest on the equator. Those perceived differences guided European assumptions about their physical bodies. They thought each body had deep connections to its native climate and suffered if removed from it. Temperate climates produced normal humans (especially Europeans). They assumed that places with extreme climates had monstrous races, with the heads of dogs or no heads whatsoever, but might also produce wondrous things, such as the gold and gems found in hot places.[5]

The most wonderful place was paradise itself, source of the spices that prolonged life. An illustration made around 1490 shows a man and woman fishing spices from a river presumed to flow from paradise. Medieval cosmographers positioned Eden to the east of the Holy Land, where it was guarded by the fiery angels who had presided over Adam and Eve's expulsion. More recently, paradise was blocked by infidels. Islamic merchants brought spices and other exotic goods into Europe via the Indian Ocean and Red Sea or else over the Silk Road. Only a handful of European merchants, mostly Italians, had the contacts with Muslim traders necessary to get spices. The tiny trade commanded high prices. In the early 1400s, merchants in Venice sold cloves at a 72 percent markup and nutmegs at 400 percent.[6]

Circumventing those Italian and Muslim traders was the main goal
for the major European voyages of exploration, including Magellan's.
But the late medieval map of the world, derived from the Greek philosopher Ptolemy, did not offer obvious alternate routes to Asia. That map
(see illustration 1) presents an Ecumene (from the Greek *oikoumene*),
the inhabited or known world. Its center is the Mediterranean, surrounded by Europe and Asia, plus the upper part of Africa, a real world
in which actual travel took place—where Polo and Battuta had done
their respective travels.

To go beyond the Ecumene would be a severe test of seamanship.
Navigation was called the "haven-finding art" because its whole point
was to find land early and often. We think of the globe as a series of
landmasses surrounded by water. But except in Polynesia and East
Asia, mariners assumed seas were landlocked, as with the Mediterranean and the known parts of the Indian Ocean. European sailors relied
on portolan charts that were surveys of coastline, not charts of open
water. Beyond the Ecumene, a great Ocean Sea flowed freely, but sailors
hesitated to enter it. They hugged the coastline, reluctant to meet the
monsters and whirlpools said to lurk beyond.

To navigate was to interpret a local ocean, using the Earth and sky.
Sailors read patterns of wind and water to determine their proximity either to deep sea or impending shoreline. They learned to memorize the
appearance of land formations along coastlines. Some of them could
make celestial readings, using the stars to measure the Earth's surface
and therefore distances between places, as if the heavens were a coded
map, reflecting and reflected in the seas below.[7]

In this way, each ocean had a pool of human knowledge about it.
Few of these pools ran into each other. The Polynesians who mastered
the Pacific did not travel beyond it. People in Asia knew a lot about
the Indian Ocean but did not share their wisdom with others. Even
at the height of Asian expansion, knowledge of the seas only went so
far. When the Chinese explorer Zheng He sailed the boundaries of the
Indian Ocean from 1405 to 1433, even to the southern tip of Africa,
he turned back; the Chinese were content with the seas and trade they
already controlled. Nor did the Norse and English mariners who found
their way into the North Atlantic and North America at the end of the
Middle Ages publicize their discoveries.

Yet some Europeans began to wonder whether two routes through the Ocean Sea could get them to Asia. Maybe somebody could sail down south past Africa's west coast to see whether there was a way into the Indian Ocean. Or somebody might sail west from Europe to Asia. Either route raised theoretical questions about the Earth. Was the globe mostly land, or equally balanced between land and water? Were seas truly landlocked, or did the Ocean Sea connect them? Were land and water distributed equally over the globe, with the Ecumene balanced by a vast southern continent?

And then the biggest question of all: how big was the world? The ancient mathematician Erathosthenes had, circa 240 BCE, calculated the globe's circumference at 252,000 stadia, between 39,690 and 46,620 kilometers. (It is now estimated at 40,008 kilometers.) But that calculation had been lost and subsequent ones disagreed with each other. Longitude divides the Earth into 360 degrees of meridian, for example, but medieval estimates of a degree along the equator ranged from $15\frac{2}{3}$ to $17\frac{1}{2}$ leagues, or 54 to 60 miles. The true size is about 69 miles. The medieval preference for flat maps over globes did not indicate belief in a flat world—it was a strategy that avoided having to give global particulars. (Celestial globes were common, in contrast.) Even rolled into a cylinder to resemble a globe, a world map did not estimate the size of the world or indicate its nature outside the Ecumene.[8]

Many a character in a fairy tale is warned, "Be bold, be bold, but not too bold." According to that classification, Bartolomeu Dias was bold, Christopher Columbus and Vasco da Gama were too, and Ferdinand Magellan was perhaps too bold. They all sailed toward the back of the rolled-up world map, but Magellan sailed through it; they all had to negotiate with populations unknown to Europeans, though Magellan most dangerously; they all underestimated the size of the world, but Magellan disastrously.

It was no accident that these four explorers worked for Spain or Portugal. Living together on the Iberian Peninsula, envious of the Italians whose own peninsula thrust a triumphant boot into the richly interconnected Mediterranean, the Spanish and Portuguese were like cousins who squabbled over a meager patrimony. That bleak treasure was their access to the Atlantic, a very unpromising part of the Ocean Sea, but an entryway, at least, to the outer world. To reach the spice-laden Indies, the Portuguese headed east and the Spanish west, though their point of intersection on the opposite side of the globe would be productively contentious.

Dias was the first to demonstrate an eastbound alternate route to Asia, though he stopped short of going there himself. His discovery extended Portugal's claims to Africa on its northern and western coasts, as well as the islands of Madeira and the Azores. In each place, the Portuguese were quick to use force, often against Muslim populations. In that spirit, Dias continued down Africa, then just around its southern tip. He was bold, though for the most part he followed the coastline, the preferred method of navigation. He returned in 1488 with portolan knowledge about the African coast and about a part of the Ocean Sea where a mariner no longer followed land south but could turn eastward into open water.[9]

The Italian mariner Christopher Columbus tried the other way to the spices, westward across the Ocean Sea. To demonstrate that route's feasibility, Columbus used a recent projection of the Earth's surface that calculated its circumference at 19,000 miles (it is actually 24,900) and therefore the distance from the Canary Islands to Japan at 2,400 nautical miles (in truth, 10,600). He also argued that Asia projected far to the east, putting the Ocean Sea on the small side. The monarchs of Aragon and Castile agreed to bankroll him. Off he went in 1492, making the first of four voyages to "another world," as he called it, insisting that the Caribbean islands and South American mainland must be parts of Asia, and close to paradise. Pausing at the mouth of Venezuela's Orinoco River, Columbus piously retreated, believing it was one of Eden's mystic rivers.[10]

The monarchs of Castile and Aragon were famous for their piety too, but they preferred to think of Columbus's discoveries as real places that happened to be unclaimed by the Portuguese. In 1493, Pope Alexander VI (an obliging Spaniard) issued a declaration or bull that drew a *raya* or line, pole to pole, 100 leagues west and south of the Azores, and granted to Spain all lands beyond the line.

Most straight lines drawn over the world are products of the human imagination, and this was no exception. The *raya* resembled a line of longitude, which lacked the naturalness of the latitudes that paralleled the equator. Nor was it clear whether the *raya* (drawn on a flat map) should run through the poles and continue down the back of the globe. And if Pope Alexander had simply wanted to keep the peace between Spain and Portugal, he failed. João II of Portugal protested that the papal bull failed to mention him. Spain and Portugal subsequently negotiated a treaty at Tordesillas in 1494 that drew a new pole-to-pole line slightly further west to guarantee that Portugal would get part of the coast Columbus had found (eventually Brazil) and retain its outposts in the Azores and Africa. This line designated spheres of interest in the Atlantic, not over the whole globe.[11]

Yet the Portuguese worried that Spain might in future be able to claim what could turn out to be part of Asia. They wanted a counter-claim based on trade routes that approached the lands of the spices from the other direction. At this point, most Europeans accepted that these lands were actual places, exotic ones, but not paradise. The new generation of explorers questioned the existence of monstrous races, for example, though they continued to associate heat with monstrous wealth, and worried that Europeans who went to such places risked ill-health if not death.

The oldest surviving world globe, made in Nuremberg by Martin Behaim in 1492, reinforced the new view that spices came from faraway places, not paradise. Behaim had lived in Portugal as royal cartographer and knew about the latest Portuguese discoveries around Africa. Those details embellish the map which he fit around a globe or *Erdapfel* (earth-apple). Columbus would not return until 1493, so Behaim supplied the ocean between Asia and Europe with a pre-Columbian wealth of islands, from Japan to the Canaries, or so-called Fortunate Isles. He then added notes. Spices came from particular Asian islands, he wrote, and were expensive because of the distance they traveled to Europe. They passed through twelve transaction zones, each with its own merchants and customs officials who added a markup—"No wonder spices for us cost their weight in gold."[12]

With their chain of outposts along Africa, the Portuguese were uniquely positioned to avoid the costly markups. Following the line that Dias had traced along Africa, Vasco da Gama headed to India in 1497. Once he ran out of usable coastline, at Malindi in present-day

Kenya, he had to venture out into the Indian Ocean, as little known to Europeans as the open Atlantic had been to Columbus. But da Gama had something Columbus had lacked: a local navigator. He hired a pilot in Malindi and that unnamed man guided him to Calicut.[13]

The separate pools of knowledge about different oceans had begun to run together, though almost always under circumstances intended to expand the boundaries of European empires, whether for settlement or for trade. We may well wonder what da Gama's pilot thought he was doing. Portuguese "trade" often resembled piracy—maritime people around the Indian Ocean regarded the newcomers with suspicion.

And, as da Gama would discover, learning a mere route from someone was inadequate without that someone's accumulated experience with winds, reefs, and currents. Somehow, da Gama never heard that sailing against the Indian Ocean's prevailing monsoons was foolish (maybe a gift of silence from his contacts), or perhaps he ignored any warning. Although monsoon winds had whisked him to India in a brisk 23 days, the recrossing back to Malindi against the wind took 123. Without adequate provisions and water, he lost over half of his 170 men, many of them to sea scurvy, fully described for the first time during his voyage. Only two of the four Portuguese ships returned in 1499.[14]

Two parts of the Ocean Sea, the Atlantic and Indian oceans, were now connected. The Portuguese established a trading empire that outperformed the older Muslim-Italian circuits. On his second voyage to India (1502–03), da Gama hauled back 1,700 tons of spices, roughly the amount Venetian traders slowly accumulated from their Arab contacts over an average year. He sold them at an estimated 400 percent profit. From Calicut, the Portuguese entered Goa, where they used a Hindu ally to undermine the Muslim sultanate and establish themselves. They also discovered, and began to tell the rest of Europe, that Asia was not a marginal and mystical place, back of beyond, east of Eden, on the edge of nowhere. Rather, India in particular was, as one economic historian has put it, "on the way to everywhere." The human worlds of the Indian Ocean were richer and more densely interconnected than anywhere else on Earth.[15]

While the Portuguese were now close to the heart of the world's riches, the Spanish seemed to be on its far fringes. It was even possible, despite Columbus's claims, that they were not in Asia. In his celebrated 1507 world map, Martin Waldseemüller said they weren't even close. He called the new lands "America" to honor the explorer, Amerigo Vespucci, who contradicted Columbus: this was not part of Asia but a new world. Waldseemüller presented Florida and the eastern coast of South America as large islands, with the West Indies floating alongside, all entirely distinct from Asia. Thrilling stuff, historically momentous, and confirmation that the Spanish settlements were nowhere near the lands of the spices. Score one for the Portuguese—for the moment.[16]

Magellan had participated in several of the conquests that moved his native kingdom of Portugal ever east. Surviving portraits of him (none done from life) show a man with dark eyes, dark beard, and the dark garb of a minor nobleman, indistinguishable from thousands of his Mediterranean contemporaries. But the record of his deeds sketches an excellent portrait of the man. It is no good wishing he were a nice person; he was more or less a ruthless thug. (The first commander of a circumnavigation who resembled a nice person would not draw breath until 1723—you will meet him in chapter 4.) Magellan had helped to subjugate parts of East Africa and India. Wounded by a javelin in Morocco, he acquired a limp without losing status as a fighter. But he also gained a reputation for prickly self-regard, on several occasions arguing for rights, authority, and remuneration beyond what was obviously his. He wanted more and became ever more creative about getting it.[17]

He saw a great opportunity in Malacca, rich and strategic possession of the sultans who ruled the Malaysian Peninsula in Southeast Asia. Magellan accompanied a 1509 mission to Malacca that was abruptly expelled. He then joined the outright invasion of 1511, led by Afonso de Albuquerque, veteran of the Portuguese invasion of India. After a six-week siege, Malacca fell. "An infinite number of Moslems, men, women and children died by the sword," Albuquerque claimed; "nobody's life was spared."[18]

It was a boast and it was a lie. The victors destroyed Malacca selectively, preserving booty, including Muslim captives. That was how Magellan gained a slave known only as Enrique, who was probably too young to have borne arms against the invaders. No image or description of Enrique survives, but he must have had some quality that made him valuable. His grasp of the region's language and geography (he was said to be from Sumatra, so perhaps knew about lands beyond Malacca) was probably what interested Magellan, who now possessed knowledge of Asia in human form.[19]

Similar knowledge was promised to the king of Portugal, including a "great map belonging to a Javanese pilot, which showed the Cape of Good Hope, Portugal and the territory of Brazil, the Red Sea and the Persian Gulf, and the spice islands." A veritable diagram of Portugal's expanding empire, the map was almost too good to be true. And it was promptly lost, along with other documents and treasure, on the *Flor de la Mar*, which sank off Sumatra. Albuquerque assured the king that he had a copy of the most valuable parts of the map, including,

> the route your ships should follow to reach the spice islands, where the gold mines are located and the islands of Java and Banda, where the nutmeg and mace come from, and the territory of the king of Siam. You will also see the extent of Chinese navigation and where they return to and the point beyond which they do not sail.

But where was that point, exactly?[20]

To find out, the Portuguese pushed farther east and the Spanish west. In 1511, Albuquerque ordered António de Abreu to lead an expedition to the Spice Islands. One of Magellan's friends and a fellow veteran of the invasion of India, Francisco Serrão, went with him. Abreu sailed first through the Java Sea and then the Banda Sea. He and his crew almost certainly glimpsed the western edge of the Pacific Ocean, but they evidently thought it was just another part of the Indian Ocean. The Spaniard Vasco Núñez de Balboa spotted the Pacific from the other side, after Native Americans guided him through the Isthmus of Panama in 1513. Because he gave the first written description of what he named the *Mar del Sur*, South Sea, Balboa gets the credit for "discovering" the Pacific. He was guessing that the body of water might be part of the Ocean Sea that lay south of the Asian mainland. If so, then it and the ocean east of Malacca were one and the same, and perhaps rather small, if the circumference of the Earth were itself on the small side. There was another *if*, a strategic *if*: if the Spanish claims to the new world gave them easy access to the Spice Islands, they might be able to keep at least part of the spice trade out of Portuguese hands.[21]

Abreu had reported that, rather amazingly (and it was true), all of the world's cloves, mace, and nutmegs came from a small archipelago that the Portuguese would call the Moluccas. These islands, south of the Philippines and east of New Guinea (part of what are now Indonesia's Maluku Islands), contained Ternate and Tidore (source of cloves) and the Banda archipelago (nutmeg and mace). Abreu had not himself reached the Moluccas, but one of his crew, Magellan's friend Serrão, was shipwrecked there. Serrão allied himself with Ternate's sultan, married several women, and generally made himself at home. When offered rescue by a subsequent Portuguese expedition, he declined, but sent at least one letter to Magellan. Now lost, the correspondence evidently invited Magellan to join him. Serrão misled his friend, however, by exaggerating the wealth of the Moluccas and their distance from Malacca, which made Magellan think that the islands lay much farther east from Europe than they actually did.[22]

Magellan decided to sail not only to the Moluccas but beyond. Sometime in late 1515 or early 1516, after his return to Portugal, he met with King Manuel I and offered to lead a new expedition. At this point, he probably meant to go to the Moluccas by the old route, via the Cape of Good Hope, which the king already controlled. Perhaps for that reason, the proposal went nowhere. Magellan made a fateful decision to turn renegade.[23]

His knowledge of Asia made him valuable to Portugal's rival, Spain. He arrived in Seville in October 1517, and sought audience with King Charles I, who was eager to enter the spice trade that Columbus had promised his grandparents yet failed to deliver. Magellan proposed to sail to the Spice Islands and claim them for Spain, presumably with the help of Serrão. Magellan knew better than to ask Charles to risk war with Portugal. Instead, he offered to prove that the passage westward to the Moluccas was less than the eastward route, which would mean that the Spice Islands lay within the Spanish sphere laid out in the Treaty of Tordesillas.

His plan depended on yet another bold hypothesis: that ships could pass through the southern landmass below the Caribbean, the place that kept cartographers busy by growing whenever someone returned with news of yet more land down under. Some of that news might have encouraged Magellan, particularly descriptions of a large river, the Río de la Plata, which at least two previous expeditions had tried to navigate to Balboa's South Sea, without success.[24]

Magellan thought he could do better. He favored estimates of a small globe and he sided with Columbus against Vespucci: there was no such place as America. The growing landmass beneath the Caribbean was surely a part of Asia that bulged east and south. Indeed, Magellan believed the ocean Balboa had seen was no South Sea but the Magnus Sinus, a partially enclosed sea at the eastern end of the Indian Ocean. And Magellan revived yet another Columbian idea, the search for places described in scripture. If he thought he could sail to paradise, he held that notion quietly to himself, but he planned to reach islands the Portuguese called the *Lequios* (Formosa, now Taiwan, and the Ryukyu Islands south of Japan). He was confident they were the gold-laden Tarshish and Ophir of the Bible. To make his case, he introduced two expert witnesses: his learned friend Ruy Faleiro, master of the latest geographic theories, and his slave Enrique, who embodied personal knowledge of the East. He and Faleiro dazzled King Charles with a world map, likely one done with a rare polar projection (with the south pole at the center), all the better to detail the proposed southern route west out of the Atlantic Ocean.[25]

Charles was convinced. He offered Magellan and Faleiro five ships and 230 men, artillery, a twentieth of the profits from the all-important spices, and the right to take title as *adelantados* or governors of any new lands they found within the Spanish division of the globe. This was the standard arrangement: an explorer offered an idea and the underwriter supplied materials and incentive. Most of the materials built or maintained sailing ships; much of the incentive was to get people onto those ships.[26]

Because we know Magellan's era as an "age of sail," we assume he and his contemporaries took to the sea gladly. They did not, and for good reason. Praise has been lavished, deservedly, on Renaissance advances in navigation, ship design, and geographic knowledge. But one thing did not improve. Ships' capacity as life-support systems was dreadful and would not get better until the late eighteenth century. That was a rather severe problem for long-distance travel, especially circumnavigations. The physical conditions of Magellan's voyage are therefore worth examining in detail, because they would be representative of what circumnavigators would endure for the next 250 years.

"Life-support system" is anachronistic as a term, but people in the Renaissance would have known what it meant, given that they thought of a sailing ship as a "wooden world," a floating microcosm of the larger world in which humans lived. The microcosms were indeed small. The vessel da Gama had taken to India and back had been a mere 178 tons. (A "ton" in this sense is not a measure of the ship's weight, but rather a measure of the volume of water it displaces.) Horatio Nelson's HMS *Victory* would be 1,921 tons and the RMS *Titanic* almost fifty times that size. Each of Magellan's five ships was even smaller than da Gama's, and only slightly bigger than Christopher Columbus's, on his first voyage to America—the Moluccas were just around the corner from there, after all. Magellan's flagship, the *Trinidad*, was probably around 110 *toneles* or 120 modern tons, and the *San Antonio* slightly larger, 120 *toneles*. The *Concepción*, *Victoria*, and *Santiago* were all smaller, with the last weighing in at 90 tons. They would each have been around 60 feet in length, as long, roughly, as a railroad car.[27]

Ships were designed for sailing, but not for sailors. The top deck of an oceangoing vessel would have been the roomiest, but the lower decks very cramped and, without air and natural light, chokingly close and pitch black. Candles and lanterns were necessary, though a constant fire risk. The bilge at the very bottom always contained some water, which rendered all parts of the ship quite damp, basically in various stages of rot. The longer the journey, the worse the conditions. Travelers expected to become ill, especially if they went to climates unlike those of Europe. They also assumed that they were meant to live on land, not at sea. Seasickness was known to affect most people, even experienced mariners. One Spaniard who vomited his way to America lamented how the sea "was not a fit place for human beings." Da Gama must have concluded the same as his men melted away from scurvy, and that disease would become the ultimate proof that human bodies simply failed at sea.[28]

However miserably undersized, sailing ships were expensive. They may not have been costly in terms of labor to build, and they may have had a free source of power in the wind, but they needed prodigious quantities of wood for their construction. Together, Magellan's five ships had probably required close to 1,000 trees. Indeed, the famous ships of European voyages of exploration represented the final demise of the forests of the Mediterranean and the beginning of an intensive exploitation of wooded parts of the Baltic, Scandinavia, Russia, and, eventually, the Americas. As if they missed the forests from which they had been chopped, wooden ships were unable to stay away from land for very long. They always needed spare lumber for repairs, yet could never carry much of it.[29]

And if the wind was a free power source, the human labor that manipulated it was not. Night and day, sailors set and struck sails to keep a ship on course. (The space belowdecks could be small because seamen were never all there at once; someone was always on watch above.) A ship's master, who took charge of provisions and cargo, had to maintain a careful balance: a bigger ship could carry more material, but would then need a larger crew—who would then need more provisions. Sailors could neither drink nor breathe the element over which they sailed, nor depend on it for food. So ships had to be crammed with as much terrestrial stuff as possible, especially food, water, and firewood.

Before the invention of canning or artificial refrigeration, methods of preserving food were limited. To delay decay, food could be dried, as with baking, or salt or sugar could be forced into it. The mainstays of a sailor's diet were ship's biscuit or hardtack (bread hard-baked to keep at sea, just about), water, and wine, supplemented with cheese, beans, rice, and salted meat or fish. The captain and other officers brought their own provisions, including delicacies suited to their station and livestock to be slaughtered en route.

The biscuit was the main thing; everything else was bound to rot or run out, and any sane captain made land as often as possible to get food and water. He also replenished supplies of wood, as well as the cloth, rope, and leather that served hundreds of purposes. Magellan knew perfectly well that his ships were like cranky unweaned babies, always needing to return and suckle at the shore. Yet he took only two years' worth of supplies, rather optimistic compared to the three-years' supply that da Gama's first fleet had carried.[30]

A sailing ship not only had to protect humans from the sea, but get them across it. For that reason, a ship was the largest and most complicated piece of machinery yet invented, with an array of internal mechanisms—rudder, rigging, and pumps—necessary to steer, adjust sails, and keep out water. A variety of other devices (cannon, compasses, spyglasses, astrolabes, sand-clocks) were loaded aboard. Human hands had to manage all this equipment. Any vessel that lost its minimum crew was doomed to drift, fill with water, or be wrecked. But any ship that needed a large crew needed greater capacity for provisions, so new developments in ship design were trade-offs. The narrow caravels that the Portugese had recently adapted from Arab ship design were built for speed and maneuverability, but they lacked a traditional carrack's carrying capacity. Expeditions often had both types of ship, to mix their benefits. Four of Magellan's ships were carracks; one, the *Trinidad*, was a caravel.[31]

It was an acknowledged paradox that a ship, that complex and expensive piece of machinery, chockablock with smaller and yet more delicate devices, was put into the hands of rabble. No one respected the common sailor. Iberian sailors earned less than skilled workers on land and came from the lowest social ranks. (Free blacks and mulattos, people one step away from slavery in European societies, were common on European ships.) Captains and officers ranked higher, but often had limited means, as with the younger sons of good but impoverished families who went to sea. And they were poorly paid. A military admiral got a mere two ducats a day; a woman in a Seville brothel could earn twice that by servicing his crew.[32]

But, in this pre-industrial world, wages were beside the point. Every free man who worked a vessel was a partner in the venture and promised a cut of the profits. For Magellan's expedition, shares were measured in the spices each man could bring back for himself, using jealously allocated cargo space. Magellan was allowed the maximum,

82 *quintales*; an ordinary sailor got 3.5 *quintales*; others had shares in between. There would have been around 46 kilos in a *quintal*, so Magellan's share was 3,772 kilos, over 4 fragrant tons. At expected market prices, a sailor's share of spices (minus taxes and required religious donations) would more than double his wages.[33]

The added cost of outfitting Magellan's ships for trade overran the royal budget. Private investors made up the shortfall. One merchant alone footed a quarter of the bill: Cristóbal de Haro, a veteran of one of the attempts to sail up the Río de la Plata and an associate of the Fugger family, Europe's most powerful bankers. Haro prepared a wealth of goods as gifts, bribes, or trade items, including textiles, bars of copper, vials of quicksilver, 20,000 small bells, 10,000 fishhooks, 1,000 looking glasses, assorted knives, and a selection of brass and copper bracelets.[34]

Assembling the ships' complements was equally important. Any ship's crew was a microcosm of Europe's social hierarchy. Captain-general Magellan presided at the top, from the flagship, with the captains of the four other ships beneath him. The five ships' masters came next. Below them were the pilots, who acted as navigators, as well as the gunners, in charge of the ordnance. Then came ordinary sailors and, at the very bottom, ship's pages, boys apprenticed to the sea. The lack of women signaled the artificial nature of shipboard society. Sailors whispered that a female on a ship was bad luck. The real problem was the sexual competition a woman might create among the men. For that reason, a woman at sea had to be under the protection of a powerful man on board, or else designated as the property of someone extremely powerful on land, as with female captives sent to a monarch. Otherwise, a woman was sexual prey. A captain might not care about her, but worried about the damage his men would do to each other by fighting for access to her. (Or to boys, who were also sexual prey.)[35]

Magellan was careful to assemble a crew he could control. Faleiro, despite plans to be the expedition's mastermind navigator, either blurted a confession that he knew everything he did *because a demon told him,* or else Magellan invented the story to discard a potential rival and select a more acquiescent pilot for his flagship. He also worried about being surrounded by Spaniards. He made special requests for more Portuguese, and in the end, about thirty-seven went, several of them his relatives. The final crew numbered somewhere between 270 and 280 men, and they were an unusually international bunch. Over 35 percent were not Spanish, including the Portuguese, but also French, Italian, German, Flemish, Irish, and Greek men, and some of African descent, plus some truly exotic characters: two Malays (including "Enrique de Malacca"), and at least one Englishman. There was the usual assortment of skilled men, including several monks to spread the gospel.[36]

Literacy and skill only sometimes indicated place in a ship's hierarchy. The ships' pilots, for instance, were of middling rank. Although they had mastered the instrumentation necessary for celestial navigation, their abilities were considered the result of mere experience at sea, similar to what ordinary sailors had. (The fact that slaves and foreigners could serve as pilots underscored navigation's low status.) Literacy and the ability to keep a record of the journey were not the preserves of high rank. Many of Magellan's men kept journals or logs, although most are lost, including Magellan's. Most were done by lower-ranking men, including several pilots. But the longest surviving narrative was written by someone of gentle birth. Antonio Pigafetta was a minor Venetian nobleman and knight of a religious order that answered to the Pope. Such was the growing fame of Magellan's expedition, even in Rome, that Pigafetta asked permission of his order to join it and arrived in Seville with suitable letters of reference. He evidently kept detailed notes and sketches, enough to tell the story of the journey.[37]

With blasts of artillery, the fleet descended the Guadalquivir River from Seville on August 10, 1519. Take a good look at them here, in their seaworthy glory, because the expedition would be the first vanishing act performed on the world stage, as its ships and men fought a war of attrition against the Earth itself. (You may follow their route on Map 1.)

On September 20, the five vessels entered the Atlantic. At Tenerife in the Canary Islands, Magellan kept following the coast of Africa south and only later swung west to Brazil. His reluctance to use the familiar route from the Canaries to the Caribbean, with its predictable westward winds, was puzzling. All of the ships' pilots were compliant Portuguese, but three of the four captains were Spaniards and one of them, not for the last time, questioned the captain-general's judgment.[38]

Once across the equator, the crew adjusted to navigating without the north star, Polaris, which indicates north in the northern hemisphere. A pilot who had lived in Brazil guided the fleet to that place, claimed by Portugal, though luckily for Magellan barely settled by his estranged nation. Magellan ordered a stop at Rio de Janeiro, and for thirteen days the crew gathered provisions from a group of Tupi or Guaraní Indians.[39]

Pigafetta's response to them mingled wonder with contempt— typical sentiments. Europeans expected non-Christians to be different, indeed inferior. Pigafetta related old tales of American cannibalism, for instance, though he failed to witness any. Yet he conceded that the Indians understood trade, a significant mark of human civility. Pigafetta lists eight words with their local counterparts, most of them relevant to trade, but also the first written instance of *amache*, a "hammock" in which a person could sleep in the tropics, cool and away from creeping vermin.[40]

Just after Christmas, the fleet continued down the coast in search of the thing essential to the mission: a way through the land. From this point on, European geographic knowledge ebbed. Magellan had recruited, maybe voluntarily, at least one Brazilian, the first of many non-European people the expedition would gather up for information. Once the fleet reached the Río de la Plata, about 35 degrees south of the equator, it was as far south as the tip of Africa, lower than most European ships ventured. After investigating some rivers that proved dead ends, the fleet wintered at a natural harbor it found on March 31, 1520, and named Puerto San Julián (now in Argentina). There, at the 49th parallel, the men hunted bizarre "geese" that could not be plucked but had to be skinned whole—penguins. For two months, they assumed the nearby land to be uninhabited until, one day, "a giant" appeared.[41]

Blame Pigafetta for that description, a last gasp of belief in the monstrous races. (He had already breathlessly claimed another Indian man was "as tall as a giant.") Certainly, everyone agreed that the man was large and marveled how he "danced, leaped, and sang," a solo performance, just for them. Magellan sent one man to approach the stranger, "charging him to leap and sing like the other in order to reassure him." Somehow, it did, and the sailor danced the stranger over to meet the others.[42]

However exciting it was to discover people, and massive ones at that, the "giants" were an unwelcome distraction for everyone but Magellan. The men fumed while the commander delayed the expedition in order to secure for Spain a land uninhabited by Christians. One giant stayed with the crew so long that the monks baptized him. Magellan also wanted to bring some giants back to Spain, as proof of a land he could claim as governor. He tricked two men into captivity by adorning them with anklets that turned out to be shackles. They roared, "loudly calling on *Setebos* (that is, the great devil) to help them." Relations deteriorated. In a skirmish, an arrow killed one of the crewmen. And yet the Europeans could not manage to shoot any of the antic giants, whose dancing now seemed a canny form of defense.[43]

Amidst it all, Pigafetta composed a Tehuelche vocabulary, a total of ninety words. Over one third of them were for parts of the body, and it is easy to imagine Pigafetta drawing aside an informant or two, pointing, listening, scribbling. As in Brazil, several words relate to trade, mostly European goods. Others describe colors, ten are verbs, one is the pronoun *we,* and yet another a command: *come here.* A subset of words address maritime affairs and navigation: *sun, stars, sea, wind, storm,* and *ship,* plus *to go far* and *the guide.*[44]

The list reveals disappointment. Unlike the Brazilians, these people, in a colder climate, had limited food and capacity for trade. One of the pilots claimed they had no concept of property. Nor did the rudimentary list of sea terms yield much information about geography or navigation. That especially must have frustrated Magellan. He gave the people the contemptuous name of *Pathagoni,* which anyone who spoke a Romance language would recognize as referring to an animal "with large paws." The assumption that the Patagonians were beasts would mar centuries of Western relations with them.[45]

Magellan tarried in Patagonia, lord of a people he disliked, for five months, ample time for unease to grow into mutiny. Why was it taking so long to get to Asia? Was the captain-general only interested in his own glory? The masters of three of the other ships conspired to kill him and sail back to Spain. This was unusual. The Spanish Empire carefully regulated its ships—mutinies of entire crews had become rare. But a long expedition, especially one headed by a foreigner, was difficult to control, particularly when the foreigner had promised a short expedition. Luckily for Magellan, news of the conspiracy leaked. He put the mutineers on trial, executing some, casting others ashore to fend for themselves, and demoting the rest. He put each ship under a Portuguese captain.[46]

Magellan ordered the smallest vessel, the *Santiago*, to keep searching for a passage. It was a bad sign that the ship did not return, and worse when debris from it began floating into San Julián. (A coincidental solar eclipse probably increased any sense of dread.) The ship was indeed wrecked, though most of the cargo and all but one man were saved, to be redistributed among the four remaining ships. Magellan placed a large cross on the tallest mountain "as a sign that the said land belonged to the King of Spain," and, at last, moved on. The expedition investigated another large river, another dead end. For two months, the men took on water, wood, and fish, then continued south.[47]

On the autumn feast day of the Eleven Thousand Virgins, Magellan discovered the most promising inlet of all, around 53 degrees south. "Found by miracle," Pigafetta declared of Cape Virgins, and the others surely agreed, over a year since their departure from Seville. Magellan sent the *San Antonio* and *Concepción* to see how far they could ascend the river. He swore he had seen, in Portugal's royal treasury, a map that showed a hidden strait from the Atlantic to the Magnus Sinus.[48]

Two days passed, a storm blew up, and those left behind feared the other ships were lost, smashed in a narrow bend or run aground. Pigafetta related how in great suspense "we saw the two ships approaching under full sail and flying their banners," already a good sign. Then the *San Antonio* and *Concepción* "suddenly discharged their ordnance, at which we very joyously greeted them in the same way . . . thanking God and the Virgin Mary."[49]

The happy discovery had come too late—the expedition had been out beyond the half-life of its supplies. Magellan asked his officers for counsel, assuring them that he welcomed any "reasons why we ought to go forward or turn back." But when they recommended a return to Spain, he ignored them. He ordered the reconnaissance ships to investigate what seemed an alternate opening to the channel. The pilot of the *San Antonio* realized this was the last chance to abandon the mission. He conspired with other Spaniards to overpower the captain, Magellan's brother. The *San Antonio* slipped off in the night and bore north. When she entered hotter latitudes, her captive Patagonian died; when she returned to Spain in May 1521, the pilot ringleader was imprisoned and the conspirators put on trial (though they were acquitted and their belief that Magellan had overstayed his time at sea vindicated). The mutineers gave the last news of the expedition that anyone in Europe would have for over a year.[50]

Meanwhile, the *Concepción* trustingly waited for the *San Antonio*, and Magellan sent a smaller boat to see where the passage ended. Within three days, the boat returned and confirmed that the strait led to another sea. Overcome, Magellan "began to weep and gave this [western] cape the name Cape of Desire, as a thing much desired and long sought." No wonder he wept—his backup plan had been to continue 20 degrees further toward the south pole, 75 degrees below the equator. (No one knew it yet, but the Southern Ocean begins to freeze around 60 degrees south.) He then learned that the *San Antonio* was missing, and ordered signals and directions placed for the stray to follow. The fleet now lacked two of its ships, including the largest, the *San Antonio*, with corresponding losses of men and supplies.[51]

It was late October when the *Trinidad, Victoria*, and *Concepción* entered the strait. The ships had to move slowly to avoid scraping the sides of the channel. It was like threading a dreadful needle, three times in a row. The strait was so deep that the ships could not anchor at night. Men in boats attached cables to the land, a very time-consuming effort, though an opportunity to get fish and fresh water. The brief nights of the southern summer enabled interesting observations. In the dark, on the starboard side of the strait, the men could see fires, evidence of people unseen, but perhaps watching them—hence one name for the place, *Tierra del Fuego*. To port, they heard roaring surf, amplified at night when their senses focused on sound, and a hint that there was not a southern landmass, only sea-lashed islands. It took thirty-seven days to navigate the passage, each ship pulling an invisible thread from Spain to . . . where?[52]

In late November 1520, the fleet emerged from the strait and into the "Pacific Sea," as the expedition named it for its apparent lack of storms. Pigafetta represented the strait on a little map, framed by the Ocean Sea and the Pacific Sea opposite. Now, onward to Asia. When he had reached the Isthmus of Panama, Columbus had thought himself nineteen days sail from Japan; Magellan's friend Serrão had probably made a similar guess. The worst of the journey was supposed to be over.[53]

Yet nothing so far compared to the agonies ahead. The Pacific is the world's largest ocean, larger than all the land surfaces of the globe put together, and covering one third of the Earth. Magellan had to admit what he had denied: an ocean quite clearly separated Asia from the Americas. He was lucky that excellent prevailing winds carried him north and west, yet unlucky not to encounter a single inhabited part of Polynesia, where he might have found food, water, and information.[54]

This was the first big surprise of the expedition: the globe was much bigger than anyone had suspected. And its southern waters might not contain corresponding land, as the roaring surf in the strait had already hinted. Pigafetta, for one, lost all faith in the existence of the great southern continent thought necessary to balance the landmasses of the global north. He shuddered to think of what might have happened had the ships headed due west from the Cape of Desire, south of the Cape of Good Hope: they would have swung around the world at the 52nd parallel, right back to the Cape of the Eleven Thousand Virgins.[55]

For three months and twenty days, the expedition continued north and west as its supplies shrank. The ships' smallness, an advantage in threading the strait to the Pacific, was now a disaster—on this part of the world's surface, they were inadequate artificial worlds. When all the real food ran out, the men ate the remaining and oldest ship's biscuit, "turned to powder, all full of worms and stinking of the urine which the rats had made on it." Next came oxhides, softened for several days in the sea, grilled, and forced down with "water impure and yellow." Every human disaster has its cheerful opportunists and those stuck out in the middle of the Pacific Ocean with Ferdinand Magellan ran the trade in tasty rats, half an écu apiece; "some of us could not get enough," Pigafetta reminisced. There was no land except for two tiny uninhabited islands that had no shootable animals, no water, and no low-growing vegetation, only birds and trees. Each islet was, moreover, in water too deep to anchor and patrolled by sharks: "Wherefore we called them the Isles of Misfortune." Those Unfortunate Isles mirrored the Fortunate Isles (Canaries) on the opposite side of the globe. The men had reached the back of the rolled-up map.[56]

When they developed scurvy, with gums so swollen they could eat nothing at all, they knew for sure that humans were unsuited to life at sea. Indeed, the cause of scurvy seemed to be prolonged removal from land. (Fatal instances of the disease ashore were too unusual or scattered to contradict that suspicion.) Pigafetta observed that men of all nations suffered—whatever their native climates, they shared an inability to live at sea: "twenty-nine of us [Europeans] died, and the other [Patagonian] giant died, and an Indian of the said country of Verzin" or Brazil. Dozens of others were too weak to work. Men of high rank and anyone connected to them (and their private stores) did better, including Magellan, Pigafetta, and Enrique.[57]

It is striking that Pigafetta's narrative would be illustrated, from this point on, with maps of islands. Only the first map, of the strait through Patagonia, hints at a larger landmass. Thereafter, starting with the Unfortunate Isles (two unpromising ovals amid a great many waves), the maps resemble Mediterranean "Island Books." These map-illustrated works presented a world of water punctuated with small bits of land. The Latin name for such a book, *isolario*, emphasizes that isolated islands were small comfort to sailors. Moreover, unlike da Gama, Magellan had no native pilot to guide him. If he had hoped that the Brazilian or Patagonian would help him, that does not seem to have happened, and then the two men died.[58]

To determine their position, the sailors looked to the skies. They could easily determine latitude from the length of the day and the Sun's position at its highest point, adjusted for season and hemisphere. Longitude could be determined on land (the Spanish had a secret method for doing that), but it was much trickier at sea because it depended on the position of other celestial bodies relative to a ship's progress, mostly guesswork. Latitudinal readings tended to be off by only a degree, but those for longitude could be 10, 20, even 30 degrees in error—as much as the distance between London and Kiev. It astonished everyone that the southern hemisphere had not only the wide-set stars of the northern hemisphere, but also "several small stars clustered together, in the manner of two clouds a little separated from one another, and somewhat dim." These clouds (later determined to be some of Earth's nearest galaxies) were fascinating to behold but not so useful for navigation. The sailors found much more helpful a "cross of five very bright stars right in the west," the Southern Cross (first described by Vespucci in 1503) that pointed toward the south pole. Each of the ships' pilots used compass and celestial readings to keep a chart on which he pricked a line to represent the ship's track, the first paper maps of the seemingly endless South Pacific.[59]

Once across the equator, Magellan veered slightly west, and fetched up in Guam. He stopped to get supplies. The crew were still getting the sails down when they noticed some people from a nearby village making off with the flagship's skiff. It was an awkward welcome, though the local people may have had good reasons to dislike outsiders and could not have known what the expedition had endured. That may have been what unnerved the Europeans: they didn't know where they were, and no one else in the world knew how far they had come. Magellan landed with an armed force that burned the village, killed at least seven people, and recovered the skiff. The Europeans declared that the island and those around it should be called the *Islas de los Ladrones*, "Islands of Thieves," and, for centuries, European mapmakers granted their wish.[60]

Guam was a rough re-entry to the inhabited world that had steadily receded, from the lonely coast of Patagonia, through the eerily isolating strait, and across the seemingly deserted Pacific. Such was the shock of encountering human societies after all that time that the "Ladrones" are the only people represented on any of Pigafetta's maps: two men row a boat of misleadingly European configuration, but with an unmistakable outrigger.

Yet the Earth, not its people, was the problem. In thinking the globe was small, Magellan had erred colossally. The Pacific had cost him time, supplies, men, and—crucially—space. Who would now believe that the Moluccas lay within the Spanish sphere? Magellan said he did not want to head immediately to the Spice Islands because he had heard they lacked provisions. More likely, he still hoped to claim lands outside Portuguese territory, which could not lie any farther west. Indeed, his northwest heading showed that he still expected to get to Tarshish and Ophir, or even to China.[61]

First, he had to feed his crew. Some of the scurvy victims begged the Guam-bound war party to bring back entrails from the enemy dead, convinced the gruesome medication would cure them. Pigafetta delicately neglects to follow that story line. Magellan was confident that mere contact with land, plus fresh terrestrial stuff, would remove scurvy—a belief that would have a long, eventful history. After making landfall in the eastern Philippines, he ordered tents pitched ashore for the sick and a pig killed to feed them.[62]

When Magellan showed the local people his samples of gold and spices and indicated his heading, they "made signs that the things which the captain had shown them grew in the places whither we were going." Unfortunately for everyone, Magellan decided to stay and claim the Philippines for Spain. That meant more delay for his spice-hungry crew. Nor were the islands likely to be easy conquests or eager trading partners. Parts of the archipelago were under powerful government, including a Buddhist kingdom and several Muslim sultanates. And most of the islands already traded with the outside world, especially the Asian mainland, Indonesia, and Japan.[63]

An accident of language may have encouraged Magellan's imperial design. At Leyte, Enrique leaned out of the *Trinidad* to hail eight men in a boat and they understood him. Malay and the Austronesian languages of the Philippines are mutually intelligible, and the Philippine aristocracy primarily spoke Old Malay. When the rajah Kolambu arrived, "the said slave [Enrique] spoke to that king, who understood him well." Magellan declined Kolambu's presents, perhaps bribes to encourage his departure, and instead offered to buy provisions, which Enrique negotiated by going ashore alone, a sign of his master's trust in him. The rajah and Magellan then displayed their respective treasures. For Magellan, that included "the marine chart and the compass of his ship," also news of "the strait [by which] to come hither, and of the time which he had spent in coming, also how he had not seen any land. At which the king marveled," perhaps relieved that the rest of Europe would not descend on him any time soon.[64]

Magellan sank deeper into the region's complicated politics, until he decided to conquer an island or two. He succeeded with Cebu, and believed he had made an ally and Christian convert of its ruler, Rajah Humabon. Magellan turned to a neighboring island, Mactan, where a local leader, Zula, reported that his overlord, Rajah Silapulapu (or Lapu Lapu), wanted nothing to do with Spain.[65]

Magellan decided to make another show of force. For reasons of his own, Zula agreed to help. But the forty-nine men Magellan led to shore at dawn met Filipino soldiers by the thousands. Disastrously, the ships' cannon could not reach the battlefield. Magellan soon realized that his firearms and crossbows were ineffective. "Do not fire, do not fire any more," he reportedly yelled at his men, but they disobeyed in order to cover their retreat back to the boats. Eventually, he was too wounded to fight. Down he went. "All at once," Pigafetta recalled, the enemy "rushed upon him with lances of iron and of bamboo and [struck] with these javelins, so that they slew our mirror, our light, our comfort, and our true guide."[66]

Pretty sentiments, but it is likely that the crew had encouraged their light and comfort's little invasion to get rid of him, after yet another frustrating dalliance in lands without spices. Rather than risk another mutiny, his men had arranged for others to kill their commander. But, really, it was the size of the globe that had brought down Fernão de Magalhães e Sousa, who might have lived longer had the Moluccas been only the year's sail from Spain he had promised.

The crew selected two new captains-general: Duarte Barbosa, Magellan's brother-in-law, and João Serrão, brother of Magellan's friend who had settled in the Moluccas. They tried to rally Enrique, who kept "wrapped in a blanket" on the pretext, Pigafetta sneered, of being "slightly wounded" in the melee. Maybe Enrique feared abuse without his master's protection. Indeed, Barbosa told Enrique, loudly enough for all to hear, "that although the captain his master was dead, he would not be set free or released" but held as property for his master's widow (Barbosa's sister) in Spain. Enrique flung off his blanket and headed for shore, where he evidently plotted with Rajah Humabon to seize the three European ships. Barbosa's threat, and the slave's reaction to it, hints at a bargain Magellan might have struck with Enrique: *get me to the Moluccas and I will free you*. Humabon's haste to conspire against the Iberians revealed an equally fragile accord: the man had suffered baptism and promised fealty only under duress.[67]

Humabon lured Serrão, Barbosa, and twenty-five others to land. Those who stayed behind heard shrieks and groans, then saw Serrão hauled out, bound and bleeding. He shouted that all but he and Enrique had been killed. His comrades dared not send a rescue boat, lest it too be taken. They abandoned Serrão, whose fate is unknown. As is Enrique's. Although Serrão had seen Enrique alive, the Malay slave was listed among those who died in the Philippines. Of course, he was not supposed to be there in the first place—Albuquerque's forces had killed him in 1511, or so the story went. Maybe he survived another Iberian boast about his demise. Maybe Rajah Humabon let him go home. Maybe he reached Malacca or Sumatra before the Spaniards returned to Seville. If so, he was the first person to go around the world.[68]

Freed from Magellan's scheming, the expedition was back on course. As Pigafetta laconically noted, "Before the captain died, we had news of the islands of Molucca."[69]

Perhaps the people on Cebu could see the smoke, though the Spanish had been careful to sail away some distance before they burned the *Concepción*. The expedition's recent losses—six or eight men on Mactan and twenty-six on Cebu—meant that only one hundred twenty men remained, not enough for three vessels. Having redistributed cargo and men into the *Trinidad* and *Victoria*, the crew destroyed the *Concepción* rather than leave her to the scavengers who were already turning up in small sailing *praus*. The fleet had lost over half its ships and crew, including many of its original leaders and its principal interpreter.[70]

Pigafetta alone was qualified to take Enrique's place. In the Philippines, he had compiled his longest vocabulary to date, Bisaya words for body parts, numbers, pepper, cloves, cinnamon, and ginger. Equipped with a regional language, and quick to learn others, he was critical to the mission and had a front seat on the action. At the Kingdom of Brunei, for instance, he got to ride on one of the silk-clad elephants that Rajah Siripada sent to collect a delegation.[71]

The delegates probably looked bedraggled, especially against those elegant elephants. While at sea, sailors (and sometimes officers) went barefoot and wore canvas trousers and jackets. In the tropics, they stripped down to their underwear, thigh-length linen shirt and drawers. When they met with dignitaries, the leaders rolled thin hose up their legs, stamped into shoes or boots, and shrugged on doublets, tailored tunics. But most of the men lacked fine clothes and all the garments must have been past their prime. When the battered ships reached Borneo, Pigafetta would complain that "the greatest labor which we had was that of searching for and cutting wood in the forests, without shoes." Pigafetta always admired any rich Asian garb he saw, even as he and his companions slipped into decrepitude, a humiliating consequence of their overstretched supply line.[72]

The expedition's leadership was equally patchy. The pilot João Carvalho, who had guided the expedition to Brazil, was now captain-general and an outright pirate. At Luzon, he seized several sailing junks. When he used his son as a hostage in a diplomatic standoff with a rajah in Borneo, and lost his boy, some of the crew believed he had sold the child to the rajah for gold. He also mistreated prisoners. Along with the cargo from the junks, Carvalho seized sixteen men and three women. To protect them from rape, the latter had been designated as gifts for the queen of Spain, but Carvalho scoffed at Her Majesty and made the women into a private harem.[73]

Maybe the other men were above rape, but not kidnapping, because it was their only way to navigate. Three island-ringed seas lie between the Philippines and the Spice Islands and it showed how lost they were that the Europeans took captives from each sea to guide them to the next. First, they crossed the Sulu Sea the wrong way, west to Borneo, where Carvalho took three men out of a *prau*. These captive pilots saved the ships (and themselves) from being wrecked on the coral reefs around Borneo. Carvalho was at this point removed from command. Gonzalo Gómez de Espinosa took the *Trinidad* and Juan Sebastián de Elcano, one of the mutineers whose life Magellan had spared, had the *Victoria*. The ships doubled back to the Philippines, then entered the next sea, the Celebes. There, the expedition seized another ship, killing seven men. One man saved himself by claiming "he knew well where Molucca was." His knowledge evidently ran out at the islands that divided the Celebes from the Molucca Sea, where the Europeans "took two [new] pilots by force that we might learn from them of Molucca." All went well until the fleet could not clear one final obstacle to the Molucca Sea, an island "very beautiful to behold" but smack in the way.[74]

Once the two ships escaped the lovely, stubborn island, one captive got his bearings. Some very high islands were the Moluccas, he said, and he was right. Such was Pigafetta's relief, after "twenty-five months less two days" since departure, that the only living plant in any of his illustrations appears on the map of the Moluccas. It is a clove tree. (See illustration 2.)[75]

Finally, *finally*, in the Spice Islands, Espinosa and Elcano were interested in nothing but cloves. On Ternate, both the sultan and Magellan's friend Francisco Serrão, whose correspondence had started the whole affair, had died eight months earlier, perhaps poisoned. But Sultan Manzor of Tidore was open for business. On Tidore, Espinosa and Elcano built a little trading house, Spain's sole outpost in Asia. There, the long-preserved stockpile of trade goods, augmented by cargo from the looted junks, was rapidly exchanged for tons of spice.[76]

The Iberians were desperate to cooperate. When Manzor asked for their Asian captives, they surrendered everyone but their useful pilots from Borneo. The sultan also demanded respect for his Muslim faith. He asked Espinosa and Elcano to kill their pigs, and they did; with his hand on his Quran, he swore an oath of friendship, which they accepted. The Europeans were painfully aware that they cut a poor figure. The sailors had offended local modesty by going ashore in their ragged drawers. They were told to wear more, a humiliating rebuke.[77]

When a Portuguese man living on Ternate visited Tidore, he confirmed that Elcano and Espinosa were out of their depth. Pedro Afonso de Lorosa told the Iberians that Manzor had kept him away in order to overcharge them, knowing they could not know the local price for fresh cloves. Oh, and they were wanted men. Word had spread, even to India and Malacca, that a Portuguese traitor had led five ships to the Moluccas. The king of Portugal had dispatched ships to the Cape of Good Hope and the Río de la Plata to intercept them, and made at least two attempts to find them in the Moluccas. The Portuguese also lost no time in fortifying their trade to the Spice Islands, and had discovered that they could sail from Malacca to the Moluccas in fifteen days.[78]

That last fact impressed Elcano and Espinosa. They must have suspected that they were more than halfway to Spain, and now they knew it. A circuit around the world would have to be done, and any pretense of never entering Portuguese territory laid aside. In a final frenzy of trade, many sailors swapped extra clothing for spices. Elcano gave the sultan firearms and gunpowder in exchange for wood; Manzor threw in some precious skins of birds of Paradise for the king of Spain. Many dignitaries turned up to wave off the two ships in late December 1521, only for the *Trinidad* to spring an unfixable leak. Fifty-three men, including Espinosa and the contentious Carvalho, would have to stay behind while the *Trinidad* underwent extensive repairs. Under Elcano, the *Victoria* departed alone with a crew of forty-seven Europeans and thirteen Asians, plus two Moluccan pilots hired to pilot the ship out of the archipelago.[79]

The *Victoria* dropped south, ultimately headed for the Cape of Good Hope and re-entry to the Atlantic. She was clearly ailing and stopped short of Timor, where the men spent fifteen days doing repairs and adding pepper to the cargo. They also "took a man to guide us to some island where there would be provisions." At Timor, Elcano needed food so badly that he kidnapped a prominent man and his son to ransom for livestock, the last provisions he would take on for two months. So poor was the outlook that some men deserted.[80]

Pigafetta, in his final map, signals access to "the Great Sea" with two small, unnamed islands. The *Victoria* began to leak dangerously and her stores dwindled to small amounts of water, rice, and putrefying meat. Having sold their caps and cloaks for cloves, many men lacked adequate clothing for cold latitudes. Some of them begged to stop at Mozambique, but Elcano did not want to risk exposure to the Portuguese. To avoid detection, he went around the Cape of Good Hope as far out as he dared, near the 42nd parallel, where the winds of what would later be called the Roaring Forties ripped off the top part of his foremast.[81]

The final run to Spain was hideous. Twenty-one men died, a disproportionate number of them Asian. It is hard to imagine that anyone on the clove-saturated *Victoria*, as they jettisoned corpses from the perfumed ship, still believed that spices preserved health. The Canary Islands were the nearest Spanish safe haven, but Elcano gave up at Cape

Verde and ventured to Santiago to buy provisions from the Portuguese. The crew pretended that the *Victoria*'s foremast had been damaged in the Atlantic, within Spanish waters. But one man blabbed enough of the truth for the Portuguese to detain a boatload of thirteen men. The *Victoria* fled, and her remaining eighteen Europeans (plus four Asians) soon reached the mouth of the Guadalquivir River, having sailed an estimated 14,460 leagues "and completed the circuit of the world from east to west."[82]

There was some confusion about how long the circuit had taken. At Santiago, the Portuguese had said it was a Thursday. The *Victoria*'s log said Wednesday, but the Spanish confirmed that the circumnavigators had somehow lost a day. When the eighteen Christian survivors processed to two shrines to give thanks for their deliverance, barefoot and wearing only their shirts, they did penance for having violated a great many sabbaths and holy days, however unwittingly, because they had crossed what would much later be called the International Date Line.[83]

Wonder of wonders, the *Victoria* was not just the first ship to go around the world, but also the first time machine. Myths and marvels may have been fading from the world—spices came from real places; monstrous races seemed unlikely. And yet Magellan's survivors had accomplished something mythic: they, unlike Phaeton, had done a daring revolution around the Earth. They had also discovered the only form of time travel that has ever been proven to work, a marvel straight out of science fiction, and a hint that something natural might divide the globe along lines of longitude, just as surely as the equator did along lines of latitude. The first around-the-world voyage would be difficult to explain, and it would be just as difficult to replicate.

Chapter 2

A World Encompassed

*G*iven *that the* Magellan/Elcano expedition had been a war of attrition against the vastness of the globe, it is amazing that, whatever else it had run out of on its long, tortured way, it never ran out of paper. Tattered he may have become, reduced to eating rats, but Antonio Pigafetta had packed so much paper that he had about half a ream to spare as gifts to dignitaries in Brunei. If the other provisions of the wooden world had kept the men alive, their journals, diaries, and charts did the same for their story. Published narratives based on those accounts would let other people go around the world by proxy, by far the safest way to do so.[1]

It is tempting to imagine the reports of around-the-world travel steadily generating better knowledge of the planet, as if a small and inaccurate globe, slowly revolving in a time-lapse display, morphs into its present-day size and configurations. But there was no such smooth integration of information about around-the-world voyages. There was not even a name for what Elcano's survivors had done; *circumnavigation* would not be used (in English) until 1625. Instead, the Spaniards were said, in Romance languages, to have *encircled* or *surrounded* the globe, and in English to have *encompassed* it. These verbs had military connotations. That was appropriate, given the deadly implications of around-the-world travel.

The most unfortunate legacy of the first circuit of the world was a vision of world domination. If the Magellan/Elcano expedition had made the globe into a real object, a global stage on which humans actually did something, it also made plans for global empire real, more than metaphors. Every suave villain in a James Bond movie who threatens that "the world is not enough" has a historic ancestor in Charles V, king of Spain, who was eager to make the most of his unprecedented expedition.

Charles had just augmented his kingdom with three imperial acquisitions: the Holy Roman Empire within Europe, Hernán Cortés's conquest of Mexico, and the outpost on Tidore in the Spice Islands. When Elcano's cargo of spices was sold, the proceeds did little more than cover the expedition's expenses. The king and his principal investor, de Haro, divided the small profit and everyone else lost everything. It is doubtful that the sailors who had gone around the world—the original eighteen plus the thirteen eventually released by the Portuguese in Santiago—received wages beyond their initial advance. But Charles's three acquisitions, if he could hold them, might string together an empire for him that would stretch most of the way around the world.[2]

To maintain his access to Asia, the king suppressed information about the size of the Pacific: the bigger that ocean, after all, the less credible Spain's claim to anything on its western side. Although Pigafetta had said that twenty-nine men died during the Pacific crossing, the official list recorded only seven, no small disparity. And although Elcano had kept a journal, Charles V ordered it destroyed to conceal the length of time it had taken the Spanish to get to the Moluccas. He paid off Elcano by promising him an annual pension of 500 ducats and by granting an augmentation to his family's coat of arms, which now featured a castle, spices, two crowned Asian kings, plus a helmet surmounted by a globe bearing the motto (and the motto said it all): *"Primus circumdedisti me"*—"You were the first to encircle me," as if the Earth, a recent virgin, sighed over her first lover.[3]

The globe emblem was the biggest compliment. Its presence on Elcano's arms compared him to the monarchs and holy figures who held orbs to indicate their divine authority over the world, or even to images of Christ holding a globe as *Salvator Mundi*, saviour of the world. Elcano was the first non-royal (and non-divine) person to use a globe to designate a similar authority in secular terms, based on his having circled the world.[4]

Charles V was not done with Elcano, however. He ordered a fleet to check on the men stranded on Tidore, via the strait Magellan had discovered, in an expedition commanded by García Jofre de Loaysa (or Loaísa), with Elcano as second in command and chief pilot. This time, the crew was huge (about 450), Portuguese were banned from it, and the fleet of four ships was better provisioned, at least at the start. The ships departed in July 1525 and headed for the strait, where bad weather wrecked one vessel (and its precious bread) and killed nine men. The ocean that had been "Pacific" for Magellan threw up storms that scattered the three remaining ships. The flagship leaked so badly that the men were exhausted by constant work at the pumps. Supplies of fish and seal meat taken in the Atlantic were inadequate to feed the crew, swelled by men from the wreck. "Each day we expected the end to come," wrote Andrés de Urdaneta, chronicler of the misery. Among the officers, there was a jumble of deaths and successions. In July 1526, Loaysa died. Elcano succeeded him for six days, then died himself. His possessions included a small globe.[5]

At least this expedition's leaders, unlike Magellan, knew where to find relief. At the Ladrones and the Philippines, they bought food and recruited new crewmen. Even so, between the strait and the Moluccas, they lost forty men. And there was no welcoming party on Tidore—the post was abandoned. When the Portuguese came to question their intentions, the Spaniards declared the Spice Islands to be on the Spanish side of the 1494 *raya* or line of demarcation. That set off an unpleasant little war, all the more desperate because the Spaniards had no idea whether they could ever get home.[6]

Neither had the Spanish whom Elcano had left on Tidore. The leaking *Trinidad* had been repaired, and in April 1522, her captain, Espinosa, leaving a handful of men to guard the trading house, had gone east, hoping to reach Spanish America via the Isthmus of Panama. It was a desperate scheme, but Espinosa worried that the *Victoria,* sailing ahead of him, might put Portuguese vessels in the Indian and Atlantic oceans on their guard. The Pacific seemed safer.

If that ocean had been difficult to cross westward, its eastward passage seemed impossible. Mariners knew that oceans had wheels of wind at their edges that could cut sailing time. Espinosa toiled up to 43 degrees, parallel to northern Japan, but failed to find any eastward wind. His men shivered and starved. After battling a twelve-day storm, he turned back and, losing a man every three days (the journey killed over half of them), he returned to the Moluccas, and the waiting Portuguese. They had already destroyed the Spanish outpost on Tidore and imprisoned its guards. They seized Espinosa and his men and beheaded Lorosa, the Portuguese man who had aided and abetted the trespassers. The Portuguese commander shrank at executing the Spaniards, however, confiding to João III that he simply "detained them in the Moluccas because it is an unhealthy place, to see if the climate would kill them." Reduced to slavery in various parts of Asia, the men steadily perished until five beaten-down warnings, including Espinosa, were finally sent home.[7]

The world's known supply of round-the-world travelers now numbered thirty-nine, less than 13 percent of Magellan's original crew. Even compared to other sea voyages over unknown routes, the 86 percent mortality rate on the Magellan/Elcano expedition was terrible. Fewer people had been lost on da Gama's first voyage to India (about 50 percent died) and on Columbus's first voyage to America (about 44 percent were left behind on Hispaniola). The casualty rate on a world-round voyage did not resemble what occurred on other expeditions; it was more like battle, another reason it may have seemed so intrinsic to the global politics of empire.

João III and Charles V arranged several diplomatic meetings to debate their competing imperial claims, which were, for the first time, designed to fit a round Earth with better-known dimensions. The Spanish used the Magellan/Elcano expedition to claim that the Treaty of Tordesillas did not divide a flat map by a line running pole to pole, but instead segmented a globe through its poles, all the better to claim their half.

As the negotiations ground on, Charles V carefully monitored initial accounts of the Magellan/Elcano expedition, lest they weaken his claim to the Moluccas. Maximilian of Transylvania, a secretary at the Spanish court, had been given privileged access to Elcano and two others from the *Victoria*. He distilled their testimony into a long letter to his father in October 1522, first published in Rome in November 1523, then reprinted several times.[8]

The letter's main point was that an around-the-world voyage was modern, entirely different from anything accomplished by the ancient Greeks and Romans, whose heroics (mythical or real) had once defined for Europeans the peak of human achievement. Maximilian declared that the *Victoria* was much worthier "of being placed among the stars than [Jason's] old Argo; for that only sailed from Greece through Pontus, but ours from Hispalis to the south; and after that, through the whole west and the southern hemisphere, penetrating into the east, and again returned to the west." That the vessel barely survived only made it more marvelous.[9]

Charles V next commissioned a history of the expedition from Peter Martyr, an Italian humanist who, settled at the Spanish court, wrote official histories of Spain's overseas expansion. Martyr interviewed the survivors, drafted what was said to be a full account, and sent it to Rome to be printed. But Charles V's complicated ambitions guaranteed its doom. His imperial troops invaded and sacked Rome in 1527, and the manuscript that extolled his nation's amazing around-the-world expedition vanished. It was a coincidence to deter any author. Martyr abandoned the project and his manuscript has never resurfaced.[10]

Tension between Spain and Portugal also conditioned the cartographic information that was made public. Each nation appointed a board of nine specialists who presented competing evidence for the two nations' geographic claims. No original maps survive from the Magellan/Elcano expedition, but at least one chart must have made it back, from which new maps of the world were prepared. The first were frauds, several of them executed by Diego Ribero, a Portuguese cartographer in Spanish employ who had made charts for Magellan. Ribero was ordered to make new maps to support Spain's case against Portugal for the negotiations of 1524 and 1526. The results invariably placed the Moluccas on Spain's side of the line of demarcation.[11]

It was all guesswork, anyway. Without a good way to determine longitude at sea, east-west distances were easy to manipulate. It seemed unlikely, however, that the Spice Islands lay closer to Spanish America than to Portuguese India. Charles V abandoned (for the moment) the idea of an Indian Ocean empire and sought instead to expand his new world empire into the Pacific, over the islands that Magellan had died trying to secure for him.

In 1529, at a final meeting in Zaragoza on their shared border, Spain and Portugal agreed to a treaty that drew yet another human-imagined line over the world. Even at that late date, the compliant Ribero prepared maps with the Spice Islands on Spain's side of the world; on one, two ships have the notation *voy amaluco*, or (roughly), "Moluccas or bust!" But Charles V backed down. The treaty drew a line 17 degrees east of the Moluccas, a presumed anti-meridian, a circumpolar mirror of the Treaty of Tordesilla's line through the Atlantic. João III paid Charles V 350,000 ducats (probably over 400 times the value of any one of Magellan's ships) to relinquish any claim to the Moluccas while promising to honor Spain's claim to the Philippines. He made a bad bargain: the Atlantic meridian, if extended through the poles, would have given Portugal not only the Moluccas but also the Philippines. The line's imprecision mattered less than its global pretensions: it was the first civil boundary drawn *around* the whole Earth, a breathtaking imperial claim to the world, premised on the ability to go around it.[12]

By 1532, news of Spain's concession reached the stranded Spaniards holding out in the Moluccas. The Portuguese had burned their ship, they were prisoners on Gilolo, where the rajah had tired of feeding them. Barefoot, ragged, they survived by hunting wild pigs. Steady attrition meant that only twenty-eight men remained. They surrendered to the Portuguese, who took them to Lisbon. The survivors had gone around the world, but not their ship. The 6 percent survival rate of the men (not quite half that of the Magellan/Elcano expedition) and the zero percent survival rate of the ships (even worse) reinforced suspicions that to try to go around the world was, but for a lucky few, a roundabout and expensive way to commit suicide.[13]

Even before the Treaty of Zaragoza, news about the Magellan/Elcano expedition had been spreading. The first uncensored world map to reflect the expedition's discoveries was done around 1525, perhaps in Nuremberg. The map consisted of twelve woodcut-printed paper gores that could be mounted on a globe. The expanse of the Pacific Ocean, broken only by the Unfortunate Isles, is especially striking. The cartographer assumed that a large southern landmass (Magellanica, or *Terra Australis*) had to offset all that water, a possibility that voyagers and mapmakers would continue to debate. The maker also drew the line of demarcation all around the world. And note how he has invented a new device, the circumnavigator's track, a dark line tracing the expedition's route, embellished in two places with a little ship. That line around the world was the first graphic representation of a global human activity: the world or bust! (See illustration 3.)[14]

With a pious sense of mischief, Hans Holbein inserted one of these globes into his painting called *The Ambassadors* (1533), now in London's National Gallery. In this allegorical double portrait, two French ambassadors pose against a high table laden with objects of learning, including the globe on the shelf below. Do the learned objects celebrate human ingenuity and knowledge? Probably not. At the bottom of the painting looms a distorted human skull, the classic warning against *vanitas*: remember you will die, and regard any worldly thing accordingly. A half-concealed crucifix at top left promises eternal salvation, and the globe is positioned with Rome at its center, just as the Catholic Church had been the center of Christian Europe for centuries.

That was intentionally ironic. Two years before Magellan had departed, Martin Luther had publicized his *Ninety-Five Theses* at All Saints' Church in Wittenberg, Saxony, demanding extensive reforms of the Catholic Church at Rome. Earlier calls for religious reform had faded, but Luther's set off a Reformation that divided Europe between Protestants and Catholics. Just when the Magellan/Elcano expedition had made spreading the Christian gospel around the world into a tangible goal, the Reformation divided Christendom against itself. On the globe within Holbein's painting, the line of demarcation, scored in red across the Atlantic, may be prominently featured in order to lament human divisions of the world, however they come about.[15]

Narratives of the expedition were more confident. Charles V had already bought the rights to Elcano's story, only to bury it. The way was open to Pigafetta, less beholden to the king, and eager to use his precious diary and other souvenirs to his own advantage. It tells us something about his survival skills that he courted multiple patrons. From Seville, he headed to the royal court at Valladolid, where he presented Charles V with "not gold or silver, but something to be prized by such a lord," meaning "a book written by my hand treating of all the things that had occurred" on the expedition. Next stop Portugal, where Pigafetta told João III "of the things which I had seen." Then to France, where he gave the queen regent "some things from the other hemisphere."[16]

To flatter these and other worthies, Pigafetta commissioned at least four handwritten copies of his diary. (Others may not have survived, including the copy given to Charles V.) All four are dedicated to the Grand Master of Pigafetta's religious order. One is in Pigafetta's native Venetian and the others are in French. Two were done simply on paper and two more elaborately on vellum, thin polished sheets of leather. The latter are elegantly written, illuminated, and gilded. One of them is especially grand, on finest white vellum, with maps that glow like jewels within their gold borders. These de luxe editions were unique art objects, possessions of the powerful. It is not impossible that other people read them—one of the paper copies is quite worn. And maybe the owners of the fancy versions on vellum liked to show them off. The illustrated versions, for example, could be appreciated even by the illiterate.[17]

But print could get out some kinds of news even better. The relatively novel technology of print produced books faster than toiling scribes could, and it used cheaper materials: soot boiled with oil to make ink; rags chopped and fermented into paper. Printed books could now make their way to more readers and places, and that would be how most people learned of Magellan's ordeal.[18]

When Pigafetta returned to Venice, he announced himself as a "Venetian Knight of Jerusalem," who had "encompassed [*circumdato*] the whole world," one of the world's most exclusive self-descriptions. On that basis, he petitioned for a copyright to enjoy profits from his printed "narration" for the next twenty years. But for some reason, his full story did not appear in his lifetime. (He probably died in the mid-1530s.) A fragment of his account was published in France sometime between 1526 and 1536. It might have left little trace had not an Italian man of letters, Giovanni Battista Ramusio, read, translated, and republished it, along with Maximilian's letter, in a 1550 collection of travel writings.[19]

That edition had legs. It was reprinted several times. Richard Eden translated it into English and published it in London, in 1555, within his collection of travel accounts. Ramusio and Eden were not mass reading, but along with Maximilian's published letter, they opened new worlds to readers of the high and middling ranks. William Shakespeare, for instance, liked the sound of the Tehuelche word *Setebos*, for the Patagonian deity, and gave it to the god that Caliban worships in *The Tempest*. *Birds of Paradise* flit about European accounts of distant parts of the world, having been mentioned, evidently for the first time in European literature, in Maximilian's report. *Canoes* begin to float through English accounts, courtesy of Eden. And on European ships everywhere, sailors adopted what they called, in variant spellings, *hammocks*—body-slings modeled on the *amaches* that Pigafetta had first described in Brazil, and so much nicer it was to snooze swaying in one of those than on any of the hard, wet, and sliding surfaces that mariners had endured for millennia.[20]

Above all, Ramusio seconded Maximilian's suggestion that round-the-world travel was modern: "The voyage made by the Spaniards round the world in the space of three years is one of the greatest and most marvelous things which have been heard of in our times; and, although in many things we surpass the ancients, yet this expedition far excels every other that has been made up till now." It was exciting to be modern, if only as a reader. And to see the world represented on pages recycled from soot and rags was much less expensive (not to mention safer) than actually sailing around the world in a wooden world—had even 5 percent of Europeans tried to do that, there wouldn't have been a tree left in Europe.[21]

If an around-the-world voyage was modern, it followed that only a modern genre, the prose narrative in a vernacular language, could tell its full story. Although that plain, just-the-facts format now seems to us the logical and inevitable one, it was a deliberate choice and at first excluded other options, notably the ballad and the epic poem.

Timing was everything. By the early 1500s, Spanish ballads had largely ceased to relate current events, as they once had done. If ballads were sung of Magellan or Elcano, they have vanished.[22] Nor were the world encompassers featured in an epic poem, the genre made famous by Homer and Virgil. However suitable for the mythic characters who wandered the Mediterranean, the epic evidently seemed an awkward choice for round-the-world travelers. The Renaissance explorers who commanded shorter voyages were comparable to the ancients (Columbus was a new Jason, da Gama was like Odysseus, and so on), but those who went around the world were not, as Maximilian and Ramusio had noted.[23]

The only epic to mention Magellan did so to denounce him. In *The Lusíads* (1572), Luís Vaz de Camões celebrates not Magellan but da Gama, reminding his primary reader, the king of Portugal, that da Gama had helped give him an empire on which the Sun barely sets:

> *You, mighty King, on whose India*
> *The new-born sun directs his first beam,*
> *Shines on your palace in mid-hemisphere,*
> *And casts his last ray on the Brazils.*

Yet the reader is later reminded to

> *. . . glance westwards*
> *To observe the exploit of a Portuguese*
> *Who, believing himself snubbed by his king,*
> *Made another voyage beyond imagining*

and gave to Spain the rest of the Sun-caressed planet.[24]

There was Magellan's epic moment, dispatched in a sideways glance. Only one other epic poet even considered him. In his *Jerusalem Liberated* (1581), the Italian author Torquato Tasso wrote of the first crusade and other episodes of Christian heroism, and, after reading Ramusio, drafted an episode that followed Magellan's passage into the Pacific. But Tasso scuttled that section in favor of a more conventional celebration of Columbus. For the first around-the-world travelers, there would be no ballads, no epics. Their glory was too distinctive.[25]

Moreover, round-the-world voyages were not, compared to other explorations, so firmly connected to specific nations or empires. Columbus "discovered" America *for Spain*; da Gama opened direct trade to India *for Portugal*. But if the Magellan/Elcano expedition went to the Moluccas *for Spain*, it went around the world for its own reasons. The motto on Elcano's augmented coat of arms had already made that point. And, as Ramusio said of Pigafetta, he "has circumnavigated [encompassed] the whole globe," for which "the ancients would have erected a statue of marble to him . . . as a memorial and example to posterity of his great worth"—no mention of Spain. It would be difficult for Spain's ambitious monarchs to reclaim the deed for themselves, though they would try.[26]

It seems fitting that news of the first around-the-world journey had an international audience. A letter in Malay from Ternate had informed the king of Portugal of the Spanish ships' arrival there, with news that the one which remained (Espinosa's) awaited twenty more from Spain. Rulers in Turkey were also interested in European expansion, and made sure to get copies of European world maps as well as globes. A locally made map of 1559 included extracts from Ramusio. And somehow, a European map with a polar projection, possibly based on the one Magellan had used to win over the king of Spain, ended up at Topkapi Palace in Istanbul. Charles V had been right to worry that strategic information about his nation's historic expedition would steadily leak out to the rest of the world.[27]

If imperialism gave world-round travel one original meaning, the element of time travel was equally arresting. Travelers had always changed their place, but only around-the-world travelers also changed time, in the mind-boggling Circumnavigator's Paradox. Although some Christian and Jewish philosophers in the Middle Ages had hypothesized that a person would gain or lose a day by encompassing the world, the phenomenon had never been widely discussed, let alone explained. One scholar suggested that to circle the globe was to race the Sun, and lose. Magellan's men had sailed west with the Sun, yet always lost time, each of their days longer than it would have been on land, extra minutes that added up to twenty-four hours. That discovery was another modern triumph. According to Ramusio: "the ancients never had such a knowledge of the world, which the sun goes round and examines every twenty-four hours, as we have at present, through the industry of the men of these our times."[28]

The idea that humans could race the Sun made them into cosmic entities, Phaetons who did not crash and burn. But were they, really?

In his book *On the Revolutions* (1543), Nicholaus Copernicus questioned whether the Earth was the unmoving center of the universe. His observations indicated that it rotated on its axis and circled the Sun—it was just another planet. It is a myth that Copernicus's book was either too arcane for general discussion or universally rejected as heresy. His revolutionary ideas had immediate impact, which means that at least some people who read Ramusio (published 1550) would have rolled their eyes at its outdated geocentric worldview. Some of Copernicus's background points about the Earth, however, were also out of date, showing that he had not read any of the accounts of the Magellan/Elcano expedition. Copernicus mocked the notion "that the [Earth's] entire body of water is ten times greater than all the land," for instance, and argued for the opposite case, citing cartographers who had pushed Asia farther east than the ancients had believed it to extend. The new Spanish and Portuguese discoveries, "especially America," had added even more land to the world. [29]

Obviously, Copernicus's knowledge of the terrestrial globe had climaxed with Vespucci and Waldseemüller. His book was both ahead of and behind the times. And because the Circumnavigator's Paradox would continue to be explained in a pre-Copernican fashion, as a race with the Sun, it too would be both an entirely new experience, yet understood in outdated terms, perhaps to blunt the shock of the new. In an engraving published in 1594, Magellan passes through the strait that would bear his name. He is guided by Apollo, the Sun god, as if the deity were acknowledging that, in his daily transit around the globe, he now had human companions.[30]

A world encompassed was a gift fit for a prince. Or so Charles V thought when, in the early 1540s, he gave a gorgeous atlas to his heir, Philip. The royal present came from the Venetian workshop of Battista Agnese. This unique edition on vellum, with illumination and gilding done to order, shows in quick succession three images of the world, indications that the globe, post-Magellan, was becoming a commonly represented object. On a special presentation page, Charles V presides in classical Roman dress as a new Caesar. Below him, Philip also wears Roman garb, but reaches heavenward for something no Caesar ever ruled: the whole globe, which God hands down to him. The next page offers a pre-Copernican view of that globe encircled by the Moon, Sun, planets, and signs of the zodiac. The first map in the atlas is an oval image of a world dominated by Spain, with the Americas at the center. A circumnavigator's track marks Magellan's route. The map somewhat disguises the breadth of the Pacific Ocean by pushing it far left, though it is big enough to emphasize Magellan's pluck, and the strait he had followed to the Pacific bears his name, as his crew had suggested it should.[31]

A second ship's track streaks from Spain through the Caribbean, over the Isthmus of Panama, and down Peru, marking yet another of Charles V's glorious acquisitions. After Francisco Pizarro had conquered the Inca empire in the Peruvian Andes in the 1530s, other Spanish explorers followed every clue to pinpoint the source of the region's abundant silver. They found it in present-day Bolivia: an entire mountain of silver at a place called Potosí. The Spanish surrounded the mountain with a mining town and Indian workers extracted tons of silver a year. Teams of mules and llamas toted the metal down to the sea where treasure fleets transferred much of it to Spain, whose glittering wealth made other Europeans dangerously envious.

Silver traced the extent of Spain's empire, not quite all around the world, but more than halfway. Magellan's stubborn invasion of the Philippines—named for Prince Philip in 1544—had given Spain its back door to the Indies. American silver would keep the door open, a constant threat to Portugal's interests there, whatever the terms of the Treaty of Zaragoza.

An expedition of 1565 established the first Spanish settlement in the Philippines. But would Spaniards have to keep making full circuits of the globe to maintain this territory? That would be tedious. The commander of the 1565 venture ordered Andrés de Urdaneta (chronicler and survivor of the Loaysa expedition) to find an eastern route across the Pacific. Urdaneta sailed north up to 36 degrees, where the eastward winds that Espinosa had failed to find wheeled him over and down to Acapulco, though at the cost of fourteen men. On arrival, the shattered survivors could barely lower the anchor.[32]

Urdaneta's route became the means by which Spain linked together its rich global holdings, and Spain had no reason to contemplate an-

other official circumnavigation for over two hundred years. At least
once a year, a galleon made the 8,000-mile run between Manila and
Acapulco, the longest active ocean crossing in the world at the time,
and then went back. The Atlantic and Pacific now carried a busy traffic
in Spanish officials, documents, settlers, and materials. Colonies spread
over Central America, down the west coast of South America, and
across the Philippines. Maps sprouted Spanish settlements in places
unknown to Europeans a decade earlier. All those places produced
something—wool, wine, lumber, grain—but above all silver. The Atlantic silver fleets were heavily guarded, but their sailing times and routes
were public knowledge. And the silver that crossed the Pacific was un-
guarded. The Spanish considered "their" ocean a Spanish Lake closed
to outsiders. Their silver-filled ships were perfect targets for pirates,
including a very bold English pirate, Francis Drake, who completed
history's second successful around-the-world expedition as a global
round of plunder.

Unlike Magellan or Elcano, Drake was painted from life several times
and described by many people. Handsome, he was nevertheless on the
short side; a lively conversationalist, he diluted his charm with constant
bragging. A small nobody from nowhere, he thought he had to remind
people that he mattered. He came from a poor family in southwest England, an extended maritime clan that launched him as a sailor—and
whatever else was said about him, everyone agreed that Drake was a
fantastic mariner. His nation barely figured on the global stage, but the
English had ambitions. Just as the Spanish and Portuguese had once
craved Italian lucre, so the English now snapped at Spanish heels.[33]

England was one of several non-Iberian nations, especially Protestant nations, to contest Zaragoza's hemispheric division. The English
defiantly claimed parts of the North American coast and Caribbean
where no other Europeans had settled, and they loudly lamented how
Indian souls fell daily into the errors of Spanish Catholicism. A religiously inflected war, sometimes hot, sometimes cold, divided Spain
and England. Although Elizabeth I was careful not to overreach her
military capacity, Drake convinced her that nips at Spanish shipping
in the Pacific might be safer and cheaper than attacks in the Atlantic.

He also promised to find places where England might trade or send colonists. Elizabeth extracted securities from Drake (to discourage his wanton plundering of Spanish territory) and gave him a commission to attack Spanish ships (he waved the document at one of his victims). He thus became an official privateer rather than a stateless pirate.[34]

Elizabeth granted Drake five ships of varying sizes and capabilities, the usual plan, though Drake's other preparations show a shrewder sense of the demands of long-distance seafaring. His crew was small, only 164 men. And his flagship, the *Pelican* (about 150 modern tons in size, with 19 guns or cannon), was brand-new, possibly custom-built, with a double lining to protect against wear and tear. The high ship-to-man ratio, and the spanking new flagship, hinted that Drake was willing to lose everyone and everything—except his precious self. Moreover, he anticipated worst-case scenarios. He carried pieces of a prefabricated pinnace, a small boat that could be rowed or rigged with a sail, good for rescue or escape. He recruited a variety of workmen to make repairs en route, packed every conceivable weapon, and traveled in style. One witness claimed that Drake ate off gold-rimmed silver plates and dined to the "music of viols," with "dainties" to nibble and "perfumed waters" to dilute any shipboard stink.[35]

Drake also kitted himself out as expedition chronicler. Like Pigafetta, he carried plenty of paper, plus ink and watercolors, and he and his nephew evidently passed the time at sea by making drawings. Lest these seem frivolous amusements, consider how one Spaniard "grieved" to see that Drake was recording, with devastating accuracy, the terrain and resources of Spain's empire. And Drake carried knowledge on paper, especially Spanish and Portuguese accounts and maps of the Americas, the Pacific Ocean, and the Indian Ocean. Before he left, he visited Lisbon, where he consulted Portuguese navigators and bought a wall-sized chart of the world. One witness claimed it was a copy of the very chart Magellan had used; others noted that Drake had an account of Magellan's voyage. Indeed, Drake inaugurated a custom among round-the-world travelers of citing their predecessors, of seeing themselves as part of a tradition. When the English reached the place "called by Magellan Port S. Julian," they found "a gibbet standing upon the maine, which we supposed to be the place where Magellan did execution upon some of his disobedient and rebellious company." It was exactly there that Drake accused one of his company, Thomas Doughty, of treason, put him on trial, and beheaded him.[36]

His ruthlessness certainly helped, but the new nature of the world was the secret of Drake's success. Ships had not improved—they still needed to nuzzle into shore for nurture—but much more of that shoreline, meaning the Spanish and Portuguese colonies, contained the equivalents of shipyards, provision shops, and personnel offices that, if a person was willing to raid them (and Drake was), provided everything (see Map 1). Although the expedition was supposed to last two years and in fact took almost three, it carried provisions only for eighteen to twenty months. Drake was the first around-the-world traveler to have the luxury of not packing what he could acquire along the way, a decision only possible because much more of the world was being developed for European use.[37]

The first thing Drake began acquiring was more ships. Cruising down the west coast of Africa, he came across Spanish and Portuguese fishing boats as well as four larger caravels, all of which he appropriated, with their cargos of fish and supplies of biscuit. He redistributed his crew into the new ships and towed the boats behind. In the Cape Verde Islands, he took two more ships and their "good store of wine." Why the large flotilla? Drake knew he was most likely to face Spanish challenges in the Atlantic—best to look big. Also, the extra vessels stored, and were, materials to be consumed as needed. Off the Atlantic coast of South America, Drake hauled ashore and burned one boat in order to extract its reusable ironwork, then ripped firewood from one of the Portuguese ships before abandoning it. Nor did he bother to maintain anything but the *Pelican* once he was in the Pacific. Drake eventually lost all but his flagship, which he renamed the *Golden Hind* to honor a patron whose arms included a hind, a speedy female deer. Beyond the Atlantic, speed was what Drake needed.[38]

The most precious resource, geographic knowledge, came in human form, and Drake stole that too. When he appropriated the Portuguese ships in the Atlantic, he released the crews but kept one pearl of a pilot, Nuño da Silva. Off Chile, he took and released some Spanish prisoners but kept back a Greek man to guide him to Lima. Near Mexico, he detained a pilot from a Spanish ship to get him to a town rumored to be rich yet little guarded. These kidnapped pilots might have found it ironic that their official status was rising. As European overseas empires expanded, navigation was becoming an art or science, a maritime skill superior to anything ordinary sailors acquired through repeat experience. Pilots who could navigate open ocean had special prestige—da Silva was a *piloto de altura,* pilot of the high seas. That was exactly why Drake reduced him to slavery.[39]

Although he benefited from a landscape settled by Spaniards, Drake denied their right to it, starting with the Strait of Magellan. At its entry, he discovered "the bodie of a dead man, whose flesh was cleane consumed"—so much for Spanish threats to guard the place. He questioned Spanish belief that the strait divided the Americas from a southern continent, suspecting the southern land was actually islands, one of which he named for Queen Elizabeth. He sailed far enough south to guess that a ship could clear the land without threading the famous passage. If his Portuguese informants had suggested the existence of an open sea, Drake was the first European to see it. His map of the area is now lost but another, done by the expedition's minister, the Reverend Francis Fletcher, shows a northern continent and southern islands.[40]

When his fleet coasted up the western side of South America, Drake marveled that it little resembled what appeared on Spanish maps, so was not "truly hitherto discovered, or at the least not truly reported," until by him. He also asserted that the Indians resented "the cruell and extreme dealings of the Spaniards." One Indian, thinking Drake was Spanish, directed him to Valparaíso in Chile, where he found his next target, a ship he commandeered along with her cargo and two of her crew.[41]

Here Drake began to raid Spanish settlements (see Map 1), each time demonstrating how much the Spanish had developed a landscape convenient to him. At Valparaíso the population had fled, so the English looted the town at leisure. To make a dubious point about the virtues of Protestantism, Drake presented the Catholic chapel's communion silver and altar cloth to the Reverend Mr. Fletcher. (His men would later desecrate a Catholic church with iconoclastic precision.) Meanwhile, his crew transferred the contents of a wine warehouse to the ship, along with fragrant boards of cedar. Lacking any market for the valuable but bulky lumber, they simply burned it, some of the world's costliest firewood.[42]

"To rifle": that was the operative verb, as in "our Generall [Drake] rifled the ship"; "we found there three small barkes which we rifled"; "Our Generall rifled these shippes." Time and again, the Englishmen were hip-deep in someone else's cargo and choosing what they needed that particular week. They were picky eaters where Magellan had starved, abandoning hardtack, throwing grain and oil into the sea. Their cargo soon included, along with gold and silver from Spanish America, porcelain and silk from China, fine linen cloth from Europe, and many items of jewelry unclasped from their original owners. The pirates admired their own greed, as if they graciously relieved Spaniards of a terrible burden, the heavy wealth from Potosí. When they came upon a Spaniard slumbering on a beach, thirteen bars of silver gleaming beside him, the Englishmen sneaked up, "tooke the silver and left the man." When he confronted a Spanish pilot who owned two silver-gilt cups, Drake teased him, "Segnior Pilot, you have here two silver cups, but I must needes have one of them," and what could the poor man say?[43]

The most hilarious prize was the biggest, the Spanish treasure ship *Nuestra Señora de la Concepción,* named for the Virgin Mary. Her men had nicknamed the heavily armed vessel the *Cagafuego,* or *Shitfire,* much as men of a later era called a fighter plane the *Spitfire.* Drake had learned of the ship from another he had rifled, and tracked her to Panama, looting gold from yet another vessel on his way. After a brief battle, Drake's men shot down the *Cagafuego*'s mizzenmast and boarded her. (They misheard the ship's nickname, which they transcribed as *Cacafuego.*) On board, they discovered over 25 tons of silver, 80 pounds of gold, miscellaneous gems, and that unlucky pilot's two silver cups. As the Englishmen packed up the spoils, the pilot's young page made free to say to Drake, "Captaine, our ship shall be called no more the *Cacafuego,* but the *Cacaplata* [*Shitsilver*], and your shippe shall bee called the *Cacafuego.*" This "pretie speech of the Pilots boy ministred matter of laughter to us," the pirates fondly recalled, "both then and long after."[44]

But look hard at the pretty booty. Some useful item (or person) is usually mixed in with the gorgeous loot. One ship offered a chest of silver, but also linen cloth. Another vessel provided gold pesos, plus a sea chart of the Spanish Pacific. A gold crucifix inlaid with emeralds came along with a cargo of ship's rope and rigging. A golden falcon with another inset emerald was accompanied by one of the captive pilots. From Guatalco in Oaxaca, the pirates extracted "the entire supply of Indian women's petticoats" as trade items (not for cross-dressing). Drake also realized that the Spanish would probably not protest some of his raids. The captured silks and porcelain, and stupendous trans-Pacific traffic in silver, revealed that the Spanish were trespassing on Portuguese trade in Asia. Complaints to Elizabeth would advertise their failure to respect the Treaty of Zaragoza. And Drake knew when to stop. After detaining a vessel that carried a governor to the Philippines, he released the ship, and "sufficiently satisfied, and revenged" on Spanish slights against his nation and himself, he stopped the plundering.[45]

Where next? Drake enjoyed spreading rumors about that—he knew the Spanish authorities were trying to track him. Several times, he showed Spanish prisoners his large world map and asked them to contemplate his possible routes home: back through the Strait of Magellan, or else another passage through the Americas, or via the Indian Ocean and the Cape of Good Hope, perhaps with a detour to China? Upon release, the captives duly contradicted each other about Drake's next move. Surely the "English Corsair" would not attempt Magellan's route over the Pacific, Indian, and Atlantic oceans, which the *Victoria* had barely survived? Wouldn't he have to go back the way he came—and be apprehended? One man suspected a bluff: Drake threatened to go to Asia just to throw the Spanish off the scent. They were all wrong. Drake meant to rival Magellan, either by going around the world or by discovering a strait through North America to mirror Magellan's strait in the south.[46]

His rapacious little tour of Spanish America had confirmed that Europeans could settle comfortably along the Pacific, and that England might make its own empire there. But Drake had come up against a geographic puzzle that the Magellan/Elcano mission had identified: if God had made the world so big, why wasn't it easier to navigate? The Almighty had a baffling sense of humor, given His crafting such an oddly configured globe, with so much water in the Pacific, so little land down below to balance all the continents at the top, and so winding a way through the water and around the land.

A Northwest Passage would be a great help. Along with the Strait of Magellan, it would provide a circular route from Atlantic to Pacific and back, without the time-consuming, ship-destroying, crew-killing exertion of going around the world. A northern shortcut into the Pacific and back appealed especially to Protestants in northern Europe, who were denied free passage over the Spanish Lake.

Drake headed north to do some reconnaissance, first releasing his European captives at Guatulco, keeping only those of African descent, two men and a woman, Maria. He clearly regarded the Europeans as potential spies but the blacks as biddable labor. His decision also reflected his failure, despite generous financial offers, to recruit a Spanish pilot who knew the route to China. None of his captives had that skill, not even Nuño da Silva, whom the Spanish authorities at Guatulco imprisoned and accused of various things, above all seeming rather helpful to his English captor, what with ushering him through the Strait of Magellan. (Smoothly switching sides, da Silva offered to show the Spanish how the strait could be fortified and supplied from Peru.)[47]

Drake's men proceeded past Spanish outposts in Baja California, and eventually into "a faire and good Baye." "The Spaniards hitherto had never bene in this part of the Countrey," they claimed, "neither did ever discover the lande by many degrees, to the Southwards of this place." Drake named the site Nova Albion, remarking that its "white bankes and cliffes, which lie toward the sea," resembled the white cliffs of Dover. To a tall post, he nailed a metal plate inscribed with Elizabeth's right and title to the land, and attached a sixpence, in effect a tiny portrait of the queen, and the one piece of silver he gave back to the Americas. His search for a passageway might have taken him as north as far as 48 degrees (roughly parallel to Vancouver Island) where he gave up and headed across the Pacific.[48]

Drake made the standard stop at the Moluccas and loaded up on spices. So gorged was the *Golden Hind* that she snagged on a submerged rock off Celebes. That was annoying, and dangerous, because the waves might grate the ship over its obstacle until the hull wore through. From eight in the evening until four in the afternoon the next day, the men jettisoned non-essentials: eight cannon, 3 tons of cloves, some grain and beans. With that and a timely shift in the wind's direction, their ship lightened, their sails filled, and they were free.[49]

Not all of them. Some of the men stayed behind with the beans; how they earned that fate is unclear. And somewhere between the Moluccas and Java, Drake stopped at an uninhabited island and "left the two negroes and the negress Maria, to found a settlement, leaving them rice, seeds and means of making fire." The euphemism of "a settlement" and the passage of over four months since Maria's abduction hint at the truth: she was pregnant, inconvenient evidence that she had been enslaved for the pleasure of the captain, and possibly other officers. Drake dumped her, an act that, more than any other, showed his instrumental attitude toward the resources of Spanish America (human or developed by humans), which had kept him alive and made him rich.[50]

Unlike Magellan, Drake had survived; unlike Elcano or Espinosa, he had a good ship beneath him; unlike the survivors of Loaysa's expedition, he was no man's prisoner. He and his crew even had the energy to admire the Cape of Good Hope, "the fairest Cape we sawe in the whole circumference of the earth." True, on the final approach to England, water had to be rationed. But 59 of the original 164 men returned, a survival rate of 36 percent, pretty bad, yet better than ever. The Englishmen confirmed Elcano's marvelous discovery, the loss of a day going westward around the world. And Drake brought with him over 10 tons of silver bullion and slightly more than 100 pounds of gold. His queen was thrilled.[51]

As Charles V had done, Elizabeth insisted that all information about the expedition go through her, to maximize its strategic value. She summoned Drake to report with "specimens of his labours." He surrendered his large chart and illustrated diary (and, later, much of the loot). The official story was that he had gone through the Strait of Magellan and up the west coast of the Americas. His foray into Portuguese territory in Asia was kept secret (perhaps to conceal his failure to find a Northwest Passage), as was his report of open sea beyond Tierra del Fuego, information of possible future use. Yet the English quickly learned that their "Country Man hath gone rounde about the whole world," as the author of a 1581 pamphlet crowed. That precis congratulated Drake for outstripping the ancients—"for greater h[e]art did never Caesar beare"—and scoffed at Virgil's and Ariosto's epic poems and the ballads of Robin Hood, those tired old celebrations of smaller achievements. Some facts about Drake's voyage would appear on a map of North America in 1582. The image announced England's dominion over portions of the Americas where the Spanish had not settled, without much detail.[52]

The rumors were enough for the Spanish ambassador to complain to Elizabeth that Drake had entered Spanish territory, especially the Pacific, without permission. The queen retorted that "the use of the Sea as of the Ayre is common to all," firm rejection of the Treaty of Zaragoza—or any global demarcation. The Spanish disagreed, and some officials suggested that "if the Strait [of Magellan] could be fortified and manned, it would remedy everything." This was the first attempt to create some kind of tollbooth to control round-the-world traffic. Because chilly Patagonia permitted little agriculture, no European colony could survive there unless supplied by relief ships, as da Silva had pointed out. How much would Spain spend to thwart the global ambitions of others?[53]

The other place Spain sought to control was Portugal, not least to silence Portuguese chatter about strategic sea routes and complaints about Spanish trade in Asia. Portugal was undergoing a succession crisis, and, disregarding the claims of one Don Antonio to be king, Philip had marched his troops into the country in 1580 and became king of a united Iberia—and united Iberian empires—in 1581. Now his dominions did go all around the world. Among other things, he acquired an elaborate tapestry originally made for the king of Portugal. In it, Atlas, the Titan from Greek legends, shoulders the heavens, represented as an armillary sphere with Earth at the center. That titanic image became a hallmark of Spain's global claims. So did Philip's assertions, struck into a medal of 1580, that he was king of Spain and the New World (on the obverse) and that NON SUFFICIT ORBIS—The world is not enough (on the reverse). The Latin motto broadcast Spain's triumph in bursting beyond the known world of the Ecumene and around the entire actual globe.[54]

Elizabeth had good reason finally to announce that Drake, her champion, had circled the globe that Spain tried to hug to itself. Yet again, the modern nature of around-the-world travel caught the imagination. In what was probably the first reference to Nova Albion (in 1582), one poet emphasized that Drake had:

> *. . . recently sailed round*
> *The vast circumference of Earth (a feat*
> *Denied to man for many centuries),*
> *To show how father Neptune circumscribes*
> *The continents, and wanders in between . . .*[55]

Even better, the queen's hero was a commander who had completed an around-the-world expedition, an Englishman who had out-Magellaned Magellan. Elizabeth arranged for the *Golden Hind* to be removed from service (a costly decision) and "preserved as a memorial" in a ship-sized house. In the spring of 1581, she knighted Drake and granted him a coat of arms, perhaps partly of his own design. The device has a shield with two stars to indicate the polar extremes of his travels, farther south than anyone before, then far north to look for a Northwest Passage. Below the stars runs the motto SIC PARVIS MAGNA— Great things from small beginnings—a nobody from nowhere was now a knight of a world power. The crest above the arms features a globe on which a monumental *Golden Hind* teeters beneath the north star. The ship is tethered to a hand emerging from clouds, the hand of God, whose beneficent bridle encircles the globe. The pious tag AUXILIO DI-VINO (divine assistance) also emphasizes that Drake had providential help.[56]

The Spanish countered any suggestion that God preferred Protestants by expanding their missionary efforts. With Spain and Portugal united under Philip II, Catholics saw an opportunity to make the Church at Rome truly universal. Catholic missionaries of this generation included the first repeat world-circler, the tough character who did it twice: Franciscan friar Martín Ignacio Martínez de Mallea, commonly called Loyola (he was grandnephew of the famous Ignatius of Loyola, main founder of the Society of Jesus or Jesuit order). In addition to making two journeys to Spanish America, the younger Loyola went around the world from 1580 to 1584, then again from 1585 to 1589, when he became the first person to circle the globe eastward.[57]

Like Drake, Loyola benefited from Spanish development of a Europeanized landscape that straddled the Atlantic and the Pacific. He was the first around-the-world traveler who (see Map 1), by traveling over the Isthmus of Panama to catch a ship in Acapulco, was spared passage through the Strait of Magellan. From Acapulco he went to the Philippines, along sea routes that, from November to January, were so gentle that Spanish sailors called them the *Mar del Damas*—the "Sea of Ladies"—as if a crew of soft-handed lovelies could cross with little effort and plenty of time for embroidery. From the Philippines, Loyola headed to the real prize, China, where he was one of the first Christian missionaries in that country, already famous for its enormous territory and population.[58]

Loyola's account of his first circling of the world appeared in Rome, as part of a 1586 collection of sources on China (translated into French and English in 1588). His story mostly follows the conventions of an around-the-world narrative, including the expanding catalog of colonized places—tamed by forest industries, agriculture, and grazing livestock—where European travelers could find provisions. His main innovation was to transfer the traditional Christian critique of the sins of a metaphorical world to a measurable globe. He estimated that he had traveled 9,040 leagues around the world, plus many others within the lands he visited. "All these leagues are full of mightie kingdomes," he allowed, yet "al, or the most part of them, are subject unto the tyranny of Lucyfer. God, for his infinite mercy, convert them, and take pittie on them."[59]

Queen Elizabeth, who feared Catholic tyranny nearly as much as Lucifer's, encouraged new strikes at the Spanish Empire, including another encompassing of the world, led by Thomas Cavendish, from 1586 to 1588. Cavendish had learned several things from Drake. He commanded only three ships (the largest, the *Desire*, was only 140 tons) and 123 men. He took workmen, tools, and a portable forge; he carried quantities of salt to preserve food along the way. And, by following Drake's route, he was able to take resources from the Spanish Empire whenever possible. Narratives of the venture would boast that Cavendish had burned nineteen or twenty ships, plus towns too numerous to list, and gleaned a "great quantitie of treasure." The expedition also mocked Spanish failures: Cavendish intercepted twenty-four starved Spaniards (of an original three hundred) who were trekking out of a ruined settlement established in the Strait of Magellan in 1583—no global tollbooth, yet.[60]

For all that his ruthlessness would rival Drake's, Cavendish embodied something else: the thinking explorer. He was the first commander of an around-the-world expedition to be college-educated (Cambridge) and he took with him another college man, mathematician and geographer Robert Hues (Oxford). This, the first fully intended and successful global circuit, would also be the first done with natural science in mind, the start of a long tradition of scientific circumnavigation.[61]

The Cavendish map of the world followed Drake's in tracing where nature was giving way to European settlement. The expedition carried two years' worth of supplies, ample, as it turned out, given that Cavendish circled the globe in just under twenty-six months. Contact with land was still necessary, however. Some details of the voyage make the ruthless corsairs sound like anxious housewives, perpetually monitoring the health of their wooden world. At Sierra Leone, the men did laundry and gathered lemons. Fresh shirts and citrus did not prevent several crewmen from developing "Scurbuto, which is an infection of the blood and the liver," and some died before they could eat the "refreshing" seagulls, penguins, and seals available on the other side of the Atlantic. Once the fleet threaded the Strait of Magellan, past the failed Spanish outpost Cavendish mockingly named "Port Famine," it encountered a Europeanized landscape that had begun to spill even beyond European settlements. Near the Bay of Quintero in Chile, the Englishmen saw herds of Spanish cattle and horses that had gone wild; Indians sold them a mix of European and New World supplies, "Spanish wheate, potatos, hoggs, hens, dryed dogfish, and divers other good things."[62]

Sometimes, the provisions were plunder. From one Spanish ship, the English stripped the sails, rope, and firewood. Off Peru, Cavendish captured two merchantmen stocked with grain, peas and beans, molasses, marmalade, clarified lard, and Castile soap. One of the ships also carried several barnyards' worth of live and clucking hens, estimated between 1,000 and 2,000—why bother to count? All those consumers who, somewhere in the Spanish Empire, had been looking forward to a soapy wash in the morning, followed by bread and marmalade and lard-fried eggs, were out of luck.[63]

Sometimes, the plunder came in human form, people held against their will to provide labor or information. The Englishmen were outraged when a Spaniard they had rescued from Port Famine fled to the Spanish in Chile, despite "deepe and damnable o[a]thes which he had made continually to our general . . . never to forsake him." Off the coast of Jalisco, Cavendish's men abducted three married couples, then released the women, who duly returned to ransom their husbands with baskets of fruit. Cavendish also appropriated foreign pilots, including "*George* a Greeke, a reasonable pilot for all the coast of *Chili*," a Spaniard from another ship when they got further north, and a native of Marseilles who seemed eager to betray the Spanish.[64]

And sometimes, the plunder was just plunder. Following a tip from his captive Frenchman, Cavendish kept watch for the *Santa Ana*, incoming from the Philippines, and early one morning, the lookout called out good news: "A sayle, A sayle." A 600-ton Leviathan, the *Santa Ana* had a crew of nearly two hundred, twice that of Cavendish's forces. Yet amazingly, she carried no arms—despite Drake's predations, Spaniards still did not expect to defend their Pacific traffic and the *Santa Ana* was quite the sitting duck. Her cargo of 122,000 silver dollars was far too much for Cavendish to carry away, but if he couldn't have it, no one could. Having set the Spaniards ashore and transferred 40 tons of their cargo to his vessels, he torched the *Santa Ana*, which sank along with the unretrieved treasure. "This was one of the richest vessels that ever sayled on the Seas," won with the loss of only two Englishmen.[65]

The booty included a bonus: the means to get to China. From the *Santa Ana*'s crew and passengers, Cavendish culled three Filipino boys and a Spanish pilot, Thomas de Ersola, who knew the route to the Philippines, plus two young Japanese men (christened Christopher and Cosmas), and a Portuguese man named Nicholas Roderigo who had lived in China. It was probably from Roderigo that Cavendish acquired a map of China, something nearly as precious as the tons of silver.[66]

Cavendish took on wood and water in California, made repairs, and, with cool nerve, went straight to the Philippines. There, he observed Manila, describing its trade with Acapulco and China. For whatever reason, he decided against a visit to China and instead went to the Moluccas, where he incited unrest among the Portuguese by claiming (and how would they know otherwise?) that Elizabeth was sheltering Don Antonio, the Portuguese claimant to the throne, and favored an independence movement against Spain. Cavendish then headed for the Cape of Good Hope and home.[67]

When he returned in 1588, Cavendish was the youngest person to circle the world, only twenty-eight, and by many measures the most successful. The longer of the two surviving narratives of the expedition lists thirty casualties; if that was the total, the 76 percent survival rate was remarkably high, though the loss of ships as appalling as ever. Yet the lone survivor, the *Desire*, was said to have sailed into the Thames with her crew clad in silk and her sails made of damask, hints of the plunder within. Unusually, a few ballads celebrated Cavendish's achievement.[68]

His return coincided with another English triumph. That year, an Anglo-Dutch force repelled the famous Spanish Armada. Because Drake lent a hand, his reputation as Protestant scourge of Spanish papists spread into northern Europe. An engraving of him watching a warship being loaded, published in Germany around 1588, describes him as encompassing the globe. He and Cavendish also enjoyed greater publicity at home. Having defeated Spain, Elizabeth finally permitted herself to brag openly about her around-the-world heroes.[69]

Prose narratives continued to be the preferred format. The first for Drake and Cavendish appeared in a 1589–90 collection of travel narratives, Richard Hakluyt's monumental *The Principall Navigations, Voiages and Discoveries of the English Nation*. Hakluyt gave a complete history of English travels in three sections, "according to the positions of the Regions whereunto they were directed," prefaced by a foldout map of the world based on Abraham Ortelius's oval world map of 1587. To the voyages into particular parts of the world were *"added the last most renowned English Navigation, round about the whole Globe of the Earth."* Hakluyt included a six-page narrative of Cavendish's voyage and a two-page summary of his map of China. But he only received permission to publish an account of Drake's expedition after the rest of the book had gone to press. He produced a brisk 10,000-word summary, an unpaginated special section tipped into only some copies of his work, to the delight of the book collectors who now love to find such things.[70]

Hakluyt emphasized the damage the world-encompassing Englishmen had done to the Spanish Empire: so many ships taken, so many towns plundered, so many Spaniards harassed. And to rub it in, his accounts stress how Drake and Cavendish cleverly used Spanish and Portuguese resources—kidnapped pilots, commandeered ships, looted stuff—against the very empires that had generated them. It made for patriotic reading, though readers in England (and elsewhere) also learned about the Europeanized landscapes spreading around the world's oceans, a somewhat less partisan point, designed to praise European colonization more generally, whoever did it.

On that level, the narratives were useful to friend or foe. Thus the account of Drake's expedition notes Cape Verde's grapevines and coconut trees (foreign transplants), but warns that the goats (also introduced) are shy and hard to catch. The narrative later recommends the mouth of the Río de la Plata for its unwary seals and fowls, and "no want of fresh water." In an account added to a later edition of Hakluyt, Cavendish's shipmate Francis Pretty extols Patagonia's delicious baby seals, "hardly to be knowen from lambe or mutton," and so much easier to kill than their elders, which needed to be thwacked over the head a tedious number of times. Pretty also recommends dried, salted iguana and notes St. Helena's abundance of water and provisions on the final approach back to Europe. In differentiating among parts of the world—colonized here, naturally abundant there, and pack ship's biscuit for the places in between—the narratives were Renaissance *Rough Guides* for European adventurers.[71]

Maps also represented a world encompassed. Their new global exploits inspired the English to import mapmakers (mostly from the Low Countries) and set up their own cartographic industry. The first map to represent Drake's voyage was executed in the very medium that had inspired the venture: silver. Michael Mercator, member of a famous Flemish cartographic family, made the palm-sized silver medallions in 1589. Each side shows a hemisphere of the world, complete with Drake's cruise track. Maybe Drake himself commissioned the medal—impossible to prove, though easy to believe he would have welcomed the shiny monuments to his expedition.[72]

More conventional maps on paper followed. The most impressive was a circa 1590 double-hemisphere world map that featured Elizabeth's portrait and Drake's *Golden Hind*. It is attributed to the Flemish engraver Jodocus Hondius, who worked in both London and Amsterdam. Hondius used a brand-new method of geographic projection, that of Gerard Mercator (great-grandfather of Michael). To represent a round globe on a flat map, the elder Mercator shaped the globe into a cylinder—pushing here, pulling there—and then unrolled it into a flat map. That method left places around the equator relatively undistorted but exaggerated the size of anything nearer the poles; Eurasia and North America bulk large on Hondius's map (as on Mercator's), and the presumed southern continent looms up like the skull in Holbein's *Ambassadors*.[73]

We will never know whether the map accurately reflected Drake's discoveries because the prototypes have vanished. When fires destroyed the Palace of Whitehall in the 1690s, the chart and diary Drake had surrendered to Elizabeth probably burned. Some details of Drake's voyage have remained mysteries. Hondius did position the Strait of Magellan between a continent and islands, as Drake had observed it. But he altered Drake's track to end its progress up Nova Albion at around 42 degrees north, erasing a line that had extended up to the 48 degrees noted on the Mercator medal, perhaps to keep the latitude of Drake's bay a secret. Indeed, it has never been definitively relocated.[74]

Other information leaked out over time. In 1592, a London merchant, William Sanderson, commissioned the first English terrestrial globe, along with a celestial counterpart. The English mathematician and instrument maker Emery Molyneux made the two globes, each 62 centimeters in diameter, slightly smaller than soccer balls. These were the largest globes yet made anywhere, with gores engraved by Hondius. Sanderson presented the globes to the queen, to win her favor. Elizabeth received the terrestrial one by remarking, "The whole earth, a present for a Prince [herself], but with the Spanish King's leave," an elaborately false courtesy toward Philip II, for whom the world was not enough.[75]

Sanderson, Molyneux, and Hondius must have been greatly relieved by Elizabeth's appreciation, because they had put forward more information about Drake's expedition than anyone before. On their terrestrial globe, "Nova Albion" has a natural harbor at 48° North named "F. Dracus," whether the actual site of Drake's bay or not, a remarkable concession to public curiosity. The globe also bears both Drake's and Cavendish's ships' tracks, though no detail about the islands below the Strait of Magellan and the open sea beyond. Yet a ship's track reveals far more on a globe than on a flat map, which can hide or exaggerate space at its edges. One mathematician measured Drake's route on Molyneux's globe and, concluding he had sailed 12,010 leagues, reckoned that total to be almost twice as much as the world's circumference, if the Earth were only 7,200 leagues (21,600 miles) around the equator (still too small).[76]

But the science of global geography was intensifying. At Sanderson's request, the English mathematician Thomas Hood wrote a scientific lecture on *The Use of Both the Globes* (1592). Robert Hues, who had sailed with Cavendish, wrote an instructional *Tractatus de Globis et Eorum Usu. A Treatise Descriptive of the Globes Constructed by Emery Molyneux.* Hues also praised around-the-world travelers for updating knowledge of the cosmos, both of the Earth and of the heavens—Molyneux's celestial globe includes the clouds of stars Magellan had discovered in the southern hemisphere, for instance.[77]

If Copernicus's news that the Earth was just another planet was still sinking in, round-the-world travelers made a similar point, and more dramatically. The globe had become, in its entirety, a place made important through human action, not because of its presumed or metaphorical place within the cosmos.

Drake went even further than Elcano by taking the globe as his personal accessory. In an engraved portrait done around 1583, a globe hangs in the background—big, fat reminder—and Drake rests his hand atop a helmet, sign of his ennobled status. In other portraits, the globe replaces the helmet. A full-length painting by an unknown artist, done around 1580 (see illustration 4), shows Drake with a simple sailor's cap, setting his right hand over a globe. In a later engraving, probably done after Drake's death, he drapes an arm on a large globe. Another posthumous image includes Drake's coat of arms (with globe) and the man himself opposite, leaning on yet another globe. Portraits of Cavendish adopted the same posture, though typically with a double-hemisphere map, which made the globe seem less tangible than the orb in images of Drake.[78]

Either way, it was a daring pose. Images of Elizabeth used the same posture, a hand or arm placed over a globe. Contemporary representations of Christ as *Salvator Mundi* were presumably familiar to the predominantly English and Low Countries artists who portrayed Drake. They were careful not to confuse the sailor with the Saviour, however. Drake's postures, holding hand or arm over part of the globe, differentiated him from a Christ who held up the entire world. Still, these cheap engravings, of a commoner possessed of the planet, showed the popularization of the idea that the globe really was a thing with which humans could interact.

Indeed, the English were globe-struck, even ordinary English people, in contrast to the somewhat narrower audiences that had learned of the Magellan/Elcano expedition. While maps and globes remained expensive, admission to the theater was not. At London's public playhouses, the "cheap seats" were not even that; a groundling paid a penny to enter the theater and stand on the ground between the actual seats and the stage. There, she or he might see the play in which Shakespeare refers to England's new images of the world, unseen by many, but evidently familiar enough that people recognized jokes about them. In *The Comedy of Errors*, first performed in 1594, a male character describes a plump kitchenmaid as "spherical, like a globe; I could find out countries in her."[79]

Then there was the entire Globe Theatre, opened in 1599 as a new venue for Shakespeare's company, the Lord Chamberlain's Men. That arena was the "wooden O" of *Henry V* (1598–99) which, somewhat like oceangoing wooden worlds, might represent all the world as a stage. Actually, a playhouse the size of the Globe cost less to make than a medium-sized ship (and used fewer trees, because plaster filled out the wooden frame), yet held far more people, up to 3,000. England's hundred-odd surviving world-circlers, if reunited in London for a day at the Globe, would have drowned in the crowd.[80]

Imagine them there, this time *enjoying* a circuit of the planet, this one imaginary. In *A Midsummer Night's Dream* (first performed in the 1590s, definitely by the Lord Chamberlain's Men, possibly in the Globe), Oberon, king of the fairies, orders his minion Puck to bring him the flower of an herb that, "on sleeping eyelids laid," will make a person fall in love with whomever she or he sees upon awakening. As he flies away, Puck promises something that only he has ever been able to do: "put a girdle round about the earth/In forty minutes."[81]

Such a timely joke—an imp circles in minutes the globe that takes mortals years to encompass—but why *forty* minutes? Did Shakespeare choose that number for its mythic sound, hinting of the forty days and forty nights that Noah rode out the Deluge on his ark, the forty days that Christ spent in the wilderness, the number 4 associated with Mercury, swift messenger of the gods? Or was the forty minutes part of a technical theory about acceleration, something the Bard had heard from a friend of a friend of a publicity-shy English mathematician who, even before Galileo's experiments with acceleration, calculated the gathering speed of an object assisted by gravity?[82]

It could be both, or neither. The forty minutes might simply signify *amazingly fast.* Shakespeare's play mentions other fairies who travel faster than the Moon, very appropriate, given the the drama's moon-struck, lunatic proceedings, and given that the Moon, in a heliocentric universe, was the only celestial body known to go around the world. Shakespeare might also have used *minute* as a pun. A minute measures an hour of time, but also a degree in space. Latitude and longitude readings come in degrees and minutes; forty minutes is a common division. The Earth's girdle is its equator. For Puck to burn an alternative equator at forty minutes (no degrees specified) might have been intended as wonderful loopy nonsense.[83]

The whirlwind jest about global travel worked all the better for its layered suggestions about speed, marvels, and modernity. By representing around-the-world travel as a joyous swoop, Puck was the anti-Phaeton.

Humans still suffered Phaeton's fate, as Cavendish learned when he set out in 1591 on a second around-the-world journey. This time, he evidently meant to get to China with a fleet of five ships and a crew of about 350 men, including his two Japanese captives. As Loaysa could have warned him, bigger wasn't better. The expedition was divided by mutiny, plagued with desertion, and culminated in Cavendish's lonely death at sea in the Atlantic, "in the moste desoluteste Cayse that ever man was lefte in." Only three of the ships returned, and only about a hundred men, and none of them had gone around the world.[84]

Would better planning yield better results? Richard Hawkins warned against that hope. After his failed around-the-world voyage in the 1590s, Hawkins wrote that one of his expedition's greatest perils had been scurvy. For that dreadful malady, "the principal [cure] of all is the air of the land; for the sea is natural for fishes, and the land for men. And the oftener a man can have his people to land, not hindering his voyage, the better it is." Given the time circumnavigators would have to spend at sea, as they wound their way around the unexpectedly large and oddly configured globe, it would be impossible to avoid the disease. The world could be encompassed, but only with casualty rates that resembled a state of war, a war against the planet.[85]

Chapter 3

Traffic

*F*rancesco Carletti had never intended to make the around-the-world voyage that ruined him. Carletti and his father, Antonio, merchants in Florence, set out in 1591 for Spain, where they obtained a license to take slaves from Africa to Spanish America. Although he was never so conscience-stricken as to free his own slaves, the trans-Atlantic trade in humans repelled Carletti. Once in America, he and his father decided to explore other options, and continued over the Pacific and into Asia, where the elder Carletti died. His son continued home, an itinerant circumnavigator who hitched lifts on others' ships, which he could do rather easily because of the expansion of maritime trade. The increased "traffick," as commerce was called at the time, created more *traffic*, congestion on the seas, which was where privateers stripped Carletti of the fortune he had acquired by going around the world. Every circle around the globe that said *mine* was, after all, an invitation to someone else to make it *theirs*.[1]

Beneath the heaving surface of national rivalries, however, a firmer foundation for world domination was emerging: a planet physically developed for the convenience of Europeans. Because they saw (and used) so many parts of the world, circumnavigators, including Carletti, were uniquely equipped to document that long-lasting and momentous transformation of the physical globe, which rival Europeans managed to do together, whatever their other, and violent, differences.

• • •

Wherever he found a market, Carletti felt at home. In the towns of Spanish Peru, he surveyed the shops and market stalls with great contentment and jotted down prices. He was no materialist, however. He loved an orderly market for itself, but also because it represented a world ordered for Christian Europeans. The loss of half a day on his arrival in Asia, where eastward-traveling Spaniards had lost track of the other half, made him confused, almost angry. When was Friday, he demanded, when Christians must not eat meat?[2]

The division of the world between the two great Iberian powers might eventually rationalize things. "Together," Carletti pointed out, "those two crowns have come to make a circle around the whole world." (The circle would be unbroken until Portugal and Spain dissolved the Iberian Union in 1640.) The global circuit was convenient to Catholic merchants like Carletti, who could "enter into that magnificent enterprise and in less than four years go around the entire universe both by way of the East Indies and by way of the West." It took four years if a man had to piece together a journey using ships with regular routes across the major oceans. Should he find someone headed for the Strait of Magellan, his tour might shrink to eighteen months. That was optimistic, meaning wrong—even Cavendish, who held the record, had taken over two years.[3]

Carletti thought any circuit, fast or slow, showed European superiority, because they knew how to navigate a round globe. He considered Japan "one of the most beautiful and best and most suitable regions in the world for making profit by voyaging from one place to another," but the Japanese lacked the ships and the seamanship to bring their goods out to the world. And while he considered Chinese maps very useful (he acquired a book of Chinese geography), he mocked the Chinese belief that their kingdom lay at the center of a flat world.[4]

Yet Carletti's world was not a single, smoothly articulated circuit, as he admitted whenever he used the word *license*. He griped that Iberian merchants had monopolies within their proprietary zones, which guaranteed Spanish merchants a 150 to 200 percent profit on trade in Chinese goods between the Philippines and Mexico. The captain-governor in Malacca cleared 70 to 80 percent by controlling the local spice trade. Carletti had to pay to enter the various trade zones. He was required to buy a license to travel from Lima to Mexico: "the annoyance was great and the cost no less." When told to purchase another license for passage from Acapulco to the Philippines, he opted instead to work his way there. In Manila, he learned that another license was required to enter Portuguese territory. He and his father decided they had been law-abiding enough. At night, they sneaked aboard a Japanese vessel bound for Japan, which neither Portugal nor Spain ruled.[5]

Nor did the licenses, ancestors of passports and visas, actually work—witness Carletti's skulking entry into Japan. Indeed, his enthusiasm for Iberian command of global trade diminished in East Asia, where Europeans were outnumbered and under surveillance. When detained in China, the stop after Japan, Carletti and his father insisted that they simply wanted "to pass over to East India for our own amusement and curiosity." They came from "a country that was independent, like Japan, and not subject to either one or another [Iberian] nation, and that moving about the world was a thing allowed to all the nations." The Carlettis were fined and released, small fry who slipped between trading zones.[6]

Though he would have resented the comparison, the slippery Carletti resembled intruders into the Iberians' round-the-world network. Despite formal Portuguese control over most of the Asian goods carried to Europe and America, "many years ago the Dutch and the English and the French took away from them, one could say, the traffic of the Moluccas." "Also ruined is the traffic with China," where the Hollanders "infest those seas and keep [the Portuguese] in continuous fear."[7]

Those warnings foreshadow Carletti's own fate at the hands of the Hollanders. Licensed or not, he had accumulated a private stock of Asian trade goods, as well as three slaves, Korean, Japanese, and Mozambican. From Malacca, he paid for passage aboard a Portuguese pepper ship bound for home. On Christmas morning of 1601, the ship departed with Carletti, his cargo and retinue, plus a hundred hens for his private table. For the Portuguese, it was a familiar route—the pilot on Carletti's ship had rounded the Cape of Good Hope "eighteen or twenty times," and made the pass again with ease. Near St. Helena, another ship hailed them with "the salute used at sea: 'A good voyage. What ship?' " The Portuguese identified themselves and learned that the other ship was Dutch. The Hollanders called again: " 'Friends, friends, do you lack nothing?' " The solicitude was a ruse to prepare an attack on the pepper ship, a floating bazaar of valuable Asian goods, and enemy property—the Dutch were no friends of the Iberian Union. For the rest of that day, the Dutchmen bombarded the Portuguese ship, aiming high to shatter rigging and masts and immobilize the prize. Night fell, along with silence, "and there, without moving, eating, or sleeping, we waited for the day fated to be our ruin."[8]

Which came with the dawn, when the Dutch shifted their fire to the pepper ship's waterline. The survivors knew they must surrender or sink. The Hollanders boarded their prey, made quick decisions about whom to take captive and whom to throw into the sea, and salvaged the cargo. (From water littered with Asian textiles and drowning humans, the privateers rescued the textiles.) Because he was not Portuguese, Carletti was spared. He spent twenty-three days in the hold of the Dutch ship, on rations of rice and water, and not much of either. Carletti confided, "I would have been in a bad way if good luck had not helped me by making me have with me one of those porcelain vases full of pears preserved in China . . . that vase did me a good turn, and with it I also maintained two titled gentlemen."[9]

On arrival in the Netherlands, Carletti found it convenient to emphasize his Florentine roots. He pleaded that he was "a vassal of a neutral prince" and that "it was a reasonable thing to be fair with merchants, the more so when they are not vassals of an enemy prince." That defense had worked in China, but failed in the Netherlands. Carletti retained some souvenirs, but never recovered his full cargo. Instead, he wrote up his story and dedicated it to a potential patron, Grand Duke Ferdinand I of Tuscany. It was a good strategy: the duke was a Medici, and Carletti spent the rest of his life in service to that powerful family. His narrative (minus the parts on the slave trade) was published in 1701, with its concluding lament for those who have been rich but lose all: "as occurred with me, who at the age of thirty years found myself to have circumnavigated [surrounded] the whole world from west to east with such fortune in my traveling that to bring it all to perfection all that was lacking was to have finished the trip safely to Lisbon."[10]

• • •

Carletti was collateral damage in a Dutch rebellion against Spain. Shortly after it had seized Portugal, Spain invaded the Netherlands. If the Portuguese had put up little resistance, the Dutch outdid themselves. From 1568 to 1648, they fought what would be called the Eighty Years' War against Spain. They had no reason to respect Spanish assertions that the Pacific and the Spice Islands were Iberian property, and they used round-the-world travel to declare their independence. Two circumnavigations, done between 1598 and 1617, would establish trade to Asia under the Dutch East India Company, a monopoly. But it says something about the Hollanders' commercial instincts that a third circumnavigation (1615 to 1617) contested the monopoly in order to sharpen the Dutch challenge to Spain, or so the rogue circumnavigators claimed.

A former tavern keeper, of all people, did the first Dutch circumnavigation. Olivier van Noort left Rotterdam in August 1598 with four ships and 248 men and crossed the English Channel to collect one of Cavendish's pilots. Several men died of scurvy and fever by the time the expedition reached Brazil; van Noort lost thirty-five others in skirmishes with Indians in the Strait of Magellan. After all that, plus an attempted mutiny, only half the crew survived to see the Pacific, and only two of the ships survived further adventures up the coasts of South and Central America. The Spanish captured one of the two vessels near the Philippines (and executed its commander and some crew), while van Noort slipped off in the *Maurits* to Brunei and Ternate, where he acquired the usual spices. He made it back to the Netherlands in 1601 with a reported forty-five men; about forty others from the seized ship returned later. The survival rate of 34 percent was similar to Drake's—at the very least, no new low record.[11]

And it was mere survival. Toward the end, the men subsisted on worm-infested ship's biscuit. They traded some precious pepper for meat and fresher biscuit from ships they passed just short of Amsterdam. Van Noort made no new discoveries and acquired little wealth. But he had lived to tell his own tale, which the excellent Dutch printing industry produced for sale only eighteen days after his return. A fuller version appeared later that year, and two amended editions the next. The multiple editions indicated that selling the story of a circumnavigation could augment any profits from it. And as usual, the big story gave other people big ideas. The Dutch East India Company (Vereenigde Oost-Indische Compagnie, or VOC) was founded in early 1602, mere months after van Noort's return. Its proprietors asserted, on the basis of his journey, that only their ships could traverse the Strait of Magellan and round the Cape of Good Hope to bring Asian goods to the Netherlands. The state-sponsored monopoly was yet another assertion of national identity, yet another challenge to Spain.[12]

Most of the time, the VOC used the old Portuguese route, via the Cape of Good Hope, coming and going. (So did England's East India Company, founded in 1600, another poke in the Spanish eye.) The Indian Ocean route seemed safer, because it did not require a long, scurvy-ridden Pacific crossing, nor risk possible entanglement with Spaniards.

But could ocean routes be owned? Recall that Elizabeth I had told the Spanish, in defense of Drake, that "the use of the Sea as of the Ayre is common to all." Underdogs always claimed this—until they became top dogs. Elizabeth's successor, James I, retreated from her stirring, universalist claim in order to assert sovereignty over England's coastal waters and keep Dutch fishermen out.[13]

So now it was the Dutch underdogs who insisted that saltwater dissolved property claims. In 1603, a VOC ship's captain seized a loaded Portuguese carrack, the *Santa Catarina*, in the Straits of Singapore. Sold in Amsterdam, the cargo's profits doubled the company's capital. When the Portuguese protested, the VOC retained the counsel of a young jurist named Hugo Grotius. In 1609, Grotius published a legal commentary on the incident, arguing that the high seas were *res communis*, common property, as Queen Elizabeth had stated. The work became known as *Mare liberum*, the "Free Sea," a classic text in the ajudication of the non-landed parts of the Earth. Indeed, Grotius presented the Magellan/Elcano circumnavigation as primal evidence for the impossibility of sovereignty over oceans: "the Castilians from the year 1519 have made the possession of the sea about the Moluccas doubtful to the Portugals." The debate was by no means settled, however.[14]

The Dutch were, if anything, a little too convinced by their own logic. During the Twelve Years' Truce (1609–21), the Dutch Republic gained de facto independence from Spain. Several Dutch adventurers decided that another round-the-world voyage would be another excellent declaration of their independence. But, oversupplied with nationalist exuberance, two Dutch circumnavigators set off with competing goals for their expeditions. They demonstrated, yet again, that it was easier to claim global sovereignty based on around-the-world travel than to maintain it.

Joris van Spilbergen set out in 1614, funded by the VOC and backed by the Dutch government. His expedition—four ships, two smaller vessels, and eight hundred crew—was a big test of a small country recently strained in its land war with Spain. Spilbergen wisely took soldiers to protect his sailors. When the Portuguese refused to allow the Hollanders to land and purchase food in Brazil, Spilbergen secured what he needed by force. He was also able to quell two mutinies, get through the Strait of Magellan without mishap, and evade or seize Spanish ships (as needed) in the Pacific. It was all going very well; the venture seemed to consolidate Dutch claims to the Iberian routes to Asia, whether east or west.[15]

But Spilbergen was being tailed. Ten months after he had departed and was blasting his way to Asia, a ship and yacht with eighty-eight men slipped in behind him. The little fleet belonged to a consortium of two families, the Schoutens and Le Maires. Brothers Willem Cornelis Schouten and Jan Cornelis Schouten captained the two vessels (the former was also the expedition's navigator). The expedition commander was Jacob Le Maire. His father, Isaac Le Maire, had created a Compagne Australe to rival the VOC, which he thought too timid in confronting the Spanish. When he told Dutch authorities that he would explore the southern ocean and the presumed southern continent, *Terra Australis* (hence his company's name), he was granted the right to enter the Pacific under the Dutch flag, though only if he discovered a new way to get there, which his company could then consider its proprietary route.[16]

Although it was an open secret that Magellan's strait might not be the only passage to the South Sea, Schouten and Le Maire did not want to risk public exposure of their plans. Only after they had crossed the equator did the crew learn their destination. Generous supplies, including over 25,000 lemons plucked or purchased in Sierra Leone, kept the men in good health. The crew accidentally burned the yacht while repairing it off Patagonia, but the men were unhurt and everyone crowded into the other vessel, the *Eendracht*. They passed the Strait of Magellan, sighted and named a new strait for Le Maire, and in January 1616, rounded "Kap Hoorn," possibly named for the lost yacht (the *Hoorn*), or else the city of Hoorn, where the venture had been conceived and the navigating Schouten brother had been born.[17]

It was a major discovery and a Dutch triumph. Schouten and Le Maire proved that the Strait of Magellan did not lie between two landmasses, South America and *Terra Australis*, but instead between South America and an appendage, Tierra del Fuego, with open ocean beyond and then, farther away than expected, the still-unseen *Terra Australis*. The new geography undermined both Spanish and VOC claims to a unique route from Europe to Asia. Sweetest of all, the Hollanders had managed to chart a part of the Spanish Empire unknown to the Spaniards—and stick a Dutch name on it (see illustration 5). Cape Horn it remains, a lasting memorial to Dutch insurgency against the Iberian colossus.

Although Jan Schouten died in the South Pacific, the mortality was otherwise low, a total of three, and the *Eendracht* proceeded to Ternate. Having secured the desired cargo of spices, Schouten and Le Maire were confident that they could get home without difficulty. They sold spare equipment—cables, anchors, scraps of the burned yacht—to buy provisions and then headed to Dutch Batavia in Java.[18]

There, as if they were regulars in the same Amsterdam tavern, the two Dutch circumnavigators ran into each other. Spilbergen had made Java in September, the climax of his scheme to unnerve the Spaniards by making them hear "the clink of our arms in places where they least expect it." When Le Maire arrived in October, the clinking got a bit too loud. The East India Company officials had given Spilbergen a hero's welcome but arrested Le Maire and his men, disbelieving their story that they had come by a substantially different route. Le Maire and Schouten, and those of the crew that did not wish to stay in Asia, were handed over to Spilbergen and sent back to the Netherlands as captive circumnavigators. Le Maire was not even that; he died in Mauritius. But because of his swift journey to Java, and Spilbergen's speedy passage home from there, Schouten's total circuit set a new record: two years and eighteen days.[19]

No wonder Isaac Le Maire wanted to preserve his claim to a new way around the world, one that avoided the slow anxiety of threading the Strait of Magellan. He sued the VOC for the expedition's surviving property, including its records, and especially the journal of his son, Jacob. And he convinced the authorities to censor the publication of any information about Cape Horn, though it is likely that sailors spread the news. As information on the new route between Atlantic and Pacific became generally known, mapmakers and engravers included it on their products. A double-hemisphere map of 1618 or 1619, for example (see illustration 6), shows the world encircled by the Le Maire/ Schouten cruise track, with some details of their discoveries in the Pacific. Alongside little insets featuring their flagships, Magellan and Schouten appear, each above a hemisphere, and each being crowned by Fame. Below them are medallion portraits of other successful circumnavigators, Drake, van Noort, Cavendish, and Spilbergen. (Poor forgotten Elcano.)[20]

Next came a triumphant cluster of printed accounts. Schouten was first to publish, in 1618, followed by Spilbergen in 1619. Isaac Le Maire had won his case against the VOC and regained Jacob's journal, printed in 1622. As with van Noort's journal, these narratives went through multiple editions and translations—Schouten's was translated into Latin, French, English, and German. Clearly, readers were not losing interest in the genre of the round-the-world travel account.

Those accounts updated the Renaissance *Rough Guides* in which English circumnavigators had presented the world as a series of resource-rich places, either wild or cultivated. Before entering the Strait of Magellan, van Noort's men slaughtered an estimated 50,000 penguins and gathered penguin "Egges innumerable." Spilbergen marveled at abundance in several places: "millions" of deer south of Acapulco, ample supplies of meat, fruit, wood, and water at Spanish ports around the Pacific, and more of the same in the Ladrones. Schouten and Le Maire found "whole forests" of citrus trees near Sierra Leone, dodged many whales, and slaughtered hundreds of penguins in the South Atlantic, too many to eat before they rotted. At the Bay of Quintero, Spilbergen saw herds of Spanish horses that had gone wild.[21]

Finding these resources, especially on land, was essential—as Hawkins had warned in relation to scurvy, sailors could not survive for long at sea. We think differently about scurvy because the disease has been defined as a deficiency of vitamin C, easily prevented or cured with certain foods or supplements. But the "scurvy" that circumnavigators described only sometimes overlapped with what a doctor, then or now, might categorize as a scorbutic condition. The special designation of "*sea* scurvy" probably included various forms of illness, beyond what a doctor today would identify as the lack of vitamin C, and not even matching what early modern doctors described as scurvy within Europe.

Plus the circumnavigators witnessed a powerful correlation: scurvy appeared at sea and went away on land. When "the Scorbute" first appeared, mid-Atlantic, van Noort vainly sought St. Helena and Ascension, but only made land in Brazil. There he set up tents for the invalids. *Terra firma*, plus "Herbes and two Trees of sower Plumbes . . . cured the sicke in fifteene dayes." Greenstuff and fruit from various Pacific islands helped them later. So too, after crossing the Atlantic, were Spilbergen's men so ill that he had tents pitched ashore for them. The restoration wore off by Acapulco, where the local meat and fruit brought "incredible joy and recuperation"; ditto for the Ladrones, where vegetables and fruit "refreshed and restored" the depleted crew. Similarly, Schouten and Le Maire noted the places in the Pacific where fish and greenery revived their scorbutic invalids.[22]

In this way, a consistent remedy for sea scurvy was emerging. Whenever possible, circumnavigators put scorbutic invalids ashore to convalesce, to steep themselves in the air, water, and natural products of land. That was what Magellan had done in Guam (in 1521), Olivier van Noort in Brazil (1598), and Joris Spilbergen also in Brazil (1614), among other examples. The narratives reveal that, over a period of a century, mariners followed the same protocol. And they represented several nations (Spain, the Netherlands, and England) as well as different statuses, from Magellan the nobleman down through van Noort the tavern keeper.[23]

Shore leave as a remedy for scurvy defied medical recommendations within Europe itself. Recall that Europeans believed the human body to be metaphorically connected to the whole world, composed of earth, air, fire, and water; individual bodies were adapted to local places, and consequently healthier there than anywhere else on Earth. Travelers risked ill health when they ventured to foreign places, and sailors who wandered around the world were at greatest risk. And yet the circumnavigators who pitched tents on foreign shores and stuffed themselves with tropical produce had come to a different conclusion: when in doubt, any land was better than no land. Scurvy was prompting a new definition of humans. They were no longer cosmic, representing all parts of the universe, nor regionally specific, products of particular places. Instead, they belonged somewhere in between: to all the landed parts of the globe, the terrestrial zone.

The desire for terrestrial places meant that the Dutch narratives traced a distinctive map of the world, a global sequence of medically necessary anchoring points. The tour went from spots in western Africa and eastern South America, to last chances to make land before Cape Horn or in the Strait of Magellan, on to lonely Juan Fernández (islands off Chile), and then to the Marquesas, the Ladrones, and parts of Asia, followed by Cape Town ("the tavern between two seas"), St. Helena, Ascension, Cape Verde, and the Canary Islands and Madeira. This guide to the world is distinctive in another way. Whatever the national rivalries that inspired circumnavigations and incited quarrels over proprietary oceans and proprietary routes, Englishmen and Continental Europeans shared information about the chain of useful landing points. Remarkably, they tended to trust the advice of their rivals, and they updated information that might help rivals. Thus, Drake confirmed Magellan's report of fresh water on islands where the Strait of Magellan opened to the Pacific; Spilbergen followed Cavendish's directions to water in Chile. Given the multiple editions of these accounts, and their multiple translations, they formed a multinational archive of information about how the world could be encompassed.[24]

The information on paper was still patchy, however, and where it failed, the Dutch circumnavigators had taken captives for information. Van Noort abducted a Spanish pilot, and when he proved unhelpful, shoved him overboard in the middle of the Pacific: "A prosperous wind happily succeeded." Van Noort detained a Chinese junk and interrogated the crew, then kidnapped a pilot off a South Asian ship. Spilbergen took some crewmen from a Portuguese vessel in the Atlantic, later releasing all but the pilot. He also seized some Spaniards in the Pacific to get information. In the Philippines, he detained six Chinese men (and possibly some Japanese) and questioned them. He released the Asians when he thought he had exhausted their knowledge, and used the Spaniards to redeem some captive Dutchmen in Ternate. Compared to all that, Schouten and Le Maire were softhearted—or wished to seem so. They admitted to taking hostages in New Guinea to ransom for pigs, but claimed to have kept only one man.[25]

It was a trade-off. Once Europeans had consolidated, recorded, and circulated their knowledge of the world, captives would be unnecessary. On the other hand, that printed knowledge would guarantee that European mariners would keep returning to the places they had learned about, having transcribed what the captives had told them and integrated it into their paper representations of the planet. Whatever quarrels they had among themselves, Europeans shared a conception of the world as useful to them, and they used that conception against a variety of non-European people they considered incapable of going around the world themselves. Deeper and deeper, the mark of empire was inscribed into the nature of circumnavigation.

Given the quickening pace of around-the-world travel, the consolidation of information about it was itself speeding up. In 1619, the year Schouten's narrative appeared in English, J.H., *"Gent.,"* considered the history of circumnavigation in verse:

> *Some thinke it true, whilst others some do doubt,*
> *Whether Capt. Drake compast the worlde about.*
> *Some say he did it in the Devils name,*
> *And none ere since could doe the like againe:*
> *But these are al decieved, why should they doubt it?*
> *They know each yeere there's some that goe about it.*

Indeed, to designate what had become a recurring feature of European history, the term *circumnavigation* was coined. In his *Mikrokosmos. A little description of the great world* (1625), Peter Heylyn referred to "the circum-navigation of the whole world" by Drake and Cavendish.[26]

That same year, an English clergyman, Samuel Purchas, offered a convenient four-volume compendium of travel literature, including the multiplying accounts of circumnavigations. Purchas noted his debt to Hakluyt's earlier such collection in his title, *Hakluytus Posthumus or Purchas His Pilgrimes, contayning a History of the World in Sea Voyages and Lande Travells, by Englishmen and others.* Purchas thought these travels were pilgrimages, forms of spiritual exercise: "Man that hath the Earth for his Mother, Nurse, and Grave, cannot find any fitter object in this World, to busie and exercise his heavenly and better parts then in the knowledge of this Earthly Globe, except in his God." Sea voyages were excellent reminders of human mortality. Any sailor knew too well "how neere in a thin ship, and thinner, weaker, tenderer body we dwell to death."[27]

If any long voyage imparted these lessons, Purchas nevertheless considered circumnavigation the ultimate test of human endurance. He was the first to print a full English translation of Pigafetta's narrative, followed by versions of Hakluyt's accounts for Drake and Cavendish, then English translations of van Noort, Spilbergen, and Schouten. By omitting the solo circumnavigators Loyola and Carletti, as well as the failed circumnavigations of Loaysa and Cavendish, Purchas helped to consolidate opinion that a circumnavigation was a distinctive type of voyage: a nationally sponsored and fully completed around-the-world expedition that was heroic and militaristic in equal measure. The frontispiece to his work showcased around-the-world travelers. This fold-out illustration bears medallion portraits of kings and sailors, beginning with classical and scriptural figures, and including the circumnavigating six: Magellan, Drake, Cavendish, van Noort, Spilbergen, Schouten. Only the English circumnavigators appear on the frontispiece's map, which bears ships marked with the names of Drake and Cavendish. The Dutch considered, however, that they were Drake's true heirs. Van Noort's tombstone (he died in 1627) features a ship atop a globe, an image probably modeled on Drake's coat of arms.[28]

It was not clear what further circumnavigations might accomplish. Europeans knew that China, India, and Japan offered rich trade, but those regions would require much greater effort for Europeans to enter, as Loyola and Carletti had noted.

A subsequent Dutch around-the-world expedition was not even intended for the East Indies, but reverted to the Drake-and-Cavendish goal of harrying Spaniards, this time to extract American territory from them and resettle it with Hollanders. A Dutch West India Company had been founded in 1621 to pursue Dutch interests in the Americas, and another circumnavigating fleet was meant to promote those interests.

Under Jacques l'Hermite's command, the Nassau Fleet left Holland in 1623: eleven large and well-armed ships, four small pinnaces, and a crew estimated at 2,500 men. Hermite's major aim was to challenge Spain's authority on the Pacific side of South America, where silver was still mined. Had he researched Loaysa's big, doomed expedition in the preceding century, he might have reconsidered the plan. In the end, it proved impossible to maintain and maneuver the fleet, the largest military force yet to enter the Pacific. Although Hermite destroyed over thirty Iberian vessels, he missed two Spanish silver fleets and he died in Peru in the summer of 1624. The remains of the Nassau Fleet crept over the Pacific and to the East Indies, where the VOC commandeered the surviving ships and men. Some of the ships straggled back to the Netherlands in 1626.[29]

The Nassau Fleet was celebrated in the Netherlands, but never quite entered the popular imagination. The published journal of the expedition, printed in Amsterdam in 1626, has a frontispiece with a world map, but that image bears no circumnavigator's track. Its outstanding feature is a bit of coastline discovered in a non-circumnavigating voyage by a different Dutchman, Dirk Hartog. The second European to set foot on Australia, in 1616 (Willem Janszoon had beat him there in 1606), Hartog named its coastline Eendracht Land, after his ship, and Hermite's frontispiece was the first printed map to show any part of Australia—all extremely interesting, but nothing to do with the Nassau Fleet. The journal of Hermite's expedition ran through at least eight Dutch editions between 1626 and the late 1660s, but only a few fragments appeared in English and French to satisfy wider curiosity about this latest attempt to get around the world. Hermite, like Loaysa, is not in the pantheon of successful circumnavigators.[30]

• • •

Five decades passed. In that half century, perhaps some individuals went around the world, maybe even a stray ship. But those voyages are historically invisible; none resulted in an extant published account, and none elevated a circumnavigator to the pantheon. Europe's national leaders were busy monitoring the trade zones they already possessed. But readers had not lost interest in circumnavigations, and authors continued to consult narratives of around-the-world travel for inspiration.

There was even interest in an old voyage, Drake's expedition, described in the Reverend Francis Fletcher's *The World Encompassed* (1628), which was published just as England began to take on truly global concerns in Asia and the Americas. Fletcher, who had accompanied Drake, praised him as the first to plow "a furrow about the whole world," which "doth not onely overmatch the ancient Argonautes, but also outreacheth in many respects, that noble mariner *Magellanus*," because Drake survived the ordeal. The published narrative included a map of the voyage and, as usual, stressed the difficulty of getting provisions and information along many stretches of the globe. Although Fletcher admits Drake's failure to find a Northwest Passage, he reminds readers that Drake discovered he could sail east through the Strait of Magellan. That knowledge might enable other ships to visit the Pacific without having "to compass the whole world," as another English ship's captain had in fact just proven.[31]

At the very end, Fletcher describes the *Golden Hind*'s arrival back in Plymouth on what the crew thought was a Sunday, but turned out to be a Monday—he, a man of the cloth, had missed the Lord's day then, and for many weeks before. It was another reminder that, however unifying round-the-world travel might seem, it also confounded universality, including universalist religions. No nation had managed to maintain global power and no religion could easily do so, whatever the ongoing rivalry over proprietary sea routes and indeed entire oceans.[32]

Perhaps because the Circumnavigator's Paradox was so uncanny, circumnavigators continued to think of themselves as bold Phaetons who lost a day because they dared to race the Sun. When Schouten's log had disagreed with the Dutch calendar in Indonesia, he and Le Maire explained the discrepancy in the old way: "we sailed westward from our country and [because we] had once circumnavigated the earth with the sun we had therefore had one night or sunset less than they" who stayed put. Purchas praised all navigators for establishing knowledge about the physical globe, but the latter-day Phaetons especially. Without them, "nor should we grow familiar with the Sunne's perambulation, to overtake him, to disapoint him of shadow, to runne beyond him, to imitate his daily journey and make all the World an Iland, to beguile this Time-measurer in exact reckonings of Time, by adding or loosing a day to the Sunne's account."[33]

At this point, the Phaeton fantasy was a Phaeton fallacy. Copernicus's argument for heliocentrism had been widely disseminated, principally by his major advocate Johannes Kepler. The Sun did not orbit the Earth. Rather, a handful of humans did, a slightly lunatic practice that they shared with the Moon.[34]

And yet the Copernican view of the cosmos troubled people because it defied their everyday experience. The Earth seemed immobile, yet it rotated; the Sun appeared to pass around the world, but didn't. Were human reason and sensations reliable guides to the cosmos? The presumed first human, Adam, contemplates the problem in John Milton's *Paradise Lost* (1667). Happily nestled in Eden, Adam nevertheless worries that "the sedentary Earth" might be no such thing. The Angel Raphael reassures him that "whether Heav'n move or Earth / Imports not":

> *What if the Sun*
> *Be Center to the World, and other Stars*
> *By his attractive virtue and their own*
> *Incited, dance about him various rounds?*

What of it? What matters is that the Creator ordered things that way: "Leave them to God above, him serve and fear." Thus are God and Copernicus reconciled, with Adam (and Eve) taught that no physical entity is the center of the cosmos—God is. Paradise is maintained.[35]

But not for long. To sow discord in the primordial garden, Satan plots a tour of Earth. He then does seven circumnavigations in a week—dark parody of the seven days of Creation—and in multiple directions:

> *The space of seven continu'd Nights he rode*
> *With darkness, thrice the Equinoctial Line*
> *He circl'd, four times cross'd the Car of Night*
> *From Pole to Pole, traversing each Colure.*

What took Puck a mere forty minutes occupies Satan seven whole nights. In the Old Deluder's defense, he, unlike Puck, is not seeking one part of the Creation. Instead, in what resembles the circumnavigators' accounts, with their attention to global natural resources, Satan surveys everything in order to find the best creature in which to conceal himself and sneak into Eden. (For that reason, he does the first polar circumnavigation.) He selects a serpent, "fittest Imp of fraud," slithers into the primal garden, and betrays first Eve and through her Adam. Seven swoops around the planet, and Paradise is lost.

History's third imagined circumnavigation is as un-Earthly as the first two, Phaeton's and Puck's. Only an imp or the Devil can do with ease what slowly killed many mortals. Yet one of the next historical world-circlers managed, in his narrative of the experience, to turn the human agony into a rough comedy. He was the first itinerant circumnavigator who, hitching lifts rather than commanding an expedition, joined the pantheon.[36]

A surviving portrait of William Dampier, executed around 1697 when he was in his late forties, shows a mild-looking fellow holding a book—not the typical portrait of a seaman, but a quiet statement of triumph for someone who had used his travels to become a noted author and expert, despite his lack of formal education. Orphaned in childhood, left no inheritance, he had been removed from school and apprenticed to a ship's master, "complying with the Inclinations I had very early of seeing the World." After shivering through some short merchant runs within the northern hemisphere, Dampier gladly headed to Java and Jamaica. He tried the logwood trade in the Bay of Campeche but, when that didn't pay, joined a crew of buccaneers, as Caribbean pirates were known. In 1678, after he had returned to England, he was offered a place on another ship headed to the Caribbean, and welcomed "the offer of a warm Voyage and a Long one."[37]

That trip metamorphosed into several others, with companions of escalating colorfulness, and Dampier would not return to England for thirteen years. After a trading venture collapsed, he joined another crew of buccaneers. These pirates ran a big operation—five ships' worth of rascals (later expanded to nine ships)—and were so well prepared that they had recruited a surgeon, Lionel Wafer, to tend their inevitable injuries. Part of the band trekked over the Isthmus of Panama, seized some Spanish ships, and raided Peru and Chile. Dampier made the return trek to the Caribbean and eventually, after some adventures on three ships, two English and one French, ended up in Virginia. There, Dampier met some old friends, including Wafer and a captain named John Cook, who convinced him to go buccaneering with them again on an appropriated French ship renamed the *Revenge*.

The seas had become rather crowded with pirates, many of them personally acquainted, and the throngs of thugs show that global trade was expanding faster than it could be regulated. There may have been as many as 1,000 renegade seamen clustered around the Atlantic and Pacific oceans, as well as the Indian Ocean and Red Sea, on any shipping route that promised loot. (By the early 1700s, their numbers approached 2,500, more than ever before, or since.) Should some freak accident destroy all records of early modern trade, yet somehow preserve evidence of piracy, historians could probably deduce the trade from the numbers and locations of the pirates. And because global trading networks interconnected, the pirate networks interconnected. Dampier kept running into the same men and the same ships—he'd sailed on the *Revenge* when she bore a French name.[38]

The coincidences continued as the *Revenge* headed south from Virginia in late summer of 1683. Cook had hired an Englishman, the tasty-sounding William Ambrosia Cowley, as his master and navigator. They acquired a Danish ship off Sierra Leone, renaming her the *Batchelor's Delight* and abandoning the *Revenge*. They decided to round Cape Horn rather than thread the Strait of Magellan but regretted it when they were blown as far south, they believed, as 60° 30′, a new record. Once they cleared the cape, they encountered two English ships that had navigated the strait—a traffic jam, indeed, for this corner of the world. Cook agreed to sail the *Batchelor's Delight* in convoy with one of the ships, the *Nicholas*, and farther north they joined still others, both English and French, in raiding Spanish settlements. At the Juan Fernández Islands, they found a Moskito Indian, William, accidentally abandoned in 1681, when Dampier's first crew of buccaneers had fled a Spanish fleet. The world was now sufficiently well traveled that William was rescued (unlikely, had Drake abandoned him), yet not so busy that the rescue came any sooner than the three years he endured alone.[39]

The Englishmen headed next to some islands that the Spanish regarded as mirages—*encantadas*, or "enchanted isles." They proved to be real, an entire archipelago of unpopulated volcanic islands later called the Galápagos. As ship's master, Cowley assumed the privilege of naming them, claiming one for himself and calling the others after Charles II, James, duke of York (the king's brother and heir), and assorted aristocratic worthies who might, if flattered by the pirates, forgive their activities.[40]

The detour to the Encantadas evidently enchanted Dampier, whose interests shifted from piracy to exploration. Another coincidence offered him a new opportunity. The *Batchelor's Delight* encountered, for a second time, the other English ship that had passed through the Strait of Magellan. This was the *Cygnet*, probably named for her captain, Charles Swan, and Dampier transferred to it. When the *Cygnet* reached the Philippines, her men bragged that the best dressed among them had come along "only to see the world." The *Cygnet* also stopped in "New Holland," alias Australia, where Dampier composed one of the first descriptions of that continent's distinctive people and natural world. Marooned on the Nicobar Islands in the Indian Ocean, Dampier and several companions reached Sumatra by canoe. From Asia, he sailed around the Cape of Good Hope and home to England in 1691.[41]

It was a bachelor's delight indeed, delightfully irregular, a far cry from the organized expeditions that usually took sailors around the world. So too did Dampier's original comrades on the *Revenge* get back to England (if at all) by jumping to other ships. Cowley, for instance, abandoned the *Batchelor's Delight* in Batavia, taking passage to Rotterdam. Others filtered into Europe as well. They never could have done it without increased traffic over the seas. Survivors of shared misadventures, crisscrossers of each other's paths, the men created one final traffic jam at London's printing presses, where each was determined to get his story out first.[42]

Stories of derring-do could pay better than doing the actual deeds. In 1678, eight years before any of Dampier's companions returned, a French author named Alexandre Olivier Exquemelin had penned a piratical thriller, *De Americaensche Zee-Roovers,* or *The Buccaneers of America*, the *Pirates of the Caribbean* of its day: gory, gaudy, and occasionally true. Subsequent editions ran well into the nineteenth century, and were translated, expanded, corrected, illustrated, and (fittingly) pirated by other authors and printers who passed off Exquemelin's material as their own.[43]

De Americaensche Zee-Roovers' runaway success was a contrast to the narrative of the Spanish Catholic priest Pedro Cubero Sebastián. Cubero did his circuit from 1670 to 1679—eastward, like Loyola, though memory seemed to have faded about that pious pioneer. An admirer described Cubero as "not happy with the direction of those who preceded" him (including Magellan, Drake, and Le Maire), but determined to travel in a "new way, against the motion of the sphere," which nevertheless allowed "a perfect rotation of the terraqueous globe." In fact, Cubero's published account of 1680 emphasized his passage through Europe and its Near Eastern neighbors. Christendom was his main concern and he was scrupulous in cataloging the documents and permissions that gave him amphibious access to places that were Catholic or he hoped would become Catholic. Compared to Cubero, the unlicensed marauders of the world seemed much more exciting; Exquemelin ran through many more editions than Cubero.[44]

A buccaneer was in a tight spot. Clearly, he had reasonable hopes of good book sales. But he risked legal prosecution if he admitted to piracy that had not been sanctioned, as privateering, against an enemy state. Dampier had returned as poor as he went out—crime didn't pay the way it used to. His scheme to exhibit for profit his heavily tattooed Malaysian slave, Jeoly, "the Painted Prince," fell through. What did he have left? His knocking about the globe, and thorough exposure to every ocean and known continent (except Africa), gave Dampier price-less expertise: he reported everything with a seen-it-all confidence that no one else could pull off. He beat everyone else into print, which helped, but—crucially—he pitched his story to a remarkable range of readers, high and low.[45]

At the high end, there was science. Philosophical inquiry into na-ture had always existed, but at the end of the seventeenth century, the philosophers gave it modern form. "I think therefore I am": René Descartes's famous statement of 1637 had asserted the priority of mind over matter, starting with the material body around the mind, then moving outward to the rest of nature, which human reason could penetrate. In England, Robert Boyle and Isaac Newton established sen-sory observation and experiments with special instruments (air pump, prism, telescope) as hallmarks of science. The Royal Society of London was England's center of learned debate on work in the natural sciences and its *Philosophical Transactions* a journal of record.

Dampier somehow knew that the Royal Society solicited informa-tion from travelers and saw an opening. He dedicated his 1697 work to Charles Montagu, first earl of Halifax, president of the society, offering his "hearty Zeal for the promoting of useful Knowledge." He even cited the *Philosophical Transactions* to show he was keeping up.[46]

Yet, right after his dedication to Halifax, Dampier inserted a preface to the common reader, promising, not science, but thrillingly rough entertainment. "As to my Stile," he warned, "it cannot be expected, that a Seaman should affect Politeness." The result was a *picaresque*, a travel tale with a lowborn hero. Since the sixteenth century, the Spanish *picaro* or rogue had taken readers on rambling tours of Europe's battlefields and hinterlands, where cunning and violence were essential to survival. Many early European novels take the form of the beguiling picaresque. Because circumnavigations were overwhelmingly maritime and usually narrated by wellborn participants, the picaresque was not an obvious genre for around-the-world travels. But Dampier, who was a traveler on land and sea, and a commoner who told his story in his own plain style, was an immensely appealing world-encompassing *picaro*.[47]

He guaranteed "a mixt Relation of Places and Actions" and "many things wholly new." He promised not to "trouble the Reader with an account of every days Run" at sea, but to "hasten to the less known parts of the World," where wonders awaited. And he didn't overdo the science. When he digressed into maritime observations that required comparisons to parts of the globe he (and the reader) had already visited, he caught himself and promised to bundle those thoughts into a special section elsewhere, so as not "to carry my Reader back again: Whom having brought thus far toward England in my Circum-Navigation of the Globe."[48]

Dampier gives a vivid sense of what going around the world *felt* like, one body against the planet. There he is, up in the rigging and holding out his coattails, part of an emergency sail composed of the crew, who catch the wind in their clothes. Next, he argues with Captain Swan about crossing the Pacific without adequate provisions, "a Voyage enough to frighten us." Later, he approaches Guam and discovers that, food having indeed run short, the crew had plotted to kill and eat Swan, then the other officers, including Dampier. ("Ah! Dampier, *you would have made them but a poor Meal*," the fat Swan teased his lean colleague.) And there is our hero in an open boat headed upriver to Hanoi, bolting for the side because his diarrhea never let him "rest long in a Place." A doctor today might diagnose parasitic dysentery, but Dampier blamed the hot climate. Even fully immersed in a cooling local river in Aceh, Sumatra, he was "sensible that my Bowels were very hot, for I felt a great Heat within me." Dampier had wished for a long, warm journey. Guts a-boil in Aceh, maybe he asked himself: *hot enough now?*[49]

If these scrapes did not exactly recommend round-the-world travel, Dampier's reassuring refrain was that all the Earth could be made useful, if not edible. Consider his contributions to the English language. The Oxford English Dictionary credits him with the first publication of seventy-nine words. The largest number, twenty-two, refer to things that help sate the human appetite, including *star-apple, tortilla, chopstick, kumquat, posole, hen-coop, cashew, sapodilla*, and *samsho* (Chinese rice spirits). The second largest group, of twenty, were mariners' terms: *catamaran, double-reef, key, sea-breeze, supercargo, kink*, and *thunder-cloud*, among others. And the third largest number (sixteen) include terms for non-edible parts of the natural world, most interestingly, *sub-species*. Five are devices (including *barbecue*, for a platform made of sticks); eight describe persons (a *valetudinarian*, for instance, is a weakling who travels to regain health); six describe places, and two, types of action. Yet even the categories of natural things stress utility, from the *dysenteric* plant that stopped diarrhea, to the *salt-ponds* harvested in Jamaica.

All this is to say that Dampier, like other circumnavigators, described a world for human use—he described sets for a world stage. He was also the propertymaster who laid out the objects that an actor in geo-drama might need. His meandering, amphibious transit had required a carefully curated survivalist tool kit, and here it is.

Compass. On the afternoon of May 1, Dampier's party of buccaneers set out across the Isthmus of Panama, using their "pocket Compasses" to bear northeast. In earlier times, only pilots and captains might have had compasses, but the devices cost less by the end of the seventeenth century. Having several compasses in a group was insurance against loss and inaccuracy; having one of his own meant Dampier could find his way even if separated from the others.[50]

When in doubt, Dampier and his men hired Indian guides and paid them in *manufactured goods*, such as "Beads, Knives, Scissars, and Looking-glasses." They secured one man's expertise with a new hatchet. Another potential guide resisted recruitment until an Englishmen "took a Sky-c[o]loured Petticoat out of his bag and put it on his Wife," who then lobbied for her husband's departure. The bright and shiny things were essential. It would have been dangerous to kidnap guides on land, surrounded as they were by friends and family anxious to retrieve them. The slaves in Dampier's party were enough of a security risk. At night, two free men kept watch while the others slept, "otherwise our own Slaves might have knockt us on the head." The slaves were more interested in escape, which all of them managed to do.[51]

A variety of other *captives* assisted Dampier and his crewmates, especially in the still-daunting Pacific. There, they took a Spanish priest and two Indian boys as pilots, abducted two old Indian women for further information, and plucked some "Spanish Pilot-books" from Spanish ships. (The books were more helpful than the captives.) Yet Dampier's party, at this point on the *Cygnet*, was still not sure of the coastline around Acapulco. They approached a village to get new guides, but the people had fled—except one mulatto woman and her three children.

The sailors took the brood and made the woman give them a tour of the area. Once they were satisfied, they put her ashore, with some clothing as payment, and all her children except one, a "very pretty Boy, about 7 or 8 Years old." The woman "cried, and begg'd hard" for her son, but Captain Swan was determined to keep him. It would have comforted the boy and his mother very little to know that scholars have declared this era to be one of great progress in European cartography. Maps were not part of Dampier's tool kit in places where he could nab knowledge from his fellow creatures.[52]

A *hat* was protection from sun and rain and, less obviously, a reminder, however ragged, of European identity. In Europe, everyone wore hats, even beggars; only animals and criminals went bareheaded. (A woman might uncover her breast to feed an infant, but never her head, not in public.) Different headgear denoted different statuses, even in the wild, where a man might have little more than the clothes on his back. Dampier told how a group of Englishmen had been intercepted by the Spanish in the Yucatán and asked which of them was captain. None of them was, but to admit that would mean they were private marauders, mere criminals. Yet had one of them claimed to be captain, the Spanish would have demanded his commission—which none of them had. "At last one John Hullock cock'd up his little cropt Hat, and told them he was the Captain," claiming his commission was on his ship. The hat-trick worked and the Spanish escorted the Englishmen to their ship, meanwhile plying Hullock with hot chocolate and other gentlemanly treats. (His messmates never let "Captain Jack" forget his little fib and temporary promotion.)[53]

A hat was also a handy head-sized container. In Central America, for instance, a thirsty and resourceful man could find a "wild Pine" bush, slice it open, and catch its liquid in his hat. That was assuming he had a good *blade*, whether ax or knife, to clear brush and cut firewood, to wield in a fight, and to butcher meat. For example, Dampier explained that sailors could find refreshment at Quibo Island (now Coimba Island on the Pacific side of Panama), which had trees to fell and deer to hunt.[54]

Of course, he needed a *gun* to hunt in the first place. On the march through the Isthmus of Panama, hungry from three days of tramping,

Dampier and his men bagged and barbecued four monkeys. Simians became their dietary staple and shooting them a pastime—they blasted several that threw sticks and dung at them, while "chattering and making a terrible Noise." Animals were there to be eaten, and rated on their edibility. Dampier thought monkey meat quite tasty, adored "sweet" manatee, and considered it a bad day when he had to eat snake.[55]

Guns also repelled creatures that sought sweet human flesh, especially alligators; his tours of the 'gator-infested Bay of Campeche convinced Dampier that "no part of the Universe is better stock'd with them." A musket was a lifesaver when an alligator clamped its jaws around an Irishman's leg and began to haul in the delicious, flailing morsel as it screamed, "Help! help!" The man had the presence of mind to pry open his captor's maw with the stock of his gun, extract his leg, and crawl away. He later recovered the tooth-marked gun and his leg healed, though he "went limping ever after." Gunshots could also locate or summon men at a distance (before the alligators got them). However essential, the loose gunpowder that went into the men's firearms was dangerous. While Wafer (the buccaneer-surgeon) was drying damp gunpowder on a silver plate, a man smoking a pipe strolled by and sparked an explosion that seared Wafer's knee down to the bone.[56]

Sailors all, Dampier and his companions not only carried *rope* but called it "line," the maritime term for ship's rope. When they came to a broad and swift river in Central America, one man waded across, line coiled about his neck, paying out a handhold the others could grope across. But midstream, the line developed a "kink" that began to strangle the man, who was overburdened with loot. He clawed off the noose but lost his footing and, "having three hundred Dollars at his back, was carried down." Two men found him drowned downriver. Necessarily unsentimental, they just left him. They also left the money, "being only in care how to work their way through a wild unknown Country," where a coil of rope and a little cropped hat were worth more than hundreds of silver dollars.[57]

Money helped elsewhere. Dampier's diary of his meals traces his progress through wilderness and into settled places where he bought victuals. Along with monkey, snake, and manatee, he dined on Mexican avocados, guavas, and prickly pears (still rarities to Europeans),

California raccoons ("a large sort of Rat"), Galápagos iguanas ("a man may knock down twenty in an hours time with a Club"), coconut flesh and juice ("I like the Water best when the Nut is almost ripe"), limes ("used for a particular kind of Sauce" in Jamaica), and breadfruit (best fresh, else it becomes "harsh and choaky"). Dampier chewed betel in the Philippines, savored the fermented fish sauce that gave many Asian dishes their characteristic flavor, and noted where to find the world's best tobacco and chocolate.[58]

He ate to recommend, not only the eating, but the transplanting of foodstuffs to construct a world for travelers who had more cash than gunpowder. He noted that the Spaniards had propagated avocados around the North Pacific and sarcastically thanked them for "stocking the West-Indies with Hogs and Bullocks" that fed the privateers who raided them. Dampier thought coconuts might be transplanted to the West Indies, both for their nuts and liquor, as well as their useful nut-shells and fiber.[59]

A joint of bamboo, stopped at both ends with wax. In this lightweight, waterproof cylinder, Dampier preserved his rolled-up journal for publication. Without that container, he would have had only a tale told from memory, rather than the richly detailed narrative bursting with new places, people, and things, and new words for them. Dampier's fellow scribblers could not compete. In 1699, for example, when Cowley's journal was published, it was condensed to fit within a collection of travel accounts; Dampier's narrative, published two years earlier, had evidently swamped the market.[60]

A New Voyage round the World (1697) made Dampier famous (see illustration 7). In English, it ran through five editions by 1705; translated into French, it appeared in Amsterdam in 1698, then spread to other parts of Europe. It had a wide influence, high and low, as Dampier had hoped it would. Just as it was influenced by picaresque novels, so it influenced others. In the *Travels into Several Remote Nations of the World . . . by Lemuel Gulliver* (1726), Jonathan Swift has monkeylike Yahoos pelt his Dampier-like hero with their dung. Daniel Defoe cribbed from Dampier to introduce a fictional traveler in 1719: "I would be satisfied with nothing but going to sea," says the young Robinson Crusoe, ignoring his father's advice to get an education and stay put. Real boys also read Dampier and dreamed of worlds beyond their local ocean. Benjamin Franklin, a lifelong Dampier fan, confessed that, after leaving "Grammar School" at age ten, he "had a strong Inclination for the Sea."[61]

Readers on the high end admired Dampier, too. He was taken up by Samuel Pepys, the famous diarist who was a patron of sailors and of scientific investigators. Pepys was Secretary to the Admiralty in the 1670s and 1680s, and, elected a Fellow of the Royal Society in 1665, served as its president from 1684 to 1686. In his diary, John Evelyn noted in 1699 that he "din'd at Mr. Pepys, where was Cap: Dampier, who had . . . printed a Relation of his very strange adventures, which was very extrordinary, & his observations very profitable." Dampier struck Evelyn as "a [more] modest man, than one would imagine, by the relation of the Crue he had sorted with." The well-spoken buccaneer had brought a manuscript chart, of "the Course of the winds in the South-Sea," that must have especially interested Pepys.[62]

To the 1699 edition of his *Travels*, Dampier added that chart and the scientific analyses he had promised not to inflict on general readers. These new chapters explained climate, weather, and hydrography (the science of bodies of water) on a planetary scale. Thus the seasons: summer and winter were climatic opposites in the temperate zones; dry and wet the counterparts for the torrid zone between. Each of the pairs had its fair and rainy months. But geography complicated the general pattern—the western sides of continents were rainer than the eastern sides, for instance, though only in the northern hemisphere. Likewise, tides were "universal; though not regularly alike on all Coasts." Currents were irregular, more like wind patterns, and probably connected to them. Dampier had confidence in his global observations: "No Place in the Ocean [is] without Tides," he declared, as if he himself had seen every inch of the seas.[63]

To represent these terraqueous patterns, Dampier worked with London mapmaker Herman Moll to create what are probably the first thematic maps, cartographic representations of phenomena that are not visible features of geography. The Dampier/Moll maps use flocks of arrows to show the prevailing air currents over oceans, for example. The trade winds could thus "be pictured, as it were, to the Eye." The map shows only the Atlantic and Indian oceans, not the whole world, but the result was far more global than circumnavigator's maps that showed specific places of discovery, as with Schouten's Cape Horn and Cowley's Galápagos. The thematic depiction of trade winds was the first significant circumnavigator's innovation in mapping since the creation of the around-the-world cruise track.[64]

It was *picaro* science, adventure and knowledge stuck together with sweat and salt spray. Dampier questioned the theoretical "Hydrographers" who had estimated the sizes of the Atlantic and Pacific oceans. Sailors considered the Atlantic to be smaller and the Pacific considerably larger. Dampier agreed, and added that the size of the Pacific made the "Terraqueous Globe" more aqueous than generally thought. All the more reason to discover that ultimate solution, a Northwest Passage. "If I was to go on this Discovery," he noted, "I would go first into the South Seas, bend my course from thence along by California, and that way seek a Passage back into the West Seas," which he assumed would lack the seasonal ice pack that had thwarted all the expeditions that had started from the Atlantic.[65]

Sometimes the *picaro* got the better of the expert. Dampier surely believed that the Earth went around the Sun, yet explained the Circumnavigator's Paradox as if the reverse were true. On arrival in the Philippines, he was "sensible of the change of time" and the confusion of universal faiths. On what they thought was Friday, Muslims went to mosques, though sailors arriving from the west considered it a Thursday. Spaniards who arrived from the east planned to worship God two days hence, on what they thought was Sunday. But Spaniards on nearby Guam agreed with the sailors that it was a Saturday. Dampier decided not to blind anyone with science and told yet another good story: "Having travell'd so far Westward, keeping the same Course with the sun, we must consequently have gain'd something insensibly in the length of the particular Days, but have lost in the tale, the bulk, or number of the Days or Hours."[66]

Promotion was a logical reward for the famous Dampier, yet a terrible mistake. The wily itinerant made a bad naval commander. By order of the Admiralty, he led an expedition from 1699 to 1701 to do reconnaissance of New Holland (Australia). Dampier managed to survey parts of New Holland and New Guinea, though his ship rotted away and his crew's confidence likewise. He lost the ship and, at a court-martial, was found guilty of mistreating his first lieutenant.

Some merchants nevertheless recruited him in 1703 to captain one of two privateers bound for the Pacific. England was at this point at war with Spain in the War of the Spanish Succession (1701–14), and the authorities issued letters of marque to privateers to harass the ships of Spain and its allies. This new venture went even worse than Dampier's mapping expedition—and all the way around the world. The men quarreled over everything, especially the division of spoils. Some were marooned for making trouble; one begged to be marooned to escape the ongoing disaster. Others were imprisoned by the Dutch in Batavia on suspicion of piracy because Dampier had mislaid his commission. Neither of the two original ships made it back, and only an estimated 28 of the 183 men, an 85 percent attrition rate.[67]

The survivors included Dampier, who returned in 1707 ready to write another best seller, only to have a crew member beat him into print with a long, unflattering account. William Funnell, self-described ship's mate, wrote the first naval circumnavigator's account that did not speak for the captain or commander. Oh, Funnell spoke *about* Dampier. Dampier's publications "have met with . . . universal Approbation in the World"; Dampier's descriptions of winds and currents are incomparable—Dampier, Dampier, Dampier, as if he were in love with the man. But Funnell put his own name where it mattered most: on the title page and—amazingly—on the narrative's world map, which is marked with "the course of Mr. Funnell's Voyage." It was the first circumnavigator's track to bear the name of a subordinate.[68]

Behold the two *picaros*, squabbling over who was most overpromoted. Funnell's account—where he and his comrades had cut wood or clubbed seals, where they starved (the Pacific, as usual), where they fought the Spanish, where they fought each other—notes Dampier's faulty observations (even on alligators) and inability to keep order. Drake the "English Hero" extracted loot from Guatulco in Mexico and Cavendish destroyed the place—Dampier can't even give accurate coordinates for it. In his *Vindication of His Voyage*, Dampier objects that

Funnell was a mere steward promoted to midshipman (apprentice), but never to mate. He had abused his apprenticeship by pillaging Dampier's library and manuscripts: "he had the advantage of perusing Draughts and Books, of which he afterward gave but a slender Account, for some he pretended were lost, and the others the Draughts are torn out of them." [69]

Clearly, a successful circumnavigation required an iron hand or golden rewards, if not both. War perpetuated both conditions, to the delight of sea rovers.

Commissioned as a privateer by the new entity of Great Britain (created in 1707), Woodes Rogers took command of the *Duke* and *Duchess* from 1708 to 1711. Rogers recruited Dampier as "Pilot for the *South-Seas*, who had been already three times there, and twice round the World." Under Rogers's command, Dampier would become history's first known triple circumnavigator. [70]

Rogers's narrative has many interesting points. He tries to avoid scurvy by laying in twenty-two months of victuals (still not enough, as it turned out) and sloshes out liquor to keep his men warm and docile. His account is the first among circumnavigations to record the ceremony of crossing "the Line": anyone who had not already crossed the equator paid a fine or was ducked overboard. He gives a potted history of circumnavigation (including Loaysa and the Nassau Fleet), though he thought Magellan's *Victoria* had been sailed home by Sebastian Cabot, who was obviously more memorable than poor old Elcano. And Rogers recruits a memorable crewman at Juan Fernández, where a landing party returns "with a Man cloth'd in Goat-Skins, who look'd wilder than the first Owners of them." He was Alexander Selkirk, the sailor who had begged to be marooned rather than continue on Dampier's blundering fleet of privateers. Alone for over four years, Selkirk had survived because the island was overrun with European animals, principally introduced goats, but also introduced cats, which Selkirk nurtured so they would keep the introduced rats from nibbling him at night. He nursed his scorbutic rescuers with greens stewed in goat broth. (His ordeal was Defoe's major source for *Robinson Crusoe*.) [71]

But Rogers's cruise is most interesting for its staggering amount of paperwork. Amidst the merry adventures are a great many Instructions,

Resolutions, Memoranda, Agreements, Letters, and Committee Reports, all done to determine the legality of the ship's operations and the division of her spoils. While crossing the Pacific, the officers even took the trouble to prepare and sign "a Duplicate of every Conclusion in all Committees since we have been in these Seas," so each ship had copies. The men nonetheless disputed the division of spoils once they nabbed two richly loaded Spanish ships. (Rogers took a bullet in the jaw during the battle, and spit out fragments of tooth and bone for months after.) It was a great "Unhappiness," he said, not of his injury, but "to have a Paper-War amongst ourselves" amidst glorious success. He returned in 1711 with prizes that various parties estimated as between £200,000 and £3 million. There was yet more paperwork and much legal activity before the money could be divided. Dampier never went to sea again and died in 1715, probably before he received his cut.[72]

To the necessity of iron rule and golden rewards, add paper by the teetering stack. The danger and costly waste of privateering circumnavigations had always been offset by their potential for great gain. But with the world divided more decisively into trading zones, principally European or Asian, the waste and danger were greater gambles. That explained Rogers's worry over legal documentation—and over provisioning his ships to survive a world more firmly settled and less abundant in wild food and free drinking water.

Precisely because the world was more clearly carved into proprietary zones, Rogers was careful about the captives and slaves he acquired. His ranking of different people showed new priorities for European empires. Although he abducted numbers of people from Spanish territories, and kept the Indians as pilots, he ransomed the Spaniards. Even in the Pacific, detection of the kidnapping of Europeans was now swifter and reprisal more likely. Rogers retained blacks and American Indians, but thought better of the latter. (His policy was to feed the black sailors less.) Those preferences, too, reflected shifts in European empire. Indians in the Americas were adept at playing Europeans against each other—to alienate them was to risk a valuable set of allies; the Atlantic slave trade, in contrast, was approaching its historic peak, and thoughts of protecting black workers consequently rare. Finally, Rogers did not dare kidnap pilots in Asia, which could impair British attempts to expand its Asian trade.[73]

The next circumnavigation, of 1719 to 1722, was also a privateering venture, this one led by George Shelvocke. His expedition had a famous afterlife because Samuel Taylor Coleridge (on the recommendation of his fellow Romantic poet William Wordsworth) read Shelvocke's narrative and refashioned one of its events into *The Rime of the Ancient Mariner* (1798). In that poem, the haunted Mariner confesses a terrible sin. Unnerved when his ship had entered eerie, icy waters around Cape Horn, he shot an albatross that accompanied the ship, though those faithful birds, which represent Christ's devotion to humanity, bring good luck. The Mariner and his companions are cursed. It is bad when the wind ceases:

> *Day after day, day after day*
> *We stuck, nor breath nor motion;*
> *As idle as a painted ship*
> *Upon a painted ocean.*

It is worse when all the Mariner's companions die; it is dreadful when they rise again. The zombie crew bring the ship home, where the Mariner seeks absolution. The poem is remarkable for its sense of dead quiet—the ship becalmed, the crew silenced—pierced only by the Mariner's compulsive telling of his tale.[74]

The historical reality was a shouting cacophony. Shelvocke had been placed under Commander John Clipperton (on the *Success*), with 101 crew on the *Speedwell,* though the only thing the two men could agree upon was to part company in the Atlantic. Six pages into Shelvocke's narrative, and a mutiny is already planned. He managed to maintain control, but alienated the crew with his self-regard and favoritism. The boatswain openly referred to the officers as *"Blood-suckers."* One of Shelvocke's favorite words was *melancholy*, which occurs eighteen times, twice on the page where the second-in-command, Simon Hatley, shoots a black albatross that haunts them in the sunless Strait of Le Maire. The cursed events that follow, however, are the usual ones: disease and death; getting lost in the Pacific. Could it get any worse?[75]

Of course it could. The climax was shipwreck at Juan Fernández. There, Shelvocke bribed the carpenter to build an emergency boat, while all the slaves tried to escape and most of the crew demanded new documents promising them larger cuts of any future loot. The island's goats proved wily and the mobs of cats (forever underfoot) had decimated the local birds, so the castaways had to catch and smoke eels as quick-cooking snacks for the escape boat. Over forty men crammed into the small vessel, sucking tiny rations of water from a musket barrel stuck into a cask, filthy as all get-out, and "everyone contending for the frying-pan." Amazingly, this was the one part of the voyage when no one died. And from that reeking, open boat the men managed to capture a Spanish ship to which they transferred. On they went.[76]

Given that he was unlikable (and dishonest), it is notable that Shelvocke betrays concern for the rest of humanity—it signaled a cultural shift that outweighed his undeniable personal viciousness. He agreed with Rogers, for example, that captives should be culled strategically. If Indians might be potential allies against the Spanish, for instance, they were to be cultivated. In Baja California, Shelvocke welcomed so many Indians aboard that they filled the entire deck. Determined to display English hospitality, he brought out pots of preserves and somehow found spoons for all. He spared neither charm nor jam on blacks, however. Twice, when he took mixed groups of captives, he released the Europeans and Indians but kept the blacks as slave labor. In contrast, he hired Asians as pilots. And, again like Rogers, he worried about provisions and tut-tutted when his men ate food he considered substandard. Rather than trust the usual small casks for his water supply, he had a wooden cistern constructed to hold 10 tons of water. (Nice idea; the tank leaked.)[77]

Shelvocke was only out of step in his views on scurvy. He concluded that a circumnavigation was bad for human health for the traditional reason, exposure to alien climates. He deplored everything that was "undergone in so great a journey as is the circumference of the Earth, where the heats and colds of so many climates are to be endur'd: to be as far from ones native soil as the circuit of the Earth will admit of." In contrast, most other circumnavigators believed that scurvy resulted from removal from land. Getting invalids ashore was the preferred remedy, as William Dampier did in the Galápagos (1684), Woodes Rogers at the Juan Fernández Islands (1709), and William Betagh on Cocos

Island (1722). After his circumnavigation (1721–23), the Dutch mariner Jacob Roggeveen stated that it took precisely "fourteen days on land" for "fresh" food and "the assisting land air" to cure scorbutic sailors.[78]

Unfortunately, it was impossible to guarantee access to land in time and in the right places. Although variations in mortality reveal that much depended on luck, poor survival rates on circumnavigations were the norm, as men died, deserted, or were abandoned:

Table 1[79]

Attrition on Circumnavigations, Magellan to Dampier

	Men: Out/back	Attrition (%)	Ships: Out/back	Loss (%)
Magellan/Elcano (1519–22)	275/39	86	5/1	80
Loaysa (1525–36)	450/9	98	8/0	100
Drake (1577–80)	164/56	66	5/1	80
Van Noort (1598–1601)	248/45	82	4/1	75
Le Maire/Schouten (1615–17)	87/21	76	2/0	100
Dampier (1703–11)	183/18	90	2/0	100

Alas, nothing as exciting as scurvy enlivens the first fictional circumnavigation to feature human characters, and quite possibly the dullest ever. Daniel Defoe's novel *A New Voyage round the World, by a Course Never Sailed Before* . . . (1725) refers to several best-selling historical circumnavigators, including Dampier, not least in its title. But the narrator begins by warning that circumnavigations, including his, were ten-a-penny; the way around the world "is now a common Road" and "the World has done wondering at it." A reader who might forgive this authorial modesty is then challenged with the promise that the novel will, at the least, be more gripping than a sailor's logbook, small praise indeed. The one imaginative touch is the "course never sailed before," which is an eastward circumnavigation, which no real sailor had yet tried.[80]

Commerce is the book's obsession. Mercantile terms—*trade, trading, merchant(s), market, exchange, goods, traffick(ing), commerce*—clog the story, as if greed were the only lust in the human heart. (The scant mentions of *woman,* five times, and *women,* twenty-five times, refer to chaste encounters.) To maximize profits, the Englishmen bring a French captain and French colors as camouflage to entice Spanish and Portuguese traders, details modeled on van Noort's subterfuge. From India to the Philippines, the disguise works and the lucre piles up. Before crossing the Atlantic, the seafarers make sure they have "rummag'd [reloaded] our Gold," lest it cause their ship to list. Ridiculous amounts of gold keep turning up, especially in Spanish America, where the story ends. (The run home is anticlimactic.)[81]

But getting rich during a circumnavigation by stealing from rivals was an old story. Defoe missed a trick by not picking up on the new story: how to protect human life while circling the big, unforgiving planet, the survivalist project that had somehow united the European rivals who, even as they created separate empires, regarded mastery of the globe's physical resources as a common pursuit, a big imperial project that framed the smaller ones within it. Going around the world thuggishly, to carve empires from oceans and from other people's territory and resources, was to be redeemed, somewhat, with a revised imperial strategy: minimizing mortality while going around the world.

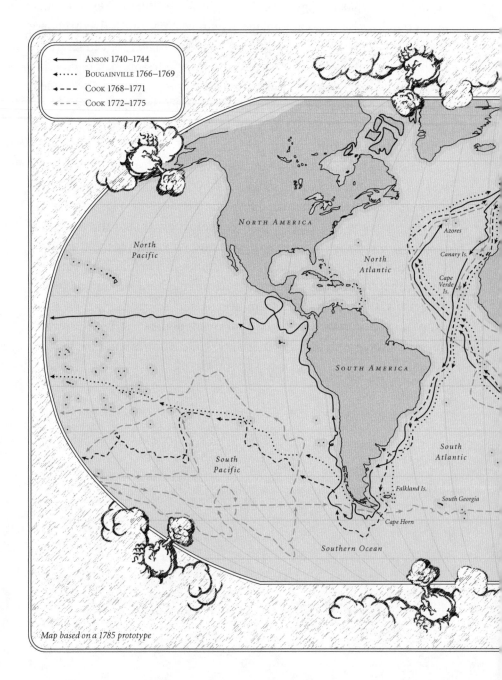

Map based on a 1785 prototype

ANSON 1740–1744
BOUGAINVILLE 1766–1769
COOK 1768–1771
COOK 1772–1775

NORTH AMERICA

North
Pacific

North
Atlantic

Azores

Canary Is.

Cape
Verde
Is.

SOUTH AMERICA

South
Pacific

South
Atlantic

Falkland Is.

South Georgia

Cape Horn

Southern Ocean

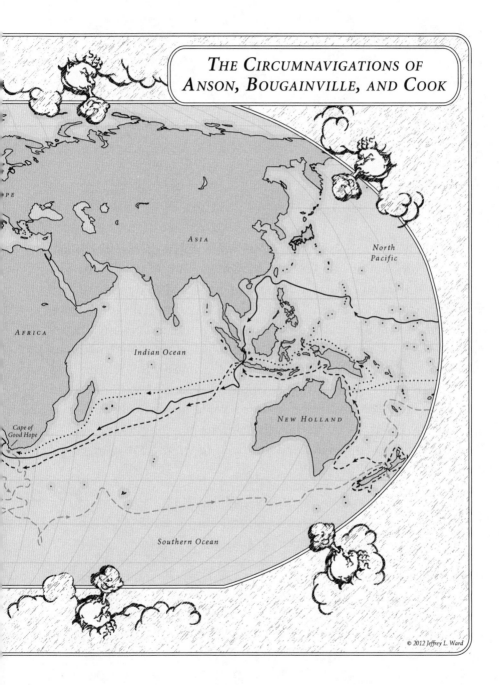

THE CIRCUMNAVIGATIONS OF
ANSON, BOUGAINVILLE, AND COOK

ASIA

North
Pacific

AFRICA

Indian Ocean

PE

Cape of
Good Hope

NEW HOLLAND

Southern Ocean

© 2012 Jeffrey L. Ward

Chapter 4

Terrestriality

*T*he War of Jenkins's Ear (1739–48) brought no decisive British victory, but one undeniable triumph. On July 4, 1744, thirty-two wagons carried an estimated £400,000 in silver and gold from a ship at Portsmouth to the Tower of London. A band of musicians led the procession and with it marched the circumnavigating sailors and officers who, under Commodore George Anson, had plucked the enormous prize from the Spanish Pacific. The men were mostly English, but also hailed from nine other European nations, plus parts of Asia, Africa, and the Americas. The wealth they guarded would be worth about £3 million today. It was, and is, the biggest maritime prize ever taken.[1]

To celebrate, a poem in London's *Daily Advertiser* acknowledged the glories of ancient Rome:

> *But round the Globe her eagle never flew,*
> *Thro' every clime is Albion's thunder hurled,*
> *And Anson's spoils are from a tribute world.*

Another rhyme, however, this from the *Daily Post*, questioned the celebrations:

> *Deluded Britons! Wherefore should you boast*
> *Of treasure, purchased at a treble cost? . . .*
> *In this attempt, count o'er the numerous host*
> *Of Albion's sons, unprofitably lost.*

The cost *was* high: of the 1,939 men who had set out with Anson in 1740, 743 had turned back and 1,051 had died. Only 145, just 7 percent, returned with him around the world. A high casualty rate is expected of a military expedition, yet only three of Albion's sons were slain in battle. The others died of natural causes, overwhelmingly from scurvy.[2]

Would it have cheered the scurvy victims—as their gums swelled, teeth fell out, skin erupted, old wounds reopened, and deaths approached—to know that their suffering would be famous? Their widely discussed agonies generated a remarkable debate over the circumnavigator's belief that scurvy proved that humans were terrestrial creatures, distinctly adapted to the earthly parts of a terraqueous planet, and unable to thrive at sea.

Anson's expedition had indeed shown that the old wooden worlds were still, two centuries after the Magellan/Elcano circumnavigation, lousy life-support systems. Europeans made celebrated efforts in the eighteenth century to command the world's oceans: better charts, improved ways to determine longitude at sea, the quest for the Northwest Passage. And yet the ships that made possible the charts, the tests of chronometers, and the searches (in vain) for the Northwest Passage, were as bad as ever at keeping humans alive over great distances, let alone the greatest distance, a circumnavigation. The mismatch among planet, ship, and human body continued to be fatal; the planet shrugged off most of those who attempted to go around it.

It was much safer to take the measure of the Earth through proxy calculations. Geodetic expeditions, which determined the planet's size and dimensions, did not go all the way around it, mercifully for their participants. Two important efforts in geodesy had been made in the 1730s, just before the War of Jenkins's Ear, to settle a debate about the shape of the Earth: was it oval (pulled toward the poles) or did it instead bulge around the equator, as Newton had predicted it would, due to gravitation? Two scientific teams departed France for Peru and Lapland, each to measure an arc of meridian, to see how these far north and equatorial units varied from measurements taken in France itself. (To measure an arc of meridian required observers to survey sections of the Earth's surface through triangulation, then determine the terrestrial distances using celestial observations of the fixed stars.) Newton was vindicated—the planet bulged around its middle. The expeditions were hazardous, even heroic, but they resulted in only one casualty, when the Peruvian group's surgeon died in a riot in Cuenca.[3]

It was easier and safer to consider the planet in the abstract, and to measure it in sections, than to experience it in its rough entirety, as circumnavigators did. Take one of Anson's suffering sailors. Let's see him off, though on a slightly different version of his tour. From England, let's drop south until we hit the equator, somewhere in the Atlantic. Say goodbye to our brave mariner, wring his hand and wipe away a tear, because we're going to use him as a measure of the planet. How many Englishmen can be laid end to end, all the way around the world? Round and round he goes, end over end over end round the equator, until he is no longer alive, and then no longer the same length as he was at the start, and finally no longer even *there*. His fragments have been absorbed into the globe that his body has been measuring.

Does this seem heartless? Well, the man was already dead when we started our thought-exercise, and had been for centuries. Yet he might have appreciated our use of him as a disintegrating unit of planetary measure, because in fact going around the world in the age of sail felt that way—that's the burden of all the descriptions of circumnavigations before about 1800: the world is enormous, putting oneself around it hurts like hell, and might prove fatal.

If anyone could have improved matters, it should have been Britain's Royal Navy, which had become one of the world's largest and most complex bureaucracies. Directed by the Board of Admiralty, the Navy Board hired sailors and, in times of war, was authorized to seize any able man—impress him into service—if he seemed to know starboard from port. The Admiralty recruited officers from the upper classes and promoted ordinary sailors who were smart, tough, and ambitious. The Navy controlled a battery of dockyards and victualling departments to build, refit, and supply ships. Masses of documents rated the men and ships that could, whenever needed, be put into service.

George Anson was a typical product of this system. Born into a wealthy family in 1697, he entered naval service around age fifteen, became lieutenant four years later, and rose to post-captain by the time he was twenty-seven. When he received a new commission in 1740—his wartime lucky break—he was promoted to commodore and instructed to "annoy and distress" the Spaniards in the Pacific by raiding settlements, seizing ships, and inciting Indians or colonists to rebellion. The Manila galleon was, as usual, the big prize, and, depending on where Anson might find it, he was authorized to return home via China or back around Cape Horn. Like Rogers and Shelvocke, Anson was a privateer; unlike them, he commanded an official naval squadron.[4]

Yet Anson's mission was secondary to British attacks on the Spanish Caribbean, and outfitted accordingly. He received six small fighting ships (three of them fourth-rate, the others smaller), plus two "pinks," tiny victualling vessels. Anson commanded the largest ship, the 1,005-ton, 60-gun *Centurion*. His men were culled from whoever could be spared from the Caribbean campaign. At least half his sailors had been impressed (kidnapped and commanded to serve), little better than slaves. To assemble the complement of five hundred marines, the armed men who maintained order among the sailors, the Admiralty collected veterans from London's Royal Hospital in Chelsea. The majority of them were over sixty years old, several had lost limbs, and some had to be carried aboard on litters. It was a sham. To give the appearance of full preparation, the 259 pensioners had been condemned to death—"Mr. *Anson* was greatly chagrined at having such a decrepit detachment allotted him." Many of the resulting crew were unlikely to survive an Atlantic crossing, let alone a voyage around the world.[5]

The ships received twenty-two months' worth of naval foodstuffs, to be served out in a standard rotation. Five days a week, Jack Tar ate pickled meat dredged from barrels of brine. The other two days were "banyan" (a word that described vegetarian Hindus), with cheese or salt fish. Beer, wine, and liquors, dried peas, oatmeal, and masses of bread (ship's biscuit or hardtack) completed the menu. Sailors got fresh stuff, such as vegetables and fruits, only in port. The stodge was plentiful— far more than most workingmen got on land—if preserved from damp, vermin, and rot. (As ever, officers ate better because they could take private stores, including livestock.) The heavily salted foods made any shortage of drink unbearable, but fresh water remained difficult for ships to store in quantity. Anson was ordered to re-water at Cape Verde and off Brazil, and take on other provisions, if necessary.[6]

Six ships, two pinks, tons of beef, beer, and bread—was it enough? Just as the Dutch who had organized the Nassau Fleet learned nothing from Loaysa's expedition, so the British Admiralty betrayed no knowledge of either of the two earlier attempts to send large fleets into the Pacific. Nor did Anson know the best time to navigate Cape Horn. He thought January; June or July are best. As he dropped south (see Map 2), his men dropped dead in scores, and not just the decrepit pensioners. Many men came down with "fevers" (probably malaria) while in Madeira, and ninety expired midway to Brazil. (Amid the agonies, a frolic: at least one ship's crew fined or ducked anyone who had not already crossed the equator.) The ships were crammed beyond their usual capacity—the men lacked even the regulation 14 inches' width per hammock. Filth accumulated below decks. One officer on the *Centurion* found it "scarce conceivable what a stench and nastyness our poor sick people had caused among each other." Already, in the Atlantic, the men fell prey to scurvy.[7]

Wind pounded the fleet as it rounded Cape Horn. Unable to take consistent celestial readings to determine latitude, let alone longitude, the *Centurion* veered dangerously off course. So did the *Wager*, which was wrecked on Patagonia. The majority of that ship's castaways mutinied and took a jury-rigged boat up to Brazil; thirty survived. Other *Wager* survivors straggled back to England via the Spanish Empire. Two other ships failed to keep up and doubled back. The depleted fleet of three managed to seize some Spanish merchant vessels and looted the Peruvian town of Paita. But the gain was small and only two ships survived to cruise the Pacific, the *Centurion* and the *Gloucester*, each substantially rebuilt from materials extracted from prize ships. Some sailors despaired: "It was the vulgar opinion amongst our people that we had sailed so far as to pass by all the land in the world," and believed that they were doomed to swirl round and round the wind-torn global south.[8]

As their wooden worlds began to fail, the men invented ways not to die. They drank blood from sea turtles. They ate "bread toasted over burning brandy to kill the numerous insects it abounded with." So many sailors were disabled by scurvy that the officers handled sails, hauled lines, and manned the pumps, tasks they had learned as young midshipmen. When the *Gloucester* leaked too badly to continue, her men transferred to the *Centurion*. The effort of crawling aboard killed three. Others, fortified with liquor they had salvaged from the sinking ship, set her afire. She burned all night, "making a most grand, horrid appearance," each of her guns discharging a farewell when the flames reached it. She blew up when the fire ignited the two hundred barrels of unretrieved gunpowder below decks. The expedition was reduced to one ship and 201 men, only about 45 of whom could work. Anson's gift for command was now apparent: there was no whisper of mutiny—no man thought he could do better.[9]

The worst was over. Stops at Juan Fernández and at Tinian in the Marianas restored the crew's strength, though the shock of being taken ashore at the first place killed fourteen men. Without finding the Acapulco galleon, Anson continued to Macao. His arrival alarmed the Chinese and panicked the British East India Company's representatives. European military ships were banned from China. The Chinese only welcomed Europeans who came to trade, on strict Chinese terms, and they considered Anson and his men pirates. When they saw any of the unaccountable visitors in the streets, they called out *"Ladrone, Ladrone,"* from the Portuguese for "Thief, Thief!" To get repairs and supplies, Anson had to bargain and bribe his way to a "chop" or license. The local British merchants secretly admired that he and his vessel, however battered, showed the Chinese something of Britain's global ambition.[10]

From Macao, Anson headed back into the Pacific, and there he made his name and fortune at last by intercepting the Acapulco treasure galleon *Nuestra Señora de Cobadonga*. Though the commodore later bragged that he had defeated a formidable adversary, the galleon was actually outgunned by the *Centurion*—easy, after all that worry and misery. A second stop in the Canton River, this time with the *Cobadonga* and captive Spaniards, further unsettled the Chinese.[11]

And then came the return to a Britain hungry for good news in the otherwise inconclusive war against Spain. Anson's officers and men added to the drama of several London plays simply by showing up in the audience, to be huzzahed and regaled with song. Two narratives of the expedition appeared within the year. Anson's track was inscribed onto world maps and globes, visible proof of Britain's global punch. Breathless notices of the venture appeared in the newest medium, the newspaper. If ordinary seamen who had gone around the world had once been nameless, the British papers now supplied names and brief fame: Mr. Maddox got married, Mr. Martin drowned while drunk, Mr. Burton was assaulted and robbed. *The Universal Spectator* published an interview—a celebrity interview—of one *Centurion* officer conducted by his "Friend":

FRIEND: Your men, I am told, dropped off like rotten sheep . . . Did not that terrify you?

VOYAGER: Not considerably. We who recovered, should have been glad that all had lived; but, as they did not, the few remaining had the greater hopes from what we might acquire.

Yes indeed, the fewer the survivors, the more loot for each, including their "brave, humane, equal-minded, prudent commander," who, along with his cut, was promoted to rear admiral and appointed to the Board of Admiralty.[12]

But the high loss-to-loot ratio was troubling. Three ships had been demolished, a dreadful expense to a deforested nation that had out-sourced much of its shipbuilding. (Together, the vessels would have required 2,800 to 4,000 trees, each at least a hundred years old.) Over a thousand men had died like rotten sheep. Not all survivors benefited equally. The Admiralty decided that the officers who had transferred from the scuttled ships to the *Centurion* were not, because not formally commissioned to serve on that vessel, entitled to a full share of its prize. That was legal, but scandalous.[13]

Anson did not publish a direct defense. Instead, with the Admiralty's permission, he collaborated with two other men to produce an official history of the voyage. Benjamin Robins (mathematician and political pamphleteer) was hired to turn the journal of the *Centurion*'s chaplain, Richard Walter, into a full narrative. Under Walter's name, *A Voyage round the World . . . by George Anson* (1748) ran through five editions within a year, nine by 1761, and fifteen by the year 1776, plus reprintings in many travel anthologies and translations into French, Dutch, Italian, Spanish, German, Polish, Swedish, and Russian. It was extremely popular. George Washington owned a copy; the Royal Library in Madrid acquired editions in French and English.[14]

Anson's circumnavigation had outdone earlier attempts to circle the globe, the official history claimed. "Notwithstanding the great improvement of navigation within the last two Centuries," it begins, "a Voyage round the World is still considered as an enterprize of a very singular nature." Much had been achieved since the first attempt. By going around "this terraqueous globe," Magellan's expedition had "demonstrated, by a palpable experiment obvious to the grossest and most vulgar capacity, the reality of its long disputed spherical figure." That of course was nonsense. Medieval folk thought the Earth was spherical, but it pleased their descendants to feel superior to them. The idea that Magellan (and Columbus) had proved the Earth to be round would stick, as if it weren't enough that they had thought the Sun went around the Earth.[15]

Given Anson's quickness to take captives, his superiority over Magellan was particularly debatable. The official history swore that captured Spanish women had suffered no "inquietude or molestation," but admitted that the male prisoners from the Manila galleon had endured terrible privation. The narrative claimed that Anson only impressed Indians and "Negroes" as workers because his crew was so depleted, and contrasted his humanity to a Spanish crew's brutality toward Indian captives. But documents from the voyage show that Anson fed captive blacks and Indians less than free sailors. In contrast, he paid for the services of Asian mariners, as with a Chinese man who piloted them to Macao and as with thirty lascars he hired there to fill out his crew.[16]

It is especially noteworthy that Anson brought back an even smaller proportion of men than Elcano had done on the very first circumnavigation. The stupendous losses incited a debate in Britain over "this important article, the preservation of the lives and health of our seamen"—at last, public concern over the fatal consequences of a circumnavigation, and especially the major killer, sea scurvy.[17]

In relation to scurvy, the official history outlined several key points, which unpublished records from the expedition's officers also stressed: doctors were useless; fatigue and deprivation made scurvy worse; provisions that were "fresh" had particular benefit; yet only land was infallible; land's specific virtues remained mysterious, however. Judgment on the medical men was swift and blunt. Anson's narrative told how the flagship's surgeon had exerted himself on behalf of the sailors but "at last declared, that all his measures were totally ineffectual, and did not in the least avail his patients." At the very least, the patent medicines of one Dr. Ward had one advantage: they made the sick no worse.[18]

The accounts emphasize that scurvy compounded the effects of other physical stress, especially cold, hunger, thirst, and overexertion. Scurvy first broke out after the fleet had fought its way around windswept Cape Horn, through cold capable of freezing the sails and "benumbing the limbs." Depleted supplies worsened matters. In the single description of scurvy that differentiated between non-Europeans and others, Anson admitted that his "Indian and Negro" prisoners developed the disease first. That was because they were on reduced rations. Scurvy spread as supplies of water and food deteriorated in the Pacific. Exertion—even the bare efforts to keep the ships on course—killed many survivors: "It was no uncommon thing for those who were able to walk the deck, and to do some kind of duty, to drop down dead in an instant." Even being shifted from one hammock to another could prove fatal.[19]

Once the disease was established, provisions were only effective if they were "fresh," a word that recurs, obsessively, to specify foods recently taken from land. One officer concluded that scurvy had only one "proper medicine: viz fresh meat and fruits." At one Pacific island, Anson's men "eagerly devoured" a boatload of mown grass and butchered seals, both items "considered as fresh provision." Conversely, heavily salted meat, the standard ration, lacked the requisite freshness. If sailors have scurvy, "to give them salt provisions before they are thoroughly recovered, would be to murder them." Indeed, the men "must inevitably perish without refreshments" derived from land, as if their sea provisions had, despite (or because of) their being salted, pickled, bottled, or confected, lost something in the translation.[20]

The commentators were not always confident that the "refreshments" yielded the full benefit of being on land—they often hedged their bets by jumbling the land's virtues (sometimes but not always its air) with those of its fresh products. At Juan Fernández, "the land" but also "the refreshments it produces, very soon recover most stages of the sea-scurvy." Relapsed cases benefit from "the smell of the earth together with the coconut milk, oranges, limes, bread-fruit and fresh meat." Sailors are "languishing" for "the land and its vegetable productions, (an inclination constantly attending every stage of the sea-scurvy)." "Nor can all the physicians, with all their *materia medica,*" one mariner explained (in another dig at the doctors), "find a remedy for it equal to the smell of a turf or a dish of greens."[21]

Anson himself believed that only land would cure scurvy. In personal letters, he described the illness as an "incurable distemper" and (somewhat more optimistically) as an "almost incurable distemper." Though it "has been generally presumed, that plenty of fresh provisions, and of water are effectual preventives of this malady," his official history declares, these measures were powerless once scurvy had settled deep within the sailors' bodies. "In some instances," the account concludes, "both the cure, and prevention of this disease, is impossible to be effected by any management, or by the application of any remedies which can be made use of at sea." On arrival at Tinian, the scurvy-struck commodore had his tent pitched ashore, "being convinced by the general experience of his people, that no other method but living on the land was to be trusted to for the removal of this dreadful malady."[22]

Why was land beneficial? Confident that they were right, most mariners did not bother to explain their claim. One of Anson's officers concluded that human beings had a terrestrial "*Je ne sais quois . . .* or in plain English, the land is man's proper element, and vegetables and fruit his only physic"—the sailor's basic belief, tricked out with a bit of French. Anson's official history attempted the first detailed answer. "Fresh air is necessary to all animal life," it stated, but polluting "effluvia" from the ocean "render the air they are spread through less properly adapted to the support of the life of terrestrial animals" than land air. That idea honored the sniffing of greenery and earth that other accounts had recommended, while gesturing toward ongoing eighteenth-century scientific debates about the nature of air. But it strains belief that Anson had provided the scientific detail. More likely, one or both of his learned collaborators, Walter the clergyman and Robins the man of science, had supplied this passage.[23]

Indeed, from the Anson account onward, sailors' belief in the healing power of earth and verdure would be translated into terms that addressed contemporary debates among landsmen, whether aesthetic, sentimental, or scientific. Each of these translations allowed the maritime belief that land cured scurvy to penetrate a broader reading public, in some ways the culmination of public fascination with this era's circumnavigators' accounts.

For the highborn officers and highly educated men who composed the known Anson narratives, land was not just useful, the sovereign remedy for the stinking horror at sea, but an aesthetic object of desire. Juan Fernández is "romantic" and has "scenes of such elegance and dignity, as would perhaps with difficulty be rivalled in any other part of the globe." Quibo is "delightful." Tinian is "charming." The phrases indicate a significant shift in views of the natural world. The eye of the educated European was becoming a sensitive instrument that detected which landscapes were morally uplifting or spiritually refreshing, even as those landscapes had (as doctors agreed) a medical capacity to affect the human body.[24]

Sailors' longing for land was also mainstreamed via the sentimental concept of homesickness. That yearning had been described as early as the sixteenth century, often among young soldiers or sailors who were far from home. In 1678, a Swiss physician, Johannes Hofer, coined the term *nostalgia* to describe a desire for home, and the word in this original meaning circulated around the European Continent, appearing in English by 1756. People debated whether the disorder affected the mind or the body. Did people miss home and then get sick? Or get sick because their bodies missed home? (The questions were particularly resonant in cases of scurvy, which was associated with melancholy.) Using those rather inchoate terms to describe sea scurvy, Anson and his officers insisted that scorbutics improved on the mere smell or sight of land. The scurvy-struck were said to rally, for instance, when carried into "the neighbourhood of land." As the fleet rounded a verdant part of Juan Fernández that featured a waterfall, the invalids "crawled up to

the deck to feast themselves with this reviving prospect." These statements did not claim that scurvy was all in a sailor's head, but implied
that, for such a serious medical condition, both mind and body must
be involved.[25]

The learned phrase about health and air in Anson's official history
also invited scientific investigators to explain why, exactly, the human
body failed at sea. A clutch of naval surgeons and British men of
science tackled the problem, essentially to debate the precise contributions of diet and air.

The best known among them was James Lind, whose recommendation of citrus fruits has generated a long tradition of regarding his
suggestion as definitive, the true cure for scurvy that was bafflingly
overlooked for several decades. And yet it is not so simple. Lind, like
the others, considered the question of human terrestriality more
broadly and he was never confident that any single land product—
including citrus—could provide the *je ne sais quoi* of land.

Several of the men of science thought that the cold, damp, and stale
air aboard ships was scurvy's main cause. They conceded that preserved
foods and inadequate water also reduced health, but warned that fresh
foods and plentiful water would have limited efficacy without proper
circulation of warm, dry air. (Invention of ships ventila-
tors was a project *du jour*.) Lind, for example, extolled fermented foods,
but especially citrus, precisely because he thought they had the capacity to overcome the effects of damp air. Another experimenter theorized
that "fixed air" (carbon dioxide) was the key because it was thought
to bind bodily tissue together. Too little fixed air and the body would
putrefy, especially when diseased, especially from scurvy. Doses of fixed
air in the form of brewer's malt and carbonated water would offset the
disease. (The malt was made into a sweet, drinkable "wort," basically
unfermented beer, an ancestor of today's non-alcoholic malted drinks.)
One dissenter declared air irrelevant, and recommended instead several
edibles, especially citrus, malt, and beer made from spruce needles, simply because they had been shown to work.[26]

But at least two of these scientific experts asserted that, should all
else fail, land was the ultimate remedy. English physician Richard Mead
said the speed of that therapy was "incredible." "Upon their being ex-

posed upon the ground," he claimed, Anson's invalids "immediately recovered." Mead related how one incapacitated sailor begged some stronger companions to dig a hole in the soil and let him inhale its restorative effluvia. "Upon doing this, he came to himself, and grew afterwards quite well." In the third edition of his work, Lind also defended the faith in land as something more than a sailor's yarn. He described an incident when some scorbutic seamen were taken ashore, stripped, and buried up to their heads in the earth. They revived, Lind related, without a hint of skepticism. He even thought that, given the melancholy that haunted the scorbutic, "the joy of being landed" was enough to begin their recovery—the mere prospect of earth began their cure.[27]

The British Admiralty would have found it inconvenient, however, to implement a plan of naval health that relied on shore leave all around the world. That would have entailed delays and (even worse) special pleas to foreigners for permission to land, in an age when imperial zones were becoming more firmly fixed. Anson had essentially followed Drake's old route (compare Maps 1 and 2), but it was dangerous to make land in enemy territory. Much better would be portable cures, either sustaining foodstuffs, perhaps combined with dietary supplements, or else technologies that could supply ventilation and pure water. In the end, the Admiralty deployed all these options, though circumnavigators would be interested in them only insofar as they promised, somehow, to bring a terrestrial freshness to ships that had to circle the entire planet.

After Anson, no expedition went around the world for twenty-one years, mainly because the major European powers were preoccupied with a new imperial conflict, the Seven Years' War (1756–63). Provisioning armies and navies over long distances was a particular challenge of that world-wide conflict, and the resulting naval reforms would be tested during the tight sequence of circumnavigations that next occurred: John Byron (1765–66), Samuel Wallis/Philip Carteret (1766–68/9), Louis-Antoine de Bougainville (1766–69), James Cook (1768–71), and Cook again (1772–75). These were nationalistic expeditions, quite often done to pursue unfinished business from the late war. France sought new territories to replace the parts of North America and South Asia it had lost to Great Britain; Britain wanted new territories to fill the Pacific gap in its empire between North America and Asia. But the violence of European rivalries was fading. Anson had led the last privateering circumnavigation; that most visible and distinctive form of European exploration, round-the-world voyaging, was at last being disentangled from European warfare.[28]

These new circumnavigations have drawn a great deal of attention, though not usually for their round-the-world character. Wallis is known for making the first European contact with Tahiti; Bougainville for helping to popularize the "South Sea" islanders as noble savages; Cook for participating in a global mission to observe the transit of Venus across the Sun, and later for mapping Australia. Conversely, Byron is often criticized for not making any new geographic discoveries. But at the time, the new circumnavigators were celebrated for something else entirely, their lowered mortality rates, starting with the much-derided Byron who, remarkably—amazingly—achieved zero mortality from scurvy, and continuing through Cook's second circumnavigation, when he duplicated Byron's results, for which he received scientific recognition:

Table 2[29]

Survival Rates on Post-Anson Circumnavigations

	Men	Deaths	Mortality (%)
Byron	153	6	3.9
Wallis/Carteret	240	31	12.9
Bougainville	200	7	3.5
Cook (I)	96	40	41.7
Cook (II)	232	6	2.6

(All original ships in these expeditions returned)

Of course, all the other tasks these circumnavigators accomplished on their voyages were important, but they were not intrinsic to an around-the-world voyage. The lowered mortality rate was intrinsic to the planet-sized voyages, however, and it was the key to everything else. Catastrophic losses of ships and men would have made charting the

Pacific and observing Venus impossible—preventing the planet from shrugging them off in the first place was the foundational accomplishment for this generation of circumnavigators. As one member of Cook's second voyage marveled, the mortality rate within the expedition was in fact statistically lower than could be found among men on land.[30]

How did *they* think they did it? The expedition narratives emphasize two goals that Anson's official history had stressed earlier: to keep the men in general good health (by maximizing rest, water, and food), and to get *fresh* food and water as regularly as possible. They also discuss two other factors, dietary supplements and fresh air below decks, that had emerged in the recent scientific debates over scurvy, though with none of the technical detail of those debates. And, offsetting these concessions to recent developments, they still stressed a fifth factor: access to any land, anywhere.[31]

All of this presupposed that sailors' lives must be treasured, and the pioneer in this regard was John Byron, born 1723, the first tender-hearted leader of a circumnavigation, the long-awaited antithesis of all the ruthless thugs who had preceded him. "Foul-Weather Jack" had survived Anson's expedition by being shipwrecked off Patagonia on the *Wager*. He had not joined the mutineers (a decision that saved his naval career) but instead made his way to the Spanish colonies, where "Don Juan's" good looks and charm let him pass the time with romantic exploits. (It ran in the family—his grandson was Lord Byron, the poet.)[32]

In 1764, Byron received command of the 24-gun, sixth-rate frigate *Dolphin*, plus a smaller vessel, the *Tamar*, with a total of 153 men. His mission was top secret and he had a secret weapon: the *Dolphin* was sheathed in copper. She was probably only the third ship to have that coating, which protected the hull from growth of marine life, the upside-down forests of seaweed that hampered speed and maneuverability. With the *Dolphin*, Byron set a speed record: around the world in under two years (July 3, 1764, to May 9, 1766).[33]

Byron's other achievement, low mortality, was also the result of careful planning and constant maintenance. He fussed when his men did not change out of wet clothes. In rough conditions, he made boat crews wear cork life jackets. He monitored diet and health and particularly extolled the one-two punch of fresh provisions and something called "portable soup," ancestor of bouillon cubes, described as resembling glue, yet thought to contain the concentrated goodness of meat without the heavy salt content of its pickled naval cousins. Byron packed slabs of the stuff. Dried peas cooked in *"Portable Soup,"* he promised, "will be as good as a fresh Meal."[34]

His only casualties were due to accidents and non-maritime diseases, and the Admiralty took note. At last, circumnavigators had learned that size definitely mattered: smaller was better. Rather than take extra ships (and men), to be lost or cannibalized along the way, the new plan, following Byron, was to take just a smallish ship or two and take precious care of them.

Ship's regimens were modified along Byron's lines, to prevent the exhaustion, thirst, and hunger that had seemed to interact so fatally with scurvy on Anson's expedition. On his circumnavigation, Wallis adopted a three-watch system: the men were divided into three groups, and each spent four hours on, then eight hours off. That schedule provided more rest and prevented what tended to happen in the watch-and-watch system, when exhausted men fell "down to Sleep upon Deck" for their allotted four hours, in cold and damp conditions. Aware of Wallis's low mortality rate from scurvy, and keen to duplicate it, Cook adopted the three-watch system and it became associated with him.[35]

Ample and fresh water was also a priority, again to maintain general good health. Carteret was determined never to ration water. He yanked the painted floor cloth out of his cabin and rigged it up as an awning, to shield those on duty and to catch rainwater. The *Swallow*'s water was never rationed and it seems always to have been purified, by the ship's surgeon, with vitriol, a concoction of sulfur. Bougainville's *Boudeuse* had a distillery that produced a barrel of fresh water a day while at sea. Vinegar disguised the staleness of stored water; red-hot cannonballs would simmer some of the slime out of it.[36]

And food was stocked in greater quantity, also to prevent general debility. That took some doing. The Spanish Monarchy was hardly likely to let British ships land and get provisions in its American territories. If anything, it was tightening control. (When Carteret needed "refreshment" in the Pacific, he made for Juan Fernández, having read Anson, who recommended the rest stop; when he arrived, he discovered that the Spanish, having read Anson, had garrisoned the islands.) Spanish authorities made it clear that even Bougainville, a French ally, could not plan to put into Spanish territory unless by prior arrangement. The only solution was to pack circumnavigating ships as densely as possible with provisions.[37]

To accomplish that, the westward-sailing Wallis/Carteret and Bougainville expeditions had store ships replenish them at the last possible moment between Atlantic and Pacific. On entry to the Strait of Magellan, Wallis's *Dolphin* was crammed so tight "that the Men was oblidged to Eate their way [through] before they could sit to Mess below." Both Wallis's and Carteret's personal cabins were "Chock full of Bread" and, contrary to custom, each man received a personal stock of ship's bread to stow. For his first voyage, westabout, Cook had chosen a vessel distinctive for her capacity. The *Endeavour* had been a collier, a ship used to transport coal. Her big belly could be stuffed with food to fill her sailors' bellies. On his second, eastward circumnavigation, Cook topped up his supply at the Cape of Good Hope.[38]

Some foods were experimental, and several had been pet projects of the medical experts. (Fans of hot bouillon or malted milk can thank the Royal Navy for championing early versions of those pick-me-ups.) The British expeditions put particular faith in portable soup, mostly on Byron's recommendation. The Wallis/Carteret expedition loaded a whopping 3,000 pounds of portable soup—dead giveaway that Byron had a hand in its planning. For his first circuit, James Cook and the Admiralty negotiated a long list of remedies, including tart essences (the juice of lemons and oranges, sauerkraut, mustard, vinegar), sugar, portable soup, malt, and a device to produce carbonated water, the latter two items meant to infuse sailors' bodies with fixed air. On his second circumnavigation, Cook subtracted the soda water and added spruce beer, salted cabbage, and carrot marmalade.[39]

Although none of the circumnavigators carried a ventilator, all established regimes meant to keep the air belowdecks as clean and dry as possible, reflecting the recent debates on air's healthful capacity. On the *Dolphin*, for example, Wallis ordered that wet clothing had to be dried above rather than below deck. Bilges were more frequently pumped, more cleaning belowdecks was done (sometimes with substances like vinegar), and fumigation was performed with gunpowder or tobacco. There was suspicion, however, that this would never generate the equivalent of fresh air. When a cook died of scurvy on his second circumnavigation, Cook stressed that the man (not on the flagship, so away from his supervision) had been "indolent & dirtily inclined" and unwilling "even to come on Deck to breath the fresh air" unavailable below.[40]

Cook's confidence in natural air was but one hint that mariners, even the officers, doubted that anything packable could generate health. The circumnavigators lacked enthusiasm for most of the new foodstuffs, for example, and were silent on the theories that championed them. Portable soup was the most popular of the new provisions, though no one bothered to say why, other than that it somehow resembled fresh food. Other preferences were likewise never explained. Wallis and Bougainville considered wine more medically beneficial than grog, perhaps because they thought fermented products (presumed to release fixed air within the body) were healthier than distilled ones, though they never spelled this out. Rather, overt statements about the preferred foods stressed their freshness. One of Bougainville's passengers believed that "Lacking fresh food, we had several sailors attacked by scurvy." Cook likewise "dreaded the Scurvy" in the absence of "refreshments" from land.[41]

A genuine man of science endorsed the seaman's faith in freshness. The chief naturalist on Cook's first expedition was Joseph Banks, the most famous scientific traveler of his generation and eventually the president of the Royal Society of London. Only once in his journal did Banks mention the latest theories about diet, air, and the body, specifically to praise the doctor who had recommended the virtues of fermentation and fixed air. Fermentation worked, Banks hypothesized, because any food that contained the "salubrious qualities" of fermented malt "would consequently acquire a virtue similar to that of fresh vegetables, the most powerfull resisters of Sea scurvy known." Banks's statement showed, not necessarily the replacement of the old maritime knowledge with something newer, but a blend of the two.[42]

Mariners' limited patience with experimental science dovetailed with their belief—still—that the ultimate remedy for scurvy was land, for its fresh foods and its general *je ne sais quoi*, as unrepentedly jumbled together as ever. The Admiralty had given Byron extraordinary permission to purchase fresh vegetables wherever needed. He got "fresh Meat" and greens in Rio de Janeiro, served "Portable Soup & Wild Cellery thickend with Oatmeal" in the Falklands, gathered wild foods in the Strait of Magellan, and welcomed "all the Refreshments necessary for scorbutick People" at several Pacific islands. When he had scurvy cases in the Pacific, Byron sent them to convalesce in tents on Tinian. In Patagonia, Wallis put his sick men ashore in tents and had others pick berries and herbs. Rather than round Cape Horn, Bougainville threaded the Strait of Magellan because it offered access to fresh food along the way. He sent the sick to live on land in Chile, Tahiti, and Malaysia (see Map 2). One of Bougainville's officers, who lamented that "fresh meat" was eventually lacking even for the scorbutic invalids, concluded, "I would be a fool if I ever have to go round the world a second time."[43]

In crafting their regimen, circumnavigators not only perpetuated their belief in their terrestriality but also deviated from medical recommendations. According to the reviled doctors, circumnavigators should not have done what they kept doing: putting invalids (or anyone) to live and sleep ashore, using foreign materials as medicines, and eating raw plants. In his *Domestic Medicine* (1769), a stunningly popular work reprinted into the 1820s, William Buchan warned against night air, medicines from abroad, and uncooked fruits and vegetables. And yet the surgeon on Cook's *Endeavour* himself prescribed "the Salutary effects of the Shore" for the sick.[44]

Ships became more like shore with the addition of domestic animals. Bellowing and squawking, livestock had always been taken to sea—to be slaughtered, usually not long after departure. Now, officers wanted milk and eggs throughout their journey, even around the world. Cats, once thought bad luck aboard a ship, were recruited to catch biscuit-nibbling rodents. (The *Endeavour*'s cat killed one of Banks's specimen birds.) And captains, especially those from the rural gentry and aristocracy, simply couldn't imagine an expedition without a dog. Byron had brought a mastiff with him, though wished he had a fast greyhound for hunting. Banks planned better, and brought two greyhounds.[45]

Animal companions could easily become meat, though that practical measure generated sadness. When Bougainville cut rations on the way home, and punished anyone who dared eat the leather patches that buffered the rigging, his men demanded that a surviving goat be killed—they needed its meat more than the officers needed its milk. Bougainville assented but confessed, "I could only pity her." Her butcher wept.[46]

It says something about sailors' new love of animals, and their continued antipathy to women passengers, that the first female known to have gone around the world was a goat. Not just a goat, the Goat, and she circumnavigated the world not once, but twice. She made her first swing around the planet with Wallis on the *Dolphin*. She must have been sturdy, docile, and a good milker, because she was pressed back into service for Cook's personal use on the *Endeavour*. Given that Cook never suffered from the dysentery that struck many others in Batavia, the Goat may have been not only a source of food but an animate water filter that shielded the captain from harm.

If one turn around the world made a man a hero, it took two for a goat. When the Goat returned with Cook, she was retired in splendor. A silver collar was fashioned for her historic neck, and the silver engraved with two lines in Latin offered by Samuel Johnson, he of the famous dictionary:

> *Perpetui, ambita bis terra, praemia lactis*
> *Haec habet altrici Capra secunda Jovis.*

Roughly: "The globe twice circled, this Goat, second only to the nurse of Jove, is thus rewarded for her never-failing milk." Put out to pasture at Cook's home just outside London, she deserved every blade of grass she munched. It was said that the Admiralty admitted her to the privileges of the Royal Naval Hospital at Greenwich. If so, she is the only animal ever to share those benefits with the usual pensioners, injured or elderly sailors. Her date of birth is unknown, and she never had a real name, but the Goat's day of death is written in history: March 28, 1772. Cook made a note of it.[47]

It had come to this: circumnavigating *pets*, anything to remind sailors of home and land. Some narrators of circumnavigations, usually from the officer class, continued to describe land sentimentally, as a tonic for the mind as well as the body. Bougainville called Tahiti the "Garden of Eden" and "New Cythera," after the place where the goddess Venus had emerged from the sea. He said another island had "verdure [that] charmed our eyes." On seeing Tierra del Fuego's towering mountains of snow, the main chronicler of Wallis's expedition knew his duty but also his limits: "it would require the pen of Milton or Shakespear to Describe this place, therefor I shall give it upp."[48]

There was still no clarity, however, about whether sick circumnavigators suffered in body or in mind. When the *Endeavour*'s crew departed New Guinea in 1770, Banks described their return to sea in sentimental terms. "The greatest part of them," he explained, "were now pretty far gone with the longing for home which the Physicians have gone so far as to esteem a disease under the name of Nostalgia." Cook never penned a languishing description of scenery, yet he too began to describe sea scurvy as a disease of more than the body. When he decided not to anchor at Timor on his first voyage, Cook thought that continuation at sea had left his men "in a very indifferent state of health and I may say mind too." One of the naturalists on Cook's second circumnavigation claimed of the prospect of landfall in New Zealand, "the hope of recruiting our wasted spirits and strength, inspired unusual chearfulness."[49]

However stirring they found scenes of wild beauty, the circumnavigators almost immediately wanted to turn them into home, to tame them with neat, pretty gardens and herds of domestic animals. Byron's surgeon sowed a garden at Tinian. Wallis had fruit stones and garden seeds planted on Tahiti, and he gave away pairs of fowls to breed and a cat "big with kitten." He regretted he had no pregnant goat (his billy goat had died at sea). Bougainville adored the "beautiful disorder which it was never in the power of art to imitate" on Tahiti, but plunked down a vegetable garden and gave the islanders sets of breeding ducks and turkeys. They should have known better. European naturalists were already describing how deforestation and introduction of European plants and animals had wrecked certain islands (which had already been "ordered," after all, and more wisely, by the inhabitants who gardened them). Goats were particularly hard on foliage, and cats on birds, as circumnavigators had repeatedly seen at Juan Fernández.[50]

The blend of sentiment and destruction—animals were companions, until they became food; wilderness was sublime, until it was gardened—was also apparent in the circumnavigators' treatment of human beings: enlightened, until it wasn't. Consideration of ordinary sailors, who now had lives that mattered, was one sign of this. The men were no longer to be used up and discarded, though they may well have thought that the dietary novelties and the nagging about dry clothing were invasive in a different way. Wallis's officers expressed surprise at the "prudent" sailors who had brought their own "tea and coffey." The men may have found that praise patronizing, and they might have wondered why the Royal Navy didn't lay on the tea and coffee. Usually, the sailors were not permitted preferences—Wallis had a man lashed for refusing to eat food because it was burned. Cook had a shrewder sense of command. When the men of the *Endeavour* declined to eat sauerkraut, Cook did not order them to do so, but instead reserved it specially for the officers until the sailors clamored for it. But whenever Cook collected fresh supplies, he simply ordered that the men eat them, whatever their personal preferences.[51]

The new solicitude for sailors' lives extended, at least partly, to the lives of others. For the first time, accounts of round-the-world ventures recorded the presence of non-Europeans who made the full circuit as free members of the expeditions, not as captives for part of the route. It had surely happened before, yet invisibly. Now, Byron noted that a black man sailed on the *Tamar*. An officer on the *Dolphin* mentioned that Wallis's cook was Malaysian.[52]

Even a woman joined the ranks of circumnavigators. Bougainville's botanist, Philibert Commerçon, had brought a servant whom the crew suspected of being a woman, though the person in question (who never disrobed, or needed to shave) claimed to have been accidentally castrated. The Tahitians were harder to fool. They greeted the supposed eunuch with the cry *"Ayen"*—girl! She confessed to Bougainville that she was Jeanne Baré (or Baret, or maybe Bonnefoy), a penniless orphan. She claimed to have deceived Commerçon. (Unlikely: she had been his housekeeper and bore a child, probably his, before their departure.) But she assured Bougainville that "she well knew when she embarked that we were going around the world, and that such a voyage had raised her curiosity." She disembarked with Commerçon at Île de France (Mauritius), where he continued his research, and after he died, she married and returned to France. In 1785, to honor the "extraordinary woman" who had accompanied France's first circumnavigation, the French navy would grant her an annual pension of 200 livres.[53]

In an equally astonishing historical shift, none among this generation of circumnavigating commanders took human captives. Byron recoiled from the possibility. Just off Cape Horn, his shore party encountered some Indians *"who gave them a Dog & offer'd them a Sucking Child if they would accept of it;* The first they brought off, but the latter they left to it's tender Parents." When he was in the Strait of Magellan, Wallis gave "strick orders not to hurt any of the Indians." After his men had been attacked in the Tuamotu Archipelago, Carteret became enraged at all Pacific islanders. Yet he never took a captive. When one Melanesian man volunteered to sail with them, Carteret pointedly named him "Joseph Freewill."[54]

It was a momentous decision, part of a new European unease over human bondage, including bondage connected to the sea. French laws that had once condemned certain criminals to be galley slaves for life were abolished in 1748, for example. The Royal Navy's practice of impressing seamen had come under attack. So had the Atlantic slave trade. The year after Cook returned from his first circumnavigation, a London judge in the *Somersett* case would state that enslaved blacks could not be removed from England by force. All in all, pulling someone aboard a ship against his or her will now seemed unsavory, to be avoided.

The circumnavigators could afford to be high-minded because, even though they still needed information about navigation in the Pacific, they could get it from volunteers like Carteret's Joseph Freewill. Bougainville navigated away from Tahiti with the help of Ahu-turu, a Tahitian man who had frequented the *Boudeuse* and asked to accompany the Frenchmen. Cook came closest to taking captives when he held hostages in New Zealand. Yet he took away only volunteers, the priest-navigator Tupaia and his servant Taiata, who joined the *Endeavour* in Tahiti. Tupaia was an invaluable translator in various islands, including New Zealand, where he helped Cook negotiate with the Maori. Those Polynesian and Melanesian men were among the most skilled sailors in the world—Europeans were wise to solicit their knowledge through negotiation, not force. To honor his informant's geographic knowledge, Cook put Tupaia's name on a printed map, "A Chart representing the Isles of the South Sea . . . collected from the Accounts of Tupaya." It was a small gesture, and could never make up for all the enslavement that had been a circumnavigating tradition. But at least it repudiated that tradition.[55]

Even as the ranks of acknowledged circumnavigators expanded to include people other than European men, those men themselves maintained a belief that scurvy revealed a terrestrial equality among all land creatures. Individuals of different social ranks, nationalities, gender, or species were described as suffering alike. On Cook's voyage, the loftiest man aboard, aristocrat Joseph Banks, panicked when his "gums swelld and some small pimples" erupted in his mouth—he "flew to the lemon Juice." When Tupaia's gums swelled, the ship's doctor gave him lemon juice as well. Two bodies, one European and one Polynesian, had the same symptoms and responded to the same treatment. It was a small statement of belief in the basic similarity of all humans.[56]

So too did mariners conclude that, like themselves, terrestrial animals got "scurvy" when taken around the world without adequate fresh provisions. On Wallis's journey, the ship's master thought that the sheep ailed once dried peas replaced their fodder and that the "poor hogs" could not "walk the Deck without falling." The Swedish naturalist Anders Sparrman said that a dog on Cook's second circumnavigation became scorbutic. On the same voyage, the crew put the starved sheep and goats ashore in New Zealand, but they did not graze because they had loose teeth and "every symptom of an inveterate Sea Scurvy." None of these animals belonged to species now thought to suffer from scurvy—sailors were not describing something to do with lack of vitamin C, but instead assimilating all examples of ill-health at sea to their concept of an innate terrestriality among all land creatures.[57]

And at least some of these creatures were female, examples of the female body at sea at a time when few women experienced travel beyond one ocean. But there were no claims that sea scurvy affected the Goat or other female animals, or Bougainville's stowaway, Jeanne "Baret," any differently than it did males.[58]

The circumnavigators' idea of a shared terrestriality may have diverged from medical opinion and commonsense views among landfolk, but it was appealing as a new European vision of their place on the planet. Although suspicion of foreign climates would continue, that prejudice was ill-suited to the expansion of overseas empires, which would transplant European bodies well beyond their places of origin. Did Europeans have to expect to suffer and die abroad? An innate terrestriality instead promised that they could move and act on the world stage as they pleased.

That displeased some of the people whose lands they coveted. When Bougainville began to settle in on Tahiti, the Tahitians demanded to know how long he would stay; he promised to leave within eighteen days. They were wise to extract that guarantee. By the time Bougainville left, his ships had consumed at least 800 of their fowls and 150 of their hogs. Worse, other European ships would keep turning up with new demands for food and trade.[59]

Muskets and bayonets killed islanders who resisted. More subtle shows of force worked, too. Bougainville set off several displays of "sky rockets" (used as signals at sea), which inspired sensations "of surprize and of horror" in the Tahitians. A British officer who observed a solar eclipse on Tahiti offered a view through the telescope to one of the island's leaders. She was duly amazed that the instrument revealed distant objects, even celestial ones. And there is a hint that some sailors may have bragged, as Dampier's men had done in the Philippines, of their unique ability to go around the world. At one tricky moment in his stay on Tahiti, Wallis suspected a plot to seize his ship, "to have made a Prize of the only ship that ever Surrounded the Globe twice." Whatever their talk of a shared terrestriality, Europeans reserved for themselves the ability to encompass the planet.[60]

Not least because they returned with most of their ships and men, the post-Anson circumnavigations received the usual forms of commemoration, published maps and narratives. Unsurprisingly, French maps featured Bougainville's track. British ones could have showcased all of the recent heroes, but usually picked Cook. His track often appeared with Anson's, to indicate British global glory, in war and peace. It was especially impressive that some British mapmakers managed to fit both tracks onto pocket globes, small orbs that could be held in one hand. A London globemaker had originated these little images of the Earth at the end of the seventeenth century. In the eighteenth century, they became popular consumer items, cheaper than the larger "library" globes that twirled in stands, and so satisfyingly grippable, the world in your hand. "A Correct GLOBE with the new Discoveries," one announces, with Cook's triumphant progress striped around it.[61]

The first-person accounts came out more slowly. Bougainville's account did appear, in France, in 1771, and sold well enough that a second edition appeared soon after. His account—scurvy, coconuts, and all—proved that, despite defeat in the Seven Years' War, France was still in the business of empire. (For that reason, it withheld specific cartographic information, including its readings for the position of Tahiti and about the probable size of the Pacific Ocean.) Like Dampier, Bougainville insisted that his observations gave direct testimony about the nature of the world: "I am a voyager and a seaman; that is, a liar and a stupid fellow, in the eyes of that class of indolent haughty writers who, in their closets [private rooms] reason in *infinitum* on the world and its inhabitants, and with an air of superiority, confine nature within the limits of their own invention."[62]

It was almost certainly without Bougainville's permission that his book was translated and printed in London in 1772. The translator was John Reinhold Forster, a naturalist based in England, who introduced his handiwork with the claim that it proved Britain's "superiority" over a "rival nation." Forster's translation is faithful, though supplied with sniping footnotes. When Bougainville describes Tahitian manners and customs, for example, Forster leaps from the bottom of the page, like an eager little dog, to assure the reader that "the English, more used to philosophical enquiries, will give more faithful accounts." It must have galled the British circumnavigators that their faithful accounts were still being discussed in the future tense, their accomplishments still underpublicized. In February 1772, when Cook looked at a friend's copy of Bougainville in French, he embellished its world map (which carried Bougainville's track) with the route of the *Endeavour*.[63]

But the price of the British Admiralty's sponsorship of its circumnavigators was a ban on publication, with the intention of producing an official history. (A few accounts by men not beholden to the Royal Navy slipped through in the meantime.) Naval leaders knew that the Anson narrative had sold well. Moreover, several collections of around-the-world travel acounts, in multiple languages, had indicated the commercial viability of a compendium, rather than a series of accounts. The plan was to produce a definitive compendium of the recent circumnavigations.

The lords of the Admiralty looked for a professional writer for the job, and someone recommended to Lord Sandwich, First Lord of the Admiralty, the author John Hawkesworth. Long forgotten, Hawkesworth was, in his day, a central figure in British cultural life. He was a playwright and drama critic, principally for the *Gentleman's Magazine*, a kind of *New Yorker* for eighteenth-century London—so just the fellow to put some life into a sailor's log. William Strahan, printer of key works of the British Enlightenment—Adam Smith's *Wealth of Nations*, Edward Gibbon's *Decline and Fall of the Roman Empire*, and part of Laurence Sterne's *Tristram Shandy*—agreed to publish the work and advanced Hawkesworth £6,000, the largest sum paid an author during the eighteenth century. Even Samuel Johnson, who believed that "no man but a blockhead ever wrote, except for money," received only £1,575 for his *Dictionary*. Arrangements were made for translations abroad. Eager readers awaited. The circumnavigating commanders were poised for new attention and flattery.[64]

It was a disaster. Hawkesworth's *Account of the Voyages Undertaken by Order of His Present Majesty* (1773) would be widely read, though to the mounting disappointment of readers. The professional writer had reduced everything to bland descriptions of individual places, without a bigger planetary sweep. Horace Walpole complained that "the entertaining matter would not fill half a volume; and at best is but an account of the fishermen on the coasts of forty islands."[65]

The worst criticisms came from the expedition leaders themselves. Hawkesworth had adopted an authorial "I" to speak for the assorted commanders and put his name, as author, on the title page. That seemed high-handed. Although Hawkesworth claimed that he had let the commanders read the manuscript, they denied it. Indeed, several of them were not pleased that their texts, or other of Hawkesworth's

sources, had been relieved of technical specifics essential to seamanship and enlivened with colorful details they didn't happen to remember. Wallis smarted at the suggestion of a dalliance between him and
Tahiti's leading woman, Carteret resented that his critical suggestions
for the Admiralty had been erased, and Cook was astonished to find
his words blended with those of Banks and another of the expedition's
naturalists, Daniel Solander. The book would be much reprinted and
widely read, even as far away as eastern Europe and North America;
that only made it worse in the eyes of its critics.[66]

It must have wounded Hawkesworth especially that the press, his
natural home, generated several scathing reviews. One London reviewer deplored how "Captains of Ships, and all superior and inferior
Seamen, and others, who undertake these Voyages at the Expence of
their Health and Hazard of their Lives, are not permitted to enjoy the
Fruits of their long Labours." Instead, the Admiralty, "with the Spirit of
a Tribunal resembling the Spanish Inquisition," demanded their journals, logs, any "Scrap of Paper." Meanwhile, an Admiralty crony had
been made rich through "the easy Business of a few Months, transacted
by a Man's own Fire-Side." "I shall not make any invidious Comparisons between the Fatigues, Sufferings, and Rewards, of the Voyagers
and the Compiler of their Labours," the reviewer concluded, having
done just that. James Boswell, the Scottish-born London man of letters,
offered sympathy to Cook upon learning of Hawkesworth's liberties
with his text: "'Why, Sir . . . Hawkesworth has used your narrative as a
London Tavern-keeper does wine. He has *brewed* it."[67]

Cook would be permitted to report on his second circumnavigation, and he would have plenty to relate. He had been promoted to
commander and was given two ships, the *Resolution* (in his charge) and
the *Adventure* (under Tobias Furneaux, making his second circumnavigation). Cook's orders were to discover the still elusive southern
continent and he was supplied with some of the new sea-clocks or
chronometers in order to do accurate readings at sea. He headed east,
which he thought would be with, not against, the prevailing wind. (Because Furneaux would get home first, in the *Adventure*, he became the
first mariner to go around the world eastabout, and the first to have
done it both ways, the long-deferred maritime equivalent of Loyola.) It

was indeed much easier to sail in that direction, and the eastabout way around the world would, along with the use of fewer and smaller ships, make many future circumnavigations much less of an ordeal. (Compare Cook's two routes on Map 2.) Past the Cape of Good Hope, Cook plunged south, spotted icebergs, and made the first recorded crossing of the Antarctic Circle. Without knowing it, he got close to Antarctica, but retreated north. Although he would cross the Antarctic Circle again, and described several islands unknown to Europeans, he found no southern continent, just chronometric readings of all the places that it didn't exist.

Cook did make land at some islands he called "Sandwich," after the Lord of the Admiralty, later renamed the Hawai'ian Islands. He had recruited another Polynesian man, Mai (called at the time Omai). With the sponsorship of Joseph Banks, Mai would be shown around London. He would return with Cook to the Pacific in 1777 and become the first Pacific islander to complete a circumnavigation. His predecessors had been less lucky. Joseph Freewill and Tupaia died on their way to Europe, while Ahu-turu, who made it to Paris, died on his way back home.[68]

One outcome of the accumulating round-the-world voyages was a sense that they contributed to a definitive, total knowledge of the world. World-circling had generated the concept of world-blanketing. Certainly, the overlapping routes and repeated observations of the eighteenth-century circumnavigators gave that totalizing impression. For the first time, and with growing force, "around the world" implied a complete vision of the planet, as if the multiple tracks that individual ships had laid around it had collectively generated the most detailed image of the globe ever seen.

That idea of global knowledge was very much at home in the eighteenth century, when the modern encyclopedia was born. Reference works had a longer history, but the big and lasting multivolume projects (the French *Encyclopédie* and the Scottish *Encyclopaedia Britannica*) were published between 1751 and 1772. A fantasy of having a reference work that contained everything in the world matched the goals of mapping all parts of the world, describing each of its plants and animals, and cataloging all of its peoples and languages. It still hasn't been done, and yet it seemed tantalizingly *possible* in the second half of the eighteenth century, as more expeditions returned with yet more information about the globe.[69]

On a slightly smaller scale, editors produced encyclopedic collections of travel voyages. Samuel Johnson nursed up one of the biggest and most popular compendiums, *The World Displayed: or, a Curious Collection of Voyages and Travels*, in twenty volumes, first published in 1759 and running through dozens of editions before the 1790s. As well, and for the first time, compilations of round-the-world voyages were done to focus specifically on that category of travel. From 1774 to 1775, David Henry produced the four-volume *Historical Account of all the Voyages round the World, Performed by English Navigators*, with the addition of a fifth volume in 1776 to include Cook. An updated six-volume collection would appear in London around 1785. A compendium in French, of ten volumes, was published in Switzerland between 1788 and 1789.[70]

The craze for going around the world made some landfolk consider doing it themselves. The French philosophe Denis Diderot read Bougainville's account as evidence that Europeans could travel the world in cosy comfort: "A ship, after all, is only a floating house . . . [a sailor] can go around the globe on a plank, just as you and I can make a tour of the universe on your floor." James Boswell considered joining Cook's second circumnavigation, having caught "the enthusiasm of curiosity and adventure" from reports of the first. Johnson, whom Boswell would immortalize, was skeptical:

> **BOSWELL:** But one is carried away with the general grand and indistinct notion of a VOYAGE ROUND THE WORLD.
>
> **JOHNSON:** Yes, Sir, but a man is to guard himself against taking a thing in general.[71]

That was the problem. The "notion" of going around the world was grand, but, as Boswell admitted, general and indistinct. No individual voyage could yield global knowledge, nor could they do it cumulatively; that placed a burden on circumnavigations they could not bear. And, as Johnson warned, getting carried away with a great notion could be dangerous.

Circumnavigators themselves knew that. Invited to join Cook's second expedition, Joseph Banks hesitated. Having done it once, he knew too much. He dreamed of another way to travel through every degree of longitude with ease: "O how Glorious would it be to set my heel upon the Pole! and turn myself round 360 degrees in a second." Failing that possibility, Banks had demanded that Cook's *Adventure* be refitted to give him more space and privacy. The weary carpenters warned that the modifications would make the ship unseaworthy. Banks was all but disinvited from the expedition.[72]

Yet it was the management of danger that was the most striking achievement of the newest circumnavigators. Out of all his contributions to science, Cook was most greatly rewarded not for his geographic discoveries, nor his successful use of a chronometer, nor his historic encounter with Hawai'i, but for his defeat of scurvy. That, after all, had made possible all the rest. The naturalist who produced an account of Cook's second circumnavigation noted that division of labor: he commented on the natural features of the lands the expedition visited while Cook narrated everything to do with "victualling or refitting the ship," the "internal oeconomy of the ship and the crew."[73]

Those duties yielded glory. Cook was elected a Fellow of the Royal Society on his return, and he consolidated his reputation in science by publishing, in the society's *Philosophical Transactions*, a short treatise on how he had fought scurvy aboard the *Resolution*. He extolled several substances provided by the Admiralty: malt, sauerkraut, and portable soup. But Cook stressed that they worked best on a well-regulated ship, one that was clean, dry belowdecks, and stocked with fresh water. Indeed, he thought these precautions would themselves prevent scurvy—a controlled environment, meant to resemble land, was the best way to support life. His word carried unique authority because, in his second journey around the world, Cook lost only four men, one to disease (not scurvy), two who drowned, and one who died after a fall: total mortality, 3.6 percent. Given that he had replicated Byron's results—no deaths from scurvy—yet on a voyage a year longer than Byron's, the achievement was particularly impressive. Cook's essay received the Copley Medal, the Royal Society's highest award, the eighteenth-century equivalent of the Nobel Prize.[74]

And yet on his last expedition, far from the Admiralty's oversight, Cook made clear his belief that land and its fresh products cured sea scurvy better than anything he had recommended in his prize-winning essay. Alexander Home, quartermaster on the voyage, in describing a brush with a poisonous plant in Kamchatka, marveled that it was "astonishing How we have Come to so little Damage in this way ... for it was the Custom of Our Crews to Eat almost Every Herb plant Root and kinds of Fruit they Could Possibly Light [upon] with[out] the Least Inquirey or Hesitation or any Degree of skill & knowledge of their Qualitys"; for the scorbutic, any land food was better than no land food, raw or cooked, familiar or alien. Cook was noted as urging his men, wherever possible, to go ashore and pluck "A Handkerchif full of greens." The vegetables were only one benefit of land. Cook would also "Order [the men] on shore in partys to walk about the Country and smell the Fresh Earth and Herbage," as he himself did, to set an "Example."[75]

This final voyage may not have been planned as a circumnavigation, and didn't go around the world, and Cook didn't return with it anyway. When some natives at Kealakekua Bay in Hawai'i stole a boat, Cook stormed ashore to get it back, as Magellan had done many years before in Guam, for the same reason. Unlike Magellan, Cook was killed. It was an ironic death: the man who had demonstrated how to keep sailors alive all around the world managed to get himself killed on one small patch of it.

There was a logic to the incident. European circumnavigators had idealized land, insisting that it provided everything, life itself. They longed for *terra firma*, convinced it would keep their teeth firm in their heads. They had decided to honor territorial claims made by fellow Europeans—they no longer raided territory for provisions or encroached on each other's trade routes. They shared information about where to anchor and gather strength, all the way around the world. But they had no respect for the territories of non-Europeans. By medicalizing their need for land, circumnavigators made their demand to it seem ethical, a matter of preserving life. It was a claim that Europeans could probably adjudicate among themselves without damage. When they insisted on their right to keep visiting the Pacific islands and other extra-

European places, however, they were invaders who seemed to South Sea islanders as piratical as Drake. Hence Cook's death.

Perhaps it consoled Cook's widow to receive a coat of arms. Her husband became the third circumnavigator, after Elcano and Drake, with the right to display the globe as his emblem. His armorial globe has the "Pacific Ocean" at its center, traced with the lines of his two circumnavigations. A human arm thrusts above, clutching the Union Jack and a banner with the words CIRCA ORBEM, "around the world." Another banner at the bottom announces NIL INTENTATUM RELIQUIT. The motto refers to Cook—"He left nothing unattempted"—but it also suggests that "Nothing unattempted remains," as if the age of extreme risk had ended, as if people could now go around the world in confidence instead of deadly fear.[76]

Could they?

First Entr'acte

The human body was as tiny as ever, compared to the planet, but Cook's defeat of scurvy gave mariners from the western hemisphere new hope that they and their ships could take on the whole world. Between 1785 and 1795, six leaders of around-the-world expeditions would try to repeat Cook's achievement of zero mortality, though only the final two succeeded, and not entirely for the reasons Cook had specified.

With two ships and 114 men, Jean-François de Galaup, comte de La Pérouse, set out in 1785. Louis XVI of France had five goals for the expedition: geographic discoveries, commercial opportunities, science, peaceful relations with other nations, and the good health of his seamen. The final point mattered no less than the others. The surgeon of the flagship was promised a state pension if mortality did not exceed 3 percent. "If we had done nothing more than an ordinary circumnavigation," La Pérouse speculated, "we would have returned to Europe without losing a single man . . . but if the length of voyages of discovery has limits one cannot exceed, it is very important to assess what they are and I believe that when we reach Europe the experiment will be completed."[1]

Patriotism did not prevent La Pérouse from modeling his experiment on what was becoming known as "the Cook plan": three watches, ample provisions, frequent fresh food, clean ships, and adequate ventilation. (The Frenchmen consulted a copy of Cook's published journal on the best way to preserve meat.) Above all, everyone was to be cheerful—to have confidence in the precautions being taken for them. Reliance on Cook's plan backfired at one point. The ship's naturalist, citing Cook on fresh water being better than anything held aboard, took a small party ashore at Samoa to fill some barrels. They were attacked and killed. La Pérouse refused to sanction any reprisal. The king had ordered arms to be used only in self-defense, and the damage was already done.[2]

La Pérouse's best asset was not anything he packed, but his access to land along the way. He noted that his ships could carry supplies for only three of the voyage's estimated four years—even the ropes wouldn't last that long. Fortunately, good relations with Spain enabled peaceful stops at Chile, California, and Manila, among other places. The Spanish mission at Monterey supplied milk, vegetables, and livestock. Even the British were welcoming. At Botany Bay in what would become Australia, La Pérouse rejoiced that "Europeans are all compatriots at such a great distance." The Russians gave similar welcomes, including a ball at Kamchatka. There, with the men in dress uniform and the dancing in full swing, the long-delayed mail arrived. Their hosts kindly allowed the Frenchmen to forget their manners and sink into chairs to read their letters. To confirm his receipt of new commands, La Pérouse ordered a Russian-speaking officer, Jean-Baptiste-Barthélemy de Lesseps, to hasten overland back to France.[3]

And then, silence. Lesseps's journey took nearly a year, but when he reached Paris, he had the only news of his commander. A rescue party went in search of the two ships, and requests for aid were issued to London and Madrid. There were troubling indications of a wreck at Vanikoro in the Solomon Islands. Several times before his death by guillotine in 1793, Louis XVI asked if there were any word of the expedition. Only much later was the wreck at Vanikoro confirmed. Lesseps was the expedition's sole survivor, and one of the few people, to date, to have made an amphibious circuit of the world.[4]

"*Resolved.* That sea letters be granted in the usual form, for the ship Columbia, burthen about 220 tons, and the sloop Lady Washington, burthen about 90 tons, bound on a voyage to the north-west coast of

America." Thus in September 1787, not four years after it had achieved independence, the United States dispatched its first circumnavigation. An upstart nation with a tiny navy, the United States demonstrated its maritime ambitions through commerce. When Robert Gray took the *Columbia* and its tender, the *Lady Washington*, around the world from 1787 to 1790, he was funded by Boston merchants who thought the fur-abundant Pacific Northwest would pay their way to China. (Buying furs from Indians to trade for Chinese tea was cheaper than acquiring the gold and silver that the Chinese usually demanded.) The Indian trade goods were so compact that the *Columbia* was able to carry the *Lady Washington*, stowed in pieces, all the way to the Pacific, there to be reassembled.[5]

Some even smaller items were more critical: the letters from Congress, plus certificates from the Dutch and French consuls in America. These functioned as passports, instruments of an emerging paper internationalism that facilitated peaceful movement among Western nations and their empires. Personal passports were rare, usually restricted to elite men on state business. (Many people carried identity papers, but usually to placate local authorities, not to travel.) Ships' passports either stated their peaceful intentions—antitheses of the old privateering contracts—or else instructed captain and crew how to treat ships they might encounter. When Cook had begun his second circumnavigation, during the American Revolution, Benjamin Franklin, acting as U.S. ambassador to France, produced documents to instruct U.S. and allied ships to let the expedition pass, as its scientific business was supposed to benefit all of humanity.[6]

Because the United States and Spain had been allies since 1779, the affidavits of Gray's U.S.-sanctioned business let him enter Spanish territory in the Pacific. Spanish Juan Fernández (off-limits to the British) welcomed Gray with fresh food, "the strongest antisc[or]butic." At Nootka Sound in what is now British Columbia, the Americans discovered four British vessels whose crews were busy building a trading fort on what Spain considered its territory. A Spanish military force came to seize the vessels and evict the British. This Nootka Crisis nearly ignited a war. The Spanish seized another American vessel at Nootka, but left Gray alone, perhaps because he carried proof of U.S. sponsorship. Warm relations with Spain were evident in a Spanish captain's gift of food and wine; cooler relations with Britain were apparent when a British captain declined to carry letters for Gray, a courtesy that ships ordinarily performed in times of peace.[7]

The journey succeeded, not least because of Spain's help. Although the Chinese profits were scant, Gray's return was news. "The Columbia and Washington are the first American vessels who have circumnavigated the Globe," announced the *Philadelphia Mercury and Universal Advertiser* in 1790, as did several other American newspapers. The official story was that Gray had lost only one man. But records from the expedition noted two deaths from scurvy and another from suicide, possibly caused by the melancholy associated with scurvy, its fatal lack of cheerfulness. Without access to Spanish territory, mortality would have been higher. The papers also noted that the Americans had exhibited an "urbanity and civility" which "secured the friendship of the aboriginals of the country they visited." That too was wishful—there had been episodes of violence.[8]

Because the expedition had secured Spanish permission to trade for furs in the Pacific Northwest, Gray set out on a second circumnavigation (1790–93), also on the *Columbia*. Scurvy continued to plague him. Despite familiarity with the Cook plan, the Americans mostly resorted to the old method of convalescing ashore. In the Northwest, a crewman related, "we buried sevrall of our sick, up to the Hips in the earth, and let them remain for *hours* in that situation [and] found this method of great service." But when the Chinese snubbed the American furs, the rationale for the voyages vanished. Maybe for that reason, narratives of the *Columbia*'s expeditions were unpublished until the twentieth century. The United States had made its around-the-world debut, though in a shy debutante's whisper.[9]

The next man to command a circumnavigation was no whisperer. He was a rising star in the Royal Navy, a talented veteran of Cook's third voyage, but who remembers William Bligh for that? He is instead notorious for the mutiny that occurred on the *Bounty*, which set out under his command, in 1787, to transfer breadfruit seedlings from Tahiti to the West Indies while making an around-the-world voyage.

Bligh wanted to equal Cook, both by doing a circumnavigation and by having zero casualties. He pronounced scurvy "a disgrace to a Ship" and thought that sailors "must be watched like Children" to be kept healthy; the first opinion was admirable but the second slightly ominous. He zealously adopted the Cook plan: three watches, cleaning and fumigation, antiscorbutic supplements, plenty of food, fresh water, and good spirits. "Mirth is absolutely necessary," Bligh declared, and made his men dance every evening in good weather. He preferred these tactics to shore leave. But arguments over food erupted on the way out. And bad weather kept the *Bounty* from rounding Cape Horn. With regret, Bligh ordered the ship to turn about and head for the Cape of Good Hope. Having exhausted his men in one attempt on Cape Horn, it would have been foolhardy to try again. There would be no circumnavigation. Plus the ship's surgeon reported signs of scurvy. Furious, Bligh called the doctor a "Drunken Sot" and dismissed the invalids' symptoms as rheumatism or heat rash. Nevertheless, he ordered doses of antiscorbutics (malt, vitriol, vinegar, mustard, portable soup) and supervised the dosing.[10]

After mere days in Tahiti, fresh food made the men "perfectly well" in body, though not in spirit—the memory of ill-health, and of Bligh's perceived indifference to it, was too bitter. The surgeon died on Tahiti, and his assistant and successor seemed less likely to contradict the captain. In a vote of no confidence, everyone squirreled away his own fresh food for the return journey, though Bligh requisitioned for common use all the hogs brought aboard and portions of the fruits and vegetables. On departure, the *Bounty* was a floating greenhouse, with 1,015 tropical seedlings everywhere, even in deck space ordinarily reserved for livestock, meat on the hoof. Did the delicate plants matter more than the hungry men? Despite daily rations of fresh pork and vegetables, other food was consumed on the sly—a refreshing coconut here, a roasted piglet there. One day, Bligh inspected his own supplies and demanded of his sailing master, "Mr Fryer don't you think those Cocoanuts are shrunk since last Night?" Convinced his men were pilfering, Bligh ordered a general search, confiscated all coconuts, and threatened to reduce the ration of yams from 1½ pounds to ¾ of a pound. Rumors spread that the captain would seize all private stocks of yams, and thus command nearly all the fresh food aboard.[11]

The mutiny began at dawn the next day: April 28, 1789. It is revealing that two of the men arrested with Bligh were the cook and the surgeon's assistant, whose duties implicated them in the captain's control
of shipboard diet and health. While Bligh waited to learn his fate, the
mutineers "oft times repeated now let the Buggar see if he can live on
three Quarters of a pound of Yams." After forcing Bligh and some others into a small launch to fend for themselves, the mutineers tossed
the pots of cosseted breadfruit into the sea. The *Bounty* had failed to go
around the world and Bligh had failed to inspire confidence among his
men that they could survive such a journey.[12]

It was just then that the Spanish planned an around-the-world expedition, their first in 250 years. During a commercial circumnavigation
he performed on the *Astrea* (1786–88), Spanish naval officer Alejandro
Malaspina saw all too clearly the increased traffic of non-Spaniards
around the Pacific. On his return, Malaspina convinced Carlos III that
an official circumnavigation would reaffirm Spain's interests around
the "Spanish Lake." In 1789, the king ordered an expedition to gather
information on his empire in the course of a grand global circuit.[13]

Malaspina commanded the expedition, though his emulation of
Cook showed a Spanish rivalry with Britain based on admiration. (His
narrative cites Cook more than any other maritime figure.) Malaspina
named his brand-new vessels, 306 tons apiece, *Descubierta* (*Discovery*)
and *Atrevida* (*Bold*), in honor of Cook's *Discovery* and *Resolution*. Above
all, Malaspina was determined to preserve his crew and make them
confident of it: "the chief means to which we had recourse to keep our
typical seaman healthy was undoubtedly peace of mind." His provisioning and work regime took hints from Cook (soup tablets and sauerkraut; three watches). But he had a significant advantage over Cook
in his access to the Spanish Empire's many ports, which had sustained
him on his earlier, private circumnavigation. Malaspina's ships were at
sea for only 40 percent of the time. In contrast, Cook had been at sea
about 70 percent of the time, a much greater test of the wooden world.
(Bligh's expedition had featured an unbearable contrast: prolonged
shore leave in Tahiti sandwiched by long stretches at sea. Gray had only
intermittent stops in Spanish territory, hence his mixed results.)[14]

In the end, Malaspina did not complete a circumnavigation, which may have seemed unnecessary, given Spain's lack of territory around the Indian Ocean. He took the comfortably populated route back the way he had come, around Cape Horn, returning in 1794 with a healthy crew. For complicated political reasons, his account remained unpublished for thirty years; it first appeared in a multivolume Russian translation, printed from 1824 to 1827.[15]

So it was a Frenchman, Étienne Marchand, who would first replicate Cook's results on a circumnavigation aboard the *Solide* (1790–92). As befit the commander of that reassuringly named vessel, Marchand wanted to make the wooden world scurvy-proof. The author of the expedition's narrative, naval authority C. P. Claret Fleurieu, also believed that science was the tool of human progress. (Not for him the poetic image of the *Solide* racing the Sun; rather, "the ship had circumnavigated the globe in the direction of the diurnal revolution of the sun, or to express myself more correctly, in the inverse direction to the diurnal revolution of the earth"—correct, if not quite as thrilling.) Marchand carried the usual assortment of antiscorbutics, and six men (12 percent of the crew) were charged with maintaining health, as surgeons, stewards or cooks, and a baker. All the officers were responsible for cleanliness and good provisioning. It worked. Only one man died, probably of a stroke.[16]

But the men of the *Solide* also benefited from landfall. True, during his longest stretch of sailing, thirteen and a half months, Marchand spent only thirty days at anchor, a remarkably low 7 percent of the time. But he timed his stops wisely, which he could do because the islands of the Pacific, and the routes between them, were now familiar. This was especially true of Hawai'i, which Fleurieu dubbed the "*caravansary*" of the Pacific. His account urges "all the nations which share the empire of the OCEAN" to share information about the seas. He criticized the Americans for not publishing accounts of the *Columbia*'s circumnavigations and the Spanish (he predicted) for consigning Malaspina's narrative to "the dusty archives of some chancery," lost to the "maritime coalition" of seafaring nations. He meant Western nations. But in his praise of Tupaia, the Polynesian navigator who had assisted Cook, Fleurieu hoped that the empire of the ocean might, someday, include other people.[17]

To his credit, Marchand had brought the rigors of an around-the-world expedition into line with those of France's geodetic surveys. The next great effort in geodesy would calculate the size of a *meter*, the core component of a universal system of measurement based on the planet itself. In 1791, the French Academy of Sciences delegated to three men the task of determining the length of one ten-millionth of the quarter meridian (the distance from north pole to equator), the unit that is the foundation of the metric system. Although some literalists wondered why an actual measure of that distance—or indeed of the equator—was not done, it was probably just as well for Jean-Baptiste-Joseph Delambre and Pierre-François-André Méchain that it was not. Their survey of a designated section of France (1792–98) would represent the full globe. The expedition was not without hazard, but both men returned alive, which would have been much less likely with a literal circuit of the planet. Their efforts confirmed that the globe was, in any case, not a true spheroid. The meter imagined a uniform planet, though it was actually lopsided in form and rugged of surface. Any clean sweep around it was still a fantasy, a far cry from the painful transits made by mere mortals.[18]

On an official British circumnavigation (1791–95), however, George Vancouver replicated Cook's scurvy-free success at sea, though again because of timely contact with land. Vancouver had served on Cook's last two expeditions and was chosen to complete the survey of the Pacific Ocean, commanding the *Discovery* (330 tons; 100 men) and the *Chatham* (131; 45). Just as Cook had retired one hypothetical entity, the great southern continent, Vancouver wanted to settle another, the Northwest Passage, the idea born when Magellan had discovered the southwest passage that bears his name. Vancouver was equally determined to show that the Cook plan would make a ship scurvy-proof.[19]

He was remarkably successful. He charted what was considered to be Earth's last unmapped section of inhabited coastline, the Pacific Northwest. He declared that "NO INTERNAL SEA, OR OTHER NAVIGABLE COMMUNICATION whatever exists, uniting the Pacific and Atlantic Oceans." He meant no passage in temperate latitudes. (Alexander Mackenzie had already made that point in 1793, in an overland expedition.) Although the Arctic still held mysteries, even a possible passageway, wooden sailing ships could not safely navigate the ice. It was a questionable triumph, given how it confirmed the globe's inconvenient configuration, in which continents blocked maritime transit from one ocean to the next.[20]

Vancouver's protection of his crew was a clearer victory. He had implemented the Cook plan, and, although he lost five men to accidents, only one died of an illness (not scurvy). There were some non-fatal outbreaks of scurvy, which a horrified Vancouver blamed on one of the cooks, who had sneaked the men "slush," tasty cooking fat, which some considered scorbutic. But with proper measures, he predicted, ships could circle the globe and return men "in perfect health," with no more missing than might have died at home.[21]

Vancouver had avoided hazard only with Spanish help, however. During most of his expedition, Spain and Britain were allies, driven together by warmongering Napoleon Bonaparte. Spanish amity permitted Vancouver to seek aid at settlements along the Pacific, places that generations of English circumnavigators had either raided or avoided. Vancouver testified that fresh food from the Spanish Empire helped preserve his men from scurvy. At Monterey, the obliging commander even allowed Vancouver to load cattle and sheep for the British colony at New South Wales, a generous (or resigned) concession to the British presence in the Pacific. The Anglo-Spanish alliance also strengthened a chain of global contacts that transmitted information. Vancouver ran into several British and U.S. ships (including the *Columbia*) that gave him news and carried his letters. Spain helped to complete the circuit: one of the Spanish missions in California received mail for the expedition.[22]

The world was being divided, not only among different nations, but also between zones arranged for the comfort and convenience of Westerners and zones that were not—or not yet. Assistance from people in places like Hawai'i and Nootka did not compare, Vancouver thought, to hospitality from fellow Europeans. Every society the Britons visited had included women, for instance, but on the way home, the Spanish-speaking world offered something that Vancouver's officers considered superior: ladies. At a ball in Valparaiso, the local gentlewomen were a treat, never mind their bad teeth and lack of English. Valparaiso's ladies marked initial re-entry to what the Britons called the "civilized world." They definitively entered that world at St. Helena, where Vancouver reset the ships' calendar by a day to represent his global circuit. Rivals for centuries, Britain and Spain settled into a friendship that would have astonished Drake or Cavendish, not to mention Anson.[23]

The amity spread. After the Congress of Vienna in 1815, Europe and its satellites entered a Hundred Years' Peace. War became the exception. Paper internationalism created what lawyers call a global condominium, in which nations agree to open their sovereign territories for certain regulated purposes, such as postal delivery, scientific investigation, or leisure travel. Subsequent cooperation among Western nations opened up still more of the globe, as when the Spanish-American wars of independence removed Spanish rule from most parts of Latin America; in parallel, the Monroe Doctrine discouraged European powers from interfering with the nations of the New World. Great expanses of the Americas and Pacific that had been off-limits to anyone but Catholic allies of Spain (if them) were now accessible nearly to all.[24]

The experiment with Cook's plan was inconclusive, though no one cared. To prove its efficacy, the plan needed to be tested without making landfall. But landfall was exactly what round-the-world voyages now achieved routinely, which voided the results of the experiment. Politics made medical precautions less necessary. Simply put, there was no antiscorbutic in the world more powerful than the Congress of Vienna.

The new shipboard regimes and the frequent shore leave together meant that a circumnavigation became rather easy. In his *Observations on the Scurvy* (1792), British physician Thomas Trotter pointed out that ships bound on peaceful exploration had an advantage over ships of war: "They are seldom long from land, and their commanders are allowed to purchase all the refreshments that they see [are] wanted." That was a historic development. In its honor, Trotter defined a new malady, "scorbutic Nostalgia," a longing for home made more painful by physical illness. This was homesickness, attachment to the one place on Earth that a traveler, by definition, had to leave behind, and not the general hunger for all terrestrial parts of the Earth, which circumnavigators no longer endured.[25]

Together, technology and political reconfiguration of the globe would change the nature of circumnavigations. In the next act of the geodrama, around-the-world travelers would be confident that they, and not the planet, had the upper hand.

Act Two

Confidence

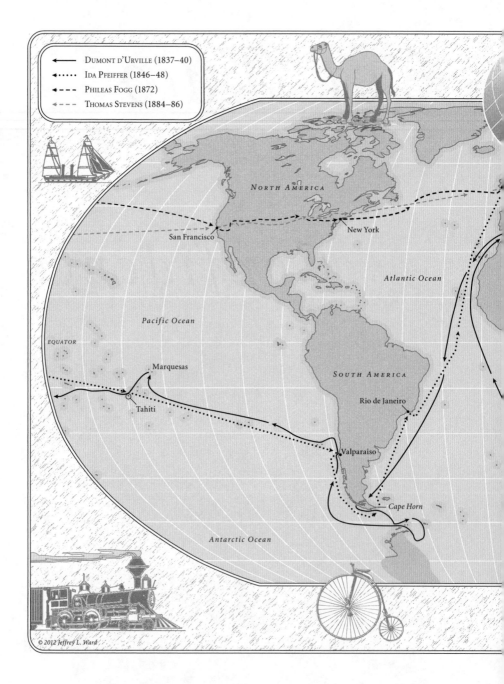

© 2012 Jeffrey L. Ward

Legend:

- Dumont d'Urville (1837–40)
- Ida Pfeiffer (1846–48)
- Phileas Fogg (1872)
- Thomas Stevens (1884–86)

NORTH AMERICA

San Francisco

New York

Atlantic Ocean

Pacific Ocean

EQUATOR

Marquesas

Tahiti

SOUTH AMERICA

Rio de Janeiro

Valparaiso

Cape Horn

Antarctic Ocean

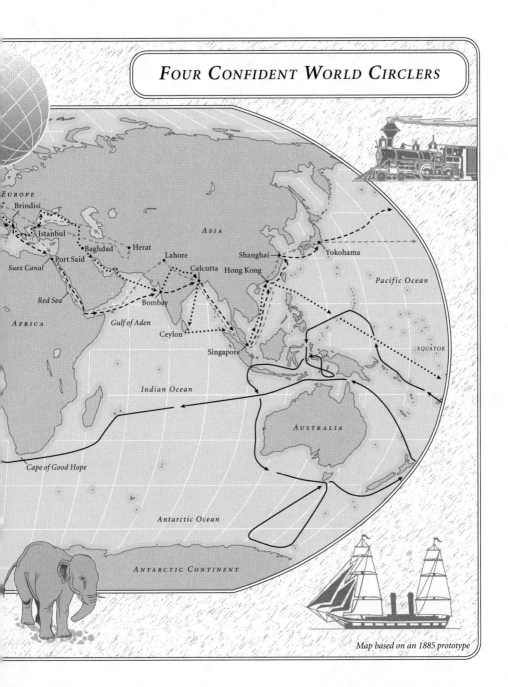

FOUR CONFIDENT WORLD CIRCLERS

EUROPE
Brindisi
Istanbul
Baghdad • Herat
Port Said
Suez Canal

ASIA

Lahore
Shanghai
Yokohama
Calcutta
Hong Kong

Red Sea

Bombay
Pacific Ocean

AFRICA
Gulf of Aden
Ceylon

Singapore
EQUATOR

Indian Ocean

AUSTRALIA

Cape of Good Hope

Antarctic Ocean

ANTARCTIC CONTINENT

Map based on an 1885 prototype

Chapter 5

A Tolerable Risk

*A*fter the pleasant bustle of departure, it was discouraging to lose the entire first week to seasickness. Charles Darwin managed to keep down his farewell meal (Champagne and mutton chops), and lay easy during his first night underway on HMS *Beagle*, but the next day revealed his lack of sea legs. "The misery is excessive & far exceeds what a person would suppose who had never been at sea more than a few days," he noted in his private diary. He resented that he never got over it. The haze of nausea became a routine contrast to his rejuvenating spells ashore. He anticipated the incredulity of landfolk: "People in general are not at all aware what a lasting misery sea-sickness is."[1]

No, they are not, and yes, it is. But it says a great deal about the shift in expectations for around-the-world travel that by the time Darwin joined the circumnavigation of the *Beagle,* from 1831 to 1836, the physical risks were tolerable. Groaning in a ship's bunk or vomiting over the rail are unpleasant, but they hardly compare to the fatal corrosion of scurvy. Darwin knew that. He and other nineteenth-century circumnavigators were confident that, unlike earlier travelers, they would probably survive a swing around the planet.

Confidence in regard to the globe would be a distinctive nineteenth-century attitude. Celebrated as the universal achievement of humanity, man over the material world, it was in fact a *proprietary confidence*, the cultural property of a small number of the world's nations. Around-the-world expeditions were done by countries that had or wanted overseas empires. The Hundred Years' Peace among the major Western powers meant that even as those powers used conquest and coercion to gain territory abroad, they were peaceful with each other. The imperial expansion, combined with the network of alliances among the major states of Europe and the Americas, offered circumnavigators more space on Earth to rest and gather strength. Once more of the planet's land was accessible to Western nations, the sea seemed tamer. The sufferings that had characterized circumnavigations were transferred to other kinds of exploration, particularly polar expeditions.

The overall sense was that mastery of the planet had been achieved, and that humanity had always been meant to achieve it. The new impulse—which still exists—is to fulfill an *us-too* ambition, to join the club of nations able to go round about the Earth, as if humanity might be united in planetary dominion.

Not that circumnavigations lost their drama entirely. As death receded into the background, the Circumnavigator's Paradox replaced it as the dramatic pivot of an around-the-world journey. And if many more imaginary round-the-world journeys featured human rather than supernatural characters—another sign of confidence—other representations memorialized the fearsome risks that had, for three hundred years, been routine during circumnavigations.

An easy way around the world, the evolving fantasy.

A swift and comfortable transit was the central activity of Xavier de Maistre's 1795 travel parody, *Voyage autour de ma chambre* (*Journey Round My Room*). In Switzerland, de Maistre was sentenced to house arrest for forty-two days after he had been convicted of illegal dueling. He passed the time by imagining his bedroom to be as big as the world and by planning to make "a forty-two days' journey" around it while wearing his dressing gown. He begins by orienting himself: the room "is situated in latitude 48 degrees east." "As we go northward," he explains, "my bed comes in sight." His mirror "offers to the sedentary traveler a thousand interesting reflections." Indeed, the "voyages of Cook" cannot compare with his own, though Satan's descent into hell (a touch of Milton here) was "one of the noblest efforts of imagination, and one of the most splendid journeys ever made, next to *the journey round my room*."²

Any earlier, de Maistre's parody would have been in bad taste. But by 1795, a circumnavigation was no longer a killing affair. People could imagine doing one with a flourish, and living to tell the tale.

Perhaps even more astonishingly, an around-the-world voyage could save lives. That was the premise of the first global medical mission. In 1803, Dr. Francisco Xavier Balmis departed Spain on a royal expedition to disseminate the strain of smallpox vaccine that Dr. Edward Jenner had developed from the disease of cowpox. (*Vaccination* was more effective than the previous remedy, *inoculation* with pus gathered from human smallpox lesions.) With no other way to keep the precious vaccine alive and active, Balmis took twenty-two orphans—small successive incubators, some late examples of captives taken on circumnavigations. After stops in the Canaries and Caribbean, the expedition divided into two groups; one team tackled the interior of South America, while Balmis's party went through the Strait of Magellan and up the coast. For the Pacific crossing, Balmis assembled twenty-four new orphans (some of them infants) from whom the vaccine could be distributed in the Philippines, China, and St. Helena in the South Atlantic. He returned to Spain in late 1806. A special supplement to the *Gazetta de Madrid*, translated and reprinted in other European newspapers, explained that Balmis kissed his sovereign's hand on September 7 of that year, having made "a voyage round the world, executed with the sole object of carrying . . . the inestimable gift of vaccine inoculation." The planet was no longer a killer, but the means by which to save lives.³

As around-the-world journeys became less risky, a great many more people did them. When the Franco-Prussian author Adelbert von Chamisso reached Hawai'i during a Russian circumnavigation (1815–18), a friendly whaler asked him: *first time around the world? Yes,* Chamisso replied, *and you?* It was the other man's tenth time. That put Chamisso in his place. The unnamed prodigy must have belonged to one of the nations—maybe the United States, possibly Britain—that could no longer keep track of its citizens' tours of the world. Others could just about maintain a tally, especially nations that had emerged from the Napoleonic Wars with something to prove. After the defeat of Bonaparte, between 1816 and 1830 the restored French monarchs (keen to make up for the La Pérouse disaster) mounted seven round-the-world expeditions. That was nothing. Russia, which had repelled a French invasion at great cost, was even more ambitious. Under Tsar Alexander I (1801–25), the Russians initiated twenty-five circumnavigations, and would complete thirty-four between 1803 and 1849. Innumerable private and commercial circumnavigations added to the total.[4]

As their ranks swelled, circumnavigators became less famous. It is hard to pick one out of the crowd, or to differentiate among their increasingly formulaic narratives—their accomplishment seemed less impressive, once more of them survived. "Honorable imitators," Jules-Sébastien-César Dumont d'Urville would call himself and his circumnavigating contemporaries, with perfect awareness of his place in history.[5]

Imitative or not, published accounts of circumnavigations continued to sell, sometimes quite well. The narratives got longer and more elaborate, continuing to fulfill the Enlightenment expectation of encyclopedic knowledge. Multivolume and illustrated editions, translations, reprintings, and collected anthologies all indicated that the reading public found the accounts interesting. Danger remained a part of the drama, though a diminishing element. Readers in this particular era were also fascinated by a new feature of around-the-world travel: safety.

The new narratives explained that circumnavigating ships were healthy places with hearty men who enjoyed many up-to-date amenities. Early nineteenth-century circumnavigators were guinea pigs for something invented during the Napoleonic Wars: canned food. More ships began to follow the British example of storing water in iron tanks, rather than the traditional wooden casks that deteriorated.[6]

Ships began to seem like home—perhaps better than many homes. Louis-Isidore Duperrey's circumnavigating *Coquille* (1822–25) had copper bathtubs for the men. On the *Mirnyi* and *Vostok*, the Russian men had clean shirts and underclothing twice a week, and dry bedding always; even the ropes were washed and dried before being stowed below, lest they spread damp and funk. Hardships still spiced the story. One Russian ship celebrated Easter with a festive double ration of water.

Another crew developed scurvy on a voyage when 70 percent of the time was spent at sea; shore leave in Port Jackson (Sydney) cured them. The risks on the first Russian circumnavigation were such that officers and crew received life pensions, and the officers were given promotion and the option of discharge; on the subsequent transit of the *Mirnyi*, the crew received eight times their normal pay.[7]

And yet circumnavigators thought most dangers could be avoided, if all went well. Amasa Delano, an American who went around the world three times on commercial voyages, said that a top-quality ship was the whole trick of the thing: "The common expression, 'I believe she will perform the voyage *well enough*,' is a disgrace to the judgment and feelings of him who uses it." Delano recommended a vessel between 200 and 400 tons, best-quality provisions, sufficient armaments, and constant care and cleaning. The last factor assumed that a crew was physically able to keep up with maintenance, which was now often true. When Auguste Bernard Duhaut-Cilly's ship returned in 1829, everything on her had been painted and polished in the Azores. People who saw her were amazed because they had assumed "that a ship returning from a voyage around the globe will be a broken wreck." Duperrey lost not a single man from the *Coquille*, mostly because he offloaded invalids along the way, another benefit of the Hundred Years' Peace. A Russian commercial circumnavigation of 1828 to 1831 also had zero casualties. In the absence of serious medical problems, one of Duperrey's medical officers gave a detailed description of seasickness.[8]

A circumnavigation could now be less exerting than life on land. Before he left, Chamisso met an ostler who, having served more than five years on part of the mail run between Berlin and Hamburg, had covered enough miles to go around the world more than once. Jacques Arago, a naturalist on Louis-Claude de Saulces de Freycinet's circumnavigation, concluded that "nothing is in fact more easy than to make the tour of the world," because all sections of the sea route were now familiar. Georg Heinrich von Langsdorff, a naturalist on the first Russian circumnavigation, noted that the expedition stayed scurvy-free, even as Russian fur-trading posts in Kamchatka and Alaska did not.[9]

That final point showed how circumnavigation's worst danger was migrating to land. Although scurvy had been associated with cold climates, including Arctic exploration, the well-publicized cases of sea scurvy during circumnavigations had outweighed such occurrences. As Europeans ventured more routinely into cold regions to trade furs or claim land, however, the balance would shift. (So too would polar expeditions begin to crowd circumnavigation out of the story of European exploration. Bellingshausen's circumnavigation had focused on the Southern Polar Sea, as the account of his expedition announced; subsequent Russian circumnavigations focused on the Arctic Ocean.) The same was true of diseases associated with hot climates, especially dysentery. Freycinet warned his men not to spend too much time out in the heat and to abstain from tropical fruits, which were thought to bring on tropical ailments. The circumnavigator's old belief that any land was better than no land gave way to a choosiness about land.[10]

The new safety of an around-the-world voyage was supposed to include anyone encountered ashore. Tsar Alexander I ordered the men of his circumnavigations never to harm landfolk except in defense. Arago said that one glory of Freycinet's circumnavigation was that "not a drop of blood has been spilt." It was true, though it did not mean that Europeans behaved that way all the time. Although Western nations mostly kept the peace with each other, their empires abroad were based on conquest. But a circumnavigation, especially a state-sponsored expedition, was now associated with the internationalism of science. The universalism looked somewhat more convincing if done peacefully; missions of war were not supposed to overlap with geographic and scientific ventures around the world, a somewhat awkward compartmentalization.[11]

Nor was the global condominium among Westerners fully worked out. During the French Revolution and Napoleonic Wars, France had demanded passports for any border crossing, and other European and American nations followed suit. The documents were not always convenient to get, however, and the highborn resented having to identify themselves to social inferiors, which border guards and port authorities tended to be. Ships were also more carefully regulated. They could now dock at a greater number of places, but only if their captains could prove them free of contraband trade goods, stowaways, or contagious disease.[12]

The paper internationalism did not always work smoothly, as circumnavigators' narratives noted. Chamisso was amused that, on his return to Europe, the English never asked to see his papers, yet the Russians constantly demanded his passport, even though he had just performed a circumnavigation for them. And until 1815, a captain could not always expect kindness from strangers. Delano's three circumnavigations unfortunately overlapped with the War of 1812. At various points in the conflict, Spanish, French, and British territories were off-limits. Even as he battled scurvy, Delano could not persuade the Spanish governor at Juan Fernández to give him so much as a basket of fruit. Toward the end of his final voyage, several captains refused to pass on any news of the war. At last, two kind souls offered help: a British captain confirmed that the United States was not at war, and an American captain handed over two items of food. Delano and his men celebrated the peace with hot buttered potatoes.[13]

As circumnavigations became less deadly, the element of time change became their new defining feature. Indeed, the issue was central to proprietary confidence. Every circumnavigation was a reminder that the world did not share a single calendar day. But where should the days be divided? Most Europeans put the prime meridian through their capitals, and let their maritime adventurers decide where the opposite line might be. Chamisso noted that the western Europeans in Hawai'i reckoned time west to east, while the Russians did the opposite. In the late 1820s, an English circumnavigator recommended two universal lines: one through Greenwich and another directly opposite. But because few ships carried chronometers, which were still expensive (and rarely accurate), it was difficult to change ship's time in the middle of the Pacific Ocean. Captains preferred to adjust their calendars in a familiar port and, if they weren't British, had no interest in Greenwich. Despite the temporal confusions, it was apparent that Europeans wanted the prime meridian to go through Europe—to that extent, their proprietary confidence united them against the peoples of the Pacific, who would hardly have agreed that they lived at the back of beyond.[14]

Once circumnavigators felt that they could entrust their lives to ships, they expected more, namely, comfort and diversion. At first, there were medical justifications for these. One medical officer warned against the "nostalgia" for home and the "black melancholy" that could afflict sailors, leading to disaffection and even physical illness. Amenities that cheered the men were medically useful, not silly indulgences.[15]

The luxuries spread fast, from the captain's cabin, to the officers and to passengers, and then to ordinary sailors. The *Rurik* of the second Russian circumnavigation (1815–18) was a mere 180 tons, yet her main cabin boasted a fireplace and overhanging mirror. Glass windows and skylights made life belowdecks brighter and less claustrophobic. A ship's library became a standard amenity. Aboard Adam Johann von Krusenstern's *Nadezhda*, the officers and naturalists could consult narratives of Anson, Cook, La Pérouse, and Marchand. The Frenchmen of the *Rhin* staged plays, complete with painted sets. As the *Nadezhda* approached Japan, the musically inclined arranged a concert called "The Voyage Around the World." The naturalists of the *Rurik* even tried to get a church organ aboard, until the captain caught them at it. He ordered the instrument ashore and dressed down the officer on watch, perhaps a music lover, who had turned a blind eye.[16]

News and letters became important diversions, and their presence or absence defined distance from a Western cultural core. Before crossing the Pacific, Chamisso savored some final newspapers from Mexico and reported a last shipment of letters back to Europe: "With them our trail vanished." Krusenstern gave his eastbound mail to an American ship headed from Baltimore to Batavia, in exchange for carrying the Yankees' westbound letters to the Cape of Good Hope. Almost home, just off Norway, he snatched a precious packet of newspapers from a Russian frigate. The increase in round-the-world traffic itself expanded the news network. At a California mission, Duhaut-Cilly greeted a deaf padre by bellowing, "I am French; I have come from Paris and I can give you recent news of Spain."[17]

With this back-and-forth, circumnavigators both experienced and expanded a proprietary world. When the *Rhin* was anchored at Sydney, the local amusements included Handel's *Messiah* and Bellini's *La Sonnambula*. Conversely, when the ship stopped at Auckland, her officers gave a dinner for their counterparts in the local 80th Regiment: "splendid dinner— glorious wines—tremendous effect," reported the *Southern Cross*. There at what they considered the end of the world, Westerners affirmed their shared identity by treating each other to European luxuries.[18]

Very occasionally, circumnavigators included non-Westerners in the club. This was particularly the case for the Russian expeditions. The steward on the *Rurik* was a Muslim Tatar, and the cook was either part Asian or else a black West Indian (accounts varied); Bellingshausen's voyage included at least one Tatar man. Four Japanese castaways, who had been picked up in the Aleutian Islands and taken to Moscow, ended up going around the world with Krusenstern, who took them home to Japan the long way. A Japanese published account of the circumnavigating castaways included a world map with the track of their route. In Japan, Krusenstern's cartographer praised the Japanese for their good "geographical knowledge," including their appreciation of an English pocket globe.[19]

It was also increasingly possible for non-sailors to go around the world. Scientists and artists were especially welcome on national expeditions. Diplomats were sometimes necessary, as with a Russian envoy to Japan who shipped with Krusenstern. Commercial agents joined some ventures. (Arago had known the commercial agent on Duhaut-Cilly's circumnavigation.) Writers also squeezed aboard. Chamisso was a famous author—settlers in Russia's far-off trading posts couldn't quite believe they were meeting him. He joked that "a trip around the world is a requisite of a scholarly education, and in England they are said to be outfitting a passenger ship to take idlers on Cook's trail for a small amount of money."[20]

"Idler" was an insult, though only in a technical sense. It was a sailor's name for someone who, lacking duties at sea, had the right to sleep all night. Boredom was their reward. (Langsdorff protested the idea that shipboard life was dull, thus confirming it was a common opinion.) The tedium seemed worse because of the limited social circle on a small ship. Arago resented that "barbarous and coarsely energetic words alone salute the ear." Hermann Ludwig von Löwenstern complained that his shipmates' bad habits and petty squabbles made his voyage an "eternity." Chamisso, who hungered for adventure, discovered that "the ship that holds a person is the old Europe from which he strives in vain to escape." The wooden world was so safe, and access to land so predictable, that a circumnavigation was nearly as dull as a tour of one's bedroom.[21]

No easy way around the world, a reminder.

Around 1780, a pocket globe produced in London showed how much the world had changed since Magellan's death (see illustration 8). On the globe's surface, European empires are expanding into the Pacific, as elsewhere. Place names in the Pacific have proliferated and a fairly complete Australia is displayed. The globe also memorializes the drama of planetary conquest. It bears Anson's cruise track, decades after the *Centurion* had returned to England, as well as the route of Cook's first circumnavigation, done years before the globe was made. Perhaps memories of Anson's spectacular casualties were fading. But even if that were so, death marks the spot: the globe identifies Hawai'i as the site of Cook's demise in 1779.[22]

Some nineteenth-century pocket globes would add the tracks of more recent circumnavigators but, remarkably, most continued to feature Anson and Cook, especially the latter, and often with a note on Cook's death in Hawai'i. Those marks on the globe show a lingering admiration for geodrama's fearful first act, with its catastrophic losses. It was an odd nostalgia, almost a rebuke to the honorable imitators.[23]

If ships now functioned as life-support systems for circumnavigating humans, they did not yet do so for animals, a crucial limitation on the ships' recent improvements. Many more creatures were being carried on circumnavigations, often as food, others as living specimens, and still others as pets to alleviate the boredom. But ships were poorly equipped for small animals (too easily knocked about or tumbled overboard), or those sensitive to changes in temperature or diet. And yet circumnavigators hated to go to sea alone. The officers of the *Nadezhda* embarked with a cat and a lapdog, which both died. In Tenerife, the men of the *Rurik* adopted a cat and a rabbit, which both died. On departing Australia, a French sailor saw a sudden roll of his ship fling a captive dingo off the deck; the dog clung briefly to a projection on the ship's side, then fell into "the abyss below." When the *Vostok* left Australia, she carried eighty-four exotic birds and "a kangaroo, which ran about loose, was very tame and clean, and often played with the sailors." The vessel headed south into Antarctic waters, where it acquired three penguins and a seal. By the time the vessel reached warmer latitudes, one of the penguins had died and the other two "grew very thin" on "a bread and meat diet." The seal refused all such food until it died. Many of the tropical birds had perished from the cold. When the survivors were brought up on deck, they "set up a many-voiced chorus," dazzled by sunshine after many dark days below. No word on the kangaroo.[24]

The categorization of shipboard animals—as meat, as specimens, as pets—established some uneasy contrasts. One Russian warned that if livestock were retained until late in the voyage (as emergency food), the men might grow too fond of them. Good milkers were especially cherished, as had been true of Cook's famous Goat. On the second voyage of the *Columbia*, one of the crewmen wrote a semi-mock eulogy:

> Between the hours of 3 and 4 PM Departed this life our dear friend *Nancy the Goat* having been the *Captains* companion on a former voyage round the Globe but her spirited disposition for adventure led her to undertake a 2d voyage of Circumnavigation. . . . At 5 PM Committed her body to the deep[.] She was lamented by those who got a share of her *Milk*!![25]

The extremes of sentiment, whimsy versus indifference, show that even as narratives of around-the-world travel had to advertise concern for humans, the equivalent for non-humans was unnecessary. Animals had taken the place of the human captives who had once been common aboard circumnavigating ships: valued, perhaps, but only up to a point. The crew of the *Nadezhda* arrived in Hawai'i in good condition, in contrast to their livestock. When some Hawai'ian women visited the ship, the desperate goats raced over to graze on the greenery they wore. One Russian naturalist passed the time at sea by torturing the specimen animals to death. The man who recorded the abuse disapproved, but confessed that, when he accidentally injured his parrot, he threw it overboard "so that I did not have to watch the misery any longer." Another Russian crew conspired to kill the ship's sow. They butchered and ate Shafekha in San Francisco, where other food was available, because they were too "ambitious" to let a beast share with them the honor of going around the world.[26]

The sailors probably felt the same way about women, but that was a battle they would steadily lose, starting in 1817, when the second woman known to have sailed around the world embarked on her journey. Rose de Freycinet was, like Bougainville's Jeanne Baret, also French and also a stowaway, but not a servant. She had married a French naval captain, Louis-Claude de Saulces de Freycinet. He and a brother, Louis-Henri de Saulces de Freycinet, had done a circumnavigation with Nicolas Baudin on the *Géographe* (1800–04). When Freycinet received command of the *Uranie* in 1817, he and Rose decided not to part, citing their mutual affection. They had probably also concluded that, after three childless years of marriage, it was unlikely that the twenty-two-year-old Rose would become pregnant on the expedition. She went aboard dressed as a man, but reverted to skirts once the *Uranie* left the Cape of Good Hope. There was official disapproval of this violation of the naval ban on female passengers, yet also widespread admiration for the young woman's intrepidity, an indication that a circumnavigation still carried risk.[27]

Rose de Freycinet herself knew the risks. Her letters home to a friend reveal that she had researched circumnavigations (she followed Vancouver's account through the Pacific, for instance). She paid a fine rather than be ducked at the equator. She complained when "science," the expedition's main purpose, required deviation from a course to the next port. She craved European supplies—Madeira in Timor, milk in the Caroline Islands. When illness struck the crew (eleven men died), she shuddered at the improper diet and tropical heat that were thought to be the culprits. She confessed her desire for land: she dreamed of fresh eggs or milk and longed to "pick a rose or a carnation." Safe in Port Jackson, she savored apricots from transplanted European trees. Freycinet also noted the officials and paperwork that, by regulating the *Uranie* and other ships, smoothed their passage and assured their safety. She met the French and Russian consuls at Rio de Janeiro, for example, and commented on the health inspector at Île de France (Mauritius).[28]

The new world of negotiated relations among Western nations saved Mme de Freycinet when disaster struck. In one letter, she is scoffing at the dangers Anson described while rounding Cape Horn; in the next, she concedes his point, because the *Uranie* had run aground at the Falklands and was wrecked. Most of the expedition's specimens and

some of its records were lost. The castaways shivered ashore in damp tents and dared not discuss what might happen when the food ran out. Pleas, threats, and promises of payment failed to persuade the captain of a passing American whale ship to help them. But when Captain Freycinet showed the whaler a "United States passport" (requesting Americans to aid the *Uranie*), this had "more impact on him than anything that he had been told until then." The man was relieved, however, when another American ship arrived and made room for the stranded French, who got home in 1820. Rose de Freycinet had survived her circumnavigation. Twelve years later, home and dry in Paris, she died of cholera, one of thousands swept away by an epidemic disease that had been transferred halfway around the world, from the Indian Ocean to the Atlantic world. (The disease would be duly added to the checklist of port authorities who examined circumnavigating vessels.)[29]

Freycinet never published her travel letters—that would have seemed immodest for a married woman of her rank—and her husband decided against publication after her death. References to her in the expedition documents, and images of her in drawings, were removed from published accounts of the venture. Knowledge that another woman had gone around the world stayed within certain French and maritime circles.[30]

Not so for another intrepid traveler, James Holman, a former sailor who did his circuit of the world from 1827 to 1832, after an illness had robbed him of his sight. He had entered naval service because he wanted to travel, and said, "I was determined not to rest satisfied until I had completed the circumnavigation of the globe," despite his blindness.[31]

Holman announced his travels and admitted his handicap in a wonderful portrait that was the frontispiece to his travel narrative (see illustration 9). The blind traveler sits at a table equipped with a special writing frame on which he kept his journal. To his left, a window opens to a seascape, with storm-tossed ship. He leans his right arm over a globe, in the proud posture of Francis Drake. And yet Holman's globe is blank, its visible features wiped away, just as they were for him when he lost his sight. The claim of the portrait, and of the four volumes that follow, is that even a blind man could encompass and experience the whole world; he had lost the vision necessary to comprehend geography, but could still act his part in a geodrama.

But how? Holman could no longer work as a sailor. His maritime connections greatly assisted in getting him berths at sea, a seat at the captain's table, and assistance ashore, but he seemed impatient with shipboard life. His account omits details of a three-week voyage from South Africa to Mauritius. Elsewhere, he complains that "a sea voyage is so extremely monotonous," forcing him into a frustrating passivity. In contrast, whenever he was on horseback ashore, he enjoyed "that independence of action common to all equestrians, my spirits rose in proportion, and my ride became again, what travelling always is to me under such circumstances, a source of exhilaration and delight." He even rode in an elephant hunt in Ceylon, where he could hear and feel the swooshing action around him. He was a pioneer in the art of echolocation, the use of noise, as with the tap of a cane, to judge his position.[32]

And Holman represented a new desire for self-propulsion, for freedom from ships. Several contemporaries wanted to walk around the world. It was not a completely mad idea—it was well known that Lesseps had survived La Pérouse's expedition by overland passage through Russia which, combined with his sea voyage, had completed his circumnavigation. John Ledyard, an American who had sailed on Cook's third expedition, attempted a fuller circuit on foot. Encouraged by Thomas Jefferson and Joseph Banks, Ledyard (who had crossed the Atlantic) planned to walk through Europe and Russia, after which he expected to ford the Bering Strait and trek across North America. He left London in 1786, spent Christmas in Copenhagen, and after a pause in St. Petersburg, continued east in 1787. Despite having a passport, essentially a letter of introduction from the U.S. Legation in London, he was arrested in Siberia on January 31, 1788, hauled back west, and deposited over the border in Poland. But he was famous as the man who *tried* to walk around the world. The next attempt was little better. John Dundas Cochrane, a Scottish naval officer, was notorious as a "pedestrian traveller" who made his way through western Europe, Russia, Asia, and Kamchatka, though the Russian authorities arrested him in their eastern territories as well. (Holman and Cochrane had met and traveled together, before they quarreled and parted in Russia.)[33]

Freedom from the regimes of the sea and the confines of the wooden world was only possible because of the global condominium. As international agreements opened more settled territory to travelers, they could carry less—a passport and minimal baggage. They began to resemble the people who go around the world today with just those things, rather than a provision-stuffed ship.

For that reason, Holman could keep a portable microcosm in his luggage: tin cup, knife, blanket, traveling cloak, soap, toothbrush. No razor; Holman had no servant, and could not shave himself, so cultivated an unfashionable beard that, he bragged, kept him warm in cold weather and shaded his neck from the sun. He traveled in a waterproof jacket and moleskin trousers, Wellington boots, and a broad beaver hat, and carried a few shirts and oddments of personal linen. That was all a man needed to see the world: "Of course, I allude to actual travellers, not to those highly perfumed idlers, who flutter from town to town throughout Continental Europe, in search of a *sensation*." Two compact devices were especially helpful. The first was a repeating (chiming) watch, with which a blind person could tell the time. The second was £100 sterling's worth of "Herries and Farquhar's circulars." That London bank had created those bills, essentially early travelers' checks, convenient wherever Holman could not draw on a reliable private bank, though those too were spreading into more parts of the world dominated by Western societies.[34]

These material conveniences gave pleasure as well as safety. Holman loved to find European high life in unexpected places. He appreciated the Champagne that one captain popped open to mark the crossing of the equator, and that another poured on Sundays. In India, where heat dulled his appetite, Holman found that "the luxury of the first glass of cool claret . . . is often the best part of the meal." Likewise, soda water was "one of the greatest luxuries of a hot climate." Those small things—bank bills, repeating watch, clean shirts, and chilled drink— made the world seem not only accessible to a blind man but comfortable as well.[35]

The effect may have made a circumnavigation seem too easy. Although admired by critics, Holman's account did not sell well. His original publisher sold 600–700 copies of the book. (The firm's most successful work, a sentimental anthology on friendship, sold about

10,000 copies a year for about twenty years.) Holman's account was reprinted, but again, sales were modest. The cool reception had a kind of logic. The point of a published circumnavigator's account was to convey a proxy experience of the world to as large an audience as possible, meaning a predominantly sighted one. Holman could not do that. People may have been interested in the *fact* that he too could act in a geodrama, but not in the details, which omitted many conventional aspects of a circumnavigation, especially the sea passages.[36]

A conventional circumnavigation, when successful, was still a marvel. Foolishly small ships, guided by mere mortals, could somehow encompass an entire planet. At the Spanish pueblo that was to become the city of Los Angeles, Duhaut-Cilly and some of his men climbed a tall hill and looked down into the harbor at their ship, *Le Héros*. "There, we said to ourselves, is the atom that carries all our hopes and of which each of us occupies only about one four-thousandth part," everything shrunk by a distance that itself shrank in comparison to the total distance the men and ship would together sail.[37]

From another perspective, *Le Héros* was no atom. The 370-ton ship had consumed something on the order of eight hundred trees. On its own, that was not much. But round-the-world traffic was adding up as never before. If the Russians alone sent twenty-three vessels around the world before 1849 and if, on average, each ship was Amasa Delano's preferred 200 tons (a low estimate), they would have required just under 10,000 trees, about an acre and a half of virgin timber. If Britain, France, and the United States sent as many ships around the world (a very low estimate), then the total would be 40,000 trees. This was a meaningful bite out of the planet. Luckily for the planet, and despite the swelled ranks of circumnavigators, only a small minority of the world's peoples went around the world.[38]

An easy way around the world, for science.

Most people circled the world vicariously, either by reading published circumnavigators' accounts or by using other media, some of which were now aimed at children. Around 1830, George Pocock manufactured (in London) a "Patent Terrestrial Globe" made of paper. The collapsible globe would, when expanded to its full 144-inch circumference, resemble a large Japanese lantern. It could simply be pulled open or else inflated with a pump to smooth the wrinkles out of the continents and oceans. It was printed with the cruise tracks of Cook (all three expeditions), La Pérouse, and Vancouver, plus those of non-circumnavigating explorers. One could also purchase *An Accompaniment to Mr. G. Pocock's Patent Terrestrial Globe* (1830), with brief geography lesson and instructions for use. The pamphlet extols the heroism of Drake, Anson, Vancouver, and "the immortal Cook." (It also claims, falsely, that Magellan's circumnavigation had proved "that our terraqueous habitation was indeed a revolving ball.") The instructions make a final suggestion: fill the globe with hot air and watch its "ascension." The rising orb would surely elicit gasps and squeals of delight, though adults were warned that, if there was any rough play, "the Globe may be injured." [39]

Consumer goods that represented a world circumnavigated would bring an awareness of the planet into many homes, perhaps especially those of the circumnavigators themselves. Charles Darwin's sister Caroline wrote him that his nephews, who followed his route, "feel proud in finding the place on the Map where their Uncle Charles is." For Darwin, himself a product of a book- and globe-rich upper-middle-class household, an around-the-world journey meant that "the map of the world ceases to be a blank." The real planet, more than any abstraction, reminded him of his smallness. In Tierra del Fuego, serving a watch over his slumbering companions, he remarked that the dark night and quiet solitude magnified awareness of "what a remote corner of the world you are then buried." The enormity of the Pacific Ocean was equally humbling: "Accustomed to look at maps, drawn on a small scale, where dots, shading, and names are crowded together, we do not judge rightly how infinitely small the proportion of dry land is to the water of this great sea." [40]

Darwin may be history's most famous "idler." He made a maritime circuit at the private invitation of naval captain Robert FitzRoy. The invitation was itself historic, because it acknowledged the old dangers of a circumnavigation, yet insisted they might be managed.

FitzRoy, a career navy man, had served on a two-ship surveying expedition that, from 1826 to 1830, had tried to map the southern tip of South America. It was probably inauspicious for the *Beagle* and *Adventure* to select Port Famine as headquarters for their work on the Strait of Magellan. Cold weather, restricted supplies, and a punishing work schedule led to scurvy. The captain of the *Beagle* fell into a depression (probably scurvy-induced), locked himself in his cabin, and shot himself. During the redistribution of officers, FitzRoy gained command of the *Beagle*. When the expedition sailed for England without completing the survey, FitzRoy hoped to return. In 1831, he again received command of the *Beagle*, now sailing solo.[41]

In addition to the survey, FitzRoy was ordered to do a series of chronometrical readings of fixed points around the world that, connected to existing readings, would give a firmer sense of the globe's lines of meridian relative to Greenwich. To perform these tasks, the *Beagle* was enlarged to 90 feet and 242 tons. The brig could have carried 10 guns and over one hundred men, but these were reduced to 6 and sixty-five to maximize space for surveying and other scientific activities. FitzRoy added nine supernumeraries and idlers, including a surgeon, a ship's naturalist, several kidnapped Indians he would return to Tierra del Fuego (yet more long-suffering Patagonians), and a few others, for a total of seventy-four. It was so snug aboard that, when offered a puma in South America, FitzRoy would turn it down, "so troublesome a companion in our crowded little vessel" a springy carnivore was likely to be.[42]

The mission was acknowledged to be hazardous enough without wildcats. The *Beagle*'s orders specified that, should the captain die, his successor was to complete the survey, but not cross the Pacific and go around the world. The Admiralty provided several items considered anti-scorbutic, including pickles, dried apples, lemon juice, vegetables, portable soup, and tinned meat. (Kilmer & Moorsom's preserved meat was ruined when salt water corroded its tins.) FitzRoy also took on fresh food and water whenever he could, as in the Falklands, a "safe 'half-way house' " for ships bound for Australia. At Montevideo, he acquired nine month's worth of food and coal, though only a month's worth of wood and water, which were easier to replenish along the way. FitzRoy insisted that the presence of provisions would keep his men hopeful and healthy. Indeed, he lost no man overboard, very few to illness, and none to diseases specific to shipboard life, despite a "long voyage, rather exceeding that of Vancouver." [43]

He also completed both the survey and the chronometrical measurements. The latter established "a connected chain of meridian distances around the globe, the first that has ever been completed, or even attempted, by means of chronometers alone." The *Beagle* carried no fewer than twenty-two chronometers, most of them of indifferent quality, but four that FitzRoy described as "good" and one as "very good." Each instrument rested on gimbals within its own box; each box was nestled into sawdust on compartmentalized shelves built into a space near the ship's center of motion, all to keep them level, whatever dance the ship did around them.[44]

FitzRoy prepared himself just as carefully. His most important precaution was to recruit a human companion. That showed an acute self-knowledge. He suffered from persistent depression and his suicidal commander's scorbutic melancholy on the earlier expedition haunted him. He circulated word among contacts at Cambridge University that he sought an educated young gentleman willing to go to sea but to be uninvolved in ship's business, a situational *friend* to fill the solitude. FitzRoy got Darwin, and both men performed for each other a truly effective mutual display of confidence in their voyage and in each other.

Darwin knew the venture was risky. His letters and journal hint at death several times, as when, in Tierra del Fuego, he referred to being "buried" in a remote corner of the world. And yet he thrilled at the prospect of becoming a circumnavigator. He had been determined not to join the expedition unless it went around the world—"till that point is dicided," he told his sister, "I will not be." Darwin knew that Alexander von Humboldt, the Prussian naturalist who made a celebrated journey to South America, had originally wanted to go around the world, and he wished to succeed where his idol had failed. He read several circumnavigators' narratives before and during the journey, including Dampier, Anson, Bougainville, Cook, Kotzebue, and Freycinet. He had to convince his family to let him go, and enlisted his uncle to reassure his father on eight possible concerns. Most of the worries were that the journey might be a waste of time. But one was about physical danger—because the official position of naturalist had not yet been filled, perhaps "there must be some serious objection to the vessel or expedition"? Darwin hopefully suggested that the Admiralty would never dispatch such an important expedition in an unsafe vessel. Once he saw the *Beagle*, he worried more about his comfort than his safety. He described his cabin as "wofully small" though "fitted most luxuriously with nothing except Mahogany." He would be as cosy as a chronometer packed in sawdust.[45]

Only unlike the chronometers, Darwin was not insulated against the ship's constant motion. Given his unceasing seasickness, it was fortunate that his work depended on observations and collecting ashore, not at sea.

And it was equally fortunate that the new internationally negotiated access to land had made his scientific work possible. This factor, along with FitzRoy's extremely high survival rate, shows, again, how historically contingent the whole expedition was. Given that FitzRoy met no opposition anywhere, it is hard to remember that he was mapping territory that did not belong to Great Britain, which would have been impossible before 1815. The new assumption was that a British cartographic mission would generate charts to benefit all seagoing nations. Chilean officials not only permitted the survey but offered to help. And when one of the *Beagle*'s surveying parties needed access to a part of Peru torn by war, Peruvian authorities arranged a cease-fire. Hostilities were suspended for a day, the survey was quickly done, and then the shooting started up again.[46]

Paper internationalism also protected Darwin on his travels. At Rio Colorado, the authorities were unpersuaded by his passport, yet a letter of reference from an official at Buenos Aires cleared the way. Darwin was only able to cross war-torn Argentina because of another supplementary document. To his delight, this paper introduced him, rather exotically, as *El Naturalista Don Carlos*, which worked like a charm.[47]

It is not too much to claim, therefore, that the theory of evolution was one particularly momentous outcome of around-the-world paper internationalism. Darwin made three of his major scientific discoveries in places that had been off-limits to English-speaking Protestants before the nineteenth century—where *El Draco* had rampaged, *El Naturalista* now rummaged. Darwin's observation of living armadillos and discovery of fossils of the armadillolike megatherium in the same part of Argentina, for instance, hinted that species could change over time. The Argentine pampas, with its distinct yet related species of rheas (ostriches), also indicated that species changed over space. The Galápagos, with its multiple types of finches, showed that the same place and time could provide slightly if significantly different habitats for closely related species.

International harmony among Western nations also kept the *Beagle* connected by mail to a larger world centered on Europe. Postal service greatly benefited from the Hundred Years' Peace. Mail had long been carried privately, by hired couriers on land or merchant ships by sea. Different nations had also established government-supported postal services within their own territories. British packet boats connected the different parts of the empire, for example. After 1815, some of the British government service was privatized, with commercial shippers receiving government contracts to carry mail on specific routes. The Peninsular Steam Navigation Company, founded in 1835, held an early contract to take British mail to and from the Iberian Peninsula. Unimpeded by war, granted access to foreign territories by treaty, mail now streamed around the world.[48]

Darwin used every available form of postal service. The Admiralty allowed him to use official British conduits to send letters and packages to London. His sister pointed out that British mail packets were faster (and cost double postage, Darwin grumbled). Sometimes, he depended on the kindness of strangers. On the lonely Galápagos, a barrel on "Post Office Bay" served as a letter drop where ships left mail for vessels going one way and took letters meant to go the other. (There was a similar arrangement in the Strait of Magellan.) A variety of other sources reported on the expedition's progress. Friends and family in England were excited whenever the *Beagle* made an appearance in the newspapers—a far cry from the silence that had fallen over earlier circumnavigations when they were too far from Europe. Several of Darwin's specimens beat him home, including fossils of the megatherium, through which "your name is likely to be immortalized," a friend predicted. His three sisters had promised that they would, in rotation, send him a monthly letter, which they did pretty faithfully. Some of the news from home was too distant to interest Darwin. But when cholera swept through Europe (where it killed Rose de Freycinet), it was a different matter. "You will escape *this* danger at least," his sister Catherine rejoiced.[49]

He knew his good fortune. "How glad I am," Darwin wrote home, that "the Beagle does not carry a years provisions; formerly it was like going into the grave for that time," whereas he could count on frequent shore leave. At Montevideo, he realized he had spent only one night aboard in the past four months. He preferred that. "How I shall long for the green plain and its galloping horses," he said, on embarking again. "The short space of sixty years [since Captain Cook] has made an astonishing difference in the facility of distant navigation," he concluded. "A yacht now with every luxury of life might circumnavigate the globe," given improvements in ships and sea provisions, but also because "the whole western shores of America are thrown" open and Australia settled. When the *Beagle* reached the Île de France, with its theater, opera, and bookshops, Darwin relaxed into the certainty of his "approach to the old world of civilization."[50]

Darwin's comments signaled, not in a subtle way, his belief that Europe was the center of the world. He and FitzRoy were among the first circumnavigators, for instance, to mock contemporaries who did not believe the world was round. In South America, FitzRoy noted that an Indian boy told him that the Sun went around the Earth, and Darwin was amazed that rural landholders, descended from Spaniards, believed the same. Near Maldonado, he remarked on his hosts' desire to see his map and compass. "I was asked whether the earth or sun moved," he marveled—"I am writing as if I had been among the inhabitants of central Africa." At another rural estate, his hosts "expressed, as was usual, unbounded astonishment at the globe being round." And yet Darwin, who would become the most noted man of science of the nineteenth century, lapsed into the circumnavigator's poetic pre-Copernicanism. When the ship's calendar was adjusted by a day in Tahiti, he said the change was "owing to our, so far successful, chase of the sun."[51]

The voyage of the *Beagle* became the new model of a circumnavigation. It kept the peace with other nations, and that peace, in turn, kept its scientific business on track. FitzRoy and most of his men had survived; so had his gentleman friend. Not that Darwin returned in top condition. In the two months after his return in 1836, his family fed 18 pounds back onto him, so he looked himself by Christmas. He contributed a volume to FitzRoy's official narrative of the expedition, and regarded that book, his very own around-the-world narrative, with particular fondness, as his first best seller. The two men thereafter drifted apart, not least because of Darwin's evolutionist theory, which FitzRoy deplored. The captain killed himself in 1865; the melancholy he had eluded around the world had, thirty years later, caught up with him.[52]

Another easy way around the world.

Even a "hard-hearted, dunder-headed, obstinate, rusty, crusty, musty, fusty, old savage" can be defeated by true love and an appreciation of geodrama. Bobby has asked Uncle Rumgudgeon for permission to marry his daughter, Kate. "Ha! ha! ha!" the "old porpoise" jeers: Bobby shall have neither the girl nor her "plum" (meaning her money; implying her luscious virginal state) until there are "three Sundays in a week." Bobby despairs. Not Kate. It is a sign of the times that no fewer than two sea captains of her acquaintance have just returned "after a year's absence, each, in foreign travel." She and Bobby invite the two men to visit on a Sunday. When a game of whist is proposed for the next day, one captain protests that he cannot possibly play cards on the Sabbath. He has gone around the world via Cape Horn and lost a day. The other captain, another circumnavigator, has rounded the Cape of Good Hope and thinks Sunday was the day before; Rumgudgeon disagrees with both. But everyone is right, says Kate, sweet reconciler of all, "and thus *Three Sundays have come together in a week.*"

It is a lovely love story and stuffed with information, most of it true. One of the captains helpfully explains that the Earth is 24,000 miles in circumference (he rounds down by about 900 miles) and that it does a full revolution, west to east, in twenty-four hours. Every 1,000 miles constitutes an hour; 24,000 of them fill a day. Sailing around the world in one direction adds a day; in the other way, a day is lost. Kate knew it all along. And thus in 1841, sometime between writing "The Murders in the Rue Morgue" and "The Masque of the Red Death," Edgar Allan

Poe knocked off "Three Sundays in a Week" for a couple of the newspapers that afforded him a stuttered income. The story is so charming, so uncharacteristically human (for Poe) that one hardly notices two of the author's obsessive touches: Kate is Bobby's cousin, and she is all of fifteen years old. But unlike Poe's other semi-incestuous, jail-bait objects of desire, Kate is neither doomed nor dead, but alive and lively. It is topsy-turvy: geodrama can give anyone, even a tender sheltered virgin, critical knowledge of the world. With that knowledge, Kate gets her object of desire, Bobby.[53]

No longer the shy debutante it had been in the 1780s when it did its first circumnavigation, the United States had, by the 1830s, begun to use around-the-world expeditions for overtly political reasons. When an American commercial vessel, the *Friendship*, was attacked off Sumatra in 1831, and her captain and several officers killed, President Andrew Jackson ordered a reprisal. ("Old Hickory" was hardly likely to forgive such a thing.) The following year, the USS *Potomac* (carrying 50 guns) did an eastabout circumnavigation, pausing to bombard the forts and town of Kuala Batee, Aceh. The town surrendered. The region's governing rajahs were warned never to attack U.S. vessels again. This circumnavigation, unlike Thomas Gray's, the first American circumnavigation, was reported in print. In Francis Warriner's 1835 published narrative of the cruise, he emphasizes the patriotism of the venture, as in shipboard celebrations of George Washington's birthday. The shelling of "Quallah Battoo" is the climax of the story, proof of American global might.[54]

A belated newspaper reference to the expedition had probably inspired Poe. On November 17, 1841, "NAVV." had written to the *Philadelphia Public Ledger* to respond to an earlier item. That previous piece, called "Three Thursdays in One Week," remarked that the phrase had been a synonym for "never," which was certainly Uncle Rumgudgeon's implication. But "circumnavigators in their voyages round the world" had proven such wonderful weeks to exist. "NAVV." agreed and related how the USS *Potomac* had crossed the anti-meridian on Sunday, the Fourth of July. The military mission therefore celebrated "*two* 4ths of July in one year; *three* Sundays within nine days, and one month of thirty-two days." Plus, "this was in 1832, which was a leap year, and consequently all on board saw *three hundred and sixty-seven* risings and settings of the sun." The parallel circumstances under which a day could be added to a calendar may have given Poe an idea. In the English-speaking world, single women could propose to bachelors on the extra day of a leap year, something Poe would note in another story, his "Thousand-and-Second Tale of Scheherazade" (1843).[55]

Leap year or Circumnavigator's Paradox—a day gained either way might permit a woman, like Kate Rumgudgeon, to maneuver a man toward matrimony. From a side story relating to his nation's budding imperial aggression, Poe plucked a wedding bouquet.

Perceived military necessity would remain a major reason for U.S. ships to circle the globe. To avenge an 1838 Malay attack on the merchant ship *Eclipse*, for example, a Second Sumatran Expedition (December 1838 to January 1839) deployed two U.S. warships that were already doing circumnavigations. Another military convoy made a circuit from 1840 to 1844. It stopped at several Asian ports, including "Quallah Battoo," and recorded encounters with ships from several friendly European nations. These circumnavigations were measures of how Americans saw themselves as global actors, commanding a spotlight on the world stage—at last: *us too.*[56]

A scientific circumnavigation admitted a nation to a more refined role in geodrama. The United States made its first serious attempt with the U.S. Exploring Expedition (1838–42), Charles Wilkes commanding. Wilkes was ordered to do extensive reconnaissance of the Pacific Ocean and to verify and chart any major landmass to the far south, following reports (since the Russian circumnavigation by Bellingshausen in 1820) of icy terrain there, which had set off a rush for the south pole. The American venture would be nicknamed, with affection, the "Ex. Ex." Wilkes was nicknamed, without affection, "Stormy Petrel," for the seabird that warned of bad weather.[57]

The Ex. Ex. was, among other things, ominously large. It included 346 men aboard six ships: three men-of-war (780, 650, and 230 tons), plus a supply ship and two tenders. The flagship, the *Vincennes*, had already been around the world twice. She was luxuriously fitted; one of her staterooms was both both curtained and carpeted, with silver candlesticks and a mirror. The mission was "not for conquest, but discovery . . . to extend the empire of commerce and science, to diminish the hazards of the ocean." Wilkes was ordered to protect his men, especially against sea scurvy and tropical dysentery, and never to order a military attack, except in self-defense.[58]

But even the earliest stages of the expedition were troubled. After traveling into the South Pacific, one vessel reported contact with a frozen coastline, though the relevant log entries are contradictory and, worse, appear to have been altered after the fact. The icy south was a cruel trial, ongoing evidence that polar zones were inheriting circumnavigation's deadly reputation. The U.S. Navy's cold-weather clothing and equipment proved inadequate and it was against the advice of his medical staff and officers that Wilkes pressed south. Stormy Petrel fell into rages and confidence in his command diminished. The international network that handed ships' letters around the world returned news of the expedition's troubles to the United States. The *Saturday Evening Post* reprinted one letter to the effect that Wilkes was "getting delirious" and "the whole expedition is a humbug." The gossip-laden papers then shot out from the United States to reach Wilkes in Honolulu. He was furious.[59]

It was even worse that the expedition violated the two post-Cook taboos against risking lives, either of sailors at sea or civilians on land. On the first count, Wilkes was not entirely to blame. The U.S. Navy, probably unaware of the fates of Anson's and Loaysa's circumnavigations, had given Wilkes too many ships. Only two of the six actually made it around the world; of the other four, one returned back around the Horn, another was sold off in Asia, one was lost though all hands saved, and the last was lost with all hands. These were hardly Anson-sized casualties, yet uncomfortably reminiscent of them. As to hurting people ashore, Wilkes violated that taboo all on his own. On a visit to Fiji, two crewmen were killed. In reprisal, not self-defense, the Americans killed almost eighty Fijians and demolished two villages, which Wilkes declared to be "just and necessary." He also captured a Fiji leader, Vendovi, suspected of killing crew from an American merchant ship, to take him back to the United States. When Vendovi died, his skull became part of the expedition's scientific collection, as if non-

Westerners were scientific objects, specimens to be collected, however interesting and companionable in the meantime, not unlike animals.[60]

It was a muddled achievement. On his return, Wilkes was court-martialed on several counts, mostly due to his treatment of his men, though he was acquitted on all but one charge. He produced a five-volume narrative that went through fourteen editions before 1865. Twenty-four other volumes of scientific material were prepared, though not all were published. A wealth of artifacts were displayed at the U.S. Patent Office in Washington, then transferred to the new Smithsonian Institution. The United States would sponsor other scientific circumnavigations—the Ex. Ex. was an excellent dress rehearsal for them.[61]

But the venture compared poorly to the contemporary circumnavigation of Dumont d'Urville, who was also engaged in the race for the south pole. The Frenchman had set out with two ships in 1837 and braved pack ice and rampant scurvy to cross the Antarctic Circle, where he made landfall and claimed the territory for France (see Map 3). His log entries on encountering what would later be named Antarctica were clearer than Wilkes's, whose claim remained tenuous until the twentieth century. (The Frenchmen had spotted one of Wilkes's ships, which fled without allowing contact.) Dumont d'Urville was also careful to preserve life when he could. He left dysentery-stricken invalids in Tasmania rather than take them to the Antarctic; he rejoiced that claiming possession of the uninhabited polar territory was "a conquest wholly peaceful." There is one jarring note, however. The expedition had taken aboard a Polynesian volunteer, Mafi. When Mafi died of tuberculosis, his body was preserved in a barrel of arrack, destined for the natural history museum in Paris. He was not a captive; but nor was he an equal.[62]

The French expedition represented the beginning of the end of scientific circumnavigations. Published reports of the mission were called *Voyage au pôle sud et dans l'Océanie sur les corvettes l'Astrolabe et la Zélée 1837–1840*—nothing about going around the world, which was secondary to the exciting perils of ice. After this venture, scurvy, circumnavigation's old hazard, was almost exclusively associated with polar exploration. Finally, Dumont d'Urville had commanded the last circumnavigation made with sailing ships. In a sign of what was to come, the fleet had departed Toulon assisted by the *Crocodile,* a steamboat. Dumont d'Urville himself, having survived two circumnavigations, would die horribly in an 1842 steam railway disaster between Versailles and Paris. His eulogy noted the irony.[63]

The planet was no longer most dangerous because of its size but because of its extreme environments. A sense that one phase of European exploration had ended was represented in the founding of the Hakluyt Society in London in 1846. Named in honor of the Elizabethan compiler of travel narratives, the society devoted itself to preserving and printing travel accounts written before William Dampier's circumnavigation. In 1847, the society took as its emblem an image of Magellan's *Victoria,* which advertised its focus on the era of European expansion its members considered to be truly heroic, meaning dangerous. The seas had been domesticated, but not the extreme regions that constituted the next frontiers of exploration, jungles, deserts, and especially polar zones. It was not yet possible to believe that commercial travelers could go to the south pole, for example, but it was now conceivable that they could go around the world.[64]

And yet another easy way around the world.

"We did not even take a carpet-bag, or a tooth-brush, or a clean collar," marveled a writer for the British magazine *Punch,* in "A Journey Round the Globe" he made for that magazine in 1851. "Our only passport was a shilling," surrendered and never returned: "We wish every passport was as easy to obtain, and as easy to get rid of." The shilling led into the globe itself. There, *Punch's* intrepid traveler discovered "small jets of gas bursting out of the Earth." The gas lights illuminated the inside-out planet, with continents and oceans represented within. The burning gas made the tour unbearably hot. Hoping the north pole would be cooler, "we steamed our panting way up there," only to find it was quite the hottest part of the world. The oddness and discomfort of the voyage were compensated for by its convenience: "the Globe in the shape of a geographical globule, which the mind can take in at one swallow."[65]

Punch had economized by sending its reviewer, not around the real

world, but into James Wyld's "Great Globe," a London entertainment
open to the paying public from 1851 to 1862. Wyld was a geographer
who helped plan London's 1851 Great Exhibition of the Works of Industry of all Nations. The exhibition showcased global humanity; Wyld
wanted to represent the physical planet. His proposed ten-foot-to-the-
inch representation of the globe was too large to be accommodated in
the Great Exhibition's venue, the Crystal Palace, so Wyld leased part of
Leicester Square. For its ten-year run, the Great Globe welcomed tens
of thousands of visitors into a sphere 60 feet in diameter, within which
they could clamber about on staircases in order to scrutinize each
ocean and continent.[66]

The Great Globe had an ancestor in a *Géorama* that C.-F.-P. Delan-
glard had opened in Paris in the 1820s. That spectacle was constructed
from fabric and paper mounted over an iron frame to form a sphere
that people entered at the south pole. A concave view of a convex
planet was a bit odd, though it provided a pleasingly seamless view of
the Earth, as one guidebook explained.[67]

And both European venues had an American cousin that was ex-
plicitly an around-the-world tour: "City of New Bedford . . . Whale
Ship in stream . . . city of Rio de Janeiro . . . Cape Horn . . . Robinson
Crusoe's Island . . . Typee Bay . . . loss of ship Essex, destroyed by a
whale . . . Island of Owhyhee, where Cook was killed . . . Cape town . . .
St. Helena . . ." These scenes from a whaling circumnavigation were
painted on cloth. The 8½ by 1,275-foot canvas was then mounted on
upright rollers and rotated by cranks, which gave spectators the sensation of transit. Two American entrepreneurs had prepared *Whaling
Voyage round the World* no later than 1848 to commemorate an actual
whaling voyage that had left New Bedford in 1841. Their product was
a panorama, a popular nineteenth-century entertainment that could
be stationary, moving, or arranged as a corral-like cylinder called a
cyclorama. Each big canvas showed some big subject: the mighty Mississippi, the yawning Arctic, the entire Battle of Tripoli. *Whaling Voyage
round the World*, a moving panorama, was a hit. A reviewer in the *Boston
Atlas* claimed that, for world travelers, "this painting must produce vivid
reminiscences; while to others it must seem almost like making for
themselves a voyage of the world."[68]

Billed, with no small exaggeration, as the world on three miles of cloth, *Whaling Voyage round the World* seems to have been the first panorama of an around-the-world voyage. In a pre-cinematic era, it displayed the world more kinetically, and totally, than any book, atlas, play, or map of any size could have. Cycloramas even made possible the Puck-like feat Joseph Banks had wanted to do: pirouette at the Earth's axis as each degree of longitude flashed by in an instant. It was easy. It was fun. And it represented how "going around the world" now allowed relatively ordinary people to make money, or to spend it.

Ida Pfeiffer spent it. Born into a wealthy family in Vienna, she inherited enough money from her parents and deceased husband to pursue a thwarted passion for travel. After limbering up with tours of the Holy Land, Scandinavia, and Iceland, she decided in 1846 to go around the world. Her two sons were by then grown, and she surely knew that, as a widow of forty-nine, she had achieved an optimal female balance between autonomy and respectability. (A married woman was expected to travel with her husband; a young woman traveling solo was suspected of being a sexual "adventuress.")

That Pfeiffer was able to do a planetary circuit in the 1840s (see Map 3) was extraordinary evidence that paying customers could pursue long-distance leisure travel. Starting in the Atlantic, commercial shipping lines were adapted to take passengers, as well as cargo and—if they had government contracts—mail. By the 1840s, many famous shipping companies (such as Cunard, Hamburg-Amerika, and Norwegian American) spanned the Atlantic, and looked to other oceans. Miserably spare conditions continued in steerage, but shippers began to provide furnished cabins and meal service for passengers with greater means. For example, the ship Pfeiffer took from Rio de Janeiro to China, via Tahiti (see Map 3), charged her $200, slightly more than a day laborer in the United States could earn in a month.[69]

Pfeiffer had guts as well as means. She packed pistols and, when necessary, wore men's clothing. At an unusual meal in South America, she pronounced the monkey "excellent" but the parrot "not quite so tender and savoury." Despite weather so rough around Cape Horn that fires could not be lit for cooking or illumination, she enjoyed the "comical positions" into which the ship hurled her and the other passengers.

She joined a caravan to Mosul (now in Iraq), 300 miles on muleback. That was hard. But the trek from Mosul to Tabriz (Iran) promised to be so much worse that Pfeiffer sent her travel diary to her sons by another route, to preserve it, if not herself. She survived. The European official who examined her passport in Tabriz found it "absolutely incredible that a woman, without any knowledge of the languages, could have made her way through such countries" without a friend or guide.[70]

She was a unique test of paper internationalism: could a white woman of good reputation be handed around the world as gently as an invalid weekending in a spa town? Yes, by the late 1840s, she could, thanks not only to good relations among European nations but also the heavy hand of European imperialism. Pfeiffer expressed her gratitude to the British authorities for arranging her safe passage from Delhi to Bombay. A letter of introduction to a British official in Rajput guaranteed that the local ruler would send refreshments and arrange a tour on an elephant. A prince in Baghdad gave her a letter of introduction to his mother in Teheran. There were occasional gaps in the paper chain. When Pfeiffer lacked a letter of introduction at Macao, she had to continue to Hong Kong. And the chain sometimes chafed. Like everyone else, Pfeiffer found Russian demands for travel papers to be endless and annoying. She knew that a man could not have scolded the male Russian officials and got away with it, whereas she could.[71]

Her narrative swarms with solicitous men. An Austrian count, no less, is her traveling companion across the Atlantic and her escort in South America. A Mr. Von Carlovitz chaperones her visit to China. Two unnamed gentlemen arrange an excursion in Singapore. When a porter in Ceylon makes her feel uneasy, she screams to some English soldiers and they come running. In India, a variety of officials and private contacts ease her passage: "I was never insulted by deed, word, or even look."[72]

She expected even more, meaning European standards of comfort, efficiency, and refinement. The opera house at Rio de Janeiro offered *Lucrezia Borgia,* "performed very tolerably." Before bad weather made cooking impossible, the steward on her ship to China set forth "boiled or roast fowls, fresh mutton and pork, ducks and geese, plum puddings or pastry, besides fruit and side dishes." In Hong Kong, Pfeiffer drank

Portuguese wine and English beer, cooled with ice from New England. Above all, she enjoyed the very latest in Western technology: steam travel. Although she took plain old sailing ships across the Atlantic and then around the Horn to China, six successive steamers sped her from Hong Kong to Baghdad. Not unlike James Holman, the blind traveler, she preferred to get the sea passages over with, anyway. She found the Pacific "excessively tedious: you very seldom meet a ship, and the water is so smooth that you seem to be sailing upon a river."[73]

Pfeiffer was at this point not unusual as a woman who went around the world, but she was a pioneer for making the journey alone and for publishing her story of it. Unlike Rose de Freycinet (or James Holman), she had no naval connections. Most other female circumnavigators were married to maritime men, including merchant mariners or whalers.[74]

Martha Smith Brewer Brown, for instance, joined a circumnavigation her husband was making on the whaling ship *Lucy Ann* (1847–49). Brown paid her keep by cooking, washing, sewing, and mending. She enjoyed the voyage, especially the sunsets, the frequent music, and jumping rope for exercise on the deck. Her husband surely appreciated having her aboard, and possibly the men for whom she did nursing and mending, and almost certainly a small white kitten, because when it fell overboard, the captain ordered the ship about and a boat lowered to rescue it. (Would he have done that before an audience only of sailors?) Brown kept a log in the form of letters to family at home, but she never completed a circumnavigation. She became pregnant and Mr. Brown put her ashore in Hawai'i to have the baby. Her incomplete log was never published, as was true for most women who accompanied male relatives to sea.[75]

Pfeiffer's greater ambition is obvious: unlike Freycinet or Brown, she wanted to be famous for going around the world. Her election to several European geographic societies recognized her public status as a world traveler. A formal portrait photograph shows her with a globe in a rotating stand. But not for her the bold posture of Drake, who clutched the sphere he had encompassed. Instead, Pfeiffer sits alongside the telltale object, shoulder-to-shoulder with the world, but looking away from it, acknowledging only with modesty her fame as the first published female circumnavigator.[76]

If Pfeiffer was a new kind of circumnavigator, the paying customer, George Coffin was one of the entrepreneurs happy to take her or his money. Coffin came from a seafaring Nantucket clan (kin to Benjamin Franklin), many of whom went whaling in the nineteenth century. Not Coffin. He considered whalers "the greatest bores afloat" because of their lonely time at sea—he claimed to have met one who "did not know the name of the President." He preferred commercial shipping and wrote an account of the business ventures that took him around the world in stages from 1849 to 1852.[77]

Coffin's bottom line depended on a mix of freight and passengers and he knew the latter, by midcentury, expected certain standards of service. In New Orleans, he spent $10,000 to refit the *Alhambra* to accommodate two hundred passengers on a voyage to San Francisco, where the Gold Rush was in full flow. The ship had staterooms around the edge of the passenger deck, each with its own ventilator, sidelight, and new mattress and linens. Other passengers were put into berths, with access to built-in tables and washstands. Coffin made it clear that he ran a respectable ship and posted rules of conduct for passengers. At meals, he gallantly seated a widow to his right, because she "had no protection." Although water had to be rationed late in the journey, there was never any real hardship. Even rounding Cape Horn was an anticlimax—the dreaded cape was a "bugbear" invented to frighten lubbers, Coffin insisted.[78]

Unfortunately, the ease of travel made for bored passengers. They passed the time by squabbling and Coffin (who had to adjudicate) must sometimes have wished for a quick little storm, nothing dangerous, just enough to render his charges limp and docile with nausea. After many hints and pleas, he managed to institute a weekly shipboard magazine, "as one means of counteracting the effects of ennui among my large family." The squabbles gave way to doggerel.[79]

And steadily, impressively, despite and through it all, Coffin made money. He was among a growing number of mariners who could turn a profit across every ocean, using newspapers and word of mouth among fellow sailors to find out what was selling where. In San Francisco, he sold the *Alhambra* for $13,500. After some time plying the coastal trade in California, he bought the *Arco Iris* and took on three passengers and $700 worth of freight; he needed astonishingly little labor, in the form of a mate, steward, and four sailors, to cross the now truly pacific Pacific Ocean. At Manila, Coffin contracted to take sugar and rice to New York, which completed his circuit. He probably agreed with his kinsman Benjamin Franklin that time was money. Speed, which translated into timely delivery of goods and passengers, was essential. He did a double gamble on speedy arrival at Valparaiso, first by sailing fast in order to beat the competition, and second by taking bets from passengers (maximum $50) on whether his or another vessel would arrive first. Coffin won.[80]

It now seemed so easy. In 1853, Jacques Arago, veteran of Freycinet's swing around the world, inventively condensed his earlier narrative. He had in the interim gone blind and, like James Holman, had to register the events of a circumnavigation differently. This Arago did by never, after the title page, using the letter "a," not even in his name, "j.cques .r.go"—a pun for readers who knew English ("go") or their classical history (the *Argo,* the ship in which Jason sought the Golden Fleece). The 6,250-word piece makes an around-the-world journey a clever act of imagination.[81]

No easy way around the world, a dissenting opinion.

"Round the world! There is much in that sound to inspire proud feelings; but whereto does all that circumnavigation conduct?" So asks Ishmael ("Call me Ishmael"), the main narrator of Melville's *Moby-Dick* (1851). The men of the *Pequod* have just learned that Captain Ahab intends to circle the globe in search of the great white whale. "Tell them to address all future letters to the Pacific ocean!" Ahab thunders to a fellow whaler, "and this time three years, if I am not at home, tell them to address them to—" He is cut off. We never learn where the hunt might go next, perhaps to hell itself. Ahab disdains the usual motive, profit. "The accountants have computed their great counting-house the globe, by girdling it with guineas, one to every three parts of an inch," whereas he goes around the world in a pure red streak of vengeance. Ishmael worries that "in pursuit of those far mysteries we dream of, or in tormented chase of that demon phantom that, some time or other, swims before all human hearts, while chasing such over this round globe, they either lead us on in barren mazes or midway leave us whelmed."

In her fateful clash with Moby-Dick, the *Pequod* is overwhelmed and sinks. Ishmael clings to a floating coffin until saved by a passing ship, he alone to tell the tale. It was an increasingly rare reminder of the implacable immensity of the planet, somewhat at odds with the newfound confidence of those who set out around the world. That was not the only reason that Melville's demanding novel sold badly, but it surely did not help.[82]

Chapter 6

Fast—faster

"*He was the* sort of Englishman who gets his servant to do the sights for him," and why not? "He was not travelling," after all, "he was tracing a circle. He was matter in orbit around the globe, following the laws of physics." The nineteenth-century Englishman as a mindless thing, shot blind into orbit, oblivious to novelty, scenery, color, pleasure, could only be a French conception, a poke at their neighbors and rivals. The fictional British gentleman was Phileas Fogg, and his French creator was Jules Verne, who, by stressing the wonders of velocity, redramatized the story of round-the-world travel.

For the first time, real circumnavigators and fictional ones were alike in their confidence that mere Earthlings could circumnavigate the planet not only safely but with unprecedented speed. Verne cited the recent historical developments that made it possible to go around the world in eighty days. But he did more than stick together a series of news items. He cleverly merged several literary genres, including the circumnavigator's just-the-facts narrative, to create the first novel that (unlike Daniel Defoe's earlier attempt) took readers around the world in breathless, blissful suspense. Even more compellingly, he made readers consider the costs, human and material, of constructing a globe capable of sustaining such fast travel.[1]

What kind of person would, by 1872, go around the world? In that year, Verne serialized *Le Tour du monde en quatre-vingts jours* in the Paris newspaper *Le Temps* and Fogg was born.

He is boring and mysterious in equal measure. His surname suggests that he is fogged into London, where he lives on Savile Row. He goes each day to his club to take his meals, play whist, and read the newspapers. But his club is the famous and important Reform, and he has gained his membership through a recommendation from the Baring Brothers, at whose bank he maintains a large deposit whose source is never explained. He is a parody of the persnickety bachelor, whose preferences are fixed with such "mathematical preciseness" that he sacks the valet who has offered him shaving water heated to 84 degrees Fahrenheit instead of the required 86. Preoccupied with precision, Fogg obsessively tracks the passage of time, so much so that he "gave the impression of something perfectly calibrated and finely balanced, like a chronometer." That is the first clue of a restless, ticking energy within the quiet man and his routine circuit around London. Indeed, he has been quite the traveler already, "in his head, at any rate."[2]

French readers would have pegged Fogg as yet another smug Briton who, ever conscious of his nation's global empire, easily imagines a much bigger world around himself, a world at his command. "Pray speak English! it's well known / Throughout my empire—now all but your own," says an extremely accommodating "Asia" in another fictional Englishman's all-in-his-head tour of the world. This was a short play with a long title, *Mr. Buckstone's Voyage round the Globe (IN LEICESTER SQUARE). A Cosmographical, Visionary Extravaganza, and Dramatic Review, In One Act and Four Quarters*, which opened in London in 1854. Mr. Buckstone was a perennial of the London stage, a Victorian Everyman who, in this particular entertainment, finds himself within Wyld's Great Globe, where he dreams of voyaging around a real globe. He is kitted out with a carpetbag, the carry-on luggage of the day, as well as a wealth of the paper instruments that assisted travelers. Cybele, world goddess, prophesies that:

> *Circular notes you'll take to my four daughters,*
> *Which will be honoured in their separate quarters.*
> *And as on Europe you may want to draw*
> *At once, I'll introduce you.*[3]

Clearly, imperialism was still a pronounced feature of Western proprietary confidence. The play manages to celebrate Britain's empire as well as cordial relations with France, and solidarity among imperial powers more generally. Over the century that would end with World War II, empires became so entrenched that it was difficult for anyone in the West to imagine a world arranged differently. Empires meant power, a power extended outward from Europe and North America, and empires meant wealth, riches extracted from abroad and flowing back to the Atlantic center. Empires also had significant cultural dimensions, among them the opportunity for Westerners to tour parts of the world they regarded as exotic. With some exceptions, notably Japan, Europeans had access to most of the world's coastlines, and in many cases controlled access to them. As James Holman's travels into parts of Africa and Ida Pfeiffer's caravans across "Arabia" had shown, the European powers were also extending their reach into continents previously inaccessible to most Europeans. As well, Holman's extended visit to China indicated another new stop on a world tour. Expansion of settlements in Canada, Australia, New Zealand, and the United States made these places into other logical destinations.

Mr. Buckstone's imaginary tour presents the world as successive scenes of local color. He is the audience through which the audience watching him delighted in whatever delights him, including a hardening set of social prejudices and racial stereotypes. A turbaned and submissive "Asia" bows before Buckstone, for example, as does "Africa," speaking the plantation dialect associated with slaves in the U.S. South and with a chorus of "pickaninnies" behind her.

Whatever else it represents, the local color shows that some of the focus of much round-the-world travel was now on the social world, on human relations and cultural variation. This was a consequence of the new ease of travel—getting around the planet was not itself dangerous, so the adventure was in the sightseeing. The costs of around-the-world travel were themselves social: access to money and to imperial circuits.

The confident folks who went around the world under these privileged conditions were able to ignore the physicality of the globe. The planet vanished. Of course it was still there. But it was no longer an obstacle, not even a very high hurdle. Instead, it was a means to whatever end a traveler might propose, a bundle of physical resources that could be used for anything—even going around the entire world. As a Russian piece of 1857 claimed, the globe could be measured by paper: an Englishman had calculated that, if all the pages were torn from the major publications and manuscripts in the world's major libraries, the planet could "be totally, if not clothed, then at least encircled."[4]

The very titles of travel accounts, from the mid-nineteenth century onward, show the easy, meandering nature of an around-the-world trip. William Hoffman did *Jottings* around the world. Thomas E. Beaumont wrote *Pencillings by the Way*. Depending on how he was translated, Alexander Graf von Hübner completed a "promenade" or "ramble" around the world. Raymond Cazallis Davis intended his *Reminiscences of a Voyage around the World* (1869) for boys, perhaps the first time that the genre was adapted for young readers, who might enjoy somewhat sanitized tales of the sea and of foreign lands, as Davis hoped they would. (He wrote for money and needed the sales.)[5]

Older gentlemen who remembered making the Grand Tour of Europe as callow youths would have recognized that a world tour was being patterned on that tradition. The European Grand Tour had followed a conventional itinerary with famous sights—the Uffizi Gallery in Florence, the Alps in winter, the Alhambra's gardens in spring—as well as experiences: Carnival in Venice, theater in London, restaurants in Paris. Now, there were global sights—the Pyramids of Giza, the Mississippi, the Taj Mahal by moonlight—and global experiences: Carnival in Rio, opera in Shanghai, wineries in South Africa. The available field of souvenirs expanded too, so that travelers now planned to buy cashmere shawls in India as well as coral trinkets from Italy. People who went around the world emphasized the time they spent *not* traveling or the time they spent at a place, not just getting to it. In a series of letters to the *New York Evening Express*, for example, the American politician James Brooks described his seven months of going "up, and down, and around the world."[6]

Other around-the-world travelers were bound on social reform. The Catholics had created the global missionary circuit and others followed suit, especially as imperialism opened access to previously protected populations. The American spiritualist James Martin Peebles traveled around the world and into what he called "heathen" countries in order to bear witness to Christianity. Toward the end of the century, members of the Women's Christian Temperance Union sang "There are Bands of Ribbon White around the World, around the World" as they dispatched emissaries on around-the-world tours to lecture on the evils of liquor. Perhaps the best known circular image of global uplift was the claim that the Sun never set on the British Empire, as if the British, like the Sun, were at any given minute enlightening someone somewhere.[7]

These are significant ideas or developments, not least for what they reveal about nineteenth-century social and political preoccupations. But their non-societal implications matter too, because they show Westerners' diminishing sense of their puny size compared to the planet. For the pencil-jotting traveler as well as the lecture circuit reformer, the globe was an easy means to a social end, no longer forbidding, or even important for itself. Only speed would make the planet itself obvious again.

"The Earth has got smaller because you can now travel around it ten times as quickly as a hundred years ago." During his usual game of whist at the Reform Club, Fogg and his companions discuss the recent robbery of £55,000 from the Bank of England. They debate the chances of the suspect being caught. Fogg observes that the odds are against the thief because the world is not as big as it used to be. The speed at which one man might encompass the globe is proof of its apparent shrinking—even with a head start, a criminal has no advantage over the authorities, who share the easy and swift means of transport now available to any paying customer.[8]

Speed may have reminded people of the size of the planet, but mainly to impart a new confidence that they could vanquish it. The old circumnavigators had always been aware of time, because it stretched so long before them as they made their slow passage across each ocean. The prolonged nature of their voyages had been their trademark. And yet round-the-world captains had also raced against each other, starting with the seventeenth-century Dutchmen who began to set speed records, and continuing with the British and French circumnavigators of the late eighteenth century who balanced quick progress toward the next scurvy-reducing bit of land against a moderate work schedule for their men.

It was captains of nineteenth-century clipper ships who made speed into a practical goal and an expectation. Several international developments in the 1840s had prompted the quick delivery of compact commodities and well-paying passengers: gold was discovered in California and Australia, Britain allowed U.S. participation in the tea trade, and the Crimean War of 1853–56 forced shippers to work around the conflict. The clipper, a narrow and shallow craft with a large spread of sails, could carry premium goods and travelers rapidly because it was built for speed, not capacity. The clipper record for the run around the Horn, from the American Northeast to San Francisco, was eighty-nine days, for example, in contrast to the two hundred days needed for ordinary ships. One of the two clippers that held that record was the *Flying Cloud*, so named because she seemed to kiss the water as fast as the air itself. From the opposite direction, clippers could go around the Cape of Good Hope from London to Sydney in a hundred days—the *Cutty Sark* set a record of seventy-two days. Another hundred days could take a clipper back to London around Cape Horn, for an around-the-world journey that took just under seven months, not counting time in port. On land or at sea, nothing else could move at such sustained speed over such distances.[9]

The clipper route, with its beautiful lean craft, was the product—the triumph—of accumulated knowledge about global ocean currents and wind patterns. After more than three centuries of round-the-world travel, new ship design had finally combined with fuller understanding of the world's oceans to make a circumnavigation fast and relatively safe. A clipper could be back home before scurvy had a chance to stir. The logs of clipper ships charted the fastest route around the world that could be done using sails and tracked adherence to "time charters," contracts that specified when a ship had to be in port, which earlier sailors would never have agreed to. Those who had or built clipper ships at the exact right moment made fortunes. And yet, by the mid-1850s, the clipper era was just about over, after a run of roughly fifteen years, even more fleeting within the history of human transport than the clippers had been upon the water. The new ships had established an expectation of promptness, but it would be another kind of transport that would make good on that promise, as Fogg and his card-playing companions note.[10]

"This is how the *Morning Chronicle* worked it out," says one of the club men who supports Fogg's idea of a shrinking world:

From London to Suez . . . by railway and steamship	7 days
From Suez to Bombay, by steamship	13 days
From Bombay to Calcutta, by railway	3 days
From Calcutta to Hong Kong, by steamship	13 days
From Hong Kong to Yokohama (Japan), by steamship	6 days
From Yokohama to San Francisco, by steamship	22 days
From San Francisco to New York, by railroad	7 days
From New York to London, by steamship and railway	9 days
Total	80 days.

Each segment of this route depends on steam locomotion (see Map 3). Steampower made travel not only quick but predictably so. Wind had been impossible to control—it was the fastest form of locomotion on Earth when it blew the right way, but useless or even dangerous if it died or changed direction. In contrast, humans could *command* a steam engine and therefore control the direction and rate of travel. Steam had been applied to stationary engines in the second half of the eighteenth century, to boats by the 1780s, and to cars on rail tracks by the early 1800s. Wood was often used to fire a steam boiler, but coal became the fuel of choice, mostly because it was more compact and transportable than wood, and because it seemed inexhaustible, compared to the world's dwindling forests. Better coal-fired engines were developed by the second two decades of the nineteenth century. These included engines for oceangoing ships, though many "steamers" also carried sail, so coal could be saved whenever the wind was right. Certainly, that was true of the New York–built paddle wheeler *Savannah* (320 tons), the first steamship to cross an ocean. During her run across the Atlantic in 1819, she used her engines for only 80 of the 663 hours she spent at sea.[11]

When Ida Pfeiffer made her way around the world for a second time

(1851–54), the conventional ships she had used on her first circuit, not a decade before, had mostly been replaced by steamers. Her second tour was much like the first: daring adventures ashore (sometimes in men's trousers) balanced by European luxuries in imperial places. (Alas, her compact baggage did not include an evening dress, so she could not attend the theater in Batavia.) She noted steam travel often, mostly to keep tally of those captains and companies that offered her free passage—by the time of her second round-the-world journey, she was the famous female traveler. The Hotel Neederland in Batavia refused payment from "so great a traveller," as did the owners of the two Batavian steamship companies. She paid a mere £3 for board on her passage from Cape Town to Singapore, and not a penny for the steamer between San Francisco and Panama. Nor did she ever pay passage within the United States. By that point, she was used to being the celebrity. She was miffed when a Canadian newspaper editor had never heard of her and irritated when British ships' captains uniformly insisted that she pay for their services.[12]

Pfeiffer's steamers were historic. That meant they were not very powerful. The ship that chugged her from Java to Sumatra had a 120-horsepower engine, comparable to a motorcycle today (or a hybrid Prius: 98 hp). The steamer that connected San Francisco to Panama was larger, 800 hp. That made it twice as powerful as a high-performance sportscar, although a sportscar generally accommodates only two people, while the ship was expected to carry up to a thousand. And yet these puffing mechanisms truly were engines of change, both in terms of the speed they offered and the resources needed to deliver it. As more powerful engines were developed, they could propel heavier ships: bigger (with greater capacity for passengers and cargo) and constructed with more durable materials, meaning metal. The thin copper cladding on wooden ships gave way to iron reinforcement, then iron construction of the entire exterior, until the tall wooden sailing ship was replaced, for everyday and practical travel, by dismasted metal vessels with engines. They were faster and tougher, if not quite so enchanting to look at.[13]

The speed of steam travel also changed the nature of sightseeing. One traveler said that his tour "would necessarily be a panoramic glance at a narrow line around the world," rather than the up-and-down diversions that constituted a typical world tour. Even those who did the world to see the sights were seduced by speed. Ludovic de Beauvoir, a French count who went around the world between 1865 and 1867, called the clipper that took him and a companion to Australia "our world," a floating home on the sea, very conventional. But by the end of his journey, he exulted in steam-powered acceleration: 3,234 miles in fourteen days from San Francisco to Panama, for instance.[14]

Altogether, steam-powered shipping, which had totaled only 32,000 tons in 1831, would rise to over 3 million tons by the mid-1870s, just after Verne had created Fogg. Most steamer traffic plunged through single oceans, or at most connected a pair. For example, beginning in 1848, the Pacific Mail Steamship Company (PMSC) connected New York to the California gold fields. After 1867, the company provided the first scheduled steamer service back and forth across the Pacific Ocean. (An early passenger marveled at the PMSC's luxury: each ship was a "floating palace," with meals fit for first-class hotels.) A person or commodity might go around the world in a series of such conveyances, though the ships they used typically did not.[15]

Coal mining and steam-driven circumnavigations rose together: 10 million tons of coal in 1800; 76 million by 1850, 760 million by 1900. In 1847, HMS *Driver* managed to steam her way around the whole world, and then the HMS *Argo* did so as part of an intended circumnavigation in 1853. When the Royal Navy sent a vessel around the world in 1871 as part of a training exercise, she required 3,461 tons of coal. It became as important to feed ships coal as to give sailors food and water. Because of the Hundred Years' Peace, access to land for re-coaling ships rarely required the violence of the piratical age of sail. Rather, the Sun never set on a chain of "coaling stations" that dotted the world, each depot a grimy reminder of the powers of imperialism and paper internationalism, not to mention a momentous planetary shift toward the exploitation of mineral resources.[16]

Coal seems darkly primitive, yet steampower seems brilliantly modern. It is an intriguing contradiction. *"Carboniferous capitalism,"* the American sociologist Lewis Mumford called it, as if the distant millennia in which seams of coal were laid into the Earth somehow merged with a coal-based industrial society—a grazing triceratops is startled by the whistle of a steam locomotive. *"Steampunk"* is a more recent name for the temporal clash: into the future, yet back in time. Although coal is the precursor of petroleum, its power seemed, in the nineteenth century, even more prodigious, more like splitting the atom.[17]

Even at the time, carboniferous capitalism was recognized as having singular drawbacks as well as virtues. Coal lighted homes, but blackened cities. Coal and iron sped more people over thrillingly greater distances, even as the means for doing so (mines, steamship companies, railroads) were concentrated into fewer hands. Steampower lifted burdens from some people and animals, yet consigned others to coal pits and factories. The humane impulse that had begun to protect some working people (including scorbutic sailors) was overlaid with a new preference: that machines do any *visible* heavy labor. On oceangoing steamships, for instance, the laboring bodies were no longer handling sails on deck but shoveling coal below, out of sight and therefore, to commercial travelers, out of mind.

Were the drawbacks temporary? Might steampower and machinery eventually liberate all of humanity? The early French socialist Henri Saint-Simon believed they would. "Steam and electricity for all tasks," the Saint-Simonians declared; "in place of the exploitation of man by man, the exploitation of the globe by mankind." Karl Marx had less faith in technology unless it, as well as all other means of production, were under workers' control.[18]

Verne was no Marxist; he followed Saint-Simon. He was confident that improved exploitation of the material world would improve lives, with new-fangled wonders spreading through the industrializing West and then out to the rest of the world.

Fogg's house on Savile Row is a steampunk paradise of newfanglement. "Coal gas supplied, in fact, all that was needed for heating and lighting," which was a luxury beyond the means of most people on Earth in 1872. It is even more unusual that Fogg's house is wired for electricity—"electric bells and speaking tubes made it possible to communicate with the suites of rooms on the ground and first floors." Fogg even owns electrical clocks, rarer still in an age of key-wound clocks, rather like having an atomic clock in one's home today. The net effect "was like being inside the shell of a snail, but a snail that had gas lighting and heating!" If a home is a world within a world, the snail hauls its home wherever it goes in the world. Again, Verne marks a tension between Fogg's desire for familiarity yet longing to escape; he underlines that tension with Passepartout's error in not extinguishing his room's gas lamp, which must burn for the length of time it takes to get around the planet. The reference to a snailshell may also have gratified readers of Verne's *Twenty Thousand Leagues under the Sea* (1869–70) with a reminder of that novel's *Nautilus*, the mechanized and luxurious submarine that is a moveable microcosm in its course (nearly) around the world. Verne may also have planted a hint, here, of a tortoiselike persistence that might win Fogg the race.[19]

"I bet £20,000 against anyone that I will go around the world in eighty days or less, in other words 1,920 hours or 115,200 minutes. Do you accept?" The offer was amazing by itself. But the men of the Reform Club who accept Fogg's wager are even more astonished when he calmly proposes setting out that very night, on the 8:45 train to Dover. They are then flabbergasted that he declines to rush off to prepare for the journey. "I'm always ready," he says. "Diamonds are trumps. Your turn, Mr Stuart." He finishes out the game, then goes home to pack. That takes minutes. Fogg decides to carry a raincoat, a travel rug, and an overnight bag with two spare shirts and three pairs of socks, plus a big stack of bank notes. The money is the main thing, being British pounds,"those splendid banknotes that are legal tender all over the world." Should Fogg smudge a shirt or lose a sock or need anything else, he can buy it along the way rather than lug it with him.[20]

Fogg also gathers up some favorite reading: Bradshaw's *Continental Railway, Steam Transit, and General Guide*, "which was to give him all the information needed for his journey." The fixed timetable, in which a train or ship promised to depart *and* arrive at specified times, depended on steampower—wind, horses, oxen, and the like, had never allowed an arrival time to be more than a rash promise, which had implications for departures as well. George Bradshaw had produced his first compilation of English railway timetables in 1839, and as his publications expanded to include the European Continent and then other parts of the world, "Bradshaw" became shorthand for any rail timetable. The guides trace the proliferation of railways, first within parts of Europe, then connecting all of Europe, then crossing lands elsewhere. Verne pointedly says that Bradshaw is "all the information" Fogg needs because his hero requires no guide to any of the places he will go, just how to get from one to the next.[21]

A normal traveler might have packed a "Murray" or a "Baedeker," the two big names in travel literature. When Mr. Buckstone enters the Great Globe in order to go around the world, the stage manager makes sure he is prepared:

> *All Murray's Handbooks for the Governor.*
> *Germany, North and South, France, Holland, Spain,*
> *Switzerland, up the Rhine, and back again,*
> *Italy, Russia, Egypt, Turkey, Greece,*
> *Some of 'em twelve or fourteen bob a-piece.*

There had been guides to the Grand Tour, and scattered guidebooks for parts of North America and western Europe appeared as early as the 1820s, but the first lasting and trusted guides came out a decade later, with John Murray (of London) publishing a guide to western Europe, and Karl Baedeker (of Coblenz) doing the same. Parts of these books would look familiar today: descriptions of the must-see sights, advice on clothing and how not to offend the locals, recommendations of hotels and eating establishments. Other warnings were unusual—where a small pistol or inflatable rubber bath might come in handy. Murray and Baedeker sold well. Their guides (and those of imitators) were updated, revised, translated, imitated, and stretched to cover places beyond western Europe, beginning with the Near East. By the end of the nineteenth century, there were guides to Asia, Canada, the United States, and New Zealand.[22]

Bradshaws and Baedekers made travel experiences that had once been the preserve of men on naval vessels and merchantmen seem attainable to the upper and middle classes. No longer was an around-the-world voyage the result of tremendous mercantile risk or enormous state effort. Verne concocts a little protest against the new, consumer-driven accessibility. After Fogg has departed, a long article in the *Proceedings of the Royal Geographical Society* "proved conclusively the madness of the undertaking." The society and its proceedings really existed. They were not quite the nest of harrumphing killjoys that Verne inserted into the story, but their fictional skepticism represents valid astonishment over a major historical shift. The long expeditionary tradition of circumnavigations that were organized by nations and learned societies seemed to be ending, and the promenading, pencil-jotting hordes have blundered into geodrama, armed with their timetables and guidebooks, priced at twelve or fourteen bob a-piece. Fogg is the hero of that historical readjustment.[23]

No travel book yet covered the world, and Fogg wouldn't have read one even if it did—his servant, after all, will see the sights for him. "Jean Passepartout, a nickname that has stuck and that I earned by my natural ability to get myself out of tricky situations." Fogg's promising new valet, who never questions his master's finicky ways, has just made himself at home in the snailshell on Savile Row only to learn that he is to go around the world. As his name suggests, the resourceful Passepartout ("go anywhere," as with a skeleton key) is much better than a paper passport. Because more people could afford international passports, they were no longer the glamorous accessory of diplomats and explorers. A British passport had cost four shillings and sixpence in the first half of the nineteenth century, but the cost was sinking, and would be only two shillings a century later. And yet, as the acrimony of the Napoleonic Wars faded, and the Hundred Years' Peace took deeper root, passports were increasingly regarded as unnecessary. Within western Europe, much travel could be done without them.[24]

"All that passports do is inconvenience law-abiding citizens and enable crooks to get away"—and this from a policeman in Verne's novel! The Reform Club worthies agree entirely. When Fogg promises that he will, on his return, surrender his much-stamped passport for their inspection, to prove he has actually gone around the world, they are puzzled.

> "Oh, Mr Fogg," replied Gauthier Ralph politely, "that's not necessary. We will rely on your word as a gentleman."
> "I prefer it this way," said Mr Fogg.[25]

Fogg has willed a self-contained world around him; he orbits in an invisible capsule. His Bradshaw tells him the succession of steam-driven microcosms he will inhabit, and they have no features that interest or detain him. He lives in his head and out of his carpetbag, confident that the planet has been configured for speed and convenience.[26]

It only works because of Passepartout, who acts out Verne's confession that human labor was, sometimes, still necessary. Technology may have been replacing labor in many sectors of the steam-driven modern economy, but only after 1900 would domestic service go into decline. Until then, it was the single largest form of employment in England. Domestic service survived because it implied privilege. European families of means wanted to be served by people rather than machines. The former practice had history behind it, intimations of feudal glory, while the latter threatened to track the dirt of the factory into the manorhouse. Machines were meant for heavy work, humans for purchased devotion.[27]

A French valet was, for an English bachelor, a sign of unusual wealth and status. Male servants were more expensive than female (and taxed accordingly), and non-British servants were rare in England. Since at least the 1840s, the English had called a valet a "gentleman's gentleman," a term which noted both the luxury and the snobbery of having a personal attendant. A gentleman and his gentleman formed an intimate world, closed to others. (For that reason, Napoleon Bonaparte observed, "no man is a hero to his valet," which P. G. Wodehouse would make into the running joke of his Wooster and Jeeves novels.) The closed world shielded a man of means from the annoyances of the real world, even when he had to go out into it. It was the valet, usually, who organized a gentleman's travel by arranging for tickets, transfers, meals, baggage, passport, visas, information—everything.[28]

And so, historically, Fogg's Passepartout exists midway between Magellan's Enrique de Malacca and Wooster's Jeeves. His circumnavigating master has no *picaro*-like characteristics. Instead, Passepartout is endowed with a Dampier-like resourcefulness. Passepartout's loyal duties to a master are of the past, while his decision to *choose* that loyalty is modern. He is not like the enslaved captives of so many circumnavigations in the age of sail, nor is he like the wage slave who mines the coal

for the steamship or shovels it into the boiler. If Fogg orbits the world
in his own little bubble, Passepartout is there to make sure the bubble
does not burst.

> The Commissioner of the Metropolitan Police had received the following
> telegraph message:
> ... Trailing bank robber, Phileas Fogg. Send arrest warrant without
> delay Bombay (British India).

Verne adds to the novel's suspense by weaving a detective story into
the geodrama. Fogg becomes the prime suspect in the Bank of England robbery. The speeding traveler acquires a nemesis, Detective
Fix, who ends up doing his own around-the-world voyage, always hot
on Fogg's trail. (Fix is convinced that his quarry "passes himself off as
an eccentric who's trying to go around the world in eighty days.") As
he shadows Fogg and Passepartout, Fix remains tethered to London
through telegraph wires. So does Fogg's reputation. The telegraph is
precisely the kind of around-the-world instrument that he himself
predicted would make it impossible for a criminal to escape justice.
As well, the telegraph is the means by which London newspapers receive information about Fogg's progress. The betting pool around the
Reform Club wager constantly expands and its participants constantly
rearrange their bets, as if the man in question were a commodity traded
on the London Stock Exchange, itself a global marketplace because of
the telegraph.[29]

The telegraph system would be the first actual lasso looped around
the planet and cinched into place. Such was the power of geodrama
over the human imagination that the telegraph was described as making a girdle around the Earth long before it would actually complete a
global circuit. By the late eighteenth century, experimenters had run
electric current through long wires, sometimes over miles, but the
charge had always faded too soon to be useful. In 1820, an experimenter noticed that an electric current moved the needle of a compass,
and subsequent manipulation of electricity's magnetic field would
make long-distance telegraphy possible. One telegraphic pioneer, the
American Samuel F. B. Morse, brushed away doubts as to the value

of his experiments in the 1830s: *"If I can succeed in working a magnet ten miles, I can go round the globe."*[30]

He was right. Over the next two decades, the globe seemed to have been visited by two species of demented spiders, strange arachnids that could extrude insulated telegraph wire at will, and with preternatural speed, though in two different patterns. One tribe wove dense local webs, over and over in the same place, producing telegraphic networks that ran around and through the major cities of the West as well as many imperial entrepots. The other variety raced crazily over long distances, leaving behind miles of single running bands that joined with some of the local snarls. The spiders were diligent. In the United States, where the telegraph spread most quickly, there were 40 miles of cable in 1846, 2,000 miles in 1848, and over 12,000 in 1850.[31]

The spiders were in truth an assortment of private companies and national postal systems, and yet the standard prediction for these different and often competing telegraph lines was that they were the beginnings of a worldwide system. It was an easy case to make because, even more obviously than the speedy movement of goods or people, any similar transmission of information required international cooperation: cargo or passengers could be taken off one ship and put on another at some national boundary, but a telegram, ideally, was sent with minimal breaks over the hard-wired web.

The laying of the Atlantic cable was the best evidence of the innate internationalism of telegraphy. Telegraph lines were speeding across continents, especially North America, but also in Great Britain, Continental Europe, and India, and they were being laid beneath the seas, including the English Channel. Wiring bigger oceans would take greater expenditure, even the smallish Atlantic Ocean, the first to be bridged by an undersea cable. It took the cooperative efforts of several telegraph companies and two nations, Great Britain and the United States, to fill the gap between Newfoundland and western Ireland. When the first cable was completed and tested in 1858, it sped national greetings between Queen Victoria and President James Buchanan of the United States. After that glorious moment, the cable failed. Another massive effort laid a successor, completed in 1866. Two hemispheres torn apart by the insurgent nationalism of 1776 were reunited in the peaceful internationalism of the nineteenth century. That was somewhat wishful thinking, given that the cable was laid during years that included the Indian Mutiny, the American Civil War, and the Crimean War, and were leading into the Franco-Prussian War, big exceptions to the rule of the Hundred Years' Peace.[32]

And yet the idea persisted that the entire globe was being physically girdled, piece by piece. In 1858, the American photographer Mathew Brady, best known for his Civil War photographs, would take two portraits of Cyrus Field. The Atlantic cable entrepreneur stands next to a world globe, over which (in the pose of an electric Drake) he holds a section of cable, anticipating the global circuit that would eventually exist. After all, an Italian periodical dizzily claimed, electricity could travel eight times around the world in a second! An 1858 *Story of the Telegraph* more soberly predicted that, with the completion of the Atlantic cable, "the whole earth will be belted with the electric current, palpitating with human thoughts and emotions." It was not yet literally true; the Atlantic cable was just about to fail, and the Pacific was not yet wired for communication. But the history pointed out that existing telegraph lines extended a total of 100,000 miles, "more than sufficient to put a quadruple girdle round the globe." At a banquet given in Cyrus Field's honor in 1866, the sixth of the many formal toasts explained that the Western Union Telegraph and a Russian extension were the "American and Asiatic links of the chain encircling the globe," even though a physically continuous chain did not yet exist. An attempt to connect Moscow to San Francisco through a Russian-American telegraph was attempted in 1865 but abandoned in 1867.[33]

Zinging on poles overhead, humming to whales beneath the seas, the telegraph circuit was making what had been ephemeral and invisible—the circumnavigator's track—material and visible. To acknowledge the older history of such dreams, a banner on Broadway, celebrating the 1858 laying of the Atlantic cable, quoted Puck: "I'll put a girdle round about the earth in forty minutes." (Commentators, including those in France and Italy, noted that electricity was actually faster than Puck.) International cooperation had made it possible. But now, that harmony among nations was thought to be the *result* of around-the-world traffic, as well as its precondition. "Speed, speed the cable; let it run / A loving girdle round the earth," urged an 1858 poem. The telegraph would nurture "that cosmopolite soul" within each human and would eventually achieve a total "unity of the race." This was the first of many predictions that some around-the-world technology or other would somehow achieve world peace. But that is getting ahead of the story. Fogg, Passepartout, and Fix have only just arrived at another monument to nineteenth-century engineering, the Suez Canal.[34]

Fogg's Bradshaw leads him straight across Europe and to the great gateway to the East, Suez. There Fix awaits him, conveniently positioned (by Verne) to admire how British ships slipped through the canal "every day, thereby reducing by half the journey from England to India compared to the old route via the Cape of Good Hope." Such is the regularity of the Suez route that the ship Fogg intends to catch, the steamer *Mongolia,* has always earned the £25 bonus that the British government gives to a ship that arrives at least twenty-four hours ahead of schedule.[35]

The Suez Canal was completed and opened in 1869, three years after the successful Atlantic cable had been laid, and the same year that the Aden-Bombay cable had been completed. Even more obviously than those telegraph wires, the canal was part of a *built environment*, where the work of the human hand nearly obscures the hand of nature. Cities and roads had been previous manifestations of the built environment, but the urban expansion and great engineering works of the nineteenth century, including the cable and the canal, were even more intensive and dramatic.

Blasting through an isthmus or two was an old dream, one that long preceded steampower. When C. P. Claret Fleurieu had described Étienne Marchand's circumnavigation in the 1790s, his vision of a "maritime coalition" suggested canals as a solution to the globe's inconvenient configuration. Absent a Northwest Passage, to get around the world in its natural state required sailing a minimum of 14,328 leagues, although a Great Circle (or Equator's worth) of the globe was only 7,200. A canal through each of the offending isthmuses would halve the time it took to get around the world, reducing sailing time to seven or eight months. It was a bold vision, and very much part of an eighteenth-century ideal of improvement, when canal-building was all the rage in Europe and North America. When he invaded Egypt in 1798, Napoleon Bonaparte had noted evidence of ancient canals and suggested modern counterparts. Neither France nor its enemy Great Britain managed to maintain direct control of Egypt, but both continued to contemplate ways to strengthen their interests in the region, and it was long promised that a canal would reduce a voyage to India by about 1,000 miles. That seemed more satisfying than the other solution, a railroad over the isthmus.[36]

It was French diplomat and engineer Ferdinand de Lesseps who parted the sands between the two seas. From a noted family (his uncle, Barthélemy de Lesseps, was the lone survivor of La Pérouse's attempted circumnavigation), de Lesseps was a Saint-Simonian (and a friend of Verne) who believed that human effort and technology could only improve the world. He used diplomatic ties to the new Egyptian viceroy, Said Pasha, to support a French canal-digging project. In 1856, the International Commission for the Piercing of the Isthmus of Suez was created, though the British resented French control of an alternate passage to India, and declined to support the venture. A subsequent private company sold stock; most of the money came from France and Egypt. At first, the project depended on *corvée* labor, workers coerced from the countryside, until the British protested that this was a form of slavery. Waged workers were substituted, but the immense human effort, and estimated fatalities in the thousands, were noted anachronisms in the age of steam. After twelve years and the excavation of 99,400,000 cubic yards of earth (more than 227 times the amount of concrete in the Hoover Dam), the new channel was fully flooded. On November 18, 1869, a French imperial yacht opened the canal, with a flotilla of forty-six ships from several nations.[37]

There was a little hoopla over the canal's helpfulness in girdling the Earth, but not much. Too many tensions were apparent, between the European and Arab worlds, and among the nations on both sides, for claims of international harmony to convince anyone. Many artists declined to immortalize the monument, though Verdi would eventually compose his Egyptian-themed opera, *Aida*, which premiered at the

Cairo opera house in 1871. A U.S. railroad spokesman rather hopefully published a guide, *Our New Way Round the World*, explaining how to do that journey using American railways and the Suez Canal. Gradually, and especially after the conclusion of the Franco-Prussian War, traffic through the canal began to take off and it became a more convincingly international means of transport.[38]

A clear result of paper internationalism, the built environment of the nineteenth century also augmented its effects. Just as the ability to make land at most of the world's coastlines had reduced the difficulty of getting around the globe, so did the physical reconfiguration of the planet. Rather than depend on a wooden world (or a series of them) to get around the world, or on time-consuming overland connections (such as Ida Pfeiffer's caravans), the world was reconstructed to let travelers pass easily from one steam-powered means of transport to the next, as with the railway connection to steamship service at Suez.

Verne's story breaks from the progress of the *Mongolia*, which has just exited the Red Sea and made Aden, to emphasize the significance of coaling stations. Britain's Peninsular & Oriental Steam Navigation Company (the P&O) paid £800,000 to lay in fuel along the canal route, "far away from large industrial centres," at a cost of "over £3 per ton." Coal-fired engines kept traffic streaming safely and on schedule. That's all Fogg wants. He pays no attention to the scenery beyond the *Mongolia*, or the music and dancing aboard, let alone the lady passengers. He takes his four meals a day (the ship's tables groan with bourgeois abundance), and gets up games of whist with a tax inspector, church minister, and brigadier general, all bound for service in British India.[39]

The Egyptian canal is a blessing for Fogg. The built environment, to this point, has ensured his privacy. Verne also scored a patriotic point by making his English character so dependent on a French creation. The Suez Canal is, courtesy of the French, the only recently constructed part of Fogg's route that actually works, as Fogg and Passepartout are about to discover.

"When he awoke, Passepartout saw to his amazement that he was crossing the Indian subcontinent in a train belonging to the Great Peninsular Railway." Of course Fogg and Passepartout have used the railroad before, to get from London to Dover and then across Europe, but Verne considered that uninteresting—there is no story, between London and Suez. Trains within Britain and Europe were old news and, frankly, they covered comparatively little territory. Much more momentous were two new railway extensions that made it possible, beginning in 1869, to cross both the Indian subcontinent and the North American continent entirely on track.[40]

Railroads increased the speed of travel and were another crucial component of the built environment of the nineteenth century. The world had 72,000 kilometers of railway track in the 1840s, but 360,000 kilometers in the 1880s. Like telegraph lines, the tracks multiplied in dense nets around the crowded parts of Europe and North America, and then hurtled in lonely lines across continents, as in Russia, the United States, and British India. Telegraph wires and railroads tended to spread together, except across oceans. Cables were amphibious; railway tracks were not.[41]

And yet railway enthusiasts made the same claim for railroads as others did for the telegraph: little by little, they would circle the world. Henry David Thoreau thought it a nightmare. In *Walden* (1854), he pointed out that "to make a railroad round the world available to all mankind is equivalent to grading the whole surface of the planet." Verne had a sunnier view. In his novel *Claudius Bombarnac* (1892), he predicted the triumph of those "iron ribbons which will eventually encircle our globe as if it were a cask of cider or a bale of cotton." This could not literally be true. And yet it was certainly the case that steam travel had an accumulating power over the globe. Some railway companies built hotels and operated steamships, which allowed them to take passengers over longer distances using amphibious means. Again, this represented convenience for consumers, many of whom now wanted to spend their time seeing the sights, not hanging around yet another ticket office.[42]

Mr. Fogg appreciates the convenience, if for different reasons. The steamer from Suez arrives two days early at Bombay. Fogg goes straight to the rail station, pauses for dinner, and boards the train without a glance at any of the sights that might have been the point of another person's journey. To emphasize that indifference yet again, Verne describes the book's fourteenth chapter as the one "in which Phileas Fogg travels the whole length of the wonderful valley of the Ganges without thinking it worth a look." He's still ahead of schedule, all set to win the bet. And yet it is in India that the story begins to go off the rails and the greater world to leak into the microcosm Fogg has willed into existence around him.[43]

The journey would have been as dull as a weekend at the Reform Club had it happened just as Fogg planned. Though he had declared at the club that "There's no such thing as the unexpected," it is precisely all the unexpected events that are the journey's true tests. Fogg learns this when the train speeding him across India stops short of Allahabad, the destination announced in all the London newspapers, on which all of his plans have depended. "What can I say, sir? The newspapers are wrong," says a railway guard. (Actually, they were right—the plot twist is invented.) While the age of steam takes a nap, Fogg learns of a nearby elephant: "Let's go and see the elephant." He buys it for £1,000 and the Europeans go swaying away on the grand beast, bound for Allahabad, the railway to Calcutta, and then the steamer to Hong Kong. They should just make it. Fogg has two days to spare—the earlier speed has made all the difference.[44]

Speed can also make a man generous, even Fogg, who has seemed to notice nothing around him but does now. Twelve miles from Allahabad, the elephant's *mahout* or driver identifies distant music as part of a Hindu ritual. A young Indian widow is to be burned alive with her late husband, as part of a traditional funeral rite. Verne (and anyone who read a European newspaper) would have known the ongoing controversy over *sati* or *suttee*, which British officials wished to abolish. Fogg and Passepartout are predictably appalled, Fogg so much so that he puts aside his mental timetable. He declares that his small party will rescue the woman. An English companion rejoices that Fogg has "feelings after all!" " 'Sometimes,' replied Phileas Fogg simply. 'When I have the time.' " The woman is rescued and she, Mrs. Aouda, joins the party on the elephant, and later the train to Calcutta.[45]

It is while they are in Asia that Fogg's party has the longest exposure to non-European peoples, and their mixed response is typical. Mrs. Aouda happens to be beautiful, to know English, and to have re-

ceived a European education, but Verne represents most of her countrymen as deplorable. By ranking people according to their support of
the new globe-girdling technologies, he picks out prejudices relevant
to geodrama. The men of the Reform Club have worried about railway
travel in places not quite secured by the Western powers, including
South Asia. Can Fogg rely on train schedules, one man wonders, with
rising hysteria, "if the natives of India or North America take up the
rails? . . . Even if they stop the trains, ransack the wagons and scalp the
travellers?"[46]

Real round-the-world travelers also implied that their global confidence distinguished them from non-Europeans. (Recall Darwin's
encountering people in Latin American who questioned the Earth's
roundness.) During her second tour, Ida Pfeiffer thought she had
scored a point against a Malaysian ruler who gave her a geography test.
He showed her a double-hemisphere map and "was much amazed that
I could immediately show him the various parts of the world" where
she had been and he had not. But those distinctions were fading. Many
non-Europeans, perhaps hundreds, had gone around the world as
crewmen on sailing expeditions. Nations beyond the western core were
joining the age of steam and the built environment; railroads crossed
Russia, steamships filled out the Ottoman navy. It was true that, until
the Russians, no nation outside western Europe or the United States
had organized a circumnavigation, and no nation farther beyond would
join the club of circumnavigators until the 1870s, just about the time
that Fogg was supposed to be on the road.[47]

Verne could not quite admit the new diversity of around-the-world
traffic. In Yokohama, when Passepartout sees many Japanese military
officers in uniform, he "joked to himself, 'Here we go. Another Japanese
delegation off to Europe.' " The delegations were no joke to the Japanese. An around-the-world mission of 1871–73 was intended to prove
that Japan was equal to any in the West: *us too*. The well-publicized tour
was a protest against Western imperial pressures, yet claimed aspects
of Western culture. That included proprietary confidence in human
command over the Earth. The Japanese thought they shared that command, as their around-the-world journey proved—Japan could bow on
the world stage with the best of them.[48]

With a nice circular symmetry, the Japanese mission was the consequence of a U.S. circumnavigation that had taken place from 1852 to 1854. After centuries of successfully regulating (and eventually banning) foreign access to their territory, the Japanese had been pulled into sustained engagement with the West by Commodore Matthew Perry. "Black Ships," the Japanese would call the four dark-painted, smoke-belching steamships that Perry anchored off Yokohama in 1853. The U.S. naval mission invited Japan into a larger world of trade and diplomacy. *"No"* was not an acceptable answer. The Japanese agreed to a diplomatic and commercial treaty with the United States. Although the Black Ships would perform an eastbound circumnavigation, Perry, who was ill, did not. He returned westward from Japan, mostly via British travel services. The fact that Perry was not aboard the flagship during her full circuit but had never relinquished command has put the expedition into a categorical limbo, a circumnavigation or not, depending on whether such a journey requires a consistent commander, or just a consistent ship.[49]

If Japan's ruling shogunate of the 1850s had resented Perry's invitation-by-duress, the restored Meiji emperor would, after 1868, seek out the West more enthusiastically. That cultural and diplomatic outreach was especially apparent with the mission named for Tomomi Iwakura, an ambassador sent on a long journey through the United States and Europe for two stated purposes—to negotiate better diplomatic and commercial treaties, and to gather information on political, economic, and cultural developments that could be adopted in Japan—and an unstated but obvious one, to impress Western nations with the acuity of the Japanese global vision. The Iwakura Mission was abroad for just under two years, 1871 to 1873, and if it failed on the first count, it did rather well on the second and third.[50]

The delegation was designed to attract attention because it was large and traveled on commercial vessels and railways. Iwakura was accompanied by 45 other government officials, plus attendants and students (including five girls and young women, one bound for Vassar), for a total of 107 people. The Pacific Mail Steamship Company's *America* transported them from Yokohama to San Francisco, in the company of several hundred other passengers. Special railroad cars then carried

them across North America. Local newspapers reported each stage of
their tour. A San Francisco paper judged Japan to be, given its recent
arrival on the world stage, "the most progressive nation on the globe,"
as exemplified in its building of railroads and steamships, among other
things. The journey would also leave a record in Japan. Iwakura's secretary, Kume Kunitake, compiled a five-volume account of the journey,
published in Tokyo in 1878. Over the next few years, it went through
four print runs and sold a total of 4,500 volumes, good intranational
advertising of Japan's international presence.[51]

Although most of the account detailed the delegation's political and
cultural work, Kunitake also explained how he and the other members
of the embassy thought of themselves as actors within a geodrama. He
described his account as "an actual record of travel around the world,"
which showed awareness of the tradition of circumnavigation. So did
the ritual, while crossing the Pacific, of observing the daily posting of
new longitudes: everyone clustered around the ship's bulletin board to
note the change, reset his or her watch with a satisfying click, and then
awaited the next day's posting. The account also admits Japanese curiosity about the physical planet. Although Copernicanism had become
the standard worldview among the educated in Japan, few world maps
had been printed there. The delegation relied on British and American
maps, and accepted, without comment, that the longitude of 180 degrees (opposite Greenwich) was the place where they gained a day in
their eastward journey. Kunitake also noted how Western technology
was daily demonstrating its Earth-girdling capacity. While in San Franciso, the delegation visited a telegraph office and sent greetings across
the continent to President Ulysses S. Grant. Kunitake later described
how the British and imperial telegraph cable system "runs roughly halfway round the globe."[52]

Certainly, the delegation's ability to travel on commercial services
made it clear that the Western powers, largely for self-serving reasons,
had nevertheless made the world accessible to travelers from other
independent nations. When the Japanese mission traveled through
the Suez Canal, Kunitake noted the conduit's value to Japan. *"An enormous amount of profit lies hidden halfway between Japan and Europe,"* and
the Japanese were, he thought, best equipped to make the most of it,

because they were the most Westernized Asian nation. But despite his praise of modern travel services, Kunitake worried that the speedy global network—tight travel connections, back-to-back events, and fitful slumber in motion—maximized human discomfort. "We wearied of eating splendid dinners," he confessed; "we wished to enjoy, just once, the simple pleasures of drinking plain water and lying with our heads pillowed on our bent elbows."[53]

From his unexpected adventures in India to the end of the story, Fogg is rarely on schedule and his capsule leaks to the point of flooding. "He continued on his scientifically calculated orbit around the world, without bothering about the asteroids gravitating around him," and yet "a 'disturbing' star" in the form of the charming Mrs. Aouda has affected him. As well as incorporating elements of the detective story, Verne's geodrama blends two novelistic genres. Within the picaresque, the old action-driven novel populated by hard-bitten men who live by their wits, the reader finds a domestic novel, the newer, quieter version, in which women and men sentimentally plot their lives together. (It is as if Daniel Defoe and Jane Austen had somehow managed to coauthor a book.) To show his polite consideration for his female companion, Fogg pauses for some shopping in Calcutta: "It was just about acceptable for an Englishman like himself to travel around the world with only one bag, but it was unthinkable for a woman to undertake such a journey like that."[54]

Fogg is thrust even deeper into the world because of Fix's machinations. In Hong Kong, the detective stupifies Passepartout with port wine and opium. The valet is unable to join his master, let alone warn him of a new sailing time for the steamer to Japan, which Fogg misses. Deprived of his factotum, and for the second time in the journey, the Englishman leaves the age of steam. He hires a 20-ton schooner, a mere sailboat, and this little anachronism gets him to Yokohama and back into the world of Bradshaw's timetables. Passepartout is reunited with Fogg and Aouda, and everyone (including Fix) piles aboard the *General Grant* for San Francisco. The ship belongs to the ever useful Pacific Mail Steamship Company and carries both canvas and a steam engine. Although Fogg's original estimate of the Pacific crossing was twenty-two days, the captain manages to shave a day off the voyage. On the fifty-second day of his journey, Fogg crosses the 180th meridian and is halfway around the world.[55]

The numbers seem to be against him. Would he not need another 52 days to cover the other half of the globe, and therefore a total of 104, rather than the requisite 80? But a final technological wonder will accelerate him toward the finish line.

" 'Ocean to ocean' is how the Americans put it—and this phrase really should be the best way of referring to the grand trunk line that crosses the United States of America at its widest point." In 1869, the Trans-Continental Railroad united several railway lines into a grand one, tapped into existence by the Golden Spike. An American achievement, and celebrated as such, it was also regarded as another segment of the international built environment that put a girdle around the Earth. Railroad travel in the United States had for some time been making the circuit shorter. When the prominent New Yorker Abiel Abbot Low went around the world in ninety-eight days—"the shortest time on record," half the time of the fastest clipper—from December 1866 to September 1867, it was noted that he could have reduced the travel time to seventy-five days had he taken the Pacific Railroad line.[56]

The Trans-Continental Railroad was not just a way across the United States, a national achievement, but it also constituted the long-awaited Northwest Passage, another strong link in a planet-spanning chain. With its completion, an American observer claimed, "a journey around the world became both easy and brief." Along with the Suez Canal and Indian railways, the U.S. railroad had made an eighty-day circuit of the planet a widely discussed goal.[57]

Verne had not invented that deadline, but brilliantly explored it. He said that newspapers and new timetables had given him the idea for an eighty-day journey. One probable source, *Le Magasin pittoresque,* had, in April 1870, announced that "thanks to the piercing of the Isthmus of Suez" and a completed railroad across North America, a person could go around the world in eighty days. The magazine laid out an itinerary very like Fogg's, except for a stop in Hawai'i that Verne omitted.[58]

And real circumnavigators had preceded Phileas Fogg on the railroad way around the world. It is perhaps not very surprising, given his surname, that the American George Francis Train was an avid traveler, especially by railroad. "I was born into a slow world," Train declared, "and I wished to oil the wheels and gear, so that the machine would spin faster." During his many careers as railroad entrepreneur, anti-slavery advocate, political radical, social reformer, and presidential candidate, Train had made several trips abroad, to Asia and Europe, and in 1870 he did a tour of the world that totaled eighty days, after subtracting some political dalliance in France. He had done his circuit on commercial travel services and dined out on the claim that he was Fogg's prototype: "I went around the world in eighty days in the year '70, two years before Jules Verne wrote his famous romance, Le Tour du Monde en Quatre-vingts Jours, which was founded on my voyage." That's as may be; a wealth of (triple-named) world tourists were possible models: Abiel Abbot Low, George Francis Train himself, and even a William Perry Fogg, the last often suspected as the model for Phileas Fogg, though Verne had originally spelled his hero's name as "Fog," like the weather. More credible is Train's claim that he had the idea to go so rapidly around the world in 1869, after "the circumference of the world had been shrunken" by the Golden Spike.[59]

Indeed, a journey that had once taken up to half a year was reduced to a mere six days on Fogg's itinerary. The passengers enjoy the benefits of a "sleeping car," an American invention, most luxurious in the version Henry Pullman developed, which would become the international model for de luxe overnight journeys: the train worked while the passengers slept. Nothing should prevent Fogg's timely arrival in New York, not even a collapsing bridge. But Sioux warriors attack the train as it passes through Nebraska and they kidnap Passepartout. A rescue could be effected, though only if Fogg is willing to leave the train and abandon his schedule. He has been generous with his time before, but only if he had any to spare. To pursue Passepartout's captors would guarantee a day's delay—Fogg would miss the steamer from New York to London.[60]

"I shall find him, dead or alive," Fogg declares. The decision to abandon the orbiting capsule, to be the mover rather than the moved, is supposed to indicate Fogg's moral progress, even as it compromises his physical progress. The valet is rescued but the train is long gone. Fix needs to get Fogg back on British soil in order, finally, to arrest him, so he locates a sled rigged with sails and the party slides over the snow to the train depot in Omaha, a third time that a pre-industrial mode of travel has saved the day. Trains speed Fogg to New York, but he has missed the scheduled steamship. He hires a small steamer, but it runs out of coal. Fogg buys the vessel and cannibalizes its combustible upper body to keep the engine turning, turning until he gets to Liverpool. There, Fix arrests him. Fogg is released—once the real thief is apprehended elsewhere—only after the essential train to London has departed. A special train is hired. It shoots to London. But Fogg arrives five minutes late.[61]

The forlorn travelers retreat to the snailshell on Savile Row. Fogg is ruined. Of his mysterious fortune, he has only the £20,000 to cover his bet. He apologizes to Mrs. Aouda for bringing her to a foreign land where her only friend will soon be penniless. She worries only for him. Will his friends and relatives not help? He has none, he replies. And then the lady does something extraordinary: "She got to her feet and offered the gentleman her hand, 'would you like both a relative and a friend? Would you like to have me as your wife?' " Fogg melts and tells her, "I love you and I am wholly yours." Passepartout is sent to the minister to arrange a wedding for the next day. He discovers that marriages cannot be performed that day because it is a Sunday, not the Monday everyone had thought it would be. The travelers have gained a day in their circuit around the world.[62]

Ah, yes, the day gained or lost. Now that people survived circumnavigations, the temporal trick had become the dramatic pivot of any round-the-world journey, real or fictional. Bret Harte's poem "The Lost Galleon" (1867) tells of a Spanish ship caught, for two hundred years, in an eerie band of territory at the 180th degree of longitude from Madrid. While the vessel has been missing for two centuries, only a day seems to have passed in the ship's log, yet even that date is wrong because a day should have been lost on the calendar:

> *Lost was the day they should have kept,*
> *Lost unheeded and lost unwept;*
> *Lost in a way that made search vain,*
> *Lost in a trackless and boundless main.*

The spell will end if the galleon can cross the fatal line on the anniversary of its original approach to it. The attempt will fail each year until 1939, three hundred years after the ordeal had begun. Harte explained the phenomenon of the calendar change in a note. He also contributed to the myth of the Dark Ages by saying that "if any reader thinks I have overdrawn the credulous superstitions of the ancient navigators," they could consult various historical narratives, including those of Anson and La Pérouse (two men from the age of Enlightenment!).[63]

Verne had dropped several hints about the calendrical puzzle. Given Fogg's preoccupation with time, his unawareness of the Circumnavigator's Paradox is ultimate proof of his fallible human nature. Verne refers to the day "unconsciously gained," using a word, *inconsciemment*, that had appeared so rarely in French before that he may fairly be said to have coined it. The implication is that the closed-in, ticking energy inside Fogg has suddenly exploded into an awareness of the larger world. Passepartout is, in contrast, consciously obtuse. Fix has tried to convince him that a watch set in London must be reset in other places:

"Me alter my watch! . . . Never."

"Well, in that case it won't be in time with the sun."

"That's too bad for the sun, sir. It's the sun that'll be wrong."

"It was an innocent fixation," Verne lies, "which couldn't harm anyone." The harm is in fact compounded at the 180th meridian. There, Passepartout remarks that his watch agrees perfectly with the ship's chronometer, as well as the Sun, when of course his watch is twelve hours off, and will lose another twelve hours before the return to London, where the total of twenty-four gained hours imparts another illusion of accuracy.[64]

But note how Verne makes his travelers learn of their extra day. Mrs. Aouda asks Fogg to marry her—that act brings into the London house the key information that the next day is Sunday. The dating of the Sabbath is a convention of the Circumnavigator's Paradox. The bonus of marriage is unusual, especially in its unification of two global characters, English bachelor and Parsi widow. Marriages between English and South Asian people existed at the time, though not as commonly as in the previous century. Mixed marriages also existed in fiction, if rarely. Some of these culturally mixed marriages are happy—though that was even rarer. But to have a woman propose marriage was rarest of all. Even Jane Eyre, who brags, "Reader, I married him," has waited for Mr. Rochester to pop the question.[65]

The reason for Aouda's boldness was appropriate to the year in which *Le Temps* serialized Verne's story: 1872 was a leap year. The tradition of women proposing marriage on February 29 was not French, but Verne's characters are not French, after all. His story resembles another that has a happy denouement in a London drawing room, and that had been inspired by the similarity between a day gained through the Circumnavigator's Paradox and a day gained in a leap year. In Edgar Allan Poe's "Three Sundays in a Week," Kate circumvents her father's threat that she shall marry her cousin, Bobby, only when a week has three Sabbaths. She invites over two circumnavigating sea captains, and she and Bobby expose their different ideas about whether it is Sunday by proposing a game of whist.

In Fogg's perpetual games of whist, beginning at the Reform Club, Verne noted his debt to Poe. He had read the story as early as 1864, when he wrote short essays on Poe's tales: "how can three Sundays exist in one week? Perfectly, *for three individuals*, and Poe shows how." From another source, Verne might, independently from Poe, have seen the parallel between circumnavigation and a leap year. He loved scatological and sexual humor and had read the French master of such things, François Rabelais. The prototype Rabelaisian hero Pantagruel is born in "the Week of three Thursdays. For it had three of them on account of the irregular bissextiles" or leap years.[66]

The topsy-turvy is essential to Verne's plot, unlike the detective-story elements, which turn out to be superfluous. But it is unlikely that Verne empowered Aouda on principle. It was even more unusual for him than for Poe to have interesting women in his stories. He wrote entire books without female characters. But an around-the-world journey upsets the usual order of things, even in the *Boy's Own* world of Jules Verne. His bold South Asian widow is the heroine, ultimately, of his geodrama, which uses the romance of the domestic novel, embedded within a wandering picaresque, to show how the Earth itself was becoming domesticated—though only so much. Aouda is the reminder that a circumnavigation is somehow contrary to nature, however fast, however new-fangled, however dependent on a planet-spanning built environment.

Geographers loved Verne's plot twist. In 1873 he was invited to address the Société de Géographie in Paris. (There, he explained that Poe's story had inspired his.) He was aware of the multiple international congresses, most recently in 1871, on the possibility of determining universal time. That system would replace local observations, which, isolated from each other, had worked for millennia until global travel and the telegraph had put them into cacophonous conversation. Verne recommended a global line of calendrical demarcation at the 180th meridian from Greenwich, Paris, or Washington, D.C. Any such line would lie in the middle of the Pacific Ocean, which Verne calls a "desert" because of its low level of population. Voyagers could pass the line "in an unconscious manner," in the absence of dense human populations that kept their own time. Verne repeats the nature of that transition: it could be done "unconsciously." It was the second time he had used that word to describe the day won or lost.[67]

His decision to omit a trans-Pacific stop on Fogg's itinerary now makes sense. Other predictions of an eighty-day circuit had included Hawai'i. But Verne omitted from his imaginary circumnavigator's track the Pacific populations who were never represented at the "international" congresses that met in Europe. Delegates at those and future such meetings assumed that the global West was the center of the world. Compared to their consensus on that point, disagreements over where the prime meridian would go were mere quibbles. In practice, just as Verne had done, most of the global West had adopted a 180th meridian as the line of calendrical demarcation because most ship's captains, including those who went around the world, had depended, since 1767, on the annual, Greenwich-based *Nautical Almanac*. None of the international congresses on universal time ever actually dictated either the adoption of Greenwich Mean Time or of the eventual International Date Line at its planetary oppposite. They didn't need to. The Circumnavigator's Paradox had made it customary for Western travelers to be "unconscious" of the other people and cultures they encountered on the way around the world. This may have been the ultimate in proprietary confidence: solar time was defined for the entire planet, without any global consensus.[68]

If an Englishman leaves London on the 8:45 p.m. to Dover, and continues around the world, mostly by steam engine, returning to London on the eightieth day, how much coal does he consume? Subtracting the miles covered with renewable energy (elephant, sailboat, sail-sled), what is his impact on the planet?

There was no concept of a "carbon footprint" in 1872. And yet just as Verne makes the passage of global time "unconscious," so he makes poor Passepartout's gas bill a semiconscious tally of the material cost of encompassing the world. The valet calculates that his "forgotten" lamp costs two shillings a day to run, sixpence more than he earns. At eighty days or 1,920 hours (the best-case scenario), he will be £2 out of pocket. Even if Fogg is paying top money for a valet, £50 per year, £2 is a considerable part of Passepartout's wages. When Verne's full novel acquired illustrations, one toward the end shows Passepartout slumped in a chair, clutching a gas bill that scrolls between his knees and down to his feet. Even better is the frontispiece, in which Fogg and Passepartout gaze up at a representation of the Earth, from which a lighted gas lamp extends (see illustration 10). To illustrate Verne's correlation between time, space, and physical resources, part of the title, "en Quatre-Vingt Jours," curls about the globe. The valet points accusingly at this extraordinary, gas-burning world.[69]

The title page, opposite, promises a happy ending. Passepartout has just rushed back into his room and is extinguishing the lamp: lights out.

Chapter 7

The Club of Eccentrics

The lights blaze over the stage of the Théâtre de la Porte Saint-Martin in Paris. Fake Hindu worshippers chant, temple dancers twirl, and the *sati* pyre smoulders. As an elephant enters (stage left), Phileas Fogg shouts for the ceremony to stop. Mrs. Aouda is saved. And most of the Parisians who were watching the 1874 premiere of the play *Le Tour du monde en 80 jours* thus saw their first elephant since the occupants of the Paris Zoo were eaten in 1870 during the siege of the city by Germans in the Franco-Prussian War.

Verne had always wanted to write primarily for the stage. After the encouraging reception of his novel, he and Adolphe d'Ennery turned geodrama into stage drama, complete with moving train, sinking steamship, snake-filled grotto, elephant, and other splendid effects. A transparent world globe rotated in front of the theater. Fearful either that he had overdone it, or maybe not done enough, Verne asked a friend whether his play might be considered a "success." Oh no, the man replied, a "fortune." The first run would last for 415 nights. The play would be staged in Paris an estimated 3,600 times before 1940, when the Germans again laid siege to the city.[1]

Verne would write many other novels about technology and travel—they became his stock in trade. His *Voyages extraordinaires,* which included *Around the World in Eighty Days,* also encompassed the circumnavigating quest of *The Children of Captain Grant* (1865–67) and the submerged near circumnavigation of *Twenty Thousand Leagues Under the Sea* (1869–70). But Fogg's story was the biggest success of Verne's career: in 1894, he estimated that the novel had generated 10 million francs. Spanish, Russian, Italian, Hungarian, and German translations were being prepared in late 1872; English was an obvious addition. There was an American edition of the novel by 1873, and British and U.S. dramatizations were produced between 1873 and 1880. Dutch, Swedish, Japanese, and Arabic translations appeared during Verne's lifetime. Other newspapers offered serializations, including *The Pioneer,* in Allahabad, which Fogg had reached by elephant. The book followed the story, around the world.[2]

It may have been a board game that realized the greatest vicarious potential of Verne's story. "Le Tour du monde" (see illustration 11) represents Fogg's voyage on a square of colored pasteboard, over a circular path that coils around and back to the Reform Club, with added adventures along the way. Players compete to circle the world first, with each square on the board representing some good or ill fortune that accelerates or delays their travel. The game's design was quite old, based on a sixteenth-century prototype, and it was incredibly simple, literally child's play. Now, even the bed-bound could spin a teetotum and perform a cardboard circumnavigation.[3]

That is the secret of the story's success: the most imaginative aspect of Fogg's imaginary voyage is that many people can imagine doing it themselves. Verne may have intended to write a satire on the dehumanizing nature of steam-age travel, but he had also issued an invitation to circle the planet for personal fulfillment, perhaps even to find true love, as Fogg did. By 1872, an around-the-world voyage existed at a particular intersection of possibility. It was out of the ordinary, *extraordinaire,* yet entirely possible. The old expeditionary model of a circumnavigation seemed defunct; private travelers were replacing the pirates, navy men, and scientists who had once done the agonizing, state-funded circuits of the globe.

Verne appreciated that this was a historic development (he helped to produce a multivolume history of exploration in the age of sail), but for many other people, the conveniences that assisted around-the-world travel had become second nature. World travelers thought of themselves as free agents, able to pick up and go on a whim, though their personal autonomy depended on the long-term, collective project to tame the planet. Their unawareness of their dependence only increased their sense of freedom. In his staged version of *Around the World in 80 Days*, Verne implied that obliviousness by making Fogg a member of a fictional Club of Eccentrics. (The members of the Reform Club had not been uniformly delighted with their starring role in Verne's novel.) Anyone determined to get around the world was odd. It was rarely *necessary* to do such a voyage. Those who did it tended to be unsatisfied with life at home, or burdened with excess money, or possessed of some complicated ambition.[4]

The most eccentric of all would shun ships and trains and the built environment more generally. That required a circumnavigator to maintain a self-willed bubble that outdid even Fogg's removal from the world he traveled. Someone who wanted to go around the world that way had to carefully assemble a travel kit from a new range of portable manufactured goods. The result was an amazing display of confidence in Western industrial society—implying the ability to carry its essence around in pieces. The technological fragments enabled ordinary travelers to be like explorers, to move further away from the Western geographic core. This remains a popular choice for adventurous circumnavigators, though it can kill them if they are oblivious of where paper internationalism can rip away and no travel kit repair it.

Globe-trotter, the English word, was coined no later than 1873 and quickly migrated into other languages. The swift uptake of the term showed yet again the new confidence in human control of the Earth and its continued association with the English-speaking might of the British Empire. Even as "globe-trotter" could criticize the casual way that the British frolicked about the world, it might also express envy, a desire to share that proprietary confidence.

Verne family legend claims that one inspiration for his novel had been an advertisement for Thomas Cook & Son's first around-the-world tour, organized for 1872. Cook was a British travel agent, a "tourist" in its original meaning as a tour director. He had begun his travel empire with railway tours of England, then moved into Europe, the Near East, Asia, and North America. His specialty was all-in-one tours that included the services that travelers (or their valets) would otherwise have to arrange themselves. By 1869, Cook, like everyone else, had realized that the Suez Canal and completed rail lines across India and the United States could provide smooth transit around the world. He arranged for continuous passage for a small, exclusive party, whom he would personally conduct on a journey of 222 days. He implied his unique qualifications by comparing his surname to "that of the great circumnavigator of the globe," Captain James Cook, repurposed as an ancestral globe-trotter. That represented a new approach to the history of around-the-world travel: a forgetting of its early agonies.[5]

Thomas Cook's party set out westward, "with the setting sun." They crossed from London to New York, took sleeping cars across North America, and then a steamer across the Pacific, at this point accompanied by a party of Japanese travelers. Everything was suitably de luxe, though Cook observed that the unique comforts of America's Pullman cars were uniquely expensive, which "does not comport with Republican equality." Crossing the Pacific was, as many now agreed, the most "monotonous" part of the journey. To conserve coal, the steamer made no more than 206 miles per day, and Cook hints that his charges found the unchanging seascape quite unsatisfactory. Asia was their reward—Cook said Japan alone was worth the Pacific crossing. From Penang, the party sent an amusing Christmas telegram to London that arrived three hours "before" it was sent. In India, they occupied a private railway carriage that they could attach or detach to passing trains as they pleased, over a total of 2,300 miles. Passage through the Indian Ocean, Red Sea, and Suez Canal returned them to Europe.[6]

It had become a formula. As on Fogg's fictional journey (see Map 3), an amphibious itinerary, done by steam, replaced the old sailing route around the two southern capes. It is noteworthy that three newspaper accounts described the formula at around the same time: William Perry Fogg in letters to the *Cleveland Leader* beginning in October 1870 (published as a book in 1872); Verne in serialized chapters in *Le Temps* in late 1872; and Thomas Cook in *The Times* of London around the same time. The three publications occurred too closely in time to have influenced each other; rather, their similarities show convergence on the same formula. It was attractive enough that other travel agencies began to organize round-the-world tours (as the German agency Stangen did in 1878). By 1892, Cook had sent twenty tour groups around the world, perhaps as many as a thousand people.[7]

The amphibious route suited the new impatience with life at sea. The daily changes in time and eventual change of date were valued distractions during the dull Pacific crossing—they had almost entirely replaced mortal danger as the momentous centerpiece of an around-the-world voyage. Canadian missionary Eugene Vetromile said of the date change at the 180th meridian, for instance, "many of our passengers could not understand this, and others would not believe it." William Perry Fogg enjoyed the changes of his watch and calendar and introduced another diversion, photography; his may be the earliest published round-the-world account to have photographs. A camera was expensive in the 1870s, another sign of the wealth needed to circle the world for pleasure.[8]

For travelers who could not afford Cook's services, printed guides explained how to assemble a world tour. E. Hepple Hall's *Picturesque Tourist* (1877), for example, gave advice on different steamship services, provided a map of U.S. railroad routes, listed departure times, and compared rates for luggage. Hall's book was subsidized by advertisements for hotels, insurance companies, railway and steamship lines, and a manufacturer of timepieces, including chronometers. Even very local travel companies sold their services as potential links in a global chain. William Perry Fogg, for example, remembered in India how he had once seen the claim " 'Round the World' " on the coupon of a ticket issued by the Erie Railway. That estimable company had advised its passengers that travel through far-off India was available via *"Allahabad Junction"*; "here I am," Fogg reflected, years later in that very place.[9]

The most striking achievement of all-in-one tours was the arrangement of nature itself. "I think I comprehend the whole of this 'business of pleasure' around the world," Cook congratulated himself, having timed the westward itinerary for maximum climatic comfort. He made sure that his party would cross North America "under the genial climate of the Indian summer." Once they reached Asia, the travelers could enjoy India in January, the coolest time, "and then we shall be just right at Suez and Cairo for the Nile," and back in Europe as springtime bloomed. Cook declared "going round the world" to be "a very easy and almost imperceptible business."[10]

That too became part of the formula. When the Australian Presbyterian minister John Dunmore Lang went around the world from 1874 to 1875, he advised would-be circumnavigators to leave the antipodes in March or April. They would then cross North America in late spring, linger in Europe over the summer, and pass through Suez and the Indian Ocean in fall. Lang contrasted that experience to an 1837 journey to Europe when some of his fellow passengers developed scurvy. (They were treated ashore with a diet of fresh vegetables and therapeutic burial in the earth.) The ability to avoid that fate, and to select the nicest seasons for travel, were reminders of the novel ability to orchestrate personal experience on a planetary scale.[11]

Given the intensifying sense of command over nature, it was appropriate that history's last great scientific circumnavigation would demystify the sea itself. The voyage of HMS *Challenger* (1872–76) was inspired by the longer tradition of scientific circumnavigations, from Cook through Darwin, and down to the most recent grand example, Dumont d'Urville, thirty-six years earlier. The *Challenger* was supported by both the Admiralty and the Royal Society, but it was the British Treasury that funded it, at the extraordinary rate of £200,000 (around £10 million today). And unlike earlier circumnavigating scientific ships, the *Challenger* was a steamer. The 2,000-ton corvette carried sail, but also a 400-hp engine.[12]

The expedition set out to gather extensive samples of water, living specimens, and seabed from the deep ocean. It did so for scientific reasons, but also to test the feasibility of laying telegraph cable between Australia and New Zealand. Indeed, the two reasons overlapped. It had long been assumed that the dark, cold, deep sea was incapable of supporting any kind of life, and uninteresting for any reason. But when repair workers in the Mediterranean had raised undersea cable, they discovered that small organisms had colonized it. The sea floor also turned out to be irregular in its configuration and composition—some of it was firm and level (good for laying cable), but other parts were oozy or plunged from underwater mountaintops to yawning abysses. Clearly, a world beneath the waves awaited discovery.[13]

The planet's oceans were no longer just the means by which to make a circumnavigation, but the very thing scientific circumnavigators should examine. Darwin himself had speculated that the seas might be governed by the same principles of species adaptation and distribution that he and his fellow evolutionist Alfred Russel Wallace had theorized for the terrestrial realm, and that oceans might harbor ancient life forms only seen as fossils ashore. The *Challenger's* samples of water, plants, animals, and seabed would form the largest oceanic database yet generated, allowing scientists to begin to gauge the extent and variation of deep-sea life. That would require stops at multiple points in each ocean, to be distributed as evenly as possible, not unlike FitzRoy's chain of chronometrical readings around the world. The expedition's equipment was as up-to-date as its scientific goals. The *Challenger's* scientists used steampower to lower and raise water samplers or dredging equipment, to keep the ship in position while sampling (sail could not do that against a brisk wind or current), even to power the ship's launch. They also used cameras to record some of their findings—the *Challenger* had a darkroom as well as laboratory space.[14]

Like its ancestors, this latest scientific voyage was supposed to be above national divisions, even as it would, one of its British participants predicted, "exalt our national reputation." The proof of the expedition's international character was evident in what the *Challenger* did not carry: weapons. All but two of her cannon were removed, and the remaining pair were mostly intended to be used for signaling. The stated reason for removing the other guns was to maximize deck space for scientific activities, but that would never have been done had trouble been expected. In contrast, each of Dumont d'Urville's ships had carried 14 guns and the circumnavigating *Beagle* had 6. Paper internationalism protected the *Challenger*.[15]

It followed that her sailors would be protected from harm. Indeed, no one aboard the *Challenger* suffered a serious illness, let alone scurvy. (When there was a hint of yellow fever at Brazil, the ship fled to Cape Town.) In the end, one sailor died in an accident with the dredging apparatus, two were lost overboard, and a zoologist died of septic shock resulting from an infection. Because these were exceptional incidents, they were treated as tragedies that might be prevented in future—zero mortality was still the goal. It was also significant that when the expedition reached the Philippines, the captain, George Strong Nares, learned that he had been recalled to England in order to head an Arctic expedition, his real ambition. His transfer represented the continuing transfer of glamorous risk from ocean to polar zones, as circumnavigations became much easier, compared to ventures into extreme climates.[16]

Frequent stops ashore made the voyage quite pleasant. The *Challenger* was always well supplied with fresh food and water, and coal for the boilers. Whenever the men anchored at a place with an observatory, they checked their chronometers. And wherever they arrived, they had a warm welcome. (Only one site, a Brazilian penal island, was off-limits.) Balls in their honor were routine. A dance at Cape Town featured the ship's dredging equipment as decoration. Other celebrations were held in Ternate, the old center of the spice trade, and the Philippines, "where Magalhaens met his death." The authorities in Melbourne gave the men free railway passes. A "Dredging Picnic" was held at Tokyo. The return to sea was hard. On leaving Australia, one man "felt, notwithstanding all the temptations of promised adventure, the full bitterness of the price we have to pay for its excitements." The long stretch from Brazil to Cape Town had been mercifully enlivened by a variety of bird life. But the 2,400 miles from Hawai'i to Tahiti were "dull, and monotonous." Access to welcoming ports was too attractive—a quarter of the ship's complement deserted. Boredom eroded morale, as the *Challenger* made stop after stop to dredge. As far as the sailors were concerned, a global collection of mud and invertebrates was beyond eccentric, truly mad, and definitely maddening.[17]

The expedition would be encyclopedically documented on its return. The scientists produced several sets of massive tomes, and many smaller works, describing the specimens and analyzing the samples of water and seabed. Two narratives of the voyage were also published, a collective work by some of the scientists and officers, in three thorough volumes, and William Spry's very readable single volume, distinct from the scientific reports in its emphasis on "the vast extent traversed in the pursuit of knowledge." Spry calculated that by the time the *Challenger* crossed her outbound path (which marked "the actual circumnavigation of the world"), she had logged 44,000 miles, and taken two hundred soundings and almost as many dredgings. (The total journey encompassed 68,890 nautical miles and 362 observing stations.) That painstaking circuit had proven "that there are laws which govern the geographical distribution of marine plants and animals, as on land." The oddball mud samplers had laid the foundation for modern oceanography.[18]

Anna, Lady Brassey, and Sir Thomas Brassey were British eccentrics of a different sort, being people who liked things just the way they liked them. The Brasseys wanted to go around the world, but even a Cook's tour would have been too common for them. (Brassey's father was an engineering contractor who had worked on railways in Europe and British India; travel by rail would have been a busman's holiday for his son.) Instead, from 1876 to 1877, the aristocrats circumnavigated the globe on their private steam yacht, the *Sunbeam*. Yachts were fast light boats meant for recreation, usually in coastal waters. To take one around the world was adventurous, though not insane, not in the case of the *Sunbeam*. She was an early example of a large yacht, over 500 tons, comparable to many of the workaday craft that had gone around the world. And the Brasseys hardly intended to work the ropes themselves. They hired a captain and full crew, and took three female servants, for a total of forty-three persons on departure.[19]

The Brasseys wanted to experience the world at leisure. Lady Brassey kept a diary, and her observations radiate confidence in her place at the heart of empire. (It is safe to say that her social observations were never enlightened.) She would publish the narrative of the journey, *Around the World on the Yacht "Sunbeam,"* following Ida Pfeiffer's example in claiming the planet for womankind. Brassey's subtitle is especially revealing in its cosy domesticity: *Our Home on the Ocean for Eleven Months.* In essence, that meant the Brasseys brought the kids, all four of them.

It was an unusual decision. Boys still served on many circumnavigating ships, but they did so as adults-in-the-making, and often were adults by journey's end. Harden Sidney Melville went to sea as a young draughtsman (sketch artist) on HMS *Fly*, for example, from 1841 to 1846. Melville's account may be the first by a "boy" circumnavigator. His narrative uses aliases to convey a sense of fun—he becomes "Greenhorn," the ship is *Bluebottle*, and the other boys, the midshipmen, have names like "Mischief" and "Rattlebrain." Just as his book presents a circumnavigation as a *Boy's Own* adventure, so Melville blames another narrative, "two gigantic volumes of *Captain Cook's Voyages*," for seducing him to sea. Maybe that edition made Cook's voyages fit for children through euphemism and omission. Melville's own narrative did plenty of that. While he admits his voyage's hazards (several sailors die), he also describes his contributions to the ship's literary magazine, the *Circumnavigator*, and antics with the ship's pets.[20]

Other youngsters went around the world without having to pay their keep. Captain Nares of the *Challenger* had brought his nine-year-old son, for instance, along with a tutor. (It was the tutor who came to harm, dying of a stroke in Bermuda.) When he was sixteen, Samuel Smiles (not the famous self-help writer of that name) embarked for Australia, later to continue around the world, because his doctor had recommended a long sea journey to cure his weak lungs. Smiles loved the voyage to Melbourne. His little cabin on the *Yorkshire* was "cheerful and even jolly"; passing ships paused for friendly swaps of newspapers, soap, and preserved milk. The passengers put on theatricals and the captain gave an astonishing final dinner. "The *menu* was remarkable, considering that we had been out eighty-one days from Gravesend"; the passengers enjoyed fresh fowl plus "hams, with lobster-salads, oyster pattés, jellies, blanc-manges, and dessert," all made possible because "the art of preserving fresh meat and comestibles must have nearly reached perfection." Australia was anticlimactic.[21]

Children's literature was also presenting round the-world travel as child's play. In the *Voyage de découvertes de Mlle Lili et de son cousin Lucien* (1866–67), translated into English in 1868, Lili plans to run away from home and go around the world, starting from the local river. She is accompanied by her cousin, her brother, and the gardener's son, who locates a rowboat. The children load it with touchingly inadequate gear: Lili's parasol, the cousin's Punchinello doll, a little food and water, a map of the world, and the family dog, Loulou. When they disembark on an island, their unsecured boat drifts away and they are "shipwrecked." A fellow "castaway" turns out to be a landscape artist and friend of the family. He scolds the pint-sized travelers and takes them home, no harm done.[22]

That was rather what the little Brasseys expected, only their "home" on the ocean really would go around the world. The *Sunbeam* was the first circumnavigating vessel organized around a nursery. In fact, it had a piano and library as well as a nursery. The children thought it a lark. Not so the yacht's captain who, accustomed to larger ships, had not realized "how near we were to the water in our little vessel." And not so the domestic servants, who were terrified.[23]

As if to prove that a yacht could be a domestic refuge for small beings, the Brasseys stocked theirs with pets. Two dogs, three birds, and a kitten set out with them. When the kitten vanished, another replaced it. Exotic animals boarded en route—parakeets, parrots, armadillos, monkeys, and a puma cub. When the oldest boy, Tab, returned to England to enter school, two lion monkeys went with him, bound for London's Royal Zoological Gardens. (The puma went to the zoo by other means, before it got too big.) By the time they left Japan, the sailors had over one hundred birds of their own. On sunny days, the deck was covered in birdcages. It was awkward when the ship took aboard food in Polynesia in the shape of two pigs: one was eaten, but the other saved itself by befriending the dogs. It was "a pity to eat such a tame creature," Lady Brassey concluded. Luckily, the ship's table could be supplied with "potted meats and vegetables, which happen to be excellent." The dogs could keep their new companion, "Beau Brummel."[24]

Lady Brassey knew quite well that earlier circumnavigators had faced great dangers, but regarded past hazards as present-day adventures. She read Dampier, Anson, and Cook, from the hair-raising days of around-the-world travel, but also Darwin, who had made an easier journey on a ship about the size of the *Sunbeam*. She found Hubner's *Promenade autour du monde* "charming," more to her taste. Such was the Brasseys' sense of safety that, in the Strait of Magellan, they amused themselves with dodging icebergs. The children were "in ecstasies at the sight of them," and the yacht towed a small berg from which the crew chipped a supply of fresh water.[25]

Not everything was fun. A number of passengers and crew caught influenza in Honolulu. "Baby," tiny Marie, was so badly congested that she turned an anoxic shade of blue. Another child was concussed when the boom smacked her head. Lady Brassey crushed her thumb in a door hinge and, after being drugged to sleep, woke to the smell of smoke and cries of "Fire!" The nursery fireplace had been improperly tended and smouldering fragments of coal had ignited the woodwork. The children refugeed in another part of the ship, all part of the adventure, as far as they were concerned. Another fire the next night was very wearing for the adults. Later, a sailor contracted smallpox and a great fuss ensued to prevent an epidemic. But, after all, such things happened on land during the nineteenth century. Unvaccinated children got many ailments, sometimes fatal ones; coal-heated households suffered frequent fires. (Impressed by their yacht's fire extinguisher, the Brasseys installed one at home.) The worst trial for Lady Brassey was something that never occurred on land, her persistent seasickness. "Nothing annoys me more," she fumed, "than to find that, having sailed tens and tens of thousands of miles, I cannot cure myself." [26]

Whatever the dangers, no lives were lost. It is significant, as well, that the *Sunbeam* never threatened other people. Yachts did not carry cannon; their design was too shallow to permit a gun-deck. At Yokohama, the Brasseys' yacht perplexed Japanese officials. (Only one yacht had arrived in Japan before, a gift from Queen Victoria to the ruler.) Why did the strange English craft have no guns? What category of ship was she, and how should harbor fees be determined? The lack of guns was the ultimate proof of the Brasseys' confidence: their status as subjects of the British Empire, and careful choice of route, made it unlikely that anyone would threaten them, on land or sea. Rather than need protection, the little *Sunbeam* herself rescued crew from a coaling ship, the *Monkshaven*, whose cargo had spontaneously combusted, a hazard of the now constant demand for coal. (Lady Brassey estimated that one in three coalers caught fire.) Fifteen men and boys were taken from the burning ship, though the Brasseys learned too late that the captain had drowned his Newfoundland dog rather than impose it on them, little realizing he was about to join a zoo afloat. [27]

In fact, the Brasseys lost many of their own animals—that remained the measure of the limits of a seaborne life-support system, even in the age of steam. The *Sunbeam* kept acquiring creatures: sheep and pigs for food, birds, more birds, endless birds, and then a monkey and gazelle at Aden to top it all off. Few of the animals could have enjoyed their out-

ing. In Penang, Lady Brassey described "a broiling day, everybody panting, parrots and paroquets dying." When an overheated pig decided
to take a swim, and resisted its human rescuers, it was "condemned
to death" and eaten. Despite their having little flannel coats or cage
covers, several monkeys and birds perished from cold on the final approach to England, and in the hurry to get home, something big fell
on Beau Brummel, crushing the pig's spine. The Brasseys honored his
memory by not eating him.[28]

That itself was a luxury. Consider another early yacht venture.
In 1884, seven years after the *Sunbeam* had returned, the *Mignonette*
was wrecked off South Africa, on her way from England to Australia,
and her survivors forced into a lifeboat. They eventually resorted to
cannibalism—shocking, though under British law not illegal, so long
as whoever was consumed was already dead or had agreed to die to
save the others. But in England, two *Mignonette* survivors were tried and
found guilty of murder for having killed and eaten a shipmate without
his consent. The *Sunbeam*'s comparatively tranquil passage around the
whole world was a comforting counter-example. Lady Brassey predicted in her published (and much republished) account that many
more yachts would go around the world, and she was right.[29]

If the *Sunbeam* was a first, the *Challenger* was a last. There would be
other around-the-world ventures done for science, and other maritime
circumnavigations, but the two models stopped coinciding. The decline in public awareness of scientific circumnavigations was apparent
by the time of Charles Darwin's funeral. When Darwin was buried in
Westminster Abbey on April 26, 1882, great statesmen and admirals
came to mourn him. So did the Darwin family cook. But no one had
thought to invite any of the still living sailors who had served on HMS
Beagle. At least one had survived to remember young Mr. Darwin,
whose nostalgia for his circumnavigation probably meant that he had
remembered them, too. But now he was dead, and everyone else had
forgotten them.[30]

For anyone who truly detested the Union Jack, it would have taken a
great deal of effort to tour the world without running across a Cook's
touring group, English yacht, or British dredging party, if not all three.

When the Spanish journalist Torcuato Tárrago y Mateos composed a fictional circumnavigation (reported for a fake "Society of Modern Travelers"), he placed it aboard a British private steamship, the *Great Devil*. The vessel's owner, Sir Richard Peen, assembles a party of English and European ladies and gentlemen, plus some Americans, and steams away. Tárrago's story, published in two volumes from 1881 to 1882, has many Verne-like touches (Sir Richard's steamship resembles Captain Nemo's *Nautilus;* there is a subplot with mysterious balloonists) but lacks Verne's gift for narrative. The characters give long historical and geographical lectures to each other—the net effect is encyclopedic. But the wonderfully named *Great Devil* shows the pan-European suspicion that the British were superhuman globe-trotters. Indeed, the majority of around-the-world accounts were authored by Britons and written in English.[31]

The non-Britons who went around the world were usually European or North American, and responded to Britain's dominance with envy and emulation. When the Spanish diplomat Enrique Dupuy de Lôme set off around the world from Madrid in 1873, he had many criticisms of British imperial politics. And yet when he embarks, "a suitcase in one hand and a Baedeker in the other," he admits that "to travel is to imitate the English, constant travellers and, by experience, masters of it." Ugo Bedinello, who served on the Italian navy's circumnavigating *Vettor Pisani,* praised the English "entrepreneurial spirit" abroad. The men of the *Vettor Pisani* benefited from British mail services, as well as access to the British ports of Hong Kong and Sydney. Ludwig Salvator, archduke of Austria, went around the world from 1881 to 1882, but his major destination was British Australia, which he admired as "the world without want."[32]

Americans were less admiring. If the British assumed they controlled the global route between Europe and Asia, Americans thought they dominated the Pacific and North American segments, and flaunted a seen-it-all confidence to rival the British. When William Perry Fogg arrived in China, he was pleased to coincide with the American statesman William H. Seward, who was also "swinging round the circle." Fogg rooted for the American who competed against an Englishman in Penang to see who could eat the most durian, the Asian fruit whose fragrance has been compared to rotting flesh, open barnyard, every dirty sock in the world. (The American won by cheating—he wadded his nose with cotton.) Andrew Carnegie had decided to go around the world on a whim, while gazing into Mt. Vesuvius alongside his chum Cornelius Vanderbilt. It was an empty triumph for the jaded Carnegie. Was it absolutely necessary, he complained, "to have seen or heard the admitted best of everything"? Of his "tour round the Ball," he wearily concluded that "yes, my dear friends, *it is round*," and warned that "If America can learn one lesson from England, it is the folly of conquest, where conquest involves the government of an alien race."[33]

And yet these were quibbles among cultural cousins. Sometimes the circling Westerners squabbled; usually, they united against everyone else. Whatever his criticisms of the British, Dupuy de Lôme thought American travelers even worse: "globe trotters, vulgar types," as if "caricatures" out of Jules Verne. But once he was aboard the steamer bound for India, Dupuy de Lôme and everyone else would gather round the piano, of an evening, and bawl out each other's national anthems. Salvator also enjoyed the nightly "music and singing [that] blended harmoniously with the sound of the waves." In another odd gesture of internationalism, Bedinello's Italian ship joined others in saluting U.S. ships on Washington's birthday—in the British harbor of Hong Kong.[34]

Much of the solidarity originated from technophilia, faith that Western technology would girdle the globe and unite all peoples. "Today the world is small," Dupuy de Lôme rejoiced, because steampower and the telegraph had "broken the proportion between time and distance." U.S. railroads and the Suez Canal had made da Gama's and Magellan's old routes obsolete. He adored the distance-killing steamship that took him to India, which resembled a "high class hotel," with a main deck "like a *boulevard*" on which to see and be seen. Fogg approvingly related how he saw telegraph cable being laid between Singapore and Hong Kong, which would put China into communication with New York and London, more globe-girdling in action.[35]

Flaws in the technology were frustrating. Bedinello's *Vettor Pisani* could never preserve her provisions in hot climates, and her 300-hp engine was not powerful enough to counteract storms in the Pacific. Vetromile found the heat of a Red Sea crossing aboard a steamer "unbearable." Young Master Smiles thought his steam passage across the Pacific rather disappointing. Though swifter than the sailing ship he had taken to Melbourne, the steady progess made travel boring, and Smiles hated "the dismal, never-ending grind, grind" of the engine. Whenever rough conditions rocked the ship enough to expose the screw propeller above the waterline, there would be an especially "horrible *birr.*" The coal-fired engine also made any space belowdecks very hot.[36]

Yet the general sense was that these were problems to solve, not reasons to renounce modernity, which only the ignorant did. Smiles tormented the single female passenger aboard his Pacific steamer because "she was so hopelessly dense." He pretended that the equator was visible and she fell for it: "by stretching a hair across the telescope glass, I made her look in and showed her the Line." Worse insults were reserved for people who resisted Western culture, especially the Chinese. Around-the-world travelers almost uniformly compared China (unfavorably) to Westernizing Japan. Carnegie was amazed by Chinese poverty and said the country wouldn't amount to much until it had railroads. Hübner said he went around the world to witness "civilization in its struggle with savage nature" and, for that reason, praised Japan for being on "the path of progress," unlike China. Thomas Cook also found Japan wonderful, but China repulsive. Thérèse Yelverton, Viscountess Avonmore, was a rare exception: she actually liked the Chinese. Most other circumnavigators thought them stubborn holdouts against the wonderful forces that were changing time and space.[37]

In 1876, Li Gui was halfway around the world and profoundly homesick. After weeks of contact only with white Americans, Li met several of his countrymen in the town of Evanston, Wyoming. They "talked intimately, like family," but too briefly. Li had another train to catch. He was making the first round-the-world voyage on behalf of China, a historic event. China had sent its first official abroad only in 1867; that man, and Li, were responses to poor treatment of Chinese migrant workers, including the "coolies" who had helped to build the Trans-Continental Railroad that Li now traveled. Li was also despatched, as the Iwakura Mission had been, to gather information on which parts of Western culture might benefit China. His *Huan you diqiu xin lu—A New Account of a Trip Around the Globe*—would be published in Shanghai (1877–78). It is a rare published narrative of a non-Western circum-navigator, and distinctive in its effort to introduce Western culture to China on Chinese terms.[38]

Li wanted to convince the Chinese that Western astronomy and geography were correct: the Earth was round and made an annual journey around the Sun. Those observations made his journey into an even more explictly geodramatic event than had been the case for the Iwakura Mission. Japan's astronomy and geography had mostly con-verged with Western traditions. China's had not. Ancient Chinese cos-mology had presented the Earth as a perfect square and the heavens in a circle around it. Debate among Chinese scholars had criticized these classical views. Western astronomy and maps began to enter China as early as the sixteenth century, though the Jesuit missionaries who were the logical conduit for them had often impeded their arrival, in order to preserve the scriptural description of the Sun going around the Earth. Theories of a round globe that circled the Sun became more common in China by the seventeenth century, though not universally accepted. A noted scholar and government official in nineteenth-century China had categorically denied that the Earth orbited the Sun. Chinese coins continued to represent the old cosmography: each round coin had a square cut out of its center, the celestial realm surrounding the square Earth.[39]

Li used his journey to promote the Western view: "The earth is shaped like a globe and orbits around the sun, while the sun remains fixed as the earth moves. Although there are a great many Chinese who understand this in principle, perhaps eight or nine out of ten do not believe it." He had doubted it himself, but "having traveled abroad by imperial edict and circled the globe, I have come to accept it." At two points in his account, he explained that North America's "position with reference to the globe is directly opposite to that of China." He dutifully reported his experience of time changes as he crossed lines of merid-ian. "If the shape of the earth were perhaps a square," he hypothesized, "and it remained stationary while the sun was in motion," how could one travel *east* without actually going *west*? "Because the shape of the earth is like a globe," he answered, "there is essentially no separation into 'east' and 'west.' " Eastern and western hemispheres do not represent two sides of a flat Earth. They exist merely because of "the inability of the Westerners to depict them as connected" on a flat map. (Li may have intended here to rebuke the Western conceit that West was best.) To illustrate the paradox, Li provided a double-hemisphere map in his published account, complete with the continuous track of his route (see illustration 12). "Readers will therefore come to realize that the shape of the earth is that of a globe," he instructed, "and will have absolutely no doubt that the sun remains fixed while the earth moves."[40]

While confident that Westerners had a uniquely truthful view of the globe, Li Gui found their Earth-girdling technology appalling. His descriptions of railway tracks and telegraph wires gave them a relentless, predatory quality. They ran in parallel all across North America, pursuing a person even into the passes blasted into mountains. The pace and sensations of modern travel seemed unnatural. American railroad engines burned 4 tons of coal a day to achieve 400 horsepower and 66 miles per hour. Passengers could not even rest while mail was collected—a hook on the speeding train snagged a waiting mailbag. By the time he reached Philadelphia for the 1876 Centennial Exposition (a patriotic response to England's Great Exhibition), Li was exhausted. He had found railcar meals "not terribly satisfying," and the constant racket of the train had affected his hearing. Everything seemed to be "dirty and greasy," he complained, "and it had all become simply unbearable."[41]

But it was unstoppable. Western railways became the primary justification for reducing time to globally standard increments, starting with the English-speaking world. With the assistance of chronometers, most British railway companies had by 1847 enforced Greenwich time as "Railway Time." A Canadian engineer, Sandford Fleming, proposed in his *Terrestrial Time* (1878) the division of the globe into twenty-four sections, and an American educator, Charles F. Dowd, recommended that the United States be divided into four zones in order to rationalize railroad timetable arrivals and departures. In 1883, U.S. railroads would adopt that system as Standard Railway Time.[42]

Li Gui could have been more critical than he was. The International

Date Line, as it would be called, may have indicated something natural, the end or beginning of an Earth day, but its position was not determined by nature. Its political foundations were revealed in the fact that it never had the straight configuration of any line of latitude or longitude. Instead, it zigzagged. Its varying position was determined in order that the Western powers which claimed various Pacific archipelagos would not be inconvenienced by having separate time zones within one colonized region. Why would anyone critical of Western imperialism, as Li Gui must have been, have accepted such a line? Why should the Chinese acquiesce in a system that measured time from a prime meridian through England, anyway? Why accept that a circumnavigation implied power over the planet and over one's global neighbors? [43]

And yet an *us-too* sequence of independent nations had accepted all of that: Russia, the United States, Japan, and China had assumed that round-the-world travels would put them in a favorable light on the world stage. The Japanese, especially, had taken to making repeated global circuits. Numerous published accounts of the world tours of prominent Japanese men in the 1880s and 1890s were ongoing demonstrations of that. The philosopher Enryō Inoue would make no fewer than three circumnavigations between 1888 and 1911, using the standard route across North America and through the Suez Canal, but also the older route around both Cape Horn and the Cape of Good Hope. [44]

Korea joined the club when it appointed Yŏnghwan Min special ambassador to Russia, a posting that took Min around the world in 1896. The account of Min's tour, unpublished until the end of the Korean War in 1959, fully accepts Western measurement of the globe and calculation of time. "Paris's noon is our Seoul's 8:15 in the afternoon," it says; "England's London's 12:11 in the morning, the Qing Beijing's 6:16 in the afternoon," and so on. "The heel and toes of Asia and America abut," the Korean narrative explains, "making the days and the nights the opposite." If two people, each holding a watch, traveled away from each other and converged at the 180th meridian, their common calendar would have to be adjusted by a day. "This can be seen as the yesterday and today combining into a single day," as if an invisible line over the Pacific Ocean had always been the natural division between America and Asia. [45]

It is not irrelevant that Japan, China, and Korea were independent nations. However much their citizens may have disliked Westerners, they could aggressively imitate them, to demonstrate their equality, as they did when they went around the world. The colonized peoples within European empires were not as acquiescent. When the Hindu missionary Protap Chunder Mozoomdar (or Majumdar) went around the world on a ten-month tour, from 1883 to 1884, he described crossing the 180th parallel without Gui's or Min's enthusiasm. "This pilgrimage round the earth," Mozoomdar concluded, "sets at their real value a great many human calculations . . . time and space are eternal, we measure them in our small poor way."[46]

For the time being, tougher tests of proprietary confidence came from confident Westerners themselves. Consider Thomas Stevens.

As this amazing circumnavigator comes into view, note that his shape is not quite human, and yet his movement unlike that of any animal. That was because he rode a bicycle or, in French, *vélocipède*, "fast foot." Cycling was faster than walking, though early bicycle designs lacked chains, so transferred the cyclist's effort only to the front wheel. For that reason, the earliest successful versions, like Stevens's "Ordinary" or "penny-farthing," had outsized front wheels and were known as "wheels." Yet the new-fangled device would prove that people could go around the world "under their own steam," as the saying went. Richard Lesclide, a French cycling enthusiast, had predicted a cycling tour of the world in his *Tour du monde en vélocipède* (1870), though his fictional "world" consists of Europe. Stevens had bigger ideas.[47]

The 1880s and 1890s were the pioneer days of stunt circumnavigators, cranks who were determined to go around the world in some unusual way: alone, faster than ever, or by the power of their own bodies. Stunt circumnavigators expressed dissatisfaction with steam-age and globe-girdling technologies, though without rejecting them outright. Each of them declared a small *me too* in relation to the big tradition of going around the world. Their proposed alternative methods of transport did not criticize the proprietary nature of Western planetary confidence, but instead expanded it to include the middle and lower-middle classes. Newspaper editors were almost always the sponsors of these odd characters, because the reading public adored them—they were just eccentric enough without being wild social radicals.

The stunts depended on the availability of new consumer goods. This was a shift as momentous as the transition from raw human effort to machine- and steam-assisted labor, a new way of projecting a confident sense of control over the material world. In earlier decades, bicycles had been toys for the rich. Now, middle- and even working-class people could buy bicycles, as well as typewriters and sewing machines. Gear once meant for military expeditions became perfectly affordable for picnics and country walks, now accessorized with maps, hard-baked biscuits, light tin cooking kits, and rubberized clothing and boots. This consumer cornucopia has been celebrated (or deplored) for changing everyday life in industrialized societies, and for transforming life elsewhere, as many goods and brand names became globally known. But the cheap goods implied an even larger sense of freedom. With the right combination of these small items, a person could tackle the world, ultimate proof that Western manufacturing had planetary power.[48]

Stevens left San Francisco on his penny-farthing in 1884, as a special correspondent for *Outing* magazine. From a working-class English family that had resettled in America, he shared the widespread optimism that bicycles were social levelers, cheaper than maintaining a horse or paying railfare, and giving unprecedented mobility to the unprivileged. It is notable that he (and others who attempted global bicycle travel) had means, but not wealth. They also had time—many people could not have left work or family for the months, let alone years, that a circumnavigation might require. (Also, Stevens was a bachelor.)

He would have to cross oceans by steamship, but intended to cycle as many miles as he could, freed from railroad timetables and close to the people of his status that richer circumnavigators avoided. The Unitarian minister Thomas Wentworth Higginson, who heard Stevens speak in Boston, praised his quest: "mechanical invention, instead of disenchanting the universe, had really afforded the means of exploring its marvels the more surely." Better to encounter the peoples of the world on a bicycle, Higginson observed, than with a rifle or bundle of do-good tracts.[49]

And yet a bicycle could not stray far from the Western societies that had created it, especially their built environments. Stevens was the Goldilocks of the Earth's surfaces: too hard, too soft, or just right. Wheeling over good pavement, he said, was like sailing a small boat over gentle waves. Europe, with its extensive paved roads, and India, with its ancient Grand Trunk Road, were wonderful. The unpaved expanses of North America, the Near East, and Asia were much tougher. Stevens had to walk his bicycle through much of the sandy American West. Unbridged waterways were even worse. Stevens learned to use his wheel to pole-vault over narrow waterways; at broader rivers, he hauled the bicycle across with rope (a cold, wet task).[50]

Stevens's preference, ironically, was to parallel the railroads he otherwise shunned. Where possible, he rode on the level cinder track alongside railroad tracks. It was particularly invigorating to try that on bridges or in tunnels, where there was rarely space for a cyclist and an oncoming train. He also followed railway tracks and telegraph lines to get from one Westernized outpost to the next. That route also maximized publicity, because the telegraph dashed updates of Stevens's progress on to newspapers ahead of him. Stevens met Rhinelanders who had read about him in the *Frankfurter Zeitung,* a South Asian schoolmaster who knew him from the local papers, and Japanese women who did likewise. Stevens was right: the socially leveling wheel had a transnational appeal that the Pullman car, for instance, did not. When cycling through Lahore, he was so pleased when a turbaned Pakistani wheelman cheerily waved to him. The greeting between equals indicated that some parts of Western culture were welcomed, not imposed.[51]

But the "mystic brotherhood of the wheel" did not run around the whole world. The Hundred Years' Peace may have opened ports and railway stations to around-the-world travelers, but interior zones beyond telegraph lines—where bicycles were as unweaned from Western culture as sailing ships had been from land—were another matter. And the good Reverend Higginson was slightly wrong about a bicycle replacing a rifle: Stevens carried a Smith & Wesson revolver. The cyclist said he only aimed to kill when animals threatened to attack him. In the case of people he found threatening, he simply displayed the weapon, or fired in the air. Still, for someone who thought that a globe-girdling journey on the leveling wheel would make a statement about common humanity, packing heat was a bit of a contradiction.[52]

Stevens might not have needed a gun had he reconsidered his route. He wanted to cross Asia come hell or high water, and both were ominously plentiful. In the Balkans, he was briefly detained for lacking a gun permit. He tried to dodge passport inspection in Turkey, uncharacteristic for him, and reflecting a possible worry that the Turkish authorities would confiscate his gun. His concern to keep the weapon showed some awareness of conditions in the Near East, where the Ottoman Empire gave only a guarded welcome to Westerners. He had to hire local men to carry his wheel, baggage, and himself over dangerous rivers, but was never confident that they took his safety to heart. As he approached Central Asia, where Russia and Great Britain vied for territory in what was called the "Great Game," matters were even more fraught. Stevens tired of being asked whether he was "Russ or Ingilis." That should have warned him to find another way to India.[53]

Instead, he was turned back. The circumstances were slightly comic, though, as with the danger, Stevens was oblivious to it. When he reaches the border of Afghanistan, the authorities say he will be killed before he gets to Kandahar. No, Stevens protests—he has a gun. The Afghanis find that hilarious. They arrest Stevens and console him, as they might a sulking child, with gifts of sugar, tea, sweets, scented soap, and Huntley & Palmers Reading biscuits (made in England; eaten everywhere). They bundle him back to Persia (see Map 3) where he will have to take a long sequence of trains and steamers to Suez, then to India. He cycles through Japan and steams back to San Francisco in 1887. The Afghan authorities had almost certainly saved his life; he is not grateful.[54]

Though he had not gone around the world the way he intended, Stevens had done it: by "circumcycling the globe" he fulfilled the dream, pursued since the would-be pedestrian circumnavigators, of encompassing the world by human self-propulsion. It was indeed amazing that a device, lighter than the human body, could take that body around the whole planet (minus its oceans). The new feat of geodrama required a new emblem, beyond the old Drakean pose with a mere hand over the world. When Stevens published his *Around the World on a Bicycle* (1887), a drawing shows him in dramatic, oversized silhouette against the globe: the lone circumcycler, whose front wheel is only slightly smaller than the world itself.[55]

Now that one wheelman had encompassed the world, cyclists who sought similar fame had to find an angle: different route, different equipment, different kind of riders. A two-man team of American college pals, William Lewis Sachtleben and Thomas Gaskell Allen, Jr., set out in 1890 to do the circle on the new "safety" bicycles, which had equal-sized wheels and a chain that transferred power to the rear. They carried another new and equally leveling device, the Kodak camera. The Kodak used a roll of thin film instead of the expensive and breakable plates of chemically treated glass that photography had previously required. (A Kodak customer sent the used film, camera and all, back to the company, which developed the pictures and returned them with a newly loaded camera.) With strength in numbers, the two men succeeded where Stevens had failed, and managed to cycle across central Asia. Their published account emphasized that historic crossing, the longest continuous self-propelled journey on record. Their "girdle about the earth" required 15,044 miles in 344 days on the road. The duo claimed no interest in setting a record, merely a wish for the "close acquaintance with strange peoples" which steam travel did not permit. Again, that warm sentiment was slightly offset by the fact that each man carried a revolver.[56]

The next circumcycler, Pittsburgh bookkeeper Frank G. Lenz, decided to use a "safety," like Sachtleben and Allen; to go solo, like Stevens; but to travel westward, unlike any of them. *Outing* magazine agreed to publish "Around the World with Wheel and Camera," and introduced Lenz's dispatches, in 1892, with an image of him atop a globe, wheeling toward a distant sun. Like circumcyclers before and after, Lenz limited his baggage to a severe minimum, assisted by the development of cheap rubber and aluminum goods. He followed railroad tracks and telegraph poles wherever he could. In case of trouble, he carried a revolver. Alone in Asia, he encountered considerable hostility and physical hardship. He had to hire men to help him through flooded Burma, and one of them drowned, circumcycling's first casualty. Lenz pressed on, and had entered Armenia when he vanished in October.[57]

His family, the U.S. government, and various sporting organizations made inquiries. Scattered testimony suggested that Lenz might have been murdered for his revolver and other manufactured goods. Concerned that the incident would tarnish cycling's reputation, Sachtleben went to investigate. While he was in Turkey, controversy over proposed reforms within the Ottoman Empire led to massacres of an estimated 10,000 Armenian civilians. To his credit, Sachtleben realized that these deaths mattered more than one lost cyclist. Abandoning the search for Lenz's body and killers, he returned to the United States and pleaded for American attention to the atrocities against Armenians.[58]

Lenz's death, and yet its international insignificance, made it clear that the Hundred Years' Peace did not actually cover the planet. The eccentric circumnavigators, those who disliked the routine tourist route, had two choices. They could venture into territories where their Western status did not protect them. Or they could tweak the familiar amphibious circuit to make it more interesting in some unusual way. The first choice tested global social relations; the second brought the planet and its physical obstacles back into visibility. In the end, most stunt travelers preferred to take their chances with the planet. By this point in history, the globe seemed more manageable than its people.

"If you do it in seventy-nine days I shall applaud with both hands," Jules Verne told "Nellie Bly" when she dropped in on her way around the world. Bly was the alias of American journalist Elizabeth Cochrane, who in 1889 wanted to circle the world in seventy-five days while filing newspaper dispatches of her progress. First she had to convince Joseph Pulitzer, her boss at the *New York World,* that she was the woman for the job. Pulitzer wanted to send a male reporter, preferably one who knew some foreign languages, which Bly lacked. He objected that Bly would need a male protector, anyway—why not just send the man? Plus, as a "lady," she would surely be burdened by luggage. Bly, who had consulted a travel agent, countered that she could travel with minimal baggage and through the United States and the British Empire, which required neither foreign languages nor chaperon. Pulitzer approved the stunt. All at once, Bly said, "the world lost its roundness and seemed a long distance with no end."[59]

Bly was not just any woman, but a New Woman, a turn-of-the-century ideal of the female adult who, freed from misogynist nonsense, could think, work, and vote as any man's equal. She could even travel independently. She did not need the clutter of trunks and hatboxes associated with a lady traveler, and—crucially—she could be confident that the rule of law would protect her.

Bly's compact baggage was proof of her freedom to move about the world. She ordered a travel dress from a dressmaker (run up in a single day) and prepared a satchel with minimal extra clothing, toiletries, writing materials, and a flask and cup; outdoors, she wore a long checked coat and a hat. Bly also carried gold coins, Bank of England notes, and U.S. currency. She made a last-minute application for a passport and declined to take a pistol, statements of her faith in paper internationalism. Her lack of proper attire meant that she could not attend a formal dinner in Hong Kong. But "the only regret of my trip," Bly claimed (sounding rather like Passepartout), "was that in the haste of my departure I forgot to take a Kodak."[60]

Only as the *Augusta-Victoria* of the Hamburg-Amerika Steamship line pulled away from New York did Bly learn that she did not have sea legs. She waved to the onlookers as long as she could, then made for the ship's rail and vomited. Unfortunately for her, the publicity Pulitzer had generated meant that the world would watch his intrepid reporter puke. In an era when bodily functions were supposed to be concealed, especially by "ladies," public vomiting was a humiliating test of a New Woman. Bly recalled of her first attack, "One man said sneeringly: 'And she's going around the world!' " During her first dinner at sea, she was sick twice, each time hastily excusing herself. On her third presentation at table, her companions cheered her with "bravos."[61]

The perpetual sleeplessness of her tight schedule was also a trial, and made other discomforts seem unbearable. In the end, her average speed around the world was 22.47 miles per hour, and only slightly higher, 28.71, if it excluded her stops, an indication that she didn't stop much. Bly made a publicity-driven visit to Jules Verne and his wife in Amiens, and the necessary leaps between trains and the Channel ferry left little time for rest. The cost of getting Phileas Fogg's creator's blessing was fourteen precious hours. Similar transfers, sometimes in the wee hours, wore Bly down. She resented any disturbance, especially from infants or small children, during precious moments of sleep. She found "happiness in its perfection" when she was alone (and prone) at sea: "Let me rest, rocked gently by the rolling sea, in a nest of velvety darkness." It was beyond endurance when water flooded the steamer between Singapore and Hong Kong. Bly lacked the sang-froid of an English passenger who donned a life jacket and bailed his cabin with his cigarette box.[62]

Bly was aware of the Western consortium that protected her, especially the power and services of the British Empire. Her travels revealed to her "how the English have stolen almost all, if not all, desirable seaports." She endured many singings of "God Save the Queen." When she saw a U.S. flag at Canton in China, she noted that it was the first since her departure. More often, however, Bly identified herself generically as a Westerner. Indeed, her sense that the world's desirable seaports must be controlled by some Western power shows her acceptance of imperialism. And she mocked a telegraph operator at Brindisi who, she claimed, didn't know where New York was. (Unlikely—probably a consequence of Bly's lack of languages.)[63]

By the time she reached the Pacific, Bly was famous as the woman who might beat Phileas Fogg, and Americans were especially invested in having her win. In Asia, people knew she was coming—the telegraph allowed local newspapers to track her progress. On the *Oceanic*, a steamer in the American-owned Occidental & Oriental Steam-Ship Line, the engineer posted signs over the ship's engines: "For Nellie Bly,/ We'll win or die." The private railroad car Pulitzer hired to cross America made a similar effort. Bly's journey from San Francisco to Chicago was the fastest on record—at one point, the train was doing over a mile a minute. Ten thousand people turned out in Topeka to cheer her on, and when she arrived in New York on January 25, 1890, "the cannons at the Battery and Fort Greene boomed." She had made it around the world in 72 days, 6 hours, 11 minutes. Monsieur Verne owed her that round of applause.[64]

Bly had beaten a rival, Elizabeth Bisland, from another New York newspaper, the *Cosmopolitan,* who was going around the world the other way. The circulation of both papers benefited from the contest. "Father Time Outdone!" the *New York World* gloated, with a cartoon (see illustration 13) of their famous globe-circler sweeping along a line of male laggards, Elcano, Drake, Cook, and Train among them. Newspaper editors fell deeply in love with around-the-world stunts, the weirder the better. Other businesses also benefited. When Bly collected her dispatches into a book, printed advertisements subsidized the publication. In one ad, Bly endorsed the Cheque Bank Limited (New York), whose checks guaranteed "ready cash in every port that I visited." Bly's forgotten Kodak was product placement, augmented by a Kodak advertisement on the back of her book. Other travelers and authors juiced up their adventures by referring to her famous feat. In his forgettable fictional circumnavigation, done in verse, "Mister Mucklemouth" rejoices that he crosses the Pacific with "*bly*some Nell."[65]

Another Victorian-era tie-in was a board game. McLoughlin Bros. based "Around the World with Nellie Bly" (1890) on Fogg's cardboard circuit. (The company would also produce a "Race Around the World" that began and ended in London, with peg holes marking a circum-navigator's track over a world map.) The Bly game's winding course from and back to New York includes trials that she had been spared, including shipwreck in the Pacific. The corners of the board represent vignettes from Bly's journey, including the visit to Jules Verne ("Good By Phileas!") and the American train that went "over a Mile a Minute." Americans probably recognized the game as a variant on "The Mansion of Happiness," a moralizing ancestor of "The Game of Life," which represented life as a stay-at-home sequence of virtues to embrace and vices to avoid. Some children might have relished the more adventurous option of racing around the world, and girls, especially, may have seen female planetary confidence as liberating. The cover of the box displays Bly atop a globe. In that new circumnavigator's pose, the New Woman looks triumphant. Also a bit tired.[66]

Her record of seventy-two days lasted mere months. George Francis Train, who had claimed to be the original Phileas Fogg, missed being in the spotlight. He struck a deal with the editor of the *Tacoma Ledger* to make it around the world in seventy days, for a payment of $1,000. Train did it in sixty-seven days, and earned several thousand dollars more from other sponsors. Two years later, he did the circuit in sixty days. He offered his achievements to show "what could be done under stress" and retired while he was ahead.[67]

Nor did Bly's other achievement, as a New Woman who circled the world alone, go unchallenged. Her principal rival was Annie Londonderry, sometimes called the first woman to bicycle around the world. That description vastly underestimates "Miss Londonderry," who was the first round-the-world con artist. She was not named Londonderry, was Mrs. rather than Miss, never cycled when she could hop on a train, and told many fibs in the course of her amiably fraudulent journey. There is no question that Anna Cohen Kopchovsky did go around the world, but that is not very interesting, given how easy it had become. What is interesting is that she was able to lie about her transit and get away with it, despite the modern media that were supposed to speed news around the world.

Kopchovsky was born in Latvia, emigrated to the United States, married in Boston, and had had three children when she decided, in 1894, to cycle around the world. She claimed to be settling a bet between two Boston sugar magnates who had disagreed, at their private club, on women's talents and capabilities. The gentlemen in question (possibly as fictional as Fogg's whist partners at the Reform Club) supposedly offered $10,000 to any woman who could cycle around the world and return with $5,000 above her expenses. That premise allowed Kopchovsky to earn money by lecturing, selling her autograph, and leasing herself and her bicycle as advertising space. The Londonderry Lithia Water Company of New England paid her to assume their name and carry it on her bicycle. "Miss Londonderry" could avoid the anti-Semitism that her maiden or married names might have prompted, and could pose as a New Woman unencumbered by family.[68]

She dispensed many other fictions during her journey, but was never exposed as a fraud—that was her achievement, and a notable one. The telegraph was supposed to have united humanity in common knowledge. When the Indonesian volcano Krakatoa had erupted in 1883, the telegraph made the disaster into what was regarded as the first global news event. Jules Verne had stated that no criminal could hide in a shrinking, interconnected world. And yet petite, charming Londonderry knew better. Only once did a newspaper use the telegraph to confirm her story. Staff at the *El Paso Daily Times* asked their colleagues at the *Boston Herald* to confirm that Londonderry had left Boston to circumcycle the world. The *Herald* affirmed that she had, and no one bothered to trace any of her other claims. Kopchovsky went home a famous woman, several thousand dollars richer, and with a new career as a journalist. Under the byline "Nellie Bly Junior," she retailed her adventures for Pulitzer's *World,* claiming that she, unlike her model, had shunned "the comforts of steamships and parlor cars." It made good copy.[69]

But the stunts were starting to get out of hand. E. C. Pfeiffer had departed Boston to walk around the world in 1894, supposedly on a $5,000 bet, but later admitted the whole thing was a joke. A year later, the *Los Angeles Times* complained that "scarcely a week passes in which some person does not turn up who is bumming his way around the world on some asserted big wager."[70]

Meanwhile, cyclists continued to make circumcycling synonymous with a petulant impatience with other human beings. Two different bicycle teams headed in different directions around the world at about the same time, to their mutual annoyance. A married couple from Chicago went westward from 1895 to 1898, and three Englishmen set out east from 1896 to 1898. They all bore firearms, were intent on setting some kind of record, and were quite put out to bump into each other in China. In their published account, the Chicagoans huffed that the Englishmen had been so very impolite. The English narrative went one better by ignoring the Yankees entirely. Their account is also better illustrated—with photographs, no less—and better written. Indeed, the narrative's directness is disarming: "we took this trip round the world on bicycles because we are more or less conceited, like to be talked about, and see our names in the newspapers." Other nineteenth-century circumnavigators had planetary confidence; this pair had planetary obnoxiousness.[71]

Joshua Slocum pioneered another—and somewhat nicer—round-the-world stunt when he did the first single-handed sailing circumnavigation, from 1895 to 1898. Although Slocum was a working ship's captain, and would present himself as an unworldly old salt, he had shrewdly sold his story to the *Century Illustrated Monthly Magazine* before he put to sea. The magazine published six installments of Slocum's passage as he went along, and one after his return. In 1900, the Century Company would publish the full and illustrated narrative, *Sailing*

Alone Around the World, paying Slocum $2,250 over the next five years, a significant amount of money for a working mariner.[72]

Slocum thought something was missing from the modern world, but wasn't sure what it was, so went to sea to find out. He commanded a wooden sailing ship that would impress people with her size, or lack of it. The *Spray* was a refit oyster sloop, not quite 37 feet, just a shade under 13 tons, and she carried only sail. But she had one astonishing virtue: she was self-steering. Slocum swore that, with a fair wind, he could simply set the boat on course, tie the helm into place, and go below to sleep soundly all night, rising at dawn to drink coffee and make any needed adjustments. He had brought a small library with him, and bragged that he could adjust the *Spray* and return to his book. Pretty much anyone else who has tried to sail alone around the world—numb from lack of sleep—has envied Slocum his incomparable *Spray*.[73]

Slocum's minimal supplies were equally amazing. Rather than pay fifteen ruinous dollars to get his watch repaired, he left it behind. He broke down later, and relinquished a dollar for an old tin clock with a smashed face. The battered timepiece was his chronometer. It eventually lost its minute hand, and sometimes had to be boiled to get the salt out, but it worked well enough for Slocum to know where he was, a small triumph in the development of cheap consumer goods that worked about as well as fancy ship's equipment.[74]

Slocum knew that "the old circumnavigators" had risked "death and worse sufferings" and considered that his "adventures are prosy and tame" in comparison. He thought that his journey's most distinctive element was its solitude. Surrounded by fog in the North Atlantic, he felt himself "drifting into loneliness, an insect on a straw in the midst of the elements." On his longest stretch at sea, he spent seventy-two days alone. He learned to make "companionship with what there was around me, sometimes with the universe and sometimes with my own insignificant self." He kept himself company by singing, for example, and convinced himself that the porpoises enjoyed the music. At the time, it was rare for people to be so alone for so long, especially if they were in motion. When Slocum met the African explorer Henry Morton Stanley, who had been a sailor, Stanley understood the stunt at once: "He looked me over carefully, and said, 'What an example of patience!' " [75]

Slocum makes all his other troubles strangely entertaining. After gathering supplies at the Azores, he lives "luxuriously on fresh bread, butter, vegetables, and fruits of all kinds." But he regrets his "high living" when stomach cramps lay him low. While "delirious" from cheese and plums, he sees a tall man at the helm of the *Spray*. The figure introduces himself as the pilot of Christopher Columbus's *Pinta,* come to guide the ship until Slocum recovers. When the Yankee got back on deck the next morning, he found the boat on course, having made ninety miles through rough weather. "I felt grateful to the old pilot," he allowed, "but I marveled some that he had not taken in the jib." [76]

It is revealing that Slocum thought the worst dangers were ashore, not at sea. Two points of peril were fresh reminders that the global condominium only went so far. Slocum had intended to go through the Suez Canal, for example, but learned that Red Sea pirates preyed on small, unaccompanied vessels. So he took the old clipper route around Cape Horn. But the Fuegians he would meet there, who had been maltreated by too many circumnavigating (and other) passersby, were only slightly less of a threat. Slocum carried a Martini-Henry, a British service rifle with good range. And another captain gave Slocum carpet tacks to scatter each night on the *Spray*'s deck, to turn barefoot intruders into alarm systems. Slocum was quick to fire in the air to frighten the Fuegians, and willing to fire across their canoes if they came closer. But he worried that they could overpower him if they realized he sailed alone. To make the *Spray* seem better manned, he rigged up a dummy sailor and pantomimed an even larger crew: he entered the *Spray*'s cabin in one set of clothes, did a quick change as he passed through, and exited at the fore-scuttle a different man. He briefly considered hiring a companion to round the Horn, but decided against it when the sailor demanded a third man, or else a dog. [77]

Having made his peace with solitude, Slocum considered company as bad as piracy. A fisherman was amazed that Slocum sailed with neither dog nor cat. Slocum considered that he could not have tolerated a dog, and that a cat would not have tolerated him. The governor of St. Helena thought he was doing a kindness by supplying a billy goat.

But Slocum had no chain and the goat could eat through any other leash. After "the beast got his sea-legs on," he snacked on the chart of the West Indies and the captain's straw hat. A sharp-clawed tree crab also favored "tearing up things generally." Two rats stole aboard, plus a centipede and a pair of spiders, but they were not sociable. Slocum preferred the independent animals that came calling—the porpoises and birds who expected nothing from him.[78]

His own independence depended, however, on help in various ports around the world, not least for food. Slocum fished along the way, but portable stores were essential, especially coffee, always with sugar and often with cream, plus salt cod, boiled potatoes, and biscuits. (An onion stewed over the oil lamp was yet more "high living.") It was essential, whenever he made land, to restock his supplies. He did not always need to spend money. Sometimes he hunted and fished. Often, he harvested gifts from well-wishers. Various women in Tasmania supplied the charismatic American with jams, jellies, raspberry wine, and an "enormous cheese." Another female admirer in South Africa baked him a large and durable fruitcake. He managed to gain a pound on the journey. Assistance came in other forms. The British Royal Mail Steamship Company paid his docking fees in Montevideo, plus repairs to the *Spray* and £20 on top. Slocum paid no port charges in Australia except at Melbourne. To cover that cost, he charged for tours of the *Spray*.[79]

Although Slocum was sorry he had missed what he considered the "poetry-enshrined" age of sail, he was not unthinkingly nostalgic. "It is a prosy life when we have no time to bid one another good morning," he said of steamers that did not stop to greet him (or each other). But he appreciated the steamship that broke into his loneliness at Tierra del Fuego, and admired the sight of her, brilliant with electric lights, set against a night dappled by the firelight that had long ago given the place its name. Most pointedly, he mocked the willed ignorance of South Africa's fundamentalist Boers, part of an intensified cult of Flat Earthers. Several Boer dignitaries insisted that Slocum's travels must prove the world to be flat. He told them to "call up some ghost of the dark ages for research." When he was introduced to South Africa's President Paul Krüger as a man going around the world, Krüger objected, "You don't mean *round* the world . . . it is impossible! You mean *in* the world." "Only unthinking people call President Krüger dull," Slocum concluded.[80]

At the end of his narrative, Slocum pondered his place in history. Unlike the explorers of old, he had discovered no new lands. Nor did "the dangers of the seas" prove him heroic. Instead, "the *Spray* made the discovery that even the worst sea is not so terrible to a well-appointed ship" and that "to find one's way to lands already discovered is a good thing." Precisely: the sea had been tamed and most of its coastlines domesticated. Moreover, "no king, no country, no treasury at all, was taxed for the voyage of the *Spray*." Slocum had used the measure of the planet to explore the limits of the individual self. Many others would find that a stunt worth doing.[81]

Chapter 8

Pure Pleasure

Slocum had a notion that the planet was a good test of personal effort, but meet the folks with a completely opposite idea, the rich and the superrich, bless them, who would at the turn of the century make encompassing the world into the work of bliss. Mud-spattered cyclists would continue to grind their way around the globe, but to do the circuit in comfort, with a cocktail rather than a revolver in hand, was the adventure that many people preferred, and still do.

The around-the-world cruise for pleasure represented the peak of planetary confidence. It cleared from the journey all the historical and material struggles that had made the voyage possible, lest the grimy effort of taming the Earth disturb the passengers. That was wonderful. But the problem was, the toilsome loners who circled the world the hard way consistently produced better stories about their ordeals. Doing the planet aboard luxury liners, in the nicest conditions history would ever afford, made for a passive experience. Rather than being authors of their own geodramas, the pleasure cruisers merely enacted what the travel brochures had scripted for them, and their contented silence constitutes a big break in the tradition of the circumnavigator's engaging first-person account.

At first, the travel brochures were nearly as hard to navigate as the planet. *Gaze's Tourist Gazette* (1895), produced by Henry Gaze & Sons of London, is representative in its resemblance to a Bradshaw. It is printed in black and white, with only a few line drawings (mostly advertisements). With dense pages of small type, and with no cross-indexing, it puts the burden of imagining the world onto the reader. The guide nevertheless has an astonishing global reach. Prospective travelers could select a route and collateral services for a particular destination (or a few) in any part of the world that was accessible to travelers with Western standards of physical comfort.

World travel was being made easier, not least by agencies like Gaze's, even as demand for it rose, especially among those with means. The global distribution of wealth and incomes was becoming more stratified, with enormous wealth at the very top, both within industrializing societies and in contrast to non-industrialized places. The greatest individual fortunes were made by manipulating the planet's raw materials or spanning its distances—mining, lumbering, railroads, shipping—or by servicing those industries, especially through banking. Physical resources were developed at an accelerating pace. From 1850 to 1900, the amount of coal being mined rose tenfold, from 76 million tons to 760 million. The built environment spread, as railway tracks and paved surfaces proliferated. World maps increasingly featured artificial networks, such as telegraph cables, shipping routes, and transcontinental railway lines.[1]

Gaze's Tourist Gazette reflects that expansion with its pages of timetables and hotels, as well as multiple options: round-trip, one-way, side trip; sleeping car versus parlor car; accommodation with meals or without. Lists of hotels where Gaze's Coupons would be accepted fill up many pages and, for places where the agency had not brokered use of its coupons, there was a separate list of recommended hotels. But the choices were overwhelming.[2]

One solution was to have the travel agency piece together anything complicated, such as an around-the-world tour, in instances where travelers did not want an all-in-one tour with a group. Gaze & Sons were eager to help. Given how "the variety of routes is so large" throughout the world, "travelers are strongly recommended to consult us personally, if possible." They also described their "Circular Tours Around the World" and offered a free brochure, by mail, for a September 1896 departure limited to twelve people. The smallness of the party implies that the Gazes' staff might select its potential members, so that the right kind would not have to rub shoulders with the wrong kind, whose manners, dress, accent, or ethnic heritage would give them away. Even an inquiry by mail for the brochure would reveal, from the writer's address, grammar, or quality of stationery, whether he or she should be among the chosen dozen.[3]

A little context may explain the desire for social exclusion. The rich may have been on the move, but so were the poor. The late 1800s and early 1900s were peak decades for transoceanic migrations done out of economic necessity or to escape persecution. Between 1846 and 1932, over 60 million people migrated from Great Britain and Ireland, Europe, Scandinavia, and Russia. Many of them went to the Americas; others scattered through the British Empire. From 1846 to 1940, at least 100 million Asians undertook long-distance travel as well. The unprecedented mobility prompted new vigilance about passport control, against the prevailing tide of permissiveness in international travel. Some better-off passengers found their counterparts in steerage inoffensive, even interesting; others resented sharing a ship with them. Andrew Carnegie went to see the eight hundred homeward-bound Chinese in his vessel crossing the Pacific and found the "voyage less satisfactory" for the experience. Many ships that connected Asian ports confined non-Europeans to a section that could be barred and bolted. That was done in case pirates, posing as passengers, tried to seize the ship. White passengers could find the precaution either reassuring or distressing.[4]

Proximity to the poor was not what leisure travelers wanted. The American social theorist Thorstein Veblen, in his *Theory of the Leisure Class* (1899), said the era's elite were obsessed with a "conspicuous consumption" that was as different from productively physical labor as possible. In some ways, a de luxe round-the-world tour was a classic leisure-class activity. It was expensive and required that months or even years be dedicated to it, impossible for anyone who had to hold down a job. But both the conspicuousness and the consuming nature of an around-the-world tour were distinctive. A luxury circumnavigation demonstrated how members of the economic elite who harvested and rearranged planetary resources could also roam freely around the planet.

That correlation was clear in the founding of the Circumnavigators' Club in 1902. The club was formed after three strangers, thrown together somewhere on the Indian Ocean, realized that going around the world represented a world of its own. Circumnavigators did essential work in creating globe-girdling circuitry and might themselves form a social network. "To be eligible," under the original terms of admission, "a man must have made a circuit of the globe, longitudinally, and be vouched for by three members." The most desirable members were "the men who, far from the noise, and lights, and gayety of great cities, make history quietly and without ostentation," by doing such essential tasks as assembling "a dredge in New Zealand, hanging a steel bridge over the upper Nile or provisioning a warship at Singapore." The club flourished; John Philip Sousa composed a "Circumnavigators' Club March" for it.[5]

A similar sense of masculine entitlement comes across in a child's moving panorama of 1905, "Voyage Around the World by a Little Frenchman." A boy (or girl) lucky enough to possess this toy could set up a brightly colored cardboard theater, about 15 inches wide, 15 inches tall, and 3 inches deep. This proscenium or frame had vignettes of global travel printed on its top and sides, with an international orchestra depicted at bottom. A roll of scenes could be cranked through the frame, each tableau depicting one of the little hero's twenty-four stereotypical adventures: fleeing wolves while dog-sledding across Siberia, resisting cannibals in Polynesia, joining a seal hunt in Greenland, and so on. A lamp could be placed behind the scrolling panorama, an effect that made the proxy geodrama resemble that newest of entertainments, the cinema.[6]

But most real travelers undertook circumnavigations for pleasure, not to be chased by wolves. They wanted to step onto a global stage that had been carefully cleaned of any dirt, strife, and toil, even when they had caused it.

Collver Tours of Boston catered to that preference when they explained, in 1908, the advantages of their *Exceptional Journeys Under Escort*. Their round-the-world option would be distinctive in "eliminating all the irritating 'commonness' of the usual traveling party." A party of social equals would be less lonely than traveling solo. It was also troublesome to meet border guards and baggage inspectors alone, or to wrangle family and luggage up and down multiple gangplanks. Rather than face that, wealthy travelers could select from Collver round-the-world tours that ranged in price from $1,750 to $4,750, at a time when an unskilled worker in the U.S. shipbuilding industry survived on an annual $497. Collver's clients could tailor their variations and side tours as they pleased. (Eastward or westward passage around the world? Rickshaw or carriage for a day trip in Asia?) On one discretionary excursion, the maharajah of an Indian state arranged for elephants to carry the tour party. Another variation offered lovers of the lovely a visit to Japan during "Wistaria and Iris time."[7]

Often, there were touches of educational uplift. Collver Tours assured clients that their itinerary included a road "used by Akbar," the sixteenth-century Mughal emperor. They also quoted President Charles William Eliot of Harvard College on his "grave doubt as to whether a college course or world travel is of the greater value." And yet it was a spotty education. Science was poorly represented whenever the tourists crossed the 180th meridian, now commonly called the International Date Line. It was becoming the custom to have ceremonies to commemorate the crossing, gentler versions of the sailors' ritual of crossing the equator. But the equator marks a natural division between warmer and cooler zones, whereas the 180th was a political designation. Yet two Americans who went around the world in 1910 proclaimed that the 180th meridian was where "the Occident dissolves into the Orient," as if nature had demarcated the cultural zones of "West" and "East." The two men also believed themselves to be following the Sun around the world. So did many others. The pre-Copernican idea of the human race against the solar ball had survived into another century.[8]

The prestige of the luxury tour appealed to non-Westerners as well. In 1908, a Japanese publishing company, Asahi Shimbun, arranged with Thomas Cook for a packaged round-the-world tour. Within five days of the tour being advertised, all fifty-six places had been filled. Social selection was as important for this tour as for any Western travel party: company officials excluded participants whom they thought might not best represent Japan, and there was much dithering over what the selected travelers should wear. Similarly, when Eiichi Shibu-sawa went around the world independently in 1902, he was concerned to convey that Japan had industrialized and Westernized to "a high-collar" level.[9]

Whether done independently or under escort, the pieced-together, steam-powered around-the-world journey was wonderful, yet tedious. Those dual responses appear in the accounts of two men who, within three years of each other, went around the world in similar conditions. One was the famous American Samuel Clemens, aka Mark Twain, and the other was the not-so-famous Frenchman Jean d'Albrey. Each man's narrative suggested improvements to the commercial circumnavigation.

Clemens was at the height of his fame in 1895 when he went around the world to make money. His writing had generated a small fortune, but after some investments had gone awry, he was broke. Banking on his famous alter ego, Twain mapped out an around-the-world speaking tour. With his wife, daughter, and agent in tow, he earned huge fees for his talks, and he would publish a narrative of his travels, *Following the Equator* (1897), that generated yet more funds.

Much of the narrative follows the Grand Tour model of an around-the-world voyage, with Twain ticking off the sights he knows he is expected to see. "For ever and ever the memory of my distant first glimpse of the Taj," he rhapsodizes, "will compensate me for creeping around the globe to have that great privilege." That's nice, though it doesn't sound like him. He is better on Hawai'i, especially its reduc-tion to a tourist paradise littered with bicycles, not to mention "rugs, ices, pictures." Twain had been to the islands before they were caressed by these luxuries, when ice had to be hauled from New England at $600 a ton. He deplores how "the ice-machine has traveled all over the world. . . . In Lapland and Spitzbergen no one uses native ice in our day, except the bears and the walruses."[10]

For Twain, the real pleasure of an around-the-world tour was escape from the bother of daily life. For that reason, the size of the planet was its best feature, especially when experienced at sea—the bigger the ocean the better. "There is no mail to read and answer," he rejoiced, "no newspapers to excite you; no telegrams to fret you or fright you—the world is far, far away." The Atlantic was the least satisfying part.

"Voyage too short, sea too rough," he complained. The Pacific and Indian oceans, larger and calmer, were perfect, day after "reposeful" day. There in the tropics, everyone sleeps in the breezy open cool of the deck. At five in the morning, the signal comes for the decks to be washed. Women and children retreat below. The male passengers go down to bathe, then return to stroll the damp deck in their pajamas, take a little fruit and coffee, be shaved, and idly watch the ship's cats groom themselves. No one has any interest in the daily postings of the ship's progress. No one wants the voyage to end—"if I had my way we should never get in at all," Twain claims.[11]

He notes the day lost in the crossing of the 180th meridian, "the center of the globe," as he thought it. He concludes that he and the other passengers will all die a day earlier than they had been foreordained, a Twain-like swipe at the long-standing Christian perplexity over the Circumnavigator's Paradox. "We shall be a day behindhand all through eternity," he predicts. "We shall always be saying to the other angels, 'Fine day today,' and they will be always retorting, 'But it isn't to-day, it's to-morrow.' " Circumnavigators might as well go to hell, given the diabolical confusion they will cause in heaven.[12]

Twain continues to note the unnaturalness of round-the-world travel. Steaming between hemispheres entailed rapid transit among very different climates. "It is odd, these sudden jumps from season to season," as if a real year were being rearranged en route. "A fortnight ago we left America in mid-summer, now it is mid-winter" in Fiji, which was actually quite warm; "about a week hence we shall arrive in Australia in the spring," with temperatures cooler than Fiji's. And yet Twain is struck that his memory moves faster than any steamer or locomotive. When, in India, he sees a servant slapped in the face, he remembers the casual humiliation of slaves in the Missouri of his childhood: "Back to boyhood—fifty years; back to age again, another fifty; and a flight equal to the circumference of the globe—all in two seconds by the watch!"[13]

That was a rare snatch of social commentary; rumors that Twain's narrative is a critique of imperialism have been greatly exaggerated. He jokes that every nation has territories that "consist of pilferings from other people's wash." And he predicts that "all the savage lands in the

world are going to be brought under subjection to the Christian governments of Europe." To which he adds: "I am not sorry, but glad." His
acquiescence may indicate his dependence on the British-run, English-
speaking empire for his tour. The Twain circumnavigator's track ran
from one Mark Twain club to the next.[14]

Twain was indeed a global commodity. Two years later and several
oceans away, Jean d'Albrey found Twain's writings very consoling during the sea travel that he (unlike Twain) detested. On his circumnavigation (late 1897 to early 1898), Albrey resigned himself to a set routine
on each ship's passage: forty-eight hours of seasickness, forty-eight
hours of enchantment with the seascape, and then weeks of boredom.
He noted that even the first-class accommodations aboard the *China*
(an American steamer that took him across the Pacific) were poorly
lit, with small portholes. The darkness, plus his seasickness, made it a
"sad" experience, with a persistent sense of "utter incarceration." Meals
were copious and service lavish, which passed the time, though Albrey
said that the food was not actually good. (Twain had less interest in
what he ate.)[15]

It helped when Albrey's fellow passengers were interesting. He enjoyed the Christmas he spent eastbound to Hawai'i, with Champagne,
plum pudding, and singing around the piano. Even better, a friend who
accompanied him part of the way helped to "kill time in lectures and
readings." Twain was a favorite, and fortunately the *China*'s library had
a rich selection of the American humorist. But without companionship,
it was another kind of incarceration. In losing its danger, the ocean had
lost "its epic grandeur." In contrast to Twain, Albrey found his Atlantic
crossing so satisfactory—faster and more luxurious than the Pacific
leg—that it merits no detail in his narrative.[16]

Whatever their different appreciations of sea travel, Twain and Albrey both hated the rigamarole of border crossings, sticky rituals in the
otherwise smooth arcs they made about the planet. The Frenchman
was openly resentful that Britain ruled the waves, perhaps because he
didn't depend on English-speaking audiences for his tour. He complained that everyone in the world knew about England but that only
the educated knew about France. He and his French companion found
American baggage handling so confusing that, for a hair-raising inter-

val in San Antonio, they thought they had lost everything. Twain likewise hated baggage transfers and inspection, though in this regard his
fame was both passport and visa. In Frankfurt, he received a few lines
from the Italian consul general on stationery bearing the Italian royal
crest. The letter guarantees a "beautiful bow" from the stationmaster at
the Italian frontier. The official barks to his underlings that they are not
to paw through the mountain of Clemens luggage. Even when Twain's
contraband German cigars tumble from his pockets, the stationmaster
steers him to untaxed safety.[17]

Twain saved for last his best zinger about the folly of going around
the world. "It seemed a fine and large thing to have accomplished,"
he concluded, "and I was privately proud of it. For a moment." Then
he read "one of those vanity-snubbing astronomical reports" that
described an astral body "traveling at a gait which would enable it to
do all that I had done in *a minute and a half.*" So much for geodrama:
"Human pride is not worth while; there is always something lying in
wait to take the wind out of it." [18]

Travel specialists could not help Twain with that, but they were
eager to erase his and Albrey's other complaints. Shipping lines could
make accommodations nicer. Since the late 1870s, passenger ships had
begun installing electricity, so that more light was available belowdecks,
along with other electrical amenities, such as fans, which rendered
sleeping on deck in warm weather less necessary. Shippers could provide more entertainment than a piano and library. They could improve
meal services, especially once electrification made refrigeration possible at sea. They could make passengers believe they were the ship's
priority—passenger ships were becoming differentiated from freighters
or mail delivery packets, more welcoming to women, families, and passengers with particular needs. (In 1908, Annabelle Kent, who was deaf,
could go *Round the World in Silence*, as she described it.) Inter-ship and
ship-to-shore telegraph services could make travel safer and more predictable. Above all, long-distance shippers could reduce the recurring
tedium of transfer from one conveyance to another.[19]

It was to meet these needs that the Frank C. Clark Travel Agency of
New York chartered the *Cleveland,* of the German-registered Hamburg-

Amerika Line, to take touring parties on circuits of the world. The *Cleveland* had been built to accommodate 700 in first class, 500 in third class, and 2,000 in steerage. But only first-class passengers were booked on its first round-the-world tour, eastbound from New York City in 1909, at a cost of $650. Clark had seen to all the little details, including arrangements for day trips. His formula worked, and he repeated it in 1910 with a westbound tour from San Francisco. The ship did not do an actual circumnavigation. Clark knew that leisure travelers did not want to round the Horn; he helped arrange transport across North America—if necessary. It was up to the passengers as to whether they wanted to claim they had made a full circuit of the world.[20]

They tended to—that was the point—and Clark would successfully repeat the tour in subsequent years, always with North America as the unfilled gap on the itinerary, and often with the Atlantic filled by another Hamburg-Amerika liner. Passengers on the 1913 tour who took that option to return to the United States did think of themselves as circumnavigators. Indeed, they welcomed the attention they received during the Atlantic crossing: "as soon as it became generally known that we had circled the globe, we were somewhat in the limelight, and some of the globetrotters were anxious to hold the center of the stage."[21]

Clark's world tours were, at the time, famous landmarks in luxury travel. Thousands of people turned out in San Francisco to wave off the *Cleveland* in 1910 and 1913. By the second tour, the ship's arrival was a known opportunity for civic boosters and local tradespeople. In Honolulu, the Hawaiian Promotion Committee sent a welcoming party of "ladies and flowers," meaning young Hawaiian women bearing leis. The international fees for such a voyage were steep. On its first tour, from New York to San Francisco, the *Cleveland* had evidently paid a $1,000 fine for violating U.S. laws against foreign vessels carrying passengers from one American port to another. She had also paid $25,000 to get through the Suez Canal, or so one passenger claimed.[22]

Another traveler described the *Cleveland* as a "small town," a familiar place to which passengers (overwhelmingly American) could cling as they passed through foreign lands. The size was about right, given the roughly 1,200 passengers and crew, though the social divisions were starker, because separated between the paying and the paid. (The captain and officers were exceptions, though their authority over passengers would have been tacit, except in emergencies.) The ship had a full range of services, including a laundry, a printing office, a darkroom, and a small German band. She also had ample space for small-town socializing, either organized by the staff or by passengers themselves. The crew held celebrations for crossing the equator, for example. On the second *Cleveland* tour, one man, a Pennsylvania photography enthusiast, founded two clubs, one for passengers from the Keystone State and the other for camera-carriers, Clark's Cruise Camera Club. The energetic Pennsylvanian also served as chairman of the Dance Committee, master of ceremonies at the Progressive Euchre Party, and secretary for shipboard meetings of the Ancient Arabic Order of the Nobles of the Mystic Shrine, the Shriners. One would have to be bored with life to be bored aboard the *Cleveland*.[23]

Then there was the food. Supplies included 26,000 grapefruit and 6,000 bricks of ice cream. To give a sense of scale, the grapefruit would have taken up three railroad cars. The *Cleveland* also carried a railroad car's worth of lemons and two of potatoes, plus 53,000 pounds of butter, 350,000 eggs, and 2 tons of sugar; 160 tons of ice kept the ice cream frozen and the drinks chilled. Everyone seemed quite satisfied with the food, once they got over any seasickness. The meal services provided physical and cultural continuity through foreign countries. After time in Japan, one man considered the "good hot dinner" awaiting him aboard the *Cleveland* "very welcome, I can tell you."[24]

It was in Japan, however, that passengers on the *Cleveland*'s 1910 voyage could peer into the wings of the stage, as it were, and see the work and resources that made their geodrama possible. That was because the ship re-coaled in Nagasaki. At six in the morning, about one hundred barges bearing coal arrived. At least 1,500 Japanese, mostly women and girls, came with the fuel to load it aboard the ship. They rigged up bamboo scaffolding on which ascending layers of them could stand. Small baskets were filled with coal from the barges, then passed up and transferred to the hold. The Japanese worked until ten that night, then came back the next morning to finish the job by one thirty in the afternoon. One passenger reported that the 8,000 tons of coal had cost

$32,000. Little of that went to the workers, who earned 15 to 30 cents per day, with women and children at the lower end of the scale. Two of the American passengers marveled that the workers accomplished it all "with the rapidity of machinery." Another passenger said that the record in the harbor was 1,500 tons of coal loaded in two and a half hours, which was faster than machinery could have done it.[25]

The process was unusual enough that all of the printed accounts of the *Cleveland*'s 1910 voyage (written by four individuals) described it, and at least two of them photographed it—the pictures appeared in their published accounts (see illustration 14). "We have seldom witnessed a more interesting sight," said one passenger. Another deemed it at least as entertaining as a fireworks display over Nagasaki Harbor. Their astonishment says something about their social status. They had probably never seen coal mined or even delivered because their servants took charge of it. Had they simply been crossing the Pacific, rather than going around the world, they might have disembarked from the *Cleveland* and missed the spectacle.[26]

Certainly, passengers on the pieced-together route noticed re-coaling only if their occupations led them to comment upon it. Li Gui, China's first round-the-world traveler, had described it, perhaps because he was interested in the status of work and workers in the West. At Aden in the Suez Canal, Li noted that re-coaling took from 12:30 to 8:30 p.m., as a gang of workers transferred the fuel bit by bit, by hand, into the ship's hold. All that coal, he added, provided not only propulsion but pleasure. The ship's engines were used "to change seawater into freshwater and to make ice." Also in the 1870s, the circumnavigating Reverend John Dunmore Lang had condemned re-coaling if done on the Sabbath. The workers in Brindisi defied the Lord's commandment. Those in Kandavu, Fiji, did not, in contrast to the "irreligious example" of white passengers who indulged in all manner of secular activities on that very day. Nellie Bly, as an investigative reporter interested in social reform, was distressed to see the "hurrying, naked people" who re-coaled ships at Port Said.[27]

The coal was not objectionable, but the labor was. Machinery was supposed to be taking over this kind of work—dominion over the Earth was supposed to have replaced dominion over people. For that reason, many around-the-world travelers disliked being conveyed by men in rickshaws and sedan chairs. Bly's first rides in Asian rickshaws gave her a "shamed feeling." She repeated the common joke that *rickshaw*, from the Japanese, meant a human-powered vehicle, or "Pull-man car," an Eastern version of the West's luxury railroad car. One American aboard the *Cleveland* said of a Tokyo rickshaw that it "didn't seem just right" to be pulled by a man who weighed less than he did. He made similar comments about sedan chairs in China. (And yet he was disappointed, in Rangoon, that steel cranes did the work elephants had once done.) Another of the passengers simply described riding in a rickshaw without comment; sedan chairs didn't bother him, either. Two Americans noted of sedan carriers in Asia that "sympathy for them spoils the pleasure or even comfort of riding." A later passenger was "ashamed" that he, at 208 pounds, was pulled by a man weighing about 130.[28]

It seemed like unfinished business. The physical burden on humanity, when still visible, was troubling. The concern was laudable, though it is doubtful that any of the *Cleveland*'s passengers went home as friends of organized labor. Usually, they concluded that American workers should be thankful to have been relieved of the kinds of tasks still done by hand elsewhere. Nor did elite passengers worry over the significant amount of planetary resources necessary to go around the world—coal, after all, might eventually power labor-saving devices. Unlike Passepartout, they accepted the gas lamp burning in the middle of the planet; but they hated to see the bodies laboring to tend the lamp, and wished them away.

They got their wish. Increasingly, long-distance travel was done with the internal combustion engine. As the name suggests, such engines consume a liquid or gas within, rather than burn fuel (usually wood or coal) outside the boiler to create steam pressure. The new engines seemed cleaner, certainly compared to coal's characteristic black smoke, and their fuel could be transported and transferred using tanks and hoses, with much less human labor. Gas engines had been used on some trains before the turn of the century (cyclist Thomas Stevens had seen one during his expulsion from Afghanistan) and diesel engines, developed in the 1890s, were adapted to ships in the 1910s and to railroad locomotives thereafter. Coal and wood would remain standard materials for heating and cooking in many parts of the world, but the age of steam was ending.[29]

That would be a great relief to the round-the-world travelers who did not like to see dirt and labor. Sir Frederick Treves's much reprinted and widely admired *The Other Side of the Lantern* (1905) expresses that preference quite clearly. Right at the start of his journey, on his way to the English Channel, he complains of the dirt, mud, and rust that, to his mind, besmirched the countryside: "The huge, beery ogre of labour, dirty and sweating from his work, has thrown himself down in the lady's garden." Treves was briefly consoled by some beauty spots in the Mediterranean, but then there was Port Said, where he found "the West at its worst" and "the East spoiled." The nadir was the re-coaling, which he describes as deadly in every sense. The coal barges arrive like "the rafts of Charon" on the River Styx. Deck chairs are piled up in a "bier" and the whole ship, shrouded in canvas and slowly infiltrated by black dust, becomes "a house of mourning." He hates the sight of the workers, whom he describes as "eager insects," a "mob," "bony witches."[30]

An around-the-world leisure traveler expected to be spared these hints of the material struggles that sullied the world, even the struggles that were necessary to make the journey in the first place. Treves did his circuit just a bit too early. Three years later, Frank Clark announced his fifth cruise around the world, aboard the "New Oil-Burner 'California,'" guaranteed to have "No Dust or Dirt."[31]

A sharp-eyed circumnavigator could still bear witness to the effort and cost of modern global transport, especially outside the industrialized parts of the globe. When he went around the world in 1925, Aldous Huxley watched part of an all-night operation to unload potatoes in Port Said. "Moving bits of matter from one point of the world's surface to another," he reflected, was "man's whole activity." He compared it to "these particles of ink, for example, which I so laboriously transfer from their bottle to the surface of the paper." The potato-movers might have disagreed.[32]

Slocum had done a circumnavigation that more convincingly joined writing to physical labor. A trio of travelers followed his lead. All were

Americans, and each chose an unusual way to go around the world; the first was a friend of the workingman, the second should have been, and the third only pretended to be.

In the early 1900s, John F. Anderson worked his way around the world. Of course, sailors and other ships' employees by definition did that. But Anderson pioneered an informal strategy for able-bodied travelers who had neither money nor maritime skills. He belonged to a Christian evangelical group, the Endeavorers, and had decided to see the wonders of God's creation throughout the world. His conviction that the world's other religions were false was softened, somewhat, by his genuine sense of fellowship with ordinary people. He wanted to see a world of "common people" and "humbler things," and deplored how most round-the-world travelers avoided exactly that. "Globe Trotters," he thought, were "seekers after notoriety who adopt various methods of making their way, and whose only purpose is the winning of a wager or the satisfactory completion of a 'freakish' undertaking." Nor did he like the alternative, tourist voyages that generated "guide-booky" narratives.[33]

Anderson had made sure to acquire skills that would be useful in both industrial and pre-industrialized societies. He could take care of people (by cutting their hair and cooking their meals), knew how to tend domesticated animals, and had a knack with machinery. These skills, rather than money or foreign languages (or a gun), would see him through. He departed Pomona, California, in a horse-drawn buggy and with eight dollars. In Seattle, he bought and broke a fresh team that took him to New York via Yellowstone and the Grand Canyon. He barbered in Brooklyn, took a bicycle tour of New England, visited Texas, then returned to New York to embark for Liverpool. By arranging to tend livestock on the ship, he paid only five dollars for his eleven-day passage.[34]

Anderson perfected a rhythm of irregular work and cheap travel. In London, he raised a little money by working in a barbershop, then cycled through Europe. He barbered in Rome and sold bicycles in Naples. He paid passage from Athens to Alexandria by cooking in the galley of a steamer—hot work, surely. He did barbering in Nazareth, then worked in the galley of a mail packet between Port Said and Bombay (even hotter). He landed a place at Bombay's "Pall Mall Hair Dressing and Shaving Saloon, the finest barbershop in Asia," which subsidized his travel in India. When he ran low on money, Anderson paused in Darjeeling to retool as a Remington typewriter salesman. This was good evidence of the transnational appeal of shiny gearwork, especially given that Anderson worked for a "Mr. Bomivetsch," possibly a Russian, from Britain's opponent in the Great Game. (In this pre–Cold War tangle of espionage, perhaps Anderson was "our man in Darjeeling"?) He earned his best money as a barber in Manila, but was nearly stranded there. He could not afford a ticket home, and Pacific steamers preferred "Oriental labor." Anderson finally convinced a U.S. transport to hire him as the third cook in an otherwise all-Asian crew and returned to California after five years, alive and well, $65 in his pocket.[35]

His account, published in 1903, is a rare memoir that discussed the labor necessary to get ships around the world. Anderson thought it good and satisfying work. His willingness to cook in small galleys with people who were from the southern Mediterranean, Near East, and Asia showed how he, unlike first-class passengers, did not mind physical contact with people of other classes and races. That was probably apparent when he worked in his Bombay and Manila barbershops, as well. As with Slocum's circumnavigation, Anderson's tour of the planet did not entirely reject the imperialist nature of the networks that sped Westerners around the world, but it stood apart just enough to reveal another set of possibilities.

That was not the case for the second American who chose an unusual way to circumnavigate the globe. Jack London was famous for his adventures. He had experienced the Gold Rush in the Alaskan Klondike, had covered the Russo-Japanese War, had written stories of extreme travel, and had identified himself with radical political criticism, to the unpopular extent of supporting socialism and the unionization of labor. He was high on the success of his novel *White Fang* in 1906 when he read *Sailing Alone Around the World* and saw in Joshua Slocum a kinsman. He dared his wife, Charmian, to contemplate a similar voyage; she called his bluff. By the time the idea grew into an actual plan, several of London's recent works had been reviewed negatively. He wanted an escape and believed he could keep up his daily quota of

1,000 words wherever he was. Moreover, he calculated, a trip through so many oceans and to so many places might keep him in material for the rest of his writing life.[36]

The rationale for the voyage that appears in London's written narrative stressed not the parts of the voyage, however, but the overall size and power of the planetary forces that he would defy. London cited Slocum on the wonder of measuring the self against the world: "It is the old 'I did it! I did it! With my own hands I did it!' " But if Slocum's measured self was his soul, the being that communed with the Creation, London's self was his body, the corpus that fought the elements. He thought of himself as "a little animal called a man—a bit of vitalized matter" that could only survive under a sorrowfully narrow set of conditions, tiny and frail against "the great natural forces—colossal menaces." To take himself around the world in a small boat defied his mortality: "I dare to assert that for a finite speck of pulsating jelly to feel godlike is a far more glorious feeling than for a god to feel godlike."[37]

It was a terrific, stirring premise for a circumnavigation. But the venture would strike a great many people as odd: America's most famous socialist built a $30,000 private yacht to sail around the world. To be fair to London, the boat should have cost less, closer to its initial estimate of $7,000, but the San Francisco earthquake of 1906 had driven up the costs of material and labor. On the other hand, he alone chose the inauspicious name of *Snark* for the vessel. Lewis Carroll's cryptic poem of 1876, *The Hunting of the Snark (An Agony in 8 Fits)*, describes a journey across an ocean using a map that is a blank sheet of paper, in order to find a creature that turns out to be able to make its pursuers vanish. If London's *Snark* were two-thirds *Spray* in her rough-tough sense of American independence, the other third of her resembled Lady Brassey's pleasure-seeking *Sunbeam*. The *Snark* had a gasoline-powered engine, to assist sailing in rough weather and to power a generator for electrical lighting and appliances, including an ice machine. The Londons also installed a bathroom and flush toilet, and they carried a typewriter, phonograph, and library of books and records. But their sharpest contrast to Slocum was their small arsenal of handguns. In the Solomon Islands, fearing attack if they tried to land, London regretted not fitting the *Snark* with a machine gun, as originally planned. That wasn't a kind way to face the world.[38]

The Londons expected to do some of the glamorous maritime duties, but not the cooking and cleaning. When they advertised for a cook, hundreds of applicants vied for the place. A great many young men recommended themselves, some of them offering to pay for the privilege of sailing with the author of *The Call of the Wild*. Grizzled sea captains were similarly enthusiastic, as was an experienced female cook of fifty. The winning candidate, a young man from Kansas, had heightened his appeal by claiming an expertise with photography and some travel experience as a ship's stowaway—"I only mention this so you will not think I have been on Cooks Excursions." The outpouring of volunteer circumnavigators heartened London. "No, adventure is not dead," he rejoiced, "in spite of the steam engine and of Thomas Cook & Son," another swipe at the tourists. He threatened to sail someday around the world on a ship that could accommodate 1,000 volunteers.[39]

The promise to go around the world *again* quickly turned into a wistful hope to do so at all. Nothing went right. After the *Snark* had been tested at sea, she was crushed between two barges in the port of Oakland. The damage, and several dozen flaws in the boat's construction, only became apparent in the middle of the Pacific Ocean. Meanwhile, Charmian London and several others fought seasickness. Seawater flooded the stores of food; the engines had every kind of problem. It took five months of repairs at Hawai'i to make the *Snark* seaworthy again, and most of the crew and servants had to be replaced. Even after that, the boat's freshwater tanks leaked. Water had to be rationed during an uncomfortably hot spell near the Marquesas.[40]

Amid it all, London was determined to teach himself navigation. He stocked the *Spray* with textbooks and instruments, and had a knowledgeable tutor in one crewman, Roscoe Eames, Charmian's uncle, who had been the foreman for the *Snark*'s construction. But it was harder than it looked. London charmingly details his ignorance of the "priesthood" of navigators. He is perceptive in pointing out that his post-industrial view of the cosmos is flawed. In making his noon observations to determine latitude, he learns that "the sun, which is the timekeeper for men, doesn't run on time," meaning the clock time of the landbound. "When I discovered this," London deadpans, "I fell into deep gloom and all the Cosmos was filled with doubt." Even "immutable laws, such as gravitation and the conservation of energy, became wobbly." His thorough explanations of his many mistakes are mischievous. Anyone who is learning celestial navigation should *not* read Jack London on it, lest they repeat his many commonsense errors. He had just about figured it all out, for example, when he remembered the mnemonic:

> *Greenwich time least*
> *Longitude east;*
> *Greenwich best,*
> *Longitude west.*

It only works if you already know how to navigate. Eventually, London did.[41]

He praised himself for knowing more about the planet than Uncle Roscoe anyway. The old fellow followed the teachings of Cyrus R. Teed, who spread the faith that the Earth was hollow. It was an old idea, though Teed embellished it with beliefs about immortality and free love, which gained him notoriety and acolytes. "Though we shall sail on the one boat, the *Snark*," London marvels, "Roscoe will journey around the world on the inside, while I shall journey around on the outside." As with Slocum confronting the Boer Flat Earthers, London put himself firmly on the side of modernity, as if stunt circumnavigators were continuing a kind of scientific investigation of the planet.[42]

So many illnesses plagued the crew, however, that London, confident that suffering could be recycled as journalism, joked he would someday write a book called "Around the World on the Hospital Ship *Snark*." In the tropical Pacific, enchanting at a distance, the heat and insects were torments. Malaria and other fevers burned the gumption from everyone. The smallest cut, scrape, or scratch became infected. The yacht's animals died off from various accidents except for a terrier, which could only move by hopping on its two unsprained paws. The resident cockroaches grew and swarmed and gnawed the finger- and toenails of sleepers. Worst of all, the skin on London's hands swelled and thickened until he could no longer grasp a rope or a pen. (His dermatological malady remains a mystery; it may have resulted from an epic sunburn in Hawai'i, where London learned to surf.) At Australia, in December 1908, he called it quits. After a year convalescing, he and Charmian sold the *Snark* and went home. In 1911, he would publish *The Cruise of the Snark*, to mixed reviews. "It is a pity that the joyfulness of the voyage is not communicated to the reader," complained the reviewer from the *Nation*, who somehow missed that the voyage was a failure, a failed circumnavigation.[43]

Harry A. Franck did make it around the world. His original plan was to do it "without money, without weapons, and without carrying baggage or supplies." He modified the plan by taking a Kodak and $104. After arranging to sell dispatches to three magazines, he set out to prove that "random wandering" could yield as much edifying experience of the world as any "personally conducted tour."[44]

It was a beautiful idea, though ugly in execution. Franck intended, like Anderson, to work his way across oceans. By tending and bunking with cattle, he paid only £5 to cross the Atlantic. But once in Europe, he could not find work. An economic slump depressed demand for labor and Franck lacked Anderson's useful skill set. He became a vagabond traveler, dressing in old clothes and tramping rather than paying train fare. In Germany, he broke down and purchased a fourth-class train ticket, jealous of the soldiers (and dogs) that cost less. In Europe, he accepted food from charitable establishments. He worked as a sailor to get from Marseilles to Port Said, then walked through the Middle East. He stowed away to get to Ceylon, where he got his best gig yet, menial tasks for a circus, then promotion to clown. He walked across the Malay Peninsula, and managed to work his way back to the United States and from Seattle home to Chicago.[45]

Except for the circus, vagabondage was a horrifying life. And yet Franck rejoiced that his journey around the world cost $113, only $9 more than originally planned. Was that admirable? As a vagabond in Europe, he took food and work away from people who were in genuine need. In colonial zones, he calculated that white people would help him stay above the level of the colonized—it was in their interest to do so. (He had no sympathy for the "coaling niggers" who worked at Port Said.) And he clung to something that set him apart from other vagabonds: his Kodak, never sold or pawned, no matter how hard up he was. Photographs of his adventures appear in the book he published in 1911. His choices show the continuing allure of measuring oneself against the planet armed only with a few oddments of industrial society. It made a good story, even better if illustrated by one of those oddments.[46]

Paper with which to tell a story was among the amenities that Nippon Yusen Kaisha steamships supplied their first-class, around-the-world passengers. Each sheet of stationery was designed to be folded into its own envelope and was printed in color, with exquisite handblock images of Japan (geisha, cherry blossoms, Mt. Fuji), plus the ship's route and the day's menu. The recipient unfolded a proxy tour. And yet the stationery undermined the tradition of the sustained circumnavigator's narrative. The story was reduced to pleasant fragments that might fit on, and match, the pretty pages. All of the work that had made the tour possible was concealed, including the recent efforts to draw Japan and the Pacific more fully into the global circuitry that was the foundation of proprietary confidence.[47]

Even as more Asian travelers were going around the world, they depended on Western agencies for transport. By the second half of the nineteenth century, most Western nations had a proprietary or favored shipping line, as with Germany's Hamburg-Amerika. The larger powers had several such services. Britain, for instance, had the P & O, White Star, and Cunard lines, among others. And the United States dominated the Pacific with its Pacific Mail Steamship Company. Directors of the Western companies were itching to fold Asia into their transport empires. In 1905, E. H. Harriman, of the Union Pacific, schemed to create an around-the-world monopoly: a railroad across the United States, steamers over the Pacific, trains over Manchuria, Siberia, and Russia, and steamers back to New York City. It didn't happen, but it was characteristic of the globe-girdling enthusiasm of the era's magnates.[48]

Western transport companies did not necessarily welcome Asian business, at least outside steerage. When Protap Chunder Mozoomdar went around the world, he endured every kind of prejudice. At the P & O office in Calcutta, the Hindu reformer was kept standing while the staff waited on a seated Englishman. Mozoomdar finally got a ticket, but was horrified to discover that his cabin had cattle on one side and their slaughterhouse on the other. The location was an insult to a Hindu forbidden to eat beef. "The fault," Mozoomdar concluded, "lay in the fact that Europeans did not like to associate with Hindus on board ship." He visited steerage on the Cunard ship he took across the Atlantic and was appalled. He knew that Cunard steamers were famous for their luxury, but made it a point, in his published account, to report that the company kept other ships that "in their dirt and discomfort rival the Arab dhow."[49]

Elite Japanese (including the members of the Iwakura Mission) were the exceptions. They were accommodated alongside white passengers, even in the first-class sections of steamships and on Pullman cars. But that, like Thomas Cook & Son's first Japanese round-the-world tour (1908), was atypical, because done for elite clients from a nation famous for its embrace of Western culture yet freedom from Western empire.[50]

The Nippon Yusen Kaisha was Japan's effort to compete with the West on its terms, including putting a girdle around the Earth. The company had its origins in a local shipper founded in the 1870s, eventually named Mitsubishi. Initially, its maritime services competed regionally, especially against Pacific Mail Steamship and the P & O. In 1885, Mitsubishi merged with another company, establishing Nippon Yusen Kaisha (NYK; Japan Telegraph Company), with combined shipping and communications interests. The corporation steadily expanded, and, like Western shipping companies, won key government contracts, especially during the Sino-Japanese War (1894–95) and the Russo-Japanese War (1904–05). In 1896, NYK took a big step forward by ordering eighteen new vessels, each over 5,000 tons, which required over one and half times the company's available capital. That same year, the company inaugurated three major services: to Europe, North America, and Australia. At this point, NYK's services were displayed on a world map, showing where its passenger and freight lines went, and where passengers could make railroad connections. By 1903, NYK was the world's seventh largest shipping company.[51]

Meanwhile, the Western powers were busy bridging the Pacific by telegraph. The Eastern Telegraph Company had laid a cable between England and Australia, via the Cape of Good Hope, finished in 1902. That 15,000-mile line brought the world's telegraph network to a total of 200,000 miles, "enough to go about eight times around the Earth," as a description of 1902 claimed. Parts of the Pacific had already been joined into local networks, and the direct England-to-Australia connection sped things up generally. But the long-imagined girdle round about the Earth still did not exist. To send a telegram from California to Japan, for instance, required cabling eastward over North America and the Atlantic, then through the British network that would carry the message to Asia.[52]

The Commercial Pacific Cable Company (CPC), a consortium of three British and American telegraph agencies, had been formed in 1901 to lay a Pacific cable from San Franciso to Honolulu, then to the Midway Islands and Guam, ending in Manila. The global circuit was completed in 1903. On July 4, an operator at President Theodore Roosevelt's private residence on Long Island, New York, sent the first telegraph code around the world. Back in the Roosevelt home, the president of the CPC Company read out the message: "Congratulations and success to the Pacific cable." The circuit took nine and a half minutes. (An attempt the day before, by the French newspaper *Le Temps*, pieced through the old routes, had taken six hours and twenty minutes.) The demonstration publicized America's global prominence on the nation's primal holiday. But it also revealed the absurdity of sending information around the planet. A human, animal, or object might be transformed by going around the world, but not a message. Nevertheless, the CPC Company proudly featured a globe on its stationery. In 1911, the *New York Times* retested the speed of an around-the-world cable. Theirs took sixteen minutes—not bad, considering that their telegram fought much more traffic than Roosevelt's.[53]

The feat of physically encircling the Earth set off a round of triumphal claims that humans were not mere Pucks, but super-Pucks. There was enough telegraph cable to girdle the globe ten times over, said one commentator; fifty times, said another. Others pointed out the equivalents in railway track: its total length could extend over nine times around the equator, or else seventeen times, depending on how you calculated it. Other items were imagined to stretch around the world, perhaps multiple times. A 1900 report on France's finances asked readers to imagine the 138 million francs recently spent by the government in terms of its planetary dimensions. Changed into silver one-sou pieces, that amount of money would form a "ribbon of silver" that could go around the world twenty-five and a half times. On Leo Tolstoy's eightieth birthday, in 1908, fans planned to "set up committees in every settlement on Earth" where people could sign copies of a birthday address on strips of parchment that "would create a ribbon that could encircle the globe" as a "symbol of the diffusion of Tolstoy's ideas."[54]

In this way, the real and imagined girdles put round about the Earth, over four centuries of them, confirmed a sense that the whole world was finally being tamed. The recent declarations of world-encompassing triumph fit, historically, between two significant milestones. The first was the announcement, from the U.S. Census Bureau, in 1890, that the American frontier was closed: there was no more territory to be taken from its original Indian inhabitants. The second was the first expedition to reach the geographic south pole in 1911. It seemed that there were no longer any geographic frontiers against which to push.

The completed telegraph circuit also meant that circumnavigators could describe two kinds of circles around the globe: the ones they made, and the others their telegrams inscribed. When the British newspaper colossus Alfred Harmsworth, first Viscount Northcliffe, went around the world from 1921 to 1922, he had to stay in touch with his business news network, which round-the-world telegraph and postal services let him do. At his stop in Singapore, he reported, "I am busy with cables and letters, which follow us steadily around." In Penang, he lay in bed, "my mosquito cage, with the electric punkah [fan] going," as if being horizontal was the only way to reconnoiter "the great sheaf of Christmas and New Year cables which have been following me about and reached me here."[55]

President Roosevelt made his nation's global ambitions even clearer with the Great White Fleet, the largest fleet of battleships ever to do a circumnavigation. This was the latest in a series of U.S. military circumnavigations. (Even the Confederates had kept up that tradition, with the circumnavigating CSS *Shenandoah*.) Roosevelt's sixteen ships carried 360 guns, 12,793 men, and about 70 animal mascots. The guns were the main thing. Over the fourteen months from December 1907 to February 1909, the fleet traveled 43,227 miles in a series of explosive (if ultimately peaceful) demonstrations of U.S. military capacity. A cartoon from the time shows the Earth "Trying on Her New Necklace," warships strung like beads around her neck. The longest published account of the expedition made clear the sailors' awareness that they acted in a geodrama. When the ships left Sydney, a passenger steamer sailed with them, "with the wide, wide ocean and arching dome of heaven as the stage, and they [the steamer] a mere dot in the centre of it all as the audience." At journey's end, Roosevelt congratulated the officers and sailors. "This is the first battlefleet that has ever circumnavigated the globe," he claimed, with more of that poor grasp (or outright forgetting) of the history of circumnavigation, now unencumbered by the existence of George Anson.[56]

And yet the fleet had little impact on global affairs. Everyone already knew that President Roosevelt was fond of guns and, if anything, the ships revealed U.S. dependence on its allies. The ships were fairly self-contained when it came to food and other supplies. Refrigeration and canning meant that they were weaned from land. But an estimated 125,000 tons of coal was necessary to fire the fleet, and 100,000 tons of that total amount had to be shipped out to it. Problems with re-coaling in New Zealand were evidence that a round-the-world fleet of sixteen ships stretched the United States beyond its supply line.[57]

Roosevelt made a more lasting contribution to circumnavigation as the man who completed a long-planned canal through Panama. Many a helmsman fighting his way around Cape Horn had dreamed of that quicker route. For much of the nineteenth century, steamer traffic delivered passengers to a railway line that crossed the isthmus between Atlantic and Pacific. (Some ships' captains still preferred to run around the Horn, especially if they were headed to destinations in the southern hemisphere.) The Trans-Continental Railroad in the United States was another bridge between the seas, though with the inconvenience of multiple ticketings, reboardings, and baggage handlings. The fact that these alternatives somehow worked was beside the point: Western enthusiasm for the built environment on a planetary scale gave schemes for another great canal the momentum of a speeding locomotive.

A French company was organized in 1879 to tackle the Isthmus of Darien (as it was still known), and after many delays and arguments about feasibility, work began in 1881 under the direction of none other than Ferdinand de Lesseps, architect of the Suez Canal. The labor proceeded with disastrous setbacks and a new French company would take over in 1894, without Lesseps at the helm. That phase of excavation failed, too. A final effort, this time with U.S. support from newly elected President Roosevelt, was initiated in 1902. The project was a global as well as regional power play. In a 1903 photograph, Roosevelt stands beside a globe positioned to show the Isthmus of Panama, the renamed U.S. gateway to the world.[58]

The canal was among the great engineering feats of the industrial era. Its cost exceeded that of the Suez Canal by four times, around $639 million. And yet the U.S. phase of the project had come in under budget by $23 million and opened six months ahead of schedule. Fewer lives had been lost than at Suez, though the toll was still profound: 25,000 individuals, 500 for each mile along the canal. The scale of the work seemed inhuman. A single canal lock, had it somehow been made to stand upright, would have been the tallest structure in the world, surpassing even the Eiffel Tower.[59]

On the afternoon of October 10, 1914, President Woodrow Wilson opened the canal. From the comfort of the Executive Building in Washington, he pressed a button that relayed a signal through telegraph wire. One minute later, the signal detonated dynamite positioned against the last of the temporary dikes in the waterway. It was a grand moment, though overshadowed by other news.[60]

War had broken out in Europe that summer. By October, the western front in France was transformed by another historic earth-moving project, the digging of the trenches that would become the hideous hallmark of World War. Many cruise ships were refitted for troop transport or as hospitals. The sinking of the British passenger liner *Lusitania* in 1915 signaled that non-combatants would not be spared the dangers of war. Not all global traffic ceased. In 1915, NYK offered westbound service around the world. By 1916 (the war still not over), the Japanese company added scheduled sailings to New York via the Panama Canal. (The NYK's freighter *Tokushima Maru* had been, in 1914, the first Japanese vessel to run through the Panama Canal and complete a full circumnavigation.) In fall of 1917, the company opened thirteen new overseas services. By this point, Japan was the third largest shipping nation in the world.[61]

Was the Hundred Years' Peace over? Imperialism, especially the global hegemony of the British Empire, survived both the war and the Treaty of Versailles in 1919, although critics of empire, including Mohandas Gandhi, were becoming unignorable. During the war, passport regulations, which had been relaxing since 1815, began to be tightened within Europe and the European empires beyond, and the trend would continue. When the Syrian Christian minister Hanna Khabbaz set out around the world from Egypt in 1917, he marveled at the density of steamship traffic through the Suez Canal, despite the war. And yet the bureaucrats who regulated travelers were denser (and slower-moving) than the ships. Khabbaz failed to get a visa to the United States. It took thirty-three days to get papers for an alternative route through Argentina. Transport through Asia on those documents was not guaranteed, however, and Khabbaz endured several delays and interrogations along the way. Although he never published a second volume that reported on the last part of his circuit, he composed what is probably the first modern round-the-world account in Arabic, a sustained complaint about the tenacity of Western dominance.[62]

Much of that old order had survived. The autumn of Western global power, combined with the internal combustion engine, two isthmus-piercing canals, and a reviving consumer culture, meant that in the two decades after 1919, the planet was organized for around-the-world pleasure travel as never before—or since.

Travel brochures sold the tours, and they were works of art. For world tours, the brochure was usually the size of a magazine, big enough to cover a global itinerary, yet light and portable, the perfect size to peruse while lounging in a deck chair, or to use as a fan while waiting for the rickshaws to assemble at the end of the gangplank. The brochures were a blend of travelogue and potted history, illustrated with colored drawings and black-and-white photographs. Above all, the text promised that an around-the-world tour would be unforgettable, an experience of time and space in which all difficulty and unpleasantness had been tidied away. Needless to say, a zero mortality rate was taken for granted.[63]

The characteristic vehicle of such travel was the luxury liner: the enormous, multi-engine, fully electrified, and grandly decorated cruise ship devoted to passenger comfort. All cabins were heated and wired for electricity, for lighting, to supply fans, and for many other conveniences—a man could press a button to fetch a maid while his wife plugged in her curling iron. The higher the class of service, the more and larger the portholes or windows in the cabin, and the greater its distance from the working decks. Likewise, the higher the passenger's class, the greater her or his access to capacious and tasteful parlors, saloon bars, libraries, and ballrooms.[64]

Such ships were thought to blend the best features of yachts and hotels. Cunard's circumnavigating *Franconia* was "a personal pleasure ship" with the "intimacy associated with private yachts." The around-the-world ships of the Dollar Steamship Lines were "like great hotels that move about." Indeed, the ships were modeled on hotels, with comparable staff and services. As in a grand hotel, given enough notice, and for a price, a ship could arrange anything from a manicure to a masked ball. Passengers who continued on the old pieced-together tours knew they were missing something. Lord Northcliffe fussed that, on ships with patchy service, he had to worry about how to get his laundry done: "after all, what is most important to the circumnavigator is washing."[65]

Meals aboard cruise ships also resembled those found in the best hotels, from the first cup of morning coffee to the climactic multi-course dinner prepared by French-trained chefs, with accompanying wines. The extensive food service required ships to install room-sized refrigerators and walls of freezer space. On United American Lines, a menu for George Washington's birthday had fifteen courses, listed as an acrostic of the American Founder's name, from "*G*-rapefruit Cocktail" to "*N*-ougat Pastry." An "Equator Dinner," to mark crossing that line, featured "Whale Oil Soup" and "Seacow Steak," probably just plain old turtle soup and sirloin. There was no fear of contracting scurvy, though a great dread of putting on unflattering pounds. Luckily, cruise ships had swimming pools and exercise rooms, plus designated areas for sport on deck, including tennis courts. "Fat in this place was certainly tabu," commented a circumnavigating doctor of his ship's busy gymnasium.[66]

The sense of traveling the world within an exclusive hotel was especially prized. The typical luxury circumnavigation went through both major canals, as with Cunard's first tourist circumnavigation, in 1923, on the *Laconia*. Passengers who glided through Panama and Suez were spared the turbulent southern passages of the clipper route, as well as the amphibious circuit's exasperating transfers from ship to shore and back. The ultimate luxury was an all-first-class ship. Cook and Cunard arranged theirs on the *Franconia*, purpose-built after the war for round-the-world tours. The 20,000-ton ship, with space for 2,000, would be restricted to 400 guests. On her circumnavigations in the 1920s, there would be no unwelcome contact with passengers in lower classes.[67]

That became the ideal: a cruise of three to five months, with several dozen stops at famous ports, and always the luxury liner as a retreat. In the 1920s, the price of such a tour began at just over $1,000 and rose to a stupendous $25,000 for the best suite provided by the United American Lines. There were cheaper tickets for children and lower rates for servants, should their employers not wish them to help at (or enjoy) special entertainments on the ship or excursions ashore. The average cost in the 1920s was probably around $2,000 (worth about $25,700 in 2011), but that didn't include tips, which ships generally calculated at 5 percent of the passage. Surcharges for special services could also substantially increase the journey's actual cost.[68]

At those prices, everything was arranged for the passenger's delight, even the weather. Cook's tours typically departed eastward from New York in January, which put passengers through the Red Sea and Indian Ocean before any hint of heat, and got them to Asia to see springtime blooming. United American Lines was one of many operators to adopt a January departure, noting that it was "delightful to sail away from the rigors of winter" and not return to them (unlike shorter escapes to Florida or the Caribbean). The Canadian Pacific Railway offered a variation: overland via Canadian Rail and the Trans-Siberian Express, and ships the rest of the way. (Dollar Steamship Lines offered rail travel as a slightly cheaper alternative to the Panama Canal.) All the Canadian Pacific brochures promised that its route traversed "the Wonder Belt of the World," where mild climates kissed sites of cultural and historical significance.[69]

Planetary imagery reminded passengers of the global scope of their travels. Anchor Line put a stylized globe on the cover of its travel booklet, circa 1925. "The long arm of C. P. R. spans the world," Canadian Pacific Railway claimed the same year, and offered a circular diagram of its "Wonder Belt" around the Earth. Hamburg-Amerika presented its 1913 round-the-world cruise of the SS *Victoria Luise* with an image of the globe wreathed in steam from the ship's smokestacks (see illustration 15). Canadian Pacific encouraged its passengers to think of themselves not on a stage but in a film. "Round the World," its brochure for 1924 exclaimed: "what a mental cinematograph these words produce."[70]

The new pleasure circumnavigators were also encouraged to consider their place in history, albeit a sanitized history. The age of sail was a heroic backdrop to the luxury cruises. The cover of a Canadian Pacific pamphlet, for a 1924 world cruise, showed both the departure in 1577 of Drake's *Golden Hind* and the imminent departure of the cruise's *Empress of Canada*. Recommended reading for Canadian Pacific passengers included the narrative of Anson's voyage and the circumnavigations in Richard Hakluyt's sixteenth-century travel compendium. Those bracing selections were balanced by Treves's *Other Side of the Lantern*. More commonly, evidence that going around the world had once been dangerous was simply omitted. The Red Star Line, for example, recommended Twain's *Following the Equator*. Brochures for Thos. Cook & Son's world tours frequently offered a potted history of Magellan's historic voyage. But even if that included the commander's death, it withheld details about the difficulties of his voyage and those in the succeeding 250 years. And Magellan's story was usually followed by the heartening history of Thos. Cook & Son's first around-the-world tour.[71]

Tour organizers also simplified the international relations that made the journey safe and its dockings uneventful. The 1929 to 1930 cruise program for the Canadian Pacific promised that its aim was "To achieve this World in comfort,/ To visit its rare places in safety"—a tall order, on a global scale. Certainly, the niggling details of international travel were reduced considerably. On the single-vessel tours, baggage was unlimited, and needed to be wrangled only on embarkation and return. A passport had to be shown only once, at the start, and then the ship's crew would make all other arrangements for entry into multiple countries. That convenience was mirrored at the end, when customs needed to be cleared only once.[72]

Although luxury circumnavigations were extraordinary experiences, they occurred frequently. Nowadays, a cruise line might offer an around-the-world option once a year, if that. In 1925 and 1926, the United American Line offered two annual tours, eastward and westward. That was on the meager side. In the mid-1920s, the Dollar Steamship Lines offered sailings from each of its around-the-world stops every other week. A first-class ticket round the world was good for two years. NYK's comparable ticket was good for a year. Within that time, passengers could disembark and re-embark as they wished. NYK emphasized that its year-long tickets were the best value. Dollar insisted that its tickets were most convenient because it was "the one steamship line affording continuous passage circling the globe," an "endless chain upon which the sun never sets." One of its customers had evidently gone around the world eleven times; another, five times. To expand its clientele, the company in 1928 introduced a discounted "First-Class Vacation Tour for College Folks" with the breathless announcement that "Collegedom wants to see the world . . . all of it . . . and will . . . by gosh!"[73]

The Great Depression did not end round-the-world cruises, though it did change how they were marketed. If anything, NYK expanded its around-the-world services during the Depression. It christened new luxury liners in 1929 and 1930. In 1929, a promotional calendar described the company as having "World-Wide Services"; by 1932, an Art Deco poster, featuring a sleekly abstract passenger ship, announced that passengers could go "around the world . . . eastward or westward with NYK line." "Make the world *yours* while costs are low," the Dollar Steamship Line urged in the early 1930s. Its new "Tourist class" had

"an 'at ease' character" particularly suited to families with small children; in the early 1930s, fares for a "Thrift Tour" began at $650, then
in 1933 dropped to $589. For its round-the-world service, NYK offered
tourist class tickets at least as early as 1934 and second-class fares
($1,000) the next year. Even in 1938, the American President Lines of-
fered around-the-world cruise departures that sailed every four weeks.
Only first-class accommodation was available, though fares ranged
from single occupancy in a suite to double occupancy in smaller cabins, and the shore programs were priced separately, for travelers now
watching their budgets.[74]

Celebrations for crossing the International Date Line were always
an affordable luxury. Passengers aboard United American Lines ships
could send telegrams home, to arrive "before" they were sent. Travelers
on Cook's tours received documents to mark the "novel sensation of
adding an extra day to our lives." Canadian Pacific distributed certificates for the "Order of the Blue." NYK gave a "Longevity Guarantee" to
each passenger who gained a day by going eastward around the world.
The net effect was surreal, in a pleasant way. On his way around the
world, one passenger, Walter Haworth, crossed the 180th meridian
while at a "fancy dress ball." In costume, surrounded by masked mer-
rymakers, and gliding over a smooth sea, "it was difficult to realize we
were in mid-Pacific, ready to cross the international date line."[75]

The paying customer had conquered the planet and enjoyed every
inch of the way around it. A cartoon for the International Mercantile
Marine Company represented that triumph with a flapper and her
dance partner (see illustration 16). The pair pivot on a luxury liner set
atop an only slightly larger globe. The Jazz Age duo are bigger than
both. Enormous elbows out, they do an unmistakable Charleston. Frozen in glee, they are forever dancing, dancing around the world.[76]

The images and advertisements for luxury circumnavigations had
required so much imagination, it seems, that tourists didn't need to
have any of their own. Instead, the brochures, with their pictures and
descriptions of the tour, told passengers the story of their experience in
advance. Cruise lines also distributed guides for "Dressing Around the
World" and for "Shopping Around the World." They gave out detailed

"log books" for passengers to use as journals. These printed forms were convenient, but suppressed spontaneity and insight. Passengers did little more than confirm that the Champagne was cold and that crossing the International Date Line was fun.[77]

In consequence, and sadly, there is no interesting published narrative of how it felt to sweep around the world on the *Franconia*, the *Empress of Australia*, or any of the other circumnavigating luxury hotels. Some preprinted "logbooks" survive with handwritten notations, mere fragments. Aboard the SS *Belgenland* in 1929, one man recorded the following:

> Kobe: "a large department store. Just like home."
> India: many beggars, "but we just *don't see them.*"
> Leaving Port Sudan: "The smoothest & bluest of water."

Many passengers wrote letters, some of them on those gorgeous NYK handblock sheets. Only occasionally might the private letters be published as a kind of narrative. Esther Leggett did that in 1938, with a standard Grand Tour story. With greater effort, Frederick James Hill wrote short poems about his round-the-world trip. He describes husband-hunters, for instance, as "Flappers smoking, gold-tipped fags— / scanning names on deck-chair tags." That is one of his better verses. Only a genuine artist could outdo the commercial descriptions.[78]

Ivan Bunin was that artist. The first Russian writer to win the Nobel Prize for Literature, in 1933, Bunin was master of the short story. His best-known is "The Gentleman from San Francisco" (1915) about a failed circumnavigation. The gentleman in question, whose name no one remembers, sets out one winter with his wife and daughter for a two-year journey. "In December and January he hoped to enjoy the sun of Southern Italy," then do a Grand Tour of Europe, "and Constantinople, and Palestine and Egypt, and even Japan,—of course, be it understood, already on the return trip" to San Francisco. "And everything went very well at first."[79]

Bunin parodies the gorgeous promotional literature of the luxury tour. The gentleman and his ladies cross the Atlantic on "the famous *Atlantida*," which resembles "the most expensive of European hotels," with "an all-night bar, Turkish baths, a newspaper of its own." Heated and padded comfort prevails within; the sea and snow rage without. Bugles and whistles signal the time, the type of meal, the enchanting activity, and the requisite costume for each part of every day. "An exquisite couple in love with each other" constantly demonstrate concentrated happiness. They are hired actors. The gentleman makes a much-anticipated stop in Capri but, dressing for dinner yet again, thinks " 'Oh, this is dreadful!' " though "without trying to understand, without reflecting, just what, precisely, was dreadful." He dies before dinner, is consigned to a soda-water crate, and recrosses the Atlantic on the *Atlantida*. The remorseless cycle of pleasure plays out for other passengers and the ship's engines are again "toilsomely overpowering the darkness, the ocean, the snow storm. . . ."[80]

For Bunin, it was quite an indictment. He loved to be in motion and had always wanted to go around the world. He regarded travel as communion with the Creation, a singular way to transcend time and space, the ideal of self-realization. His nameless Gentleman from San Francisco is tragic not because he cannot achieve these goals, but because he only begins to do so. One third of the way around the world, he has a dim, initial epiphany, and is then reduced to cargo.[81]

Another artist, Jean Cocteau, was much more cheerful about the touristic uses of the planet. His 1936 circumnavigation was a reenactment, of sorts, of Verne's dramatic story, which Cocteau reclaims for France. The Verne novel is for "French youth" what *Robinson Crusoe* is for English speakers: "Bound in red and gold, it figures as a school-prize." Even better, "the red-and-gold curtain of the Châtelet Theater has risen on the play based on the book," and the play is what Cocteau remembers best. His friend Marcel Khill suggested redoing the voyage in (somewhat belated) honor of the centenary of Verne's birth. The editor of *Paris-Soir* agreed to publish Cocteau's dispatches and, if he did the circuit in eighty days, to pay his expenses. Cocteau is a new Fogg; Khill is his Passepartout.[82]

Cocteau repeatedly considers the world's globeness. On the train between Rome and Brindisi, he said, "I pictured the track that lay before us coiling round the globe like the snake on which the Virgin sets her foot" in some Christian art. He buys "Passepartout" a wristwatch in which a magnet moves two balls that indicate time. "It was not the immortal timepiece of Jules Verne's Passepartout," yet still wonderful: "Better than normal watch-hands, the mercury balls conveyed the lapse of time, the movement of the globe." As his Japanese ship glides over a waveless sea between Singapore and China, "it was impossible not to realize that we were moving round a sphere hung in the void." He reflects that "this globe round which we travelled was once a globe of fire." Now, the cool planet circled a fiery orb, and that explains the two Tuesdays that Cocteau experiences while crossing the Pacific. He compares his experience to Poe's "Three Sundays in a Week" and says that, while science can explain the phenomenon, it has for him "the glamour of the occult."[83]

He also compares his real geodrama to Verne's stage drama, sometimes peevishly. "Jules Verne never mentions heat or sea-sickness," he complains. But "the mournful hoots of steamers" in Malaysia are perfect, exactly like the sound effects "that stage-hands reproduce in the Verne play by blowing down lamp-chimneys." He admires the clothing in Singapore and wonders: where has he seen it before? "Why of course, it was on the stage" in Verne's play. "The producers were evidently much traveled men; they had got the scenery and costumes exactly right."[84]

At several points, Cocteau describes, in contrast to his free circle around the world, cages that confine a buzzing energy. At first glance, the distant Parthenon in Athens looks like "a little broken cage, very long and low, like the grasshopper-cages children weave of grass-halms." The electric fans on the old steamship that takes him from Calcutta and Rangoon are "shimmering overhead like small caged clouds." He himself is a whirring dragonfly, "out to circle the globe in leaps and bounds." He acquires a suitable companion in Japan, "a caged cricket" that Passepartout names Microbus. The buzzing on the page becomes louder and closer—boundary-busting instead of metaphorical. At the Parthenon, "an aeroplane (or was it Mercury, the god of commerce?) soared above the open cage, vanished into the blue." Later, in San Francisco, Cocteau, Passepartout, and Microbus board a propeller plane that takes them to Los Angeles. His contract with *Paris-Soir* did not rule out air travel, Cocteau explained, "provided we did not cheat, and use it to catch up on wasted time." They release Microbus to fly away in California, and they fly to New York on TWA.[85]

"Passepartout summed up his views on flying in three words. *"Man has guts!"* Exactly. There went the zero mortality rate, along with the sense of confidence it had fostered over the past hundred years.[86]

Second Entr'acte

*A*t *last, humans* could fly. A circumnavigation done in the air—perhaps eventually in orbit—would be unprecedented, a break with the past history of geodrama. But there were fictional precedents: Phaeton's burning crash to Earth, Puck's forty-minute sprint, and Satan's searching and joyless swoops around the planet. Obviously, aviators hoped to be Pucks, though aviation's uneven achievements would raise the first significant doubts that humans could continue to master the planet. Would the international order be capable of fostering the unfettered global movement that airplanes promised? Would flight above the Earth, or in space around it, demonstrate continued dominion over the planet? Would faster and more frequent circumnavigations take too much out of the world to be environmentally sustainable? As the worries accumulated, the globe, in its implacably huge physicality, came back into sight, as if to rebuke the confidence of circumnavigators during the extended nineteenth century.

The early centuries of attempted flight were dismal. A variety of gliders, propellers, and parachutes had been tested in nearly every human culture that could generate a surplus of building material. There is little evidence that any of those heavier-than-air devices worked, and abundant proof that they killed or crippled their experimenters. Not until the 1780s were humans, meaning the French, able to keep themselves in the air for hours and return in safety. These first aeronauts

used lighter-than-air vehicles: balloons. It is hard to find negative testimony about balloon flight—most witnesses were awestruck. When a
rare skeptic wondered aloud what good a balloon might be, Benjamin
Franklin famously retorted, "what good is a newborn baby?" Only time,
he implied, would reveal the practical use of the marvelous floating
orbs.[1]

Franklin was somewhat ingenuous, given that flight machines were
deadly in two senses: they were dangerous and they were associated
with war. Balloons were lighter than air because filled with flammable
gases. Fiery accidents were inevitable—and horrendous. Crash landings
were also common. Because balloons could not be steered, they easily
drifted to places where safe landing was impossible. And aerial devices
had long been described as potential weapons, able to penetrate enemy
lines in the perfect sneak attack.

Subsequent experimenters either developed navigable heavier-than-
air vehicles or tried to make balloons steerable. The relatively light
internal combustion engine was necessary for both. By the early twentieth century, it was Americans (especially Orville and Wilbur Wright)
who were associated with completing the first task and Europeans
(particularly Count Ferdinand von Zeppelin) with the second. Steerable
rigid airships (zeppelins) were expensive, products of great independent
wealth or national resources. Single-engine, fixed-wing aircraft were
comparatively cheap, increasingly so over time. Indeed, they were not
unlike automobiles, another twentieth-century consumer item that
allowed more people to exult in the speed of an internal combustion
engine, as with Henry Ford's Model T's, nicknamed "Flivvers" and "Tin
Lizzies."[2]

Motorcars and airplanes roar through F. T. Marinetti's "Futurist
Manifesto" (1909). In a night-darkened prologue, Marinetti and two
friends leap into automobiles—fast, darting reproaches to ponderous
steamers and locomotives. Marinetti brakes just short of two bicyclists,
"wobbling like two lines of reasoning . . . what a bore! Damn!" His
quick stop flips his car into a ditch, but his restless "shark" is fished
out and he zooms to the finish line, meaning the eleven points of the
manifesto. These include Marinetti's notorious declarations in favor
of war and against feminism, but above all a celebration of masculine

quickness. "We affirm that the beauty of the world has been enriched by a new form of beauty: the beauty of speed." Velocity, and the male hand that controls it, will measure the planet: "We intend to hymn the man at the steering wheel, the ideal axis of which intersects the earth, itself hurled ahead in its own race along the path of its orbit." The "oscillating flight of airplanes, whose propeller flaps at the wind like a flag and seems to applaud a delirious crowd," will also carry man into the future. And "it is from Italy that we are flinging this to the world, our manifesto of burning and overwhelming violence, with which we today establish Futurism."[3]

That Italian pedigree, and the Futurists' affinity for fascism, seem historically particular to one nation. But the overwhelming desire for speed was bigger.[4]

Aviation was one of two airborne technologies designed to dissolve boredom and borders. The other was radio. Since the 1870s, innovators had been manipulating the electromagnetic waves that had been discovered to oscillate all around the Earth. Different inventors (the most famous were Guglielmo Marconi and Nikola Tesla) developed ways to capture the aerial waves and transform them into signals, typically audible ones. In contrast to telegraphy, this communication was *wireless*. The built environment of telegraph cables, railway lines, and shipping docks had shackled the world into usefulness, just as empires held the world into a particular configuration. Radio was different, less bound to Earth and to the powers of the Earth. It seemed to match the social worlds that were straining against imperial bonds.

But would airplanes soar as freely as radio waves? An individual possessed of an internal combustion engine could tear up timetables and just get going, flicking aside bicycles, breakfasting on one side of a border and dining on the other. For that reason, aviation was the first form of transport for which there was an immediate international market. Not by coincidence, few early pioneers of long-distance aviation were British. The truly famous long-distance aviators included the Frenchman Antoine de Saint-Exupéry, the Americans Charles Lindbergh and Amelia Earhart, and the Australian Charles Kingsford Smith. The privilege of easy movement about the world that white Britons had enjoyed as a consequence of empire could now be achieved by other means.

And yet the ability to cross borders meant that aviation was the first transport industry regulated even before it existed. Delegates gathered at two European conferences (in 1900 and 1910) to define how flights into foreign territory might be controlled. Although the world was still configured by a Europe-centered paper internationalism, aviation was helping to create another that consisted of a broader network of shifting negotiations. This was a much more international scenario, the result of constant interplay among many nations, not a few accords that favored Westerners.[5]

The turn of the century was probably the high tide of that morphing and voluntary internationalism. A record number of non-governmental and intergovernmental organizations were created, from the International Olympic Committee (1894) to the International Broadcasting Union (1925). International movements generated a profusion of other global outposts. Some of the most visible at the time were international workers' organizations. Others included Esperanto clubs, Boy Scout troops, and branches of the YMCA and YWCA.[6]

Those organizations and networks were useful to long-distance aviators, because early aircraft needed frequent support from land. Flying machines were livestock on tethers, not free-range creatures. Weight constraints kept fuel tanks small; engines needed constant repairs; updated information about flying conditions was crucial. Travel services were still geared to telegraphy and steam locomotion, which followed established routes around the world, but aviation only sometimes followed these tracks.

The early flight machines were the smallest vessels (so far) capable of bearing people over long distances. Bicycles of course were smaller, but they were no match for oceans, as aircraft were expected eventually to be. And yet one of the early single-engine airplanes could have nestled into the deck space of Slocum's *Spray*. And in contrast to Slocum, who could amble and sprawl at his ease, a pilot or navigator on an airplane would be clenched into position for the duration of the flight. Their open-cockpit aircraft barely protected them from the elements. Early aviators wore their characteristic goggles, leather jackets, and gloves for protection, not jaunty effect. Flying through rain, one American aviator commented, felt like "lead shot being thrown into our faces and against our hands." Ignore the wings on one of these planes and think of it as a kayak. The ratio of water to air is lower, but whatever advantage that gave an aviator, he, unlike a kayaker, could not keep warm through exercise.[7]

Given the vulnerability of small aircraft and their pilots, and the astonishing lack of ground support (none at first, and then fitful radio impulses, sometimes), accidents were frequent. Flying in dark, fog, or storms was particularly hazardous. In his best book, *Night Flight* (1931), Saint-Exupéry shows an aviator preparing for the dark by feeling over his instruments, "training his fingers for a blind man's world." The darkening air is friend and enemy. The aviator "felt himself softly heaved up, as by a giant shoulder," even as "heavy clouds were extinguishing the stars." Full darkness has a "startling fatality" made terrifying by storm conditions:

> But now the instrument needles in front of him began oscillating wildly, growing increasingly difficult to follow. Misled by their erratic readings, he lost altitude. Slowly but surely he was sinking into a dark morass, a murky quicksand. The heading on his altimeter was now "500 metres"— the height of the hilltops beneath him. He could feel them heaving up their towering breakers towards him.

A desperate climb for altitude brings the plane above the storm and into moonlight. Reflected by the clouds below, the lunar radiance is unreal. "Too beautiful," worries the pilot, who is about to run out of gas and plunge back through the clouds, to his death.[8]

In short, early aviation offered all the experiences of Phaeton, Puck, and Satan. Lucifer's survey of what lies below him on his flights is dull, but only that kind of obsessive reconnaissance allowed aviators to enjoy the freedom of Puck without risking a Phaeton-like crash.

Even before an around-the-world flight was feasible, surface speed records were accelerating. After George Francis Train beat Nellie Bly's round-the-world record in 1890, with a 67-day trip, the time dropped to 60 days (1901), then 54 days (1903), then 40 days (1907), and then 39 days and a bit in 1911, this time by a one-legged Frenchman. The

New York writer and theatrical producer John Henry Mears managed to scrape off four more days in 1913, and he was the first circumnavigator to use an airplane, if only for forty miles. The flight was the most theatrical part of the stunt: Mears perched on the wing of the plane, hugging one of its struts.[9]

The first tests of whether a single internal combustion engine could circle the entire planet were done not with airplanes but automobiles. For a long time, historic first circumnavigations had been made by men, with women later repeating the feats. By the early twentieth century, it was about time for a woman to just be first. And so it serves F. T. Marinetti right that the very first person to drive around the world was a woman. Worse, she was a *lady*: a rich, middle-aged New Jersey industrialist, Harriet White Fisher. Over thirteen months in 1909 and 1910, Fisher motored around the world in a 40-hp Locomobile roadster, with a chauffeur, a maid, a cook-cum-secretary, and a Boston bull terrier named Honk-Honk. After a trial run in the United States, the Locomobile and travelers took a steamer for France, where Fisher paused to order a tent at Louis Vuitton. The party packed the Locomobile solid (every running board was heaped with lashed-down luggage) and ran down to Fisher's villa on Lake Como. Refreshed by that visit, they used P & O to get to Bombay, drove overland through Asia, took another steamer across the Pacific, and headed back to New Jersey, having covered 20,000 miles of the journey by car.[10]

They knew what they were doing. The car had an extra-large oil tank and could carry enough gas to cover 400 miles in good conditions. Their supplies included car parts and tools, plus planks and rope to manipulate the automobile through mud and over rivers. (On the hairpin turns over a Japanese mountain, the two men of the party walked behind while holding ropes tied to the rear fender, to prevent the car from skidding out over the edge.) They stayed in top-end hotels, where available, and, in places without accommodation, used camping gear: guns for hunting, tinned food, and the Vuitton tent. Fisher refused to use weapons against people; surrounded by bandits in India, she bluffed her way out. Her confidence was her noteworthy feature. She pleads and threatens authorities when she lacks necessary papers (such as driving license and registration in Ceylon) and usually gets away with it. And she recommends a car for making travel and sightseeing seamless: no need to transfer from one conveyance to another, except for ocean passages. Manufactured goods let her achieve personal control throughout her travels, down to the two small pillows from home she takes around the world—the redoubtable Mrs. Fisher is the patron saint of all travelers today who board their flights clutching their favorite pillows.[11]

On November 3, 1910, several months after Fisher's party had returned, a team of men set out in a Hupmobile, product of the Hupp Motor Car Company. The Detroit-to-Detroit journey tested a new model of Hupp touring car. The marketing gambit was also a gamble. Wide publicity for the journey meant that failure would surely hurt sales. The Hupp tour was more painful than Fisher's because the Hupmobile was completely open to the weather. (The Locomobile had a windshield and convertible top.) Clad in leather and furs, the Hupmobile team got plenty of exercise on the winter return to Detroit, when the car had be dug out of snow almost constantly. The Hupmobile did have headlights, however, which meant it could motor at dusk or even night, which Fisher never attempted.[12]

Fifteen months later, the dinged-up Hupp returned to Detroit with 47,777 miles on its odometer, over one and two thirds of the Earth's circumference. The Hupp Company squandered its opportunity by failing to produce a full account of the journey. It published a pamphlet on the Detroit-to-Manila leg, but did nothing else until it advertised its 1912 model, the "Globe-girdling Hupmobile." The car's rate of travel was significantly slower than what steam locomotives could have achieved; in perfect conditions, the Hupmobile probably topped out at 40 mph. At the likely average of, say, 25 to 35 mph, the car covered 25 to 30 miles per gallon of gas. Gas cost about 11 cents a gallon. That put the total fuel budget somewhere between $175 and $210, though of course the total journey cost more in oil, repairs, supplies, ocean passages, and the purchase of a Hupmobile in the first place. (The globe-girdling model would be priced at $900; an unskilled worker in the auto industry earned about $504 per year.)[13]

Still, it already seemed likely that the automobile, not the bicycle, might become, practically and symbolically, the technology of socially inclusive round-the-world travel. As Mrs. Fisher had realized, an automobile was distinctive in giving female circumnavigators a sense of protection and privacy. Clärenore Stinnes, daughter of a German coal-mining magnate, drove around the world from 1927 to 1929. (Unlike Fisher, she did actually do the driving, though she did not publish the journal of her journey.) By the early 1930s, the National Refining Company of Cleveland, Ohio, would sell "Around-the-World-Motor Oil" in cans that showed automobiles circling Earth on a roadway extending around the equator on a Saturn-like ring. It is an arresting image of belief that the human built environment would encompass the world.[14]

And yet, because automobiles could not cross oceans, aircraft were unique in their capacity to make people imagine a satisfyingly unamphibious global circuit, even before it happened. A children's story of 1910 claimed it could be done in seven days. The Bureau of Aeronautics of the Panama-Pacific Exposition Company, which would organize a World's Fair in San Francisco in 1915, announced in 1914 that it would sponsor an eastabout aerial circumnavigation to leave San Francisco on May 15, 1915, with $100,000 for the winner. *Aero and Hydro* magazine offered baleful encouragement: "even if the hazardous voyage is tried several times, with loss of life and consequent failure, it will still be worth trying." World War I intervened, however. The race was put off.[15]

It was during the war that aircraft fulfilled their deadly potential and established their lingering association with masculine violence. Airplanes were mostly deployed in aerial combat against each other. Manfred von Richthofen, the "Red Baron," was the most famous of the deadly flying aces. Zeppelins were effective for strategic bombing, as German air raids into France and England quickly proved. The airship casualties were small compared to what other weapons achieved, but the huge, humming balloons were despised as threats to civilian populations. Germany was punished for weaponizing the air; after the war, it was forbidden to build new aircraft and all existing air vehicles were transferred to civilian authorities.[16]

Like the war itself, post-1919 regulation of aircraft represented the end of the Hundred Years' Peace: suspicion replaced reciprocity among the Western powers. Although Hugo Grotius's doctrine of the free sea

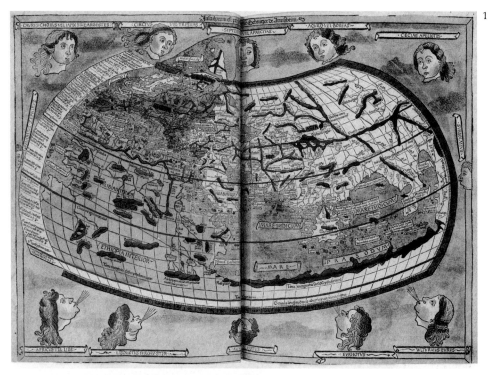

An inward-looking,
pre-circumnnavigated
world

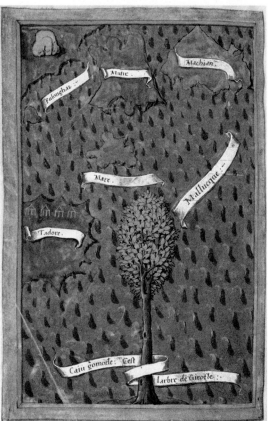

The Spice Islands,
Magellan's intended destination

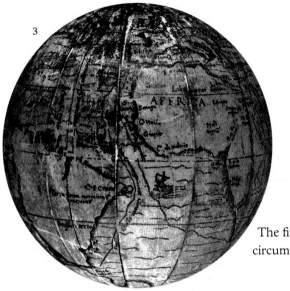

3

The first world image marked with a circumnavigator's track, made c. 1525

4

Francis Drake grasps the world he encompassed (portrait c. 1580)

5

THE RELATION OF

a Wonderfull Voiage made by WILLIAM CORNELISON SCHOVTEN *of Horne.*

Shewing how South from the Straights of *Magelan*, in *Terra Del-fuogo*: he found and difcouered a newe paffage through the great South Sea, and that way fayled round about the world.

Defcribing what *Iflands, Countries, People,* and ftrange Aduentures he found in his faide Paffage.

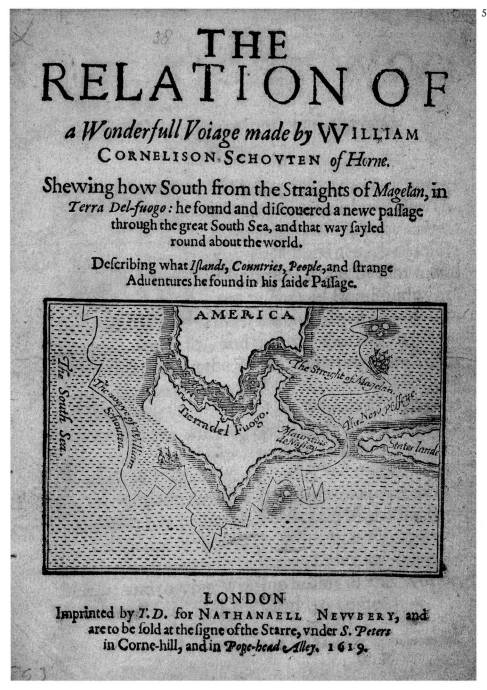

LONDON

Imprinted by *T.D.* for NATHANAELL NEVVBERY, and are to be fold at the figne of the Starre, vnder *S. Peters* in Corne-hill, and in *Pope-head Alley.* 1619.

Cape Horn: another way around the world

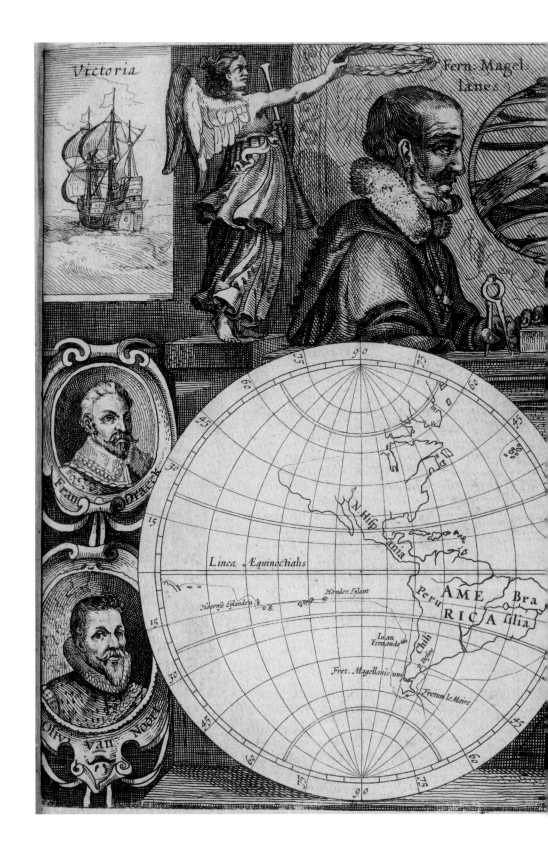

Victoria

Fern. Magel
lanes

Fran Draeck

Oliu van Noort

75 90 75
60 60
45 45
30
15
Linea Æquinochalis
N. Hisp ania
Hoernse Eylanden Honden Eylant Peru AME Bra
15 Iuan RICA silia
Fernando Chili
30 Fret. Magellanicum P. Desire
45 Fretum le Maire 45
60 60
75 90 75

6

Willem
Schouten

Eendracht

The circumnavigators' pantheon, c. 1618

7

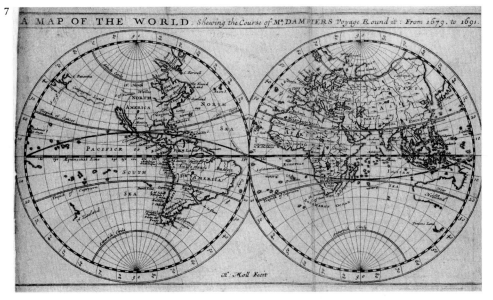

William Dampier's route round the world, 1679–91

8

"Owhyhee here C[apt.] Cook was Kill'd"

James Holman, blind circumnavigator, claims the world (1834 portrait)

Fogg and Passepartout marvel at the gas-burning planet

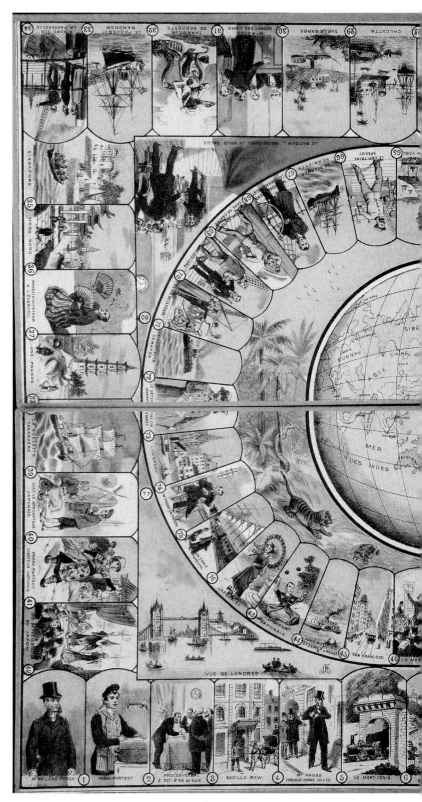

"Around the World in 80 Days," the board game (c. 1915)

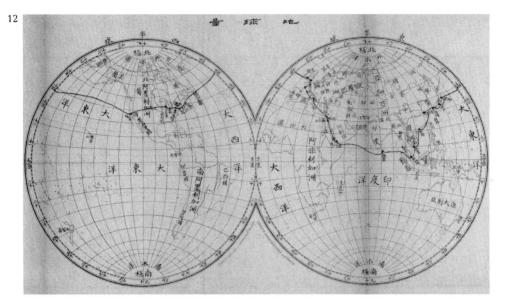

Li Gui's circumnavigator's track, 1876

Nellie Bly is introduced to the circumnavigator's pantheon in 1890

Coaling the SS *Cleveland* in Nagasaki on her way around the world (c. 1910)

Steamships circle the globe for pleasure

Jazz Age circumnavigators

17

The Baron with his Cat, "Felix."

F. K. Koenig-Warthausen, first to fly around the world solo (1929–30), and companion

Faster than the Moon (1928): Mears, Collyer, and Tail-Wind on a brief stop in Paris, greeted by flying ace Captain Alexander von Bismarck

18

19

Laika, first Earthling
to orbit the Earth (1958)

20

Explorer 1,
pre-launch
(1958)

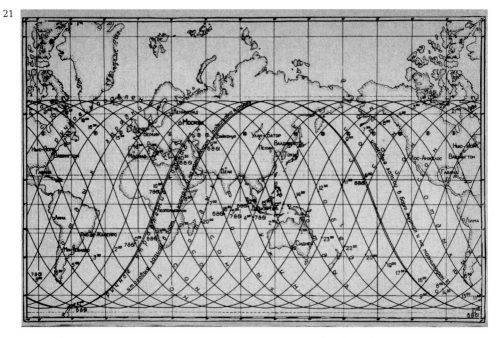

Gherman Titov's seventeen circumnavigators' tracks aboard *Vostok* 2 (1961)

The first round-the-world commercial flight, by Qantas (1958)

might have been a precedent for airspace, it was not clear whether the realm above the Earth was an extension of land, property of nations or individuals, or whether it was comparable to oceans, open to all. Legal experts had anticipated the problem—the debate began in 1900, before sustained flight was really possible, but without resolution. Adjudication was even more difficult after World War I. The Paris Peace Conference of 1919 resulted in a Convention Relating to International Air Navigation which declared that any nation had sovereignty over its airspace. Most European nations ratified the convention, but the United States did not, mostly because it opposed the League of Nations, also proposed in Paris.[17]

There was not, and never has been, an aerial equivalent to Hugo Grotius's free seas. Instead, individual nations have sovereignty over their airspace. Flight over the oceans would resemble maritime traffic, but flight over land was different. That was the upshot, as well, of subsequent conventions on airspace.

All of this is to say that, more than other vehicles, aircraft have carried a heavy burden of international suspicion mingled with hopes for world peace. An around-the-world flight would be a unique achievement, because it demonstrated both power and peaceful intention. Doing it in the first place would require a truce between two slightly warring interests. Military leaders in different nations wanted to continue to build up airpower, while other people wanted aircraft to be rebranded as instruments of peace, the means to deliver mail and carry goodwill missions. Three British aviators (and their cat, Mike) made one round-the-world attempt in 1922, determined that it should be done "by a British crew on British machines before foreigners stepped in," but they ditched in the Bay of Bengal. The U.S. Army Air Service wanted to perform the feat both to determine the feasibility of "an airway around the globe, and incidentally to secure for the United States, the birthplace of aeronautics, the honor of being the first country to encircle the world entirely by air travel."[18]

Everyone else had the same idea. In 1924, three years before Charles Lindbergh would cross the Atlantic, aviators from France, Italy, Great Britain, Argentina, Portugal, and the United States set out to fly around the world first. Most of these teams consisted of a single plane with two men, pilot and mechanic. The planes, either seaplanes or amphibious, were all open cockpit. They followed different routes, depending on preferred sites for ground support and on weather forecasts. Stuart MacLaren, with a pilot and mechanic, left England on March 25. Brito Pais and Sarmento de Beires left Lisbon on April 2; another Portuguese team departed later. Eight men of the Army Air Service, paired up into four Donald Douglas Cruisers, departed Seattle on April 6 and headed west. Georges Pelletier d'Oisy, with a mechanic, left Villacoublay in a Breguet 19-A2 on April 24. Antonio Locatelli, with a co-pilot and two mechanics, left Pisa and headed west in a twin-engine seaplane. Finally, Pedro Zanni (an Argentine) flew with a navigator and pilot, leaving Amsterdam on July 26 in a Fokker C.IV. The punishing circuit was all the more hair-raising because it would require crossing the Pacific, which had never been done by aircraft.[19]

Some good news first: everyone survived. The loss of machinery, however, was spectacular. The French crashed near Shanghai. The Italians capsized into the sea near Greenland. The British lost their plane east of Kamchatka. The Americans wrecked one plane during a crash landing in Alaska; a second had to be abandoned in the sea near Iceland. The Argentines had crashed near Hanoi, though they transferred to another plane and pressed on. But when their new plane flipped over in rough water in Osaka, they quit. The Portuguese gave up near Macao. They were not cowards. Rather, it was becoming clear that the Americans, with their four planes and eight men, had devised an ongoing backup system—the loss of a plane or two made little difference. Given the rate of mechanical failure, the American superabundance of men and machinery turned out to be the best plan for flying around the world.[20]

The Yanks were probably going to do it. But just in case things went badly, the U.S. Army had selected fliers whose deaths would have the least impact on their families, because they were bachelors.

Act Three

Doubt

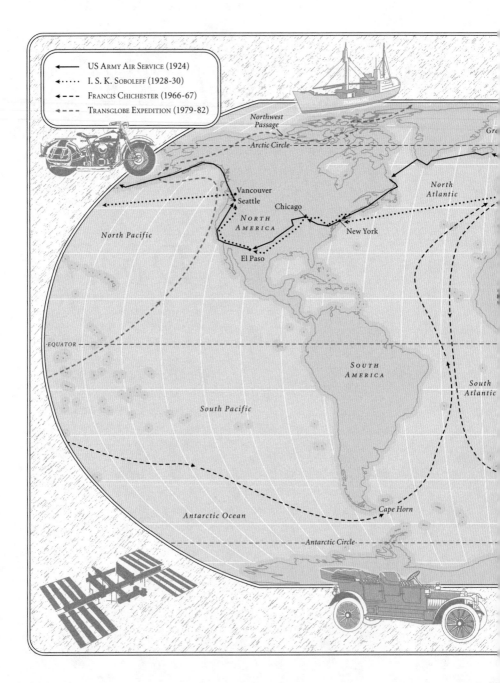

US Army Air Service (1924)
I. S. K. Soboleff (1928-30)
Francis Chichester (1966-67)
Transglobe Expedition (1979-82)

Northwest
Passage

Arctic Circle

Gre

North
Atlantic

Vancouver
Seattle
Chicago
*North
America*
New York
North Pacific
El Paso

Equator

South
America

South
Atlantic

South Pacific

Antarctic Ocean

Cape Horn

Antarctic Circle

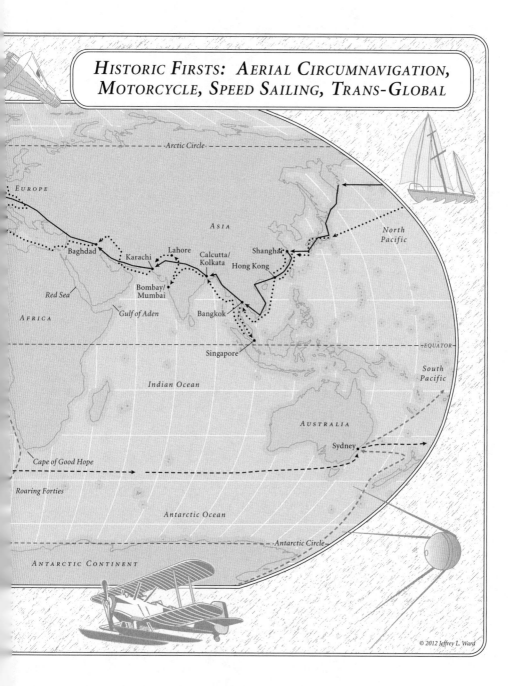

Historic Firsts: Aerial Circumnavigation, Motorcycle, Speed Sailing, Trans-Global

Arctic Circle

EUROPE

ASIA

North Pacific

Baghdad

Karachi

Lahore

Calcutta/ Kolkata

Shanghai

Hong Kong

Bombay/ Mumbai

Red Sea

Bangkok

AFRICA

Gulf of Aden

Singapore

EQUATOR

South Pacific

Indian Ocean

AUSTRALIA

Sydney

Cape of Good Hope

Roaring Forties

Antarctic Ocean

Antarctic Circle

ANTARCTIC CONTINENT

© 2012 Jeffrey L. Ward

Chapter 9

Flight

"**S**ure, *I'll fly* to Mars if you say so," said Leslie Arnold, on being invited to join the U.S. attempt to fly around the world. Many men had volunteered for the mission. The Army chose four pilots, Fredrick L. Martin, Lowell H. Smith, Leigh Wade, and Erik H. Nelson, from a field of 110 applicants. The pilots selected mechanics from a pre-approved pool of eight. Arnold was a last-minute and overjoyed substitute for another mechanic. "I'm ready now," he said. "Let's go." He joined Henry H. Ogden, John T. Harding, and Alva L. Harvey for a total team of eight.[1]

They were characters, possibly chosen for that reason. If they didn't already have nicknames, they acquired them: "Houdini," "Smiling Jack," the "Flying Viking." Their official biographies highlight their Americanness. Smith had maintained Pancho Villa's air force of five planes; Ogden kept a pet raccoon. They were smart and experienced. Several had some college education, though not so much that they might strike the public as eggheads. They were outward-looking and thought big. Nelson, a naturalized citizen, had already been around the world, twice, as a sailor (he was the "Flying Viking"). Smith and a friend had pioneered the intra-air refueling of planes, key to later experiments with long-distance flight. It didn't hurt that the boys photographed well, nor that a couple of them had theatrical experience, because they would be highly visible U.S. representatives to the world.[2]

For that reason, the eight aviators had something else in common, lack of combat experience. Several had served in the Great War, but—crucially—not in the line of fire. Flying aces were as welcome on a goodwill flight as homewreckers at a wedding. The circumference of the planet was supposed to measure a new internationalism for a world no longer unified under the imperial condominium of the Hundred Years' Peace. Each round-the-world plane was loaded with several million tons' worth of incoherent and contradictory hopes—heavier than air, indeed.

Although their work as a team was an ongoing demonstration of their nation's democratic character, the U.S. flyers did not represent a cross-section of American society, let alone the world. All of them were white and from Christian families. Their military status evoked the old expedition model of a circumnavigation: expensive, dangerous, and inaccessible to private citizens, completely unlike the commercial globe-trotting represented by Phileas Fogg, among others. The expedition advertised U.S. wealth, after World War I, compared to other nations. (The British aviator in the race mostly funded himself; the Frenchman intended to set a record for a half-circuit of the planet, which was all that France could afford.)

But the U.S. mission needed international support in the form of permission to cross and land in foreign territories. All of the 1924 teams required ground support at fixed points where materials had been cached, no farther apart than their planes could safely travel. Each engine on the Americans' planes, for instance, could fly only fifty to seventy-five hours before it needed to be replaced with a fresh one. Above all, there was the internal combustion engine's non-negotiable craving for petroleum products. Of the many good-luck charms pressed upon the Americans on their stops, none was more welcome than "plenty of gas and oil."[3]

The United States had to arrange flight over twenty-eight different countries, including colonized territories and protectorates controlled by twenty-five different political entities (see Map 4). Each of the U.S. aviators' passports was supplied with two extra pages to accommodate all the anticipated stamps. It was a tricky business. No German-speaking country could afford to put a team in the race, yet they were expected to concede flying and landing rights. The Japanese, fuming over the terms of two recent U.S. naval treaties (and recovering from the devastating three-day earthquake and fire of September 1923), conceded access only at the last minute. And the Bolshevik Revolution that had created the Soviet Union had negated Russia's earlier treaties of recognition and alliance. Because the United States had not recognized the Soviet government, the Army Air Service devised a route around the bulking USSR—no mean feat.[4]

Connection to the ground was also apparent in the airplanes' resemblance to the wood-and-canvas vessels from the age of sail. The Americans' biplanes, for example, were constructed of painted canvas over a wooden frame, with wings attached to the body by wire. The propellers were made of wood. Wooden boxes that held spare parts were themselves material for emergency repairs. Emergency precautions were modeled on shipwreck scenarios. The fliers' light and compact food supplies included malted milk tablets and concentrated beef essence, updated versions of Captain Cook's malt and portable soup. The men had guns and fishing tackle, to get food if downed beyond easy help. (The two men who were wrecked in Alaska would shoot ptarmigan and rabbits to eat. They also burned pieces of their plane to keep warm.)[5]

Nor were the planes so different from land vehicles. Many of their indicators, including fuel gauges, could be found on automobiles. The American planes had only four navigational instruments: compass, altimeter, bank-and-turn indicator, and airspeed indicator. Like automobiles, they lacked the communications systems that had become standard aboard commercial and naval vessels. None of the U.S. planes had radio equipment—too heavy. (To send a simple message, the pilots dipped or "waggled" their wings. For anything complicated, they wrote a note, put it into a special bag, and tried to drop the bag in the right place, either on land or a ship's deck.) When one of the American planes made a forced landing at a fairground near San Francisco, and the aviators needed a battery with more juice, they swapped with a man who happened to drive by in a truck.[6]

Aviators also used the built environment on Earth to guide themselves, much as around-the-world cyclists had done. Over India, the Americans followed a railway line. When a sandstorm reduced visibility, they dropped to within 50 feet of the ground and "groped along" the track. After one of the planes began to falter near Karachi, its pilot made sure to follow railroad track, where rescue might come. When they got to the Faroe Islands in the North Atlantic, they used telegraph lines to navigate.[7]

Despite the international rivalry among the aerial teams, demonstrations of international cooperation were rife. Although the Americans could not land in Russia, they came alongside a Soviet ship near the Komandorski Islands, and the Russians were hospitable, until ordered to move the Americans along. Antonio Locatelli agreed to fly part of the Atlantic route with the Americans. When the British aviator Stuart MacLaren crashed in Burma, the U.S. team came to the rescue. The expense of bringing his spare plane from Japan was too great for the postwar British government and MacLaren had exhausted his private funds. The Americans arranged to have a U.S. destroyer bring him his spare plane, so he could stay in the race. The British officer who supervised MacLaren's land reconnaissance was astonished: "That's the finest bit of sportsmanship I've ever heard of!" He reciprocated by telling the Americans that they could use MacLaren's caches of supplies, if necessary. And the British were able to help the Portuguese team when it crashed in India, where a spare British plane was available.[8]

International the flights may have been, but not in a liberationist sense. When one U.S. plane made a forced landing on a lagoon in French Indochina (now Vietnam), they depended on France's imperial might. The airmen spoke only a little French, and no Asian languages, and had no idea where to get food and water, let alone mechanical help. Vietnamese lake-dwellers approached the plane in canoes, one man furious that the Americans had moored their plane to his fishtrap; the aviators worried that the canoes might damage the plane's pontoons. The local people exchanged food for cigarettes, and French settlers and officials arranged to get a replacement engine. Within seventy-one hours, the Americans were back in the air. Later, 5,000 Parisians turned out to welcome the Americans, even though the French no longer had a team in the race.[9]

The tide of Western goodwill was the result, at least in part, of the new media that publicized the mission. Newspapers continued to cover round-the-world events, and would eventually sponsor aerial versions. But radio broadcasts and cinematic newsreels were overtaking anything in print. The crowing rooster emblem of newsreel titan Pathé News was a global icon. Recorded interviews and moving pictures would together make circumnavigators more vividly present than print media had done.

The attention showed that geodrama remained newsworthy, but also that the general public took the conquest of the planet for granted. Few appreciated the marathon nature of circumnavigation. Almost everywhere, the aviators were welcomed with official dinners and formal engagements that cut into their time to rest. In effect, they were always onstage, even if on the verge of collapse. When they had reached Unalaska Island, a Pathé News cameraman was there to film them in all their grime and exhaustion. By the time they landed in Paris, they had learned to "shoot back" on their Kodaks. Oddly, they could only be offstage when watching someone else perform. They saw their chance at the Folies Bergères and conked out as soon as the lights went down and the curtain went up. The world's most alluring showgirls were no match for the strain of traveling around two thirds of a smallish planet.[10]

That night, the Americans placed hand-lettered signs on their hotel room doors:

PLEASE DO NOT WAKE US

UNTIL NINE O'CLOCK TOMORROW MORNING

UNLESS THIS HOTEL IS ON FIRE

AND NOT EVEN THEN

UNLESS THE FIREMEN HAVE GIVEN UP ALL HOPE!!

One hopes the French staff got the message.[11]

But without publicity, there was no point to the journey. The resigned aviators had taken aboard, at Calcutta, an Associated Press representative who had been tailing them since Japan. They claimed he was a stowaway—impossible, given how tiny their planes were. In Boston, the first stop back in the United States, a new menace appeared in the form of the live broadcast, the first attempt to represent geodrama in real time. A reporter "shoved a radio microphone" into Low-ell Smith's face. "I looked at it dumbly," he recalled, "and then asked: 'What am I supposed to do with this?' " The blurted question consti-tuted the aviators' "first words to the American public." The Army Air Service thereafter worked up a three-page crib sheet, so the men would be ready for the most obvious questions reporters might lob at them.[12]

The excitement soared to dangerous heights. In New York, souvenir hunters had to be kept from cutting pieces of canvas from the planes and snapping off bits of equipment. A crowd of 35,000 received the men in San Diego. A mob in Santa Monica swept aside the fences, police, and soldiers between them and their darling heroes, who had to fight them away from the aircraft. The fans resorted to taking relics from the aviators, by snipping off their buttons and any loose edges of clothing. The scrum was so tight at one welcome that even though Smith could see and feel a woman slicing at his collar with a penknife, he couldn't do anything about it. He feared for one of his ears; Leigh Wade emerged with three cracked ribs. When the race ended in Seattle, an estimated 50,000 people roared and applauded during a 21-gun salute. Douglas, which had made the expedition's aircraft, promptly adopted "First Around the World" as its company motto.[13]

It was a success built on excess. Only two of the four original planes had actually gone around the world, and only six of the men. (The crew wrecked in Alaska had never caught up, but the pair who lost their plane off Iceland received a replacement.) All told, the aircraft wore out fifteen engines, forty-two sets of wheels, fourteen sets of pontoons, and an unknown number of propellers; almost 27,000 gallons of liq-uid gas and 2,900 of oil were drained from the caches of 91,800 and 11,650 gallons, respectively. It had taken 175 days to cover the 26,345 miles (average speed 72.5 mph), though most of that time was spent on the ground, with only 15 days, 3 hours, and 7 minutes in the air. The journey had lasted longer than many pleasure cruises, with few of their pleasures.[14]

It was common to praise the achievement in historical terms. *The Times* of London noted that "the challenge thrown down by the sphericity of the earth was bound to be answered by airmen as it was generations ago by seamen." The description of the feat as a "circumnavigation" made an even stronger historical claim. So did calling the airmen "Magellans," as the knife-toting souvenir hunters did.[15]

Lowell Thomas made the most of that claim. Thomas was an American journalist and a celebrity himself. (With his reports from the Middle East, he had helped make T.E. Lawrence into "Lawrence of Arabia.") He interviewed the U.S. airmen en route and wrote the official history of the expedition, one that emphasized its planetary scale. Embossed onto the book's cover, and repeated on the title page, was a biplane superimposed over a globe turned to display North America. (Oddly, the plane is headed in the wrong direction.) Inside, two of the photographic portraits of the fliers show them with globes; in a group portrait, they all lean in to examine a big library globe. The men were "the Magellans of the Air"; just as "Magellan's journey around the world opened up the sea-routes and civilization of today," so the aviators "shall bring the nations closer still."[16]

The claim was not new—the telegraph had incited similar enthusiasm—though its circumstances were. It showed the ongoing memory lapses about circumnavigations in the age of sail, when Magellan had died during an expedition that was meant neither to unite nations nor to go around the world. It also revived the claim that aircraft need not endanger humanity. "The airplane was destined to be the agency that would bring the races of the world into such an intimate contact with each other" that wars would cease, said one of the American airmen. "The airplane is destined to have an immense influence on the peace of the world," said another.[17]

That was a hypothesis draped over a fact. The internal combustion engine was indeed becoming a global force—whether it would be a force for peace was another matter. The same year the U.S. team made it around the world, the Texas Company printed a series called "Around the World with Texaco" for its company magazine, *Texaco Star*. The articles, later published as a book, formed a travelogue garnished with information about consumption of petroleum products. "Korea is not an important outlet for lubricating products," one learns, "but kerosene moves in large volume." The world is beginning to be bound by a chain of Texaco outlets: "its principal [New Zealand] office is in the city of Wellington"; "wherever one travels over Africa's immense distances Texaco Products are known favorably to the virile people who are building a great country out of the Darkest Continent." At the end comes the predictable claim that "the sun never sets on the Texaco Star." The suggestion, and not just here, was that a global traffic in oil would finish the job that empires had begun, of transforming the Earth into a place uniformly fit for Western consumers.[18]

Subsequent flights around the world were presented as harmless: zero mortality for all, whether in the air or on the ground. Preferred passengers included women, pets, and journalists, whose presence was thought to prove that an aerial circumnavigation was perfectly safe. "A Flight around the World" (1928) made it into child's play. The Anglo-German company J. W. Spear & Sons (inventors of Scrabble) produced that rather earnest board game, which mixed a cardboard race around the world with a variety of "Educational facts" about geography. It captures aviation's sense of autonomy by giving each player his or her own teetotum, which is the spinning propeller on a card printed with a colored airplane, one per person.[19]

Around-the-world flights became media-sponsored stunts. Linton Wells, the reporter who had hitched along with the American team in 1924, decided to set a new speed record in 1926 with Detroit industrialist Edward S. Evans. Neither man knew how to fly—they were not aviators but travelers, though travelers with a famous prototype: "who is there that has read Verne's fantastic tale of travel and adventure and not pictured himself in the rôle of Phileas Fogg," the non-expeditionary model of a round-the-world traveler. Wells and Evans put it even better when they said they wanted "to draw in the waist-line of the world." Puck's new girdle was not a belt that happened to fit, but a foundation garment that would give ample Gaia a svelte figure to match the human world of aerodynamic travel.[20]

The two men set their record: 28 days, 14 hours, 36 minutes, and 5 seconds, slightly more than the 27 days, 12 hours that Wells had predicted. It required airplanes to cross landmasses, steamers to navigate the Atlantic and Pacific, a few gap-filling trains, and an automobile to dash over the finish line in New York City. Some hitches added suspense. It turned out to be illegal for the duo's French aircraft to fly over

Germany—lingering ill-will from the recent war. Later, the two men split up in China, where the only available plane seated one, leaving the other to travel by express train. They proved not only their ability to set a speed record, but that observers were fascinated with the attempt. They estimated that they had been interviewed by over one hundred reporters in Japan alone.[21]

What is most notable about their voyage is that it represented two and a half tours of the globe, not just one. To work out everything in advance, Wells made a preliminary circuit, which took him seven months, several thousands of dollars, and approximately double the 20,000 miles of the official journey. And then there was the half-circle that some of their clothing made independently. Wells and Evans wore travel gear and carried minimal luggage, but sent ahead to Yokohama a complete wardrobe, including evening clothes, for all the occasions they expected to endure in their honor. Again, there is that Satanic excess, the multiple swoops around the world that showed the super-abundant resources that people in the industrial world, and particularly white Americans, had at their disposal.[22]

The French made a more generous international gesture with the aerial circumnavigation of Air Force Reserves captain Dieudonné Costes and Navy lieutenant Joseph Le Brix, from October 10, 1927 to April 14, 1928. They set out to demonstrate the feasibility of greater aerial cargo service to Latin America, as well as a route from Tokyo to Paris. Their published account conveyed flight's rapidity by recording events down to the minute, quite a contrast to earlier circumnaviga-tors' accounts, for which the day and hour were sufficient. With a sense of drama, they described the Paris–Buenos Aires leg as "the first act," South America to North America as the second, and Tokyo to Paris as "the third and last act." In the second act, they visited almost every country in Latin America, a grueling schedule broken by many sand-wiches and the brief companionship of a baby crocodile, supplied in Baranquilla. The reptile hated flying and Le Brix, as navigator, wasn't crazy about sharing his legroom; the crocodile went no farther than Mexico City.[23]

The promise of greater air service to and within Latin America guar-anteed that the two Frenchmen received rapturous welcomes. The choreographed frenzy—Champagne, the *Marseillaise*, motorcycle escorts, cheering crowds, press conferences—became so routine that the flyers called it *le brouhaha*. Still, they were moved when a crowd in Guatemala shouted *"Viva la Francia!"* and *"Viva la raza latina!"* for what seemed like a thousand times. The two men made sure to emphasize that they came in peace. They observed Armistice Day in Buenos Aires and laid a wreath on the Tomb of the Unknown Soldier in Washington, D.C.[24]

Given the cost of aircraft, most civilians could experience only short flights. When Cunard, with Thomas Cook & Son, arranged its de luxe 1929 round-the-world cruise on the *Franconia*, a small plane added a premium touch to a day trip over the Holy Land and Arabian Desert. Travel on the ground there would be uncomfortable, through heat and over bumpy roads. It was also dangerous, given Arab uprisings against the disintegrating European empires. Airplanes sailed above it all, perpetuating a sense that Westerners dominated the region, if not the world.[25]

Having a private airplane was the ultimate luxury. By the late 1920s, a small plane, rather like a yacht, signaled membership in a global elite. And if going around the world on a whim was the ultimate in human mastery over the planet, then travelers with wings were the most whimsical and masterful yet. But because aircraft could carry so little, even their richest owners had to leave servants behind, another instance of technology replacing human labor.

Consider a trio of aristocrats. Two of them, Jacques, vicomte de Sibour, and his American wife, Violette, liked to hunt, and wanted to try a tiger shoot in India, so over cocktails one evening in 1928, they decided to fly around the world in their private plane. *Safari* was a de Havilland Gipsy Moth, with two seats, one engine, small gas tank, and hardly any room for provisions or luggage. The plane, as trim Violette pointed out, weighed less than nine times what she did. Jacques was a flying ace and war hero, but neither of the de Sibours was a mechanic. Nor did Violette, as navigator, have much more than school geography. They flew low enough to follow rivers and coastline, and spot major towns and cities. (They never flew at night.) The whimsy grew thin after many forced landings and much engine trouble. "I often wonder which is the most tiresome, an optimist or a pessimist," Violette grumbled, after the umpteenth mechanical failure.[26]

But their amateur status was supposed to prove their peaceful intentions. Violette would later learn to fly, as many women did in the 1920s and 1930s. They were encouraged to do so for the simple reason that female aviators relieved flying of some of its associations with warriors, battle, and death.

The other flying noble was Baron F. K. Koenig Warthausen (Friedrich Karl Freiherr Koenig von und zu Warthausen). Bored with his motorcycle, Warthausen got a pilot's license in early 1928 when he was twenty-two. His father gave him a Klemm Daimler: 560 pounds, with a 20-hp engine that could maintain a cruising speed of 70 mph. The light aircraft was built as a glider. (A handkerchief pasted on with egg white could patch a hole in a wing.) If his engine failed, Warthausen could glide as much as twenty-eight miles before he had to come down—a good margin of safety. With the engine and a suitable wind, he could do even better, which extended his gas mileage. After only seventeen hours of solo flying experience, Warthausen decided to head east, alone, perhaps inspired by Lindbergh's solo crossing of the Atlantic one year earlier.[27]

Although he would win Germany's Hindenberg Prize for his non-stop flight from Berlin to Moscow, Warthausen was an amateur who felt his way as he went. He followed railroad tracks to Baku, in Azerbaijan, for example, where he picked up spare parts he had shipped ahead. (Later, he would race the Twentieth Century Limited train over North America.) When Warthausen came down on a road outside Teheran, a motorist in a Ford pulled alongside to offer help. The British consul at Bandar Abbas, near Teheran, advised Warthausen to follow the telegraph lines to Karachi. Should he be forced down, he was to climb a telegraph pole and cut the lower line, which would signal his location to rescuers.[28]

As that advice revealed, the international order was on Warthausen's side—and the de Sibours'. As civilians, none of them needed much documentation and all received friendly assistance. Warthausen arranged visas with the Polish and Soviet governments. When he decided to continue into Asia, the British Air Forces helpfully provided maps for air routes between Calcutta and Singapore. The vicomte lost his passport early in the trip, but it didn't matter. The de Sibours depended on the complaisance of Western officials, as well as the "wonderful freemasonry among flying people." The prince of Jodhpur, for instance, offered the use of his private aerodrome in India. Such was the out-pouring of welcome in Rangoon that the couple hired a secretary to manage their invitations and correspondence.[29]

Warthausen and the de Sibours were delighted to fall across each other in Bangkok. The couple left before Warthausen, but he "flew out a little way with them to bid them good-bye." He was consoled by a different companion, a Siamese cat named Tanim. Confined to a box in the plane's fuselage, Tanim emerged from his first flight as "wobbly" as a sailor back on land. When he and Warthausen flew over the equator, they had a "Neptune celebration, and were both christened." Inter-estingly, only after he had Tanim did Warthausen decide to continue around the world (see illustration 17). A small passenger may have lessened his loneliness or sense of vulnerability. He would be the first person to fly solo around the world, from August 9, 1928, to November 26, 1929, though, like the de Sibours, he loaded himself and his plane onto steamships to cross the Atlantic and Pacific.[30]

The hardships of long-distance air travel could not disguise its privileges. "Oh, the joy of shedding one's oily clothes, having a luxuri-ous bath, and finally appearing in complete evening attire—diamond earrings and all," said Violette, at the Ritz in Barcelona. Less joyful was the vicomte's emergency appendectomy in Teheran. And outright de-spicable was his use of *Safari* to attack a group of Arabs near Shaibah, now in Iraq. The men had been accused of shooting at some traveling Americans, killing one. (The skirmish was part of a series of Arab re-bellions against any foreign presence, including the British, who had extracted Iraq from the Ottoman Empire during World War I.) The authorities had organized a retaliation; with a gun and smoke bombs, Jacques set out on a personal war mission. Violette's published narra-tive includes a photograph of Arab casualties.[31]

Like their passports, their airplane identified the couple with West-ern control over the globe. Even though they, as French citizens, had no immediate interest in the British Empire in the Middle East, they thought it imperative to protect Western interests there, or anywhere. A world of empires was the stage, after all, on which they were the center of attention. As Violette put it, on embarkation from Yokohama to Seattle, *Safari* was packed into the ship's hold, "and the curtain was going up for the last act of the show." [32]

The closest thing to a weapon that Warthausen carried was a signal pistol and, given postwar anti-German sentiment, he knew better than to fire it at anyone. He was careful to extol the kindness he experienced in nations that had recently defeated his own. He too enjoyed the "freemasonry" of the air, as when he met Leigh Wade, of the 1924 U.S. around-the-world flight. He let Americans sign his plane, a custom that had replaced ripping souvenirs from an aircraft. (Fifty people signed in Oklahoma City alone.) And Warthausen was cleverly armed with a disarming mascot. Everyone loved Tanim, the wonderful flying cat. In Buffalo, the cat was filmed and recorded for the newsreels. Americans couldn't believe that a famous cat wouldn't be called Felix, after the famous cartoon of the day. Warthausen obligingly renamed his pet. [33]

The unimpeded transit of the three flying aristocrats showed how petroleum had become a global commodity, even more than when Mrs. Fisher had motored around the world. Warthausen suggested that appetite for oil might level off. Airplane engines, including his, were becoming more fuel-efficient. In the roughly 20,000 miles he flew, he used less than 1,000 gallons of gasoline, worth about $300; oil required another $70. He predicted that a small plane would "become the 'flivver' of the air in a few years." Actual flivvers were not necessarily more economical. Warthausen spent $700 on taxicabs to get to and from airports. Nor were automobiles safer. When the de Sibours drove overland to Teheran, conditions were so rough that they wore their flying gear. Warthausen's only serious accident was not in his plane, but in a head-on collision in a taxicab outside El Paso. He went through the car's windshield and came to in a hospital. (Tanim was found yowling in a tree.) It took the aviator two months to recover. [34]

Whatever its drawbacks, the gas-powered engine looked like a permanent fixture of daily life in the West. Jacques de Sibour went to work for one of the oil companies created in the division of Standard Oil. Was he an optimist in accepting that petroleum was the future and Warthausen a pessimist for wishing to reduce demand for it? Or was de Sibour the pessimist and Warthausen the optimist? Which is worse, an optimist or a pessimist?

• • •

There would always be suspicion that *flying* around the world was cheating, that surface travel was more challenging. In 1928, the centenary of Jules Verne's birth, the Danish newspaper *Politiken,* in partnership with the *Stockholm Tidning*, advertised that it would send a boy, age fifteen to seventeen, around the world to commemorate Verne's famous novel. The boy had to be in good health, needed permission from his family, must speak English and German (to do interviews), and was not allowed to fly. The Canadian Pacific Railway arranged the travel and the newspapers would pay any remaining expenses.[35]

Over one hundred boys mobbed the newspaper office in Copenhagen. (A girl telephoned to protest her ineligibility.) The *Politiken*'s staff eliminated all but the fifteen-year-olds. An essay contest identified two finalists, both of them Boy Scouts. They drew lots. The winner, Palle Huld, a freckle-faced redhead, went home to tell his mother that he had a week in which to pack and get his vaccinations.[36]

Huld was required to keep a diary and either mail or telegraph a report at every stop. Another obligation was to meet and charm the public, which made the boy into as much of a roving commercial advertisement as Annie Londonderry had been. A book narrating the journey appeared in the following year. It is hard to tell how much of it is Huld's own prose. His comment on meeting the press in London— "the reporters were awfully witty and we had a lot of fun together"— hints that the newsroom pros back in Copenhagen had given Huld's story its final shape, with a compliment for their English colleagues.[37]

But for the forty-four days he went around the world, Huld was the star of the show. The premise of the journey was that a circumnavigation was the ultimate boy's adventure, a good but not dangerous test of his character. That emphasis was reinforced by the introduction to the English translation of Huld's book, written by a grown-up: "Around the world! Every one of us has made the voyage many times—in our imaginations." The introduction made the usual nod to Magellan, threw in Drake for good measure, but fast-forwarded to Verne, whose posthumous reputation was beginning its descent to that of children's author. Verne's stories were challenging, yet safe, much as a circumnavigation was now thought to be. Huld's narrative claimed his mother had read Verne to him when he was twelve, also that the novel was the only book he took around the world, as if it were the circumnavigator's Bible.[38]

The *Boy's Own* element of the story was present, above all, in Huld's membership in the international fraternity of Boy Scouts. The year 1928 marked a peak in early enthusiasm for scouting; that Huld and the other finalist for the adventure were both Scouts was probably not happenstance. "A Scout is never without a home," the narrative concluded—"He is sure of being received wherever he goes in the entire world." Indeed, Huld was the guest of the Scoutmaster of Japan, and met groups of Scouts everywhere, from Tokyo to Warsaw. Other aspects of his journey likewise emphasized his youth. When he arrived back in Copenhagen, two policemen had to hoist him through the crowd and into the newspaper office.[39]

The juvenile drama climaxed after Huld's return, during subsequent visits to England and France. In London, he attended a gala luncheon with the head of Canadian Pacific Railway and, even better, met Sir Robert Baden-Powell, founder of the Boy Scouts. In Paris, Huld saw *Around the World in 80 Days*, still playing at the Châtelet. He watched a copy of the novel being printed expressly for him, bound in red (like one of Cocteau's remembered school prizes), with his name embossed in gold on the cover. Huld then met Jules Verne's grandson, who escorted him to *grand-père*'s grave in Amiens. There, surrounded by local Scouts, he laid a wreath with a message: IN MEMORY OF JULES VERNE, FROM HIS GREATEST ADMIRER, PALLE THE DANE."[40]

Adult world-circlers also avoided aviation in order to make some kind of point about their place in the world. Cyclists who were not from the Western imperial powers, for instance, began to rebrand the bicycle as a peaceful *me-too* way to see the world. Haru Kichi Nakamura, for example, circumcycled the world from 1901 to 1904, gathering admiring newspaper accounts as he did so. Because he did not publish his own narrative of the journey, he remained better known within Asia than beyond.[41]

Nakamura was later pleased to welcome to Japan three fellow Asian cyclers, Adi B. Hakim, Jal P. Bapasola, and Rustom B. Bhumgara, who did a world tour to show India's equality with other nations. The three young Parsi men, members of a Bombay weightlifting club, had left home on bicycles in October 1923, returning in March 1928, having covered 44,000 miles and demonstrated that the "sons of Mother India were as able, as enterprising, and as courageous as the children of any other nation in the world." Affidavits from assorted admirers—Benito Mussolini, Jawaharlal Nehru, Chang-Taolin (warlord of Manchuria), Lord Birkenhead (Secretary of State for India)—made it clear that they had succeeded.[42]

In making a point about Mother India, the three men revealed the several kinds of global society that assisted them. The first was the British Empire, not an obvious choice, but the cyclists were anxious to make clear that their British passports and a letter of introduction from the British governor of Bombay had been critical to their passage into and through Europe. Whatever their private feelings about the Raj, they saved their criticism of imperialism for French Indochina, where they claimed to encounter racism unparalleled in any other part of the world. They routinely stayed at branches of the YMCA, the equivalent, for grown men, of Huld's Boy Scouts. And they were cheered on by enclaves of Indians, and especially Parsis, who constituted a South Asian diaspora over most of the globe, a consequence of empire, and a kind of counterweight to it.[43]

A different diaspora, and yet similar manifestations of internationalism, supported I. S. K. Soboleff on his slightly later surface tour of the world. Soboleff came from a privileged Russian family, but that was no help when he found himself on the losing side in the Russian civil war. As a "White" Russian stranded in China, Soboleff was a man without a country, so destitute that he made his way to Shanghai in a mix of men's and women's castoff clothing. He obtained a Nansen passport, a document that the League of Nations had begun to issue to stateless refugees (initially Russians) in 1922, a first step in the development of international refugee law and policy. (The Nansen International Office for Refugees would win the 1938 Nobel Peace Prize.) Soboleff yearned to rally members of the non-Bolshevik Russian diaspora and he wished a Russian could do something akin to Lindbergh's inspiring flight across the Atlantic. In 1928, he decided it was up to him to do a proudly tatterdemalion equivalent: to go around the world by bicycle (see Map 4).[44]

Luckily, he didn't have to do that. Soboleff departed Shanghai on a battered secondhand bicycle, but upgraded to a new bicycle in Bangkok, then to a battered secondhand motorcycle in Singapore. A benefactor gave him a brand-new Ariel motorcycle in Karachi, plus a letter that guaranteed parts and assistance from Ariel offices around the world. Soboleff also thanked the worldwide services of the YMCA, Shell Oil, and the Firestone Company, and depended on the global availability of gasoline, oil, and tinned food—the array of industrial goods and services now spread almost everywhere. Like the circumcycling Parsis with their South Asian diaspora, Soboleff made his transit with the encouragement of many scattered White Russians. Above all, there was his Nansen passport, for which he was an around-the-world ambassador. The document raised eyebrows at the start of his journey, but once it bore an impressive succession of transit stamps, officials stamped it without suspicion. He arrived back in Shanghai on November 7, 1930, "just two years after the day that I set out with my Nansen passport, no visas, a broken-down old bicycle, and twenty Mexican dollars in my pocket." He fulfilled a promise to continue to Karachi, making a full global circle on the same motorcycle.[45]

It was not clear, however, that the satisfying internationalism of ground circumnavigations—linked up with Scout troops and Nansen passports—would have an aerial counterpart. Instead, round-the-world flights continued to be undertaken to set speed records, not to see the world and meet its peoples.

In the summer of 1928, the same year Huld and Soboleff set out, the New York writer and theater producer John Henry Mears recovered his record as fastest around-the-world traveler. (Mears was the fellow who had perched on the wing of a plane for forty miles.) This time, he bought a plane and hired a pilot, Charles B. D. Collyer. The goal was to circle the planet in less than twenty-seven days and seven hours, the time of the Moon's orbit. The human duo would not cross oceans in their plane—island-hopping was slow and crossing the open ocean would entail waiting for perfect weather conditions. Mears's plane, the *City of New York*, had wings that could be folded upward, and the origami-aircraft stowed in a ship's hold for crossing the Atlantic and Pacific. When told that what the flight needed most was a good tail-wind, Mears procured a white terrier puppy, named him "Tail Wind," and took him along as a mascot.[46]

In surviving photos (see illustration 18), Mears and Tail Wind look delighted with themselves, while Collyer looks like a man daily diminished by torture. He had to be alert at every moment in the air, with little time to rest on the ground. Formal dinners and public appearances were agony. (Ocean travel meant recuperation, a spa at sea.) Flight over the Soviet Union offered the most direct route, though also a grueling number of airmiles, and it required Soviet permission. It helped that Mears flew as a private traveler, not with U.S. sponsorship. The Soviet government could not directly approve his request to enter and land, but Mears negotiated with the Amtorg Trading Corporation, which maintained offices in New York. Amtorg's representatives indicated that their nation looked forward to his visit. Soviet citizens indeed proved to be excellent hosts. The two men (though not Tail Wind) were sick of caviar by the time they reached China. While "congratulating the travelers," *Pravda* was pleased to report that "all the [Western] newspapers remark on the assistance and hospitality given to them in the USSR."[47]

At a cost of $29,407, mostly for the plane, Mears set his new record. "The difference between my two records reflects the speeding up of the world in a period of fifteen years," a Verne-like claim that paralleled Mears's professed enthusiasm for Verne's novel and for "the scientific and mechanical wonders of our time." The full title of Mears's book says it all: *Racing the Moon (and Winning): Being the Story of the Swiftest Journey Ever Made, a Circumnavigation of the Globe by Airplane and Steamship in 23 Days, 15 Hours, 21 Minutes and 3 Seconds by Two Men and a Dog.* The whimsical details of the journey (adorable dog, cute folding plane) should not detract from the scientific precision of Mears's achievement. He did not win a race against a fanciful pre-Copernican Sun, but against the entirely real Moon. He predicted that only a plane or dirigible capable of crossing oceans could do better.[48]

One year later, a zeppelin proved Mears right by going around the world in twenty days. The world tour was an opportunity to demonstrate world peace, plus the commercialization of a planetary circuit done in the air. Dr. Hugo Eckener had taken over the Zeppelin company on Count Zeppelin's death in 1917. He knew that commercial expansion of German airship service would require rebranding the vehicles as something other than bombers, and so in 1929 prepared the new *Graf Zeppelin* for an around-the-world journey.[49]

Eckener asserted that the voyage would test air flight through the greatest variety of atmospheric and meteorological conditions. It was a gamble—no airship had ever covered the proposed 19,500 miles. Each passenger ticket cost a stupendous $9,000 for a journey that lasted from August 15 to September 4, at $3,000 per week, or $429 a day. The sixty-one passengers came from ten different nations, including the USSR. The journalists and foreign dignitaries surely did not pay full price; they subsidized their transit by generating publicity. (U.S. newspaper magnate William Randolph Hearst, a sponsor of the flight, dispatched two of his reporters.) Dr. Jerónimo Megías, from Spain, was present in case anyone needed medical attention, though this was unofficial. Two mascots, a baby alligator and a terrier puppy named Happy, boarded as well.[50]

"The whole world literally held its breath," *National Geographic* would claim. Certainly, the airship was front-page news. Koenig-Warthausen pridefully watched its passage over Texas. The polar exploring party of Richard E. Byrd followed its progress by radio. The flight segments over Siberia and the Pacific were in their own way expeditionary, not least because they were out of radio range—"the Russian posts were mute and deaf" and the Pacific offered prolonged, perilous suspension over open water. Eckener was duly described as the "Magellan of the air."[51]

Exhaustion crept through crew and passengers. Obsessed with filing stories, the journalists were dull company. Léo Gerville-Réache, who wrote the best account, teased his fellow members of "the Tower of Babel of Information." Hunched over their typewriters, they pecked away with their carbon paper–smudged fingers while some of the world's most fantastic scenery rolled beneath them. Perhaps to commemorate one way the passengers cut the tedium, one of the journalists, Max Geisenheyner, created an around-the-world card game: the winning player collected the greatest number of four matching cards, from a total of twelve sets that represented segments of the journey. The relentless schedule was worst for Eckener. At the Los Angeles stop, he checked into the Ambassador Hotel's Lindbergh Suite. "Let no one disturb me," he told the staff, "for no purpose, at no time, under no circumstances." The public never knew that; the physical strain of circumnavigation remained invisible.[52]

In 1931, Wiley Post and Harold Gatty outdid the zeppelin in an airplane. The two men, farm boys from Oklahoma and Tasmania, had come of age as aviation was becoming more affordable—just about. Post traded an eye for his first plane. When an industrial accident damaged his left eye, which had to be removed, he used the insurance money to buy and repair a damaged aircraft. Then he was hooked. That was why he disliked the *Graf Zeppelin*'s round-the-world record: a plane should have done that. One of the Oklahoma oilmen for whom Post was a pilot had a Lockheed Vega named for his daughter, *Winnie Mae*. The Vega was a small single-engine model, 420 hp, sturdy, and fast, built to eat up distance. Post would fly it around the world, and Gatty, who had trained in the Royal Australian Navy, would navigate.[53]

Post and Gatty depended on a raft of commercial sponsors and media outlets. They sold their story to the *New York Times*. The *Winnie Mae* carried radio gear, plus a Pathé camera; a painted Pathé rooster crowed from the side of the plane. Gatty and Post thought they might garner upward of $300,000 in prizes, endorsements, and speaking fees. They took stacks of stamped souvenir envelopes, to be canceled at various points and sold at journey's end, an increasingly common philatelic feature.[54]

It was the first well-publicized round-the-world flight in a plane with a closed cockpit. The Vega was not pressurized, but its mere enclosure was protection from the elements. Instead of flying gear, Post and Gatty wore ordinary suits, though their decision not to wear hats (and ties only sometimes) gave them, by contemporary standards, a sporty look. The closed cockpit meant that they (like other aviators at this point) were increasingly relying on instruments rather than sensory observations.[55]

Their goal of a circumnavigation in eight days (and their secret desire to do it in a week) meant extraordinary bodily discipline. Post trained by adopting an irregular cycle of sleeping and eating. That helped, though only so much. When interviewed by a *New York Times* reporter in Berlin, as per contract, Post kept nodding off. Fatigue probably accounted for the only serious injury. In Alaska, when Gatty swung the *Winnie Mae*'s propeller while Post revved up, the exhausted navigator lacked the quick reflexes to jump clear when the engine started. A propeller blade "smacked" his shoulder and threw him to the ground. The tight schedule in a small plane raised a delicate issue (never publicly discussed): excretion. Both men ate and drank lightly before any flight. A stash of wax-coated ice-cream cartons could provide relief when absolutely necessary, then be tossed overboard, airborne gardy-loo.[56]

As always, ground support was crucial. Post and Gatty had sought and received permission from all the countries over which they would fly, including the USSR. Through Amtorg, the Soviets agreed to reciprocate U.S. support of Soviet test flights over the north pole by assisting the Americans. Prepaid gas and ground staff awaited the men all along their route—except for Newfoundland, where they went broke buying sandwiches. Old and impromptu technologies saved them more than once. Post followed railroad tracks over the Soviet Union. Horses pulled the *Winnie Mae* out of the mud in rural Russia. After a propeller blade became bent in Alaska, Post banged it into shape, sort of, with a wrench and a big rock. Another danger was human. Gatty hated the mobs that greeted them in the United States, especially after somebody in Cleveland tore off his coat pocket.[57]

The most unusual experience was a new version of the Circumnavigator's Paradox. The narrative of the flight was the first description of an "acute fatigue" associated with "variance of time," later called time-zone fatigue, a precursor of jet lag. The *Winnie Mae* had four chronometers set to different times: the next destination, New York, Greenwich, and sidereal (the time it takes the Earth to rotate relative to the fixed stars, somewhat less than the twenty-four-hour solar day). The four clocks made Post and Gatty aware that time passed with confounding speed—their mental comprehension rasped against their physical awareness of multiple time zones. Given the precision of the timekeeping, the zigzag configuration of the 180th meridian was especially bizarre. At their point of crossing, any of their chronometric readings would suddenly be one hour and thirty-two minutes "late."[58]

Once across the Pacific, they were most of the way home, and back they came, to screaming crowds and a ticker-tape parade down New

York's Fifth Avenue. Their subsequent book was ghostwritten (and not free from error), but it has a suitably cocky tone, with a first-person voice that alternates between the two men: "Wiley had $34, and I had what the old vaudeville joke calls 'some money' . . . the air behind 450 horsepower certainly can get hot . . . I swung the prop and felt the compression on each cylinder." Post's good friend, the Oklahoma humorist Will Rogers, offered an introduction. Rogers explained that the aviators had "carried no Parachutes, or rubber life boats, they simply made it or ELSE."[59]

It wasn't enough for Post. In 1933, he set an unbreakable record: around the world *solo* in slightly under eight days, or 186 hours and 44½ minutes. Post could do it because he had an unusual companion, a "robot." That was how the *New York Times* reported his decision to fly with an autopilot, using the word Karel Čapek had introduced in 1921 to describe humanoid automata. Post named his robot "Mechanical Mike." Only with its help could he catch the catnaps necessary to keep going. As with a closed cockpit and increased instrumentation, the robot represented another remove from the natural world that earlier aviators had so vividly sensed as they flew.[60]

It was a less exuberant flight than the one in 1931. The deeper economic depression made it hard to find sponsors; Post focused on Oklahomans and on aeronautical and petroleum companies, including Texaco. He may have been relieved not to court newspapers and radio stations. (There is no published narrative of his flight.) He deployed his training regimen of irregular eating and sleeping, planned to follow much the same route he and Gatty had flown, and again flew the *Winnie Mae*. Another man threatened to beat him, but bowed out after he crashed in Siberia. By that point, Post was busy flying, catnapping, flying, landing to refuel, flying—and so on. Without a companion, the threat of deep sleep was real, and potentially fatal. The autopilot and the radio were Post's aid and comfort. Just over halfway across the Atlantic, he picked up station G2LO of Manchester, England, which was giving a special broadcast for him. That was terrific, much appreciated.[61]

Post made it back with his new record. In New York, a crowd of 50,000 watched him land, the first man to fly around the world more than once. And his just-under-eight-day circuit was unbeatable. To circle more slowly wouldn't be impressive. Nor would doing it faster, which would, after all, be less of an endurance test for the pilot. Post may have hit the planet's sweet spot. When Howard Hughes, with a crew of four, flew around the world faster in 1938, he knew better than to gloat. His circuit of 3 days, 19 hours, and 8 minutes only proved, he graciously admitted, that the equipment was getting better.[62]

In 1934, Bradshaw's published its first *International Air Guide,* modeled on its railway timetables, though considerably shorter. The guide showed that commercial air services were not evenly spread over the globe. They clustered where wealthy customers or governments were willing to pay for flight.[63]

Long-distance commercial aviation was an imperial specialty, a way to connect Europe with its overseas colonies. Since 1918, French aviators had been flying "the Line," the celebrated Aéropostale service which included an imperial mail route that ran south to North Africa, across the Atlantic, and through Latin America. It was dangerous work for the intrepid. (Saint-Exupéry and Costes flew the Line.) For many years, KLM (founded in the Netherlands in 1919) maintained the longest route in aviation, the 9,000 miles between Amsterdam and Batavia (now Jakarta). Britain's Imperial Airways (1924) covered slightly shorter distances, including routes from London to Cape Town or Karachi, done in segments and, on the extra-European sections, with "flying boats," four-engine aircraft that landed and took off from water.[64]

The densest aerial services buzzed within continent-sized countries, especially Australia and the United States, where folks were keen to conquer distance. Qantas (Queensland and Northern Territories Aerial Services) had been founded in 1920 by four Australians, three of whom had begun to fly during the Great War. Their company grew from a rugged outback service to national dominance. In 1934, the company formed a partnership with Imperial Airways, as Qantas Empire Airways. With the introduction of flying boats in 1937, the "Kangaroo Route" from Brisbane to London became the world's longest aviation path. Pan American World Airways (PAA) likewise began locally, with air service between Key West and Havana. The company was reorganized and gained its formal name in 1928, and grew steadily. PAA specialized in long-distance flights over water, especially the Pacific. It named its flying boats "Clippers," for the historic clipper ships.[65]

Because airlines remained closely tied to national governments, passenger services were not a high priority. Pioneer air fleets generated most of their earliest revenue by carrying government-subsidized mail. The philatelic cover, an envelope or postcard meant to be a collectible item, had quickly become a staple of historic flights, including around-the-world flights. Air mail was almost the sole foundation of aviation in the Soviet Union. Even though the USSR was larger than either Australia or the United States, it did not develop civil aviation as quickly. Political instability prevented that until the 1930s, when Aeroflot more than caught up. (By 1939, Aeroflot was handling a greater volume of air freight within the Soviet Union than all airlines did in the United States.) The cost of petroleum kept airmail expensive, and passenger service even more so, usually 25 percent more than first-class passage by sea.[66]

To compete with luxury liners, airline operators modeled their long-distance passenger service on those ships. The bigger the aircraft, the better the illusion. Zeppelins and flying boats were noted for their attention to food, drink, and accommodation for sleeping. Airlines also copied from cruise ships the ceremony of crossing the International Date Line. A surviving PAA certificate for Celeste Briggs noted that, at 3:15 p.m. on August 22–21, 1937, she had entered the "Domain of Phoebus Apollo / Ruler of the Sun and Heavens." Aboard "Our Flaming Chariot," the *Hawaii Clipper*, "the Today of mortals at once becomes Tomorrow and all is confusion."[67]

Flying all the way around the world was rare, and that was, at least initially, its value. "This business of circling the globe, now such a prosaic occurrence," needed a jolt, some kind of boost, as the U.S. politician Emil Hurja concluded in 1935. He traced the history of notable circumnavigators—Magellan, Fogg, Bly, Train, and Mears—and prophesied many aerial successors. "The stratosphere is still to be explored," Hurja noted, "and they predict four days around the world."[68]

Aerial speed was a welcome novelty for some people, enabling them to feel part of the pantheon of famous, fast globe-circlers. The

American journalist H. R. Ekins, for example, set a record for travel on commercial services in 1936—eighteen days plus assorted hours and minutes. (For luck, he carried Nellie Bly's pocket flask.) He placed himself within the usual potted history of circumnavigations, though he accurately noted Magellan's death and speculated that Enrique de Malacca was the first person to go around the world, unusual touches. To distinguish himself from Ekins, Bolivar L. Falconer announced, also in 1936, that he was the first *paying* passenger to fly around the world. (Ekins was subsidized by newspapers and possibly by air services that welcomed the publicity.) Falconer wanted to girdle the globe in thirty days; journalists dutifully compared him to Phileas Fogg.[69]

Air travel was sufficiently unfamiliar, and expensive, that Falconer and Ekins had little company. On his flight from Penang to Hong Kong, Falconer was the only passenger. His total bill explained why: $3,354.66 for an all-in-one tour arranged by American Express. Airlines justified their high prices by offering first-class service almost exclusively. Everyone loved the Atlantic Zeppelin service for that reason. Falconer and Ekins compared it to the best luxury liners: it had an asbestos-lined smoking room; "the wines were perfect, the service suave." Smaller planes offered other comforts. Over the tropics, Ekins noted, KLM aircraft had cool air pumped into them. Cruise ships were not yet air-conditioned, but their sprawl was never to be replicated aloft. Instead of the beds and bathtubs of ocean liners, Zeppelins had narrow berths and shower-baths. Falconer concluded that "No one will be apt to travel by air unless his time is limited or he desires the novelty." For "a pleasure trip around the world by air," the whole point would be to maximize access to famous places. Ekins agreed.[70]

The speed and novelty of flight were exactly what tempted Mr. and Mrs. Charles H. Holmes to set out from Brisbane in 1936. They took the RMA *Adelaide* of Qantas, on the first stage of the Kangaroo Route "to London Town!" The *Adelaide,* a four-engine de Havilland 86, typically seated ten passengers, plus pilot, radioman, and steward. Its limited lifting power restricted each passenger to 35 pounds of luggage, even less for the stout. (Ample Mr. Holmes could carry only 33 pounds.) It was thrilling for anyone who had never been aloft, as the Holmeses discovered 10,000 feet over Timor, midway between land

and sea. Mr. Holmes records their joy, and the fact that they were the only two people on the flight:

> "Happy?" I asked the other occupant of the saloon.
> "Rather! It feels so . . . tremendous!"[71]

Aviation's to-the-minute schedule required airlines to play nanny with their charges. Whenever they checked into a hotel, the Holmeses, Falconer, and Ekins received an airline information card with the next day's schedule: predawn wakeup, breakfast, baggage collection, transport to the airport, meals, and final landing (place, time) for the day. Occasionally, the schedule meant sleeping aboard a plane, as the Holmses did between Basra and Baghdad. It was efficient. But it made the most expensive way of going around the world miserably similar to the conditions of a military expedition, especially compared to travel on ships and trains, with their long periods of idle transit. It was a trade-off. Every dawn departure, or fitful night aboard a plane, meant more time aiming the Kodak at the Taj Mahal.[72]

Sleep deprivation was now permanently inscribed on round-the-world travel. Ekins was eloquent on the combined effects of pinched sleep and time-zone fatigue. He wore three wristwatches set to New York, local, and Greenwich time. But he found the results too confusing really to comprehend. At Athens, he complained, "seven hours had dropped out of my life, and how I could have used them!" By the time he was in Iraq, he "experienced a sensation which was not weariness, but much akin to it. . . . The explanation, of course, was speed." He regretted seeing all the great sights "through a haze of speed." The globe was conquered, but at a price.[73]

Then there was the question everyone was eager to ignore: were aircraft safe? When his Passepartout had exclaimed that *"Man has guts,"* Jean Cocteau concluded: "He was right." Bolivar Falconer pooh-poohed reports that airships were dangerous, citing their frequent and uneventful trips. But on May 6 of the year his book appeared, the arriving *Hindenburg* exploded over Lakehurst, New Jersey. Thirty-five of the ninety-seven passengers died, plus one person on the ground. It was a final blow to a company that was ailing for many reasons. Passenger airship service ceased, as it had in other countries that had unsuccessfully tested airships.[74]

Cocteau criticized the removal from humanity that aircraft represented, even when they didn't crash. When his Los Angeles–bound plane took off in San Francisco, he marveled how "the machine parted with its shadow and, gaining height by gradual stages, drew level at an altitude from which no vestige of man's presence could be seen and the earth looked barren and inhuman as a relief map." Of the takeoff from Los Angeles, "once more that feeling of 'dehumanization' and utter loneliness, that curious impression that no worlds exist inhabited by man, came over me. All trace of humanity was vanishing from the earth. People were the first to go; then animals, then cars."[75]

And yet, as humanity became invisible, other life came into focus. "How can we not marvel at the way that life, so fragile yet so obdurate, has made itself at home in every corner of the globe?" That response might have resulted from any airline flight, but Cocteau thought it all the more striking because done during a tour of the globe: "Without this bird's-eye view of the epidermis of the earth, our adventure would have lacked completeness."[76]

Airplanes were not unique in fostering a sense of detachment. When he went around the world, Aldous Huxley deplored how "the automobile had placed the whole world at the mercy of the machine." "Muscular effort" was necessary for a traveler to see things. Automobile-induced sloth, while "a new and genuine pleasure," also induced a "mild tipsiness" or "passive idleness," not unlike the state induced by the cinema. And, between them, airplanes and automobiles created some of the world's ugliest built environment. In New York, Cocteau complained, the taxi-infested space between the airport and the city was "a warren of gasometers, a region of scrap-iron, factories and bridges."[77]

Some travelers liked airplanes precisely because they avoided the world's protests and problems, especially the crumbling of the imperial order that had been so convenient to confident globe-trotters. The Australian Holmeses benefited from the global reach of a British passport, as they announced in the title of their narrative: *A Passport Round the World*. As an American, Ekins found "the blanket British Empire visa" most useful. The British had been reluctant to allow him entry, as a reporter, to restive India and Palestine, until he assured them that his object was to pass through as quickly as possible, which aircraft certainly facilitated.[78]

As if to reinforce the claim that nations in the northern hemisphere still dominated the world, many of the maps that illustrated accounts of round-the-world flights used a polar projection, with the north pole at the center. Because no one flew around the two southernmost capes, that projection was sufficient, even if much of the southern hemisphere was offstage. A reader could see the total route at a glance, which was harder with a foldout world map done on a Mercator projection, and impossible if such a map were divided between a book's endpapers.[79]

The new popularity of polar-projection world maps indicated a rising dissatisfaction with the Mercator projection, but also hard-won knowledge of the north pole. The USSR mounted all the significant expeditions above the Arctic Circle: twice, in 1937, Soviet pilots established distance records for aviation by crossing the pole from Moscow to the United States; in the same year, the USSR was the first nation to land aircraft at the north pole. By relying on polar projections, around-the-world aviators were not only registering a hemispheric prejudice but also tugging back some of the glamour of extreme danger that circumnavigators had ceded to polar explorers a century earlier.[80]

But the polar projections also revealed that some aerial circumnavigations had fallen short of a "Great Circle," the equivalent of the equator, just under 25,000 miles. It wasn't universally true. The Army Air Service flight had covered 26,345 miles. In separate journeys that went around the world between 1928 and 1930, the Australian aviator Charles Kingsford Smith had crossed the equator at least three times (depending on which of his flights are counted). Ekins's commercial voyage also crossed the equator twice and, at 25,794 miles, had "exceeded in distance the circumference of the earth at its fat middle." But many of the famous circuits fell short, including the Post-Gatty and solo Post flights, about 15,500 miles apiece. They had flattened Gaia's chest, not nipped in her waistline. A single flight that stuck to an equatorial route had yet to be done.[81]

That was what Amelia Earhart proposed in 1937: "a circumnavigation of the globe as near its waistline as could be." Earhart was a veteran of several long-distance flights. Like her friends Wiley Post and
the de Sibours, she was a celebrity pilot, member of a new Club of
Eccentrics who were resigned to the stunt flights and media appearances necessary to fund sport aviation. To make an equatorial circuit,
she wrangled use of a new airplane, a twin-engine Lockheed Electra
that could, she calculated, give her a base speed of 100 mph. Jacques
de Sibour, still working at Standard Oil, agreed to arrange the necessary
fuel caches. But it showed some public weariness with stunt circumavi-
gations that Earhart and her husband, the publisher George P. Putnam,
had some trouble selling the idea. They announced that Earhart's plane
would be a "flying laboratory" for something or other, but dropped the
scheme when response was cool.[82]

Earhart had intended a westward circumnavigation, but her takeoff
from her first stop, in Hawai'i, went wrong and the plane was damaged.
Earhart hauled it back to California for repairs and announced a new
departure. Money was even harder to raise for this second attempt.
Putnam managed to sell Earhart's dispatches to Harcourt, Brace &
Company; he and Earhart mortgaged their home, and wealthy friends
made contributions. A stack of philatelic covers, to be sold later, would
also help. Earhart pared her crew from four down to one, navigator
George Noonan, and set a new course: eastward. With a California
point of origin, Earhart was going to do the hardest segment (the Pacific) not first but last, when she, Noonan, and the aircraft would be
most fatigued. She departed on May 21, 1937.[83]

Unlike Wiley Post, Earhart disliked flying with instrumentation. She
missed the days when she could smell what was on the ground. She resented the "over a hundred dials and gadgets" in her Electra's cockpit,
fifty on the dashboard alone. The flight seemed positively "chatty" with
radio signals and conversation, and "pretty sissy" with all the gadgetry.
She skimped on her radio training. As air traffic thickened and crossed
greater distances, pilots had to coordinate flight plans and pinpoint
their position through radio signals with other aircraft and posts on the
ground. Yet neither Earhart nor Noonan knew Morse code, the emerging standard for aviation. She barely learned to use her Electra's radio

and, after the Hawai'i fiasco, removed a 250-foot trailing wire antenna, which she deemed a nuisance.[84]

Many of Earhart's dispatches lament that the places on her route went by so fast. She went from one "hasty visit" to the next, each place marked with her "calling cards," drums of gas labeled MISS AMELIA EARHART. By the time she reached Lae, New Guinea, with 22,000 miles covered in nearly forty days of flying, she had lost weight and looked exhausted. She and Noonan had 7,000 miles to go. Unfortunately, the very next segment of the flight would be one of the most difficult, because it meant finding and landing on tiny Howland Island.[85]

For over six hours, Earhart signaled the radio operators on Howland, and the USS *Itasca*, offshore, that she needed a bearing to land. For some reason, she could not receive the directions they sent. Had her radio, repaired in Australia, malfunctioned again? Or was her training with it too rudimentary? Did crossing the IDL (one of its confusing zigzags was close to Howland) befuddle her and Noonan, who had been unable to set the aircraft's chronometers? Were their out-of-date charts the problem? As she ran out of gas, Earhart's decision not to fit the plane with pontoons for a water landing seemed ill-judged, as did jettisoning spare parts, flares, and smoke bombs in Lae. After more than twenty hours since departure from Lae, it was obvious that the plane had come down somewhere. On July 2, Fleet Air Base in Pearl Harbor, Hawai'i, officially announced that Earhart was missing.[86]

How and where Earhart and Noonan went down is still unknown. The U.S. Navy called off its search on July 18. Putnam would have Earhart declared dead in 1939, when he remarried. He had gone ahead with her book, published in 1937 under the title *Last Flight* rather than *World Flight*. It includes a photograph of Earhart in front of a world map marked with her proposed circumnavigator's track. Head tilted, she looks over her right shoulder and at the flight segment over the Pacific she never completed. Of all aerial circumnavigators, Amelia Earhart most resembles Ferdinand Magellan, because the Pacific, and the size of the world, defeated her. Her failure with her aircraft's radio left her fatally unmoored to the planet—too much like Puck, not enough like Satan, and all too sadly, just like Phaeton.[87]

• • •

By 1939, commercial airlines had created nineteen world routes. None of them went around the world. Instead, each strategically connected a home country to selected parts of the world. To some extent, that was because maritime shipping services could still outcompete long-distance air transport. Another factor was at work in the United States, where anti-monopoly laws prohibited the largest airline, PAA, from flying over North America. When PAA selected a new emblem in 1941, it chose a globe bearing a set of bird wings and the motto: "Wings over the World." It was all they could claim.[88]

And yet, as with the telegraph, people imagined aerial services going around the world even before they actually did, renewed proof that humanity was united in a project to span the planet. "Puck is still busy girdling the earth with air routes for passengers and mails in spite of the preoccupation of airmen with the more horrid purposes to which their skills and daring can be directed," *The Times* of London declared in August 1939, not two weeks before Hitler would order the invasion of Poland.[89]

At this, the eleventh hour, the Japanese would do their first aerial circumnavigation, and the last non-commercial world flight to carry the internationalist burden of world peace. Two newspapers, the *Osaka Mainichi* and the *Tokyo Nichi Nichi*, pooled funds to send the *Mainichi*'s twin-engine Mitsubishi airplane, the *Nippon* ("Japan"), on a goodwill flight. Sumitoshi Nakao was the pilot, with a co-pilot, two mechanics, two radio operators, and the chief of the two newspapers' aviation departments. *Nippon* rose into the air on August 26, 1939, all set to make the longest journey yet attempted by a single airplane and crew, on a month-long series of visits to thirty nations.[90]

But because the journey began just before armed conflict in Europe did, the *Nippon*'s progress had to be reported along with extremely bad news. At first, the *Mainichi*'s front page featured the flight alongside the impending conflict. By September 2, however, all the front-page news was about Hitler's invasion of Poland; the "globe-girdling goodwill plane" was, on page two, reported as being in Seattle. An extra edition of the newspaper for September 4 announced that Britain had declared war on Germany (page one) and that the *Nippon* had reached Oakland, California (page two). Two days later, the plane received a robust welcome in Los Angeles. Matsuya department store in the Ginza district of Tokyo opened a panorama display of *Nippon* circling the world and offered kimono fabric patterned with globes and airplanes. But the *Nippon's* flight path had to be re-routed through Africa to avoid "the European war." When the plane returned to Japan on October 21, the chairman of the Osaka Mainichi Publishing Company extolled its demonstration of "international comradeship" but also noted that it had come close to "the thrill and danger of war," statements that didn't quite match.[91]

The technology to fly around the world existed, but the political will to do it in peace did not. Instead of leading to a greater international harmony that would prevent another Great War, aerial circumnavigations had established a base of expertise for another global conflict. The airborne warfare of World War I seemed as nothing compared to the grim achievements of World War II. At what point did a kimono printed with globes and goodwill planes seem like a bad joke? After the bombing of Pearl Harbor? The Blitz on London? The firebombings of Dresden and Tokyo? Or the dropping of atomic weapons on Hiroshima and Nagasaki?[92]

Even before the United States entered World War II, it did a military aerial circumnavigation. Under the terms of Lend-Lease, in May 1941, the U.S. Army Air Corps established a Ferrying Command to fly aircraft from American factories to the British. That fall, two B-24s flew the North Atlantic route to Scotland, then to Moscow to help negotiate U.S. support of Joseph Stalin's war against Hitler. One B-24 returned over the Atlantic. The other was piloted by Alva L. Harvey, one of the two aviators downed in Alaska during the 1924 round-the-world attempt. Harvey flew eastward from Moscow over the South Pacific in order to help piece together landing sites for aircraft that were not, unlike Flying Boats, amphibious. He returned to Washington, D.C., on October 30, 1941, with information on strategic Pacific landing sites, just in case. Harvey had at last flown around the world, not as part of a peacetime demonstration, but to prepare for war.[93]

Later that year, a PAA pilot was the first to circle the world in commercial aircraft, a notable event in the history of unintended circumnavigations. Captain Robert Ford was in command of a scheduled round trip to New Zealand that left San Francisco on December 1, 1941. After the Japanese attack on Pearl Harbor on December 7, Ford was warned not to return. PAA needed the aircraft, however, so on December 14 it ordered Ford to strip off the plane's identification and get it back to the United States—the long way—avoiding the war as best he could. "GOOD LUCK," concluded the orders. Ford and a skeleton crew flew westward, over the Indian Ocean and Near East, then across the Atlantic Ocean and North America, for a total of 31,500 miles. It was the first around-the-world flight done mostly near the equator. The event got lost amidst other news and the full story would not be told until the 1990s. At that point, no one among the surviving crew could remember whether the full circumnavigation had been made on the same aircraft; the original may have been swapped out in Honolulu, en route to New Zealand.[94]

Once war spread from Europe into the Atlantic and Pacific, around-the-world travel for pleasure became unthinkable. World War II made the Pacific as dangerous to cross as it had been in the early age of sail. And yet circumnavigations had to be done for military reasons. While the Axis divided the Pacific and Atlantic between Japanese and German areas of concentration, the Allies fought in both. The British and Americans, in particular, operated within an active military zone that stretched around the planet.

"American Wings Soar Around the World," *National Geographic* proclaimed in July 1943. The article outlined the ground reconnaissance that formed a wartime chain linked by aerial traffic. The "Epic Story of the Air Transport Command of the U.S. Army," the essay claimed, "is a Saga of Yankee Daring and Doing." "Planes must have bases," they pointed out, "and ATC had to construct those bases" from Arctic to desert. Somewhere in North Africa, a "young officer in charge" negotiated with "the local sheik to hire all the camels in the territory." Each of the estimated 1,000 camels carried 35 gallons of fuel to awaiting airplanes. The head of Air Transport Command (ATC) stated that "the tremendous aerial expansion under war's relentless pressure" had advanced aviation by at least a generation, not least "in the far-flung extension of routes and bases" around the world. In 1945, *Popular Science* explained how a U.S. link between Calcutta and New Guinea had completed the chain, and permitted a group of senators to be flown around the world in 1943. The Royal Air Force, meanwhile, made round-the-world flights, eastward or westward, for a total of 24,706 miles in nine days, with thirteen stops on the way.[95]

Both articles were propaganda. They withheld strategic details about the RAF and ATC, including accurate maps of their round-the-world routes, lest that encourage attacks. The short essays, with their can-do spirit, were designed to lift civilian morale. Easily translated into German or Japanese, they warned the Axis that the Allies had the planet surrounded.

That was how aviation's peacekeeping promise was altered: to do the job, the technology had to be kept in the right hands. When Wendell L. Willkie, a political adviser to President Franklin D. Roosevelt, went around the world in 1942, he did so in a refitted bomber, *Gulliver*, and to meet leaders of countries that had agreed to join the United Nations, proposed that year. Willkie's narrative, *One World* (1943), regarded a circumnavigation as useful for political reasons—there is little about the planet or globe-girdling technology in it. His 31,000-mile trip had proved that "the world has become small and completely interdependent . . . there are no distant points in the world any longer," hence the need to unify on peaceful terms.[96]

This too was the message of an exhibition that opened at the Museum of Modern Art in New York in late 1943. "Airways to Peace: An Exhibition of Geography for the Future" predicted how the world would look at war's end. The show was organized with Willkie's help, and again emphasized the international need for global circuits. The curators used globes and polar-projection world maps to argue that, because most of the populous countries clustered around the north pole, within easy flying distance of each other, the Germans should have known better than to attempt world domination—"the world is small and the world is one."[97]

The statement was a version of the old confident view, faith that the human ability to comprehend and command the globe would improve over time. But one diorama admitted that humans had physical limitations. Even a soaring aviator was restricted to the bottom layer of the Earth's atmosphere. Man was "a deep-air mammal," comparable to a deep-sea fish, able only to survive at the lowest level of terrestrial atmosphere. So far, that kept him flying quite close to the planet.[98]

For the Allied powers, postwar recovery included the ability to go around the world in peace, either for leisure or as part of some goodwill effort. The two motives were never quite distinct.

Idealists like Willkie called for a new internationalism, once-and-for-all agreements that would harmonize relations among the world's peoples, as with the United Nations. One of the many critics of this kind of talk called it "globaloney," and indeed it failed to create any new global consortium. At an international conference held in Chicago in 1944, Hugo Grotius's doctrine of the freedom of the seas was again considered as a model for the air, and again rejected. Airlines, air forces, and individual aviators would still have to seek permission to enter foreign airspace and to land.[99]

Internationalism was more obvious in the consumer economies that began to dominate everyday life in the West, especially in industrialized nations outside the Soviet sphere of interest. U.S. corporations worked with the government to market their products abroad, and their goods and services were parts of what has been termed "soft power," the demonstration of the advantages of democracy and capitalism. As with the conspicuous consumption of the Gilded Age, around-the-world travel was the perfect demonstration of the planetary plenty available to ordinary citizens, at least in theory, because it required so many resources and it exhibited consumption on a world-sized stage. It is striking, therefore, that postwar descriptions and testimonies of around-the-world travel stood aside from contemporary debates over whether planetary resources would be adequate for a world population that was growing both in size and in material demands.

Civilian air services were eager to show they could take on the world. In 1945, PAA published a booklet called *Ten Thousand Times Around the World*, a conservative estimate of PAA's achievement. "Ten thousand times around the world *at the equator* is 249 million miles," and yet "Pan American Clippers passed that figure in 1943." A parallel press release, "Around the World Today and Yesterday," placed the airline's achievements in a tradition that had begun with "the famed Portuguese explorer, Ferdinand Magellan." "Circumnavigation of the globe has caught the imagination and stirred the efforts of man ever since he first became conscious of the true shape of the earth," PAA declared, continuing the tradition of regarding people in the past as geographically ignorant yokels.[100]

The aerial circle was quickening. In 1936, PAA founder Juan Trippe had used his and other air services to go around the world in just under two months; in 1947, he did it on PAA planes (not counting the United States) which took fourteen days, one for each of the fourteen airports through which PAA's "round-the-world" service passed. That year, PAA introduced a new slogan, announcing itself as "The Round-the-World-Airline." It wasn't true—the airline was still forbidden to provide domestic airline services. PAA had to transfer passengers to affiliated airlines to cross the United States. More truthfully, BOAC (Imperial Airways' successor) announced its "Speedbird Routes across the World." A Soviet postage stamp of 1949 showed Aeroflot's contribution with a globe dominated by "CCP" and a circle of overlapping letters ringing the world.[101]

Journalist Allene Talmey and photographer Irving Penn made the PAA circuit in 1947 to report on the state of the world for *Vogue* magazine. Restricted to 66 pounds of luggage, Talmey gamely described how she projected American chic around the world for thirty days with only two pairs of underpants and a can of flea powder. She surely laundered her lingerie on the way, though in what? Clean water was often unavailable. Depending on where they were, Talmey and Penn brushed their teeth with boiling water, hot tea, or Coca-Cola. Rather than complain, they noted that the world had bigger concerns, including "food shortages, housing shortages, the Russians, and the lowered hem-line."[102]

The ravening consumer economy—PAA, Coca-Cola, hemlines, and so on—would represent another historic transformation of expeditionary circumnavigations into personal circuits of the planet. But that can wait. "The Russians" were the story to watch. They were about to prove that humans, like Puck, could put a girdle round about the Earth in a matter of minutes. The cost of that feat would be the revival of death as the hallmark of a circumnavigation, as well as, briefly, the use of captive circumnavigators.

Chapter 10

The Outer Limits

*H*er *heart raced with terror,* not Puckish glee. On November 3, 1957, at 5:30 a.m. Moscow time, Laika, hero dog of the USSR, became the first creature sent into space. As she rocketed above the Earth, her pulse rate tripled, briefly slowed, then reaccelerated as she swung into orbit. Harnessed into place, Laika was also tethered to the Earth via radio and television signals and by an electronic probe connected to her carotid artery, all meant to gather evidence that animals could survive space flight. The gadgetry dutifully recorded the subject's misery: Laika barked, she strained at her bonds, and she sent an unanswerable SOS of panicked heartbeats back to ground personnel in Kazakhstan. Her untested capsule, Sputnik 2, overheated as it made its first pass around the world. By the second orbit, the heat must have been unbearable; by the third, medically damaging. Sometime into her fourth circumnavigation, Laika died.[1]

Newspaper headlines in the world below her announced that Laika's ordeal represented the start of an unprecedented conquest of the cosmos, a new age, the space age. Yet the very first circumnavigators in the age of sail might have saluted the small doomed dog as their comrade in a grim old business. Like her, many of those maritime pioneers were expected to die; like her, their suffering exceeded expectations.

Despite its jet engine, rocket-powered, right-stuff popular image, the space age did not launch a clear vindication of the nineteenth century's proprietary confidence. Instead, even more than aviation had done, it circled back to the original doubts about the human ability to withstand a planetary voyage. That is sobering. Many people prefer to think of space as a final "frontier." Their opinion is interesting as an artifact of geodrama's second and confident act, but they are wrong. Unlike the terrestrial frontiers of imperial expansion, space has physical conditions that kill immediately.

Voyages in orbit do however resemble maritime circumnavigations, the old, bad, fatal kind that everyone had been trying to forget. Concern that spacecraft might not preserve humans in space, away from Earth, looks like nothing so much as the old fear that the wooden worlds could not sustain human life at sea, away from land. Because humans have bodies ill-suited to the outer limits, they have depended on animals and robots for their orbital explorations, much as maritime circumnavigators relied on human captives. The non-human travelers send messages back to Earth that are unlike human testimony.

For that reason, orbital travel has been novel, not so much in itself, but because it has created qualitatively different circumnavigators' narratives. All of the generic variations among previous round-the-world accounts—novel or poem, picaresque or scientific narrative, illustrated or not—were as nothing compared to the big difference between human accounts and those generated by animal or robotic surrogates. The surrogates, and the data they generate, represent a human interaction with the planet that is mediated rather than immediate, and this is an enormous rupture within geodrama, the story of conscious and deliberate human interaction with Earth.

Fantasies of space travel are too old to have a beginning. Theories of how it might be done are more recent, dating from the late seventeenth century, and the liquid-fuel jet propulsion that would make it possible originated in the early twentieth century. All along, humans have never been certain whether the fantasies, theories, and technology would deliver them entirely from Earth. In 1728, Isaac Newton conjectured that a projectile fired at immense speed would "describe the same curve over and over" around the world. Being trapped in orbit was not quite an escape from the planet, however. The preface to Newton's treatise describes what lies beyond as "an immense ocean of celestial space," unearthly, yet somehow like Earth's seas. Fictional accounts of human travel into space, including Jules Verne's *De la Terre à la Lune* (*From the Earth to the Moon*, 1865), imagined extraplanetary travel even before suitable technology existed. For the premier Russian theorist of rocketry and space travel, Konstantin Tsiolkovskii (1857–1935), that act of imagination was essential. The Earth may have been the "cradle of reason" but "we cannot live forever in a cradle." The point was to blast away from Gaia, however unreasonable that might seem.[2]

Could the body survive it? In 1929, J. D. Bernal, the British scientist and Marxist writer, proposed space colonies as zones of liberation from everything that held people down and to the Earth. He thought humanity would in future follow two paths. Conventional humans would stay to tend the globe while an engineered version of "man" could go into space—"if he could get rid of the major part of his body." In the 1930s, the British philosopher and science fiction author Olaf Stapledon pointed out that because interstellar exploration would take so long, the creatures who finished the task might barely resemble those who had initiated it. But that bare survival might matter, should the Sun begin to fade, and life on Earth begin to die.[3]

Much of this was speculative, though aviation had already warned that humans, as "deep air" creatures, could not survive the outer limits. High altitudes did not supply enough air pressure for them to inflate their lungs. Fly too high, and they began to black out, which made it impossible for them to enter the thin, fast-moving layers of air later called "jet streams." During the 1930s, inventors and pilots developed helmets and body suits that could be pressurized. A Leningrad engineer, Evgenii Chertovskii, designed the first such devices in 1931. But the first practical pressure suit was invented by Wiley Post, the one-eyed, multitalented pioneer of flight. (His 1934 prototype looks like something Jules Verne might have dreamed up with Ned Kelly, the Australian outlaw who wore homemade armor.) Post's invention allowed him to survive flights as high as 50,000 feet—he was the first person to ride the jet stream. Pressure suits remained undeveloped during World War II, when most flights were done at lower altitudes. Cold War demands for high-altitude reconnaissance—spying—were the spur for their adoption and refinement. Aviation required only partial-pressure suits and aircraft, but that technology was the basis for the full-pressure suits and pressurized cabins necessary for spacecraft.[4]

Nothing could go into space, however, except at a tremendous rate of speed. Orbital velocity, the minimum needed to put an object into Earth's orbit, was calculated as 17,000 mph, far beyond the capacity of the piston-driven internal combustion engine, which is restricted by propeller efficiency. (As a propeller's blade tips approach the speed of sound, efficiency declines.) Another technology, generating a jet of gas to thrust something upward, had been used as early as the thirteenth century, as with Chinese fireworks. By the eighteenth century, Europeans used Asian-style rockets as military signals or, in barrages, as weapons. They were highly inaccurate, however, and their power limited. Jet propulsion within a guided system first occurred with aircraft—working jet engines were adapted to airplanes by 1939, just in time for the war.

Jet propulsion was more difficult to adapt to large rockets. Several theorists, including Tsiolkovskii in Russia, Robert Goddard in the United States, and Herman Oberth in Germany, banged away at the problem early in the twentieth century. Each man predicted that liquid-fuel rockets would achieve enormous speed and range, and Goddard launched a small prototype. German scientists developed the first large rocket-powered missile, the V-2 (called the A-4 in Germany), used in the Blitz on London. When they struck, the rockets were devastating, but their accuracy was terrible, mercifully for the English.

The technology was deadly in another way. On April 12, 1945, the men of the 104th U.S. Infantry Division liberated a pair of concentration camps near Nordhausen in Germany but found few survivors. Forced to produce A-4 rockets, the slave workers had been worked to death—5,000 human bodies surrounded the abandoned missile technology. The Americans extracted the hardware and documentation before the Soviets arrived. At a test site in New Mexico, twenty-five of the missiles would be launched in top secret Project Hermes in 1946. The Soviets harvested other information for themselves, and both they and the United States recruited German rocket scientists, offering amnesty in exchange for nationally subsidized work on the new technology. Allies during the war, the United States and Union of Soviet Socialist Republics were now rivals.[5]

Just as imperial rivalry had catalyzed the first circumnavigations, the Cold War shaped the space race, with the two contentious eras book-ending the Hundred Years' Peace. In some ways, technology made the second era of conflict more dangerous. The possibility that rockets could carry weapons of mass destruction over oceans and continents was a real threat. But considered in a longer historical perspective, the new imperial contest was safer. The expectation that nations might work together, not least for science, was now ingrained. More than two centuries separated Magellan's rapaciously nationalistic mission from Cook's promise that he served science as well as George III. During the space race, in contrast, science always promised to bridge national divisions, and actually did so within decades. And, for the first time, scientific circumnavigations questioned the imperialism that had subordinated large sections of the world's populations. Non-Western powers could now participate in two linked projects: around-the-world travel (in orbit) and scientific comprehension of the Earth and its atmosphere.

The British science fiction writer Arthur C. Clarke predicted, in 1945, that a rocket-launched "artificial satellite" would have extraor-dinary peacetime applications. (In his "Venus Equilateral" stories, published from 1942 to 1945, George O. Smith had similarly explored the idea of satellites orbiting the Sun in Venus's wake as relay com-munication stations.) In two pieces published in *Wireless World*, Clarke posited that satellites orbiting Earth could be "extra-terrestrial relays" that would be "invaluable, not to say indispensable, in a world society." Three orbiting satellites, all solar-powered using mirrors, could relay signals sufficient for world service. A satellite could even grow into a "space-station," with successive rockets delivering materials for labora-tories and living quarters.[6]

It was at an informal meeting of scientists in Maryland in the spring of 1950 that James A. Van Allen proposed an international effort to study the outer limits. The activities would take place over a period modeled on the International Polar Year (actually, two of them, 1882–83 and 1932–33) that had examined the Arctic and Antarctic regions. Yet again, round-the-world travel would recover some of the deadly thrills it had ceded to polar exploration.

Van Allen suggested multiple and overlapping efforts: observers and sensors on the ground, balloons in the atmosphere, and rockets farther above. Scientists in western Europe seconded the proposal. At a subsequent meeting in Rome, held in October 1954, Soviet delegates assented to a U.S. plan to orbit artificial satellites as part of the investigations. The International Geophysical Year (IGY) actually lasted eighteen Earth months (July 1957 to December 1958) in order to include a period of unusual solar activity. The International Council of Scientific Unions (ICSU) would oversee the funding and activities of the IGY, which eventually involved sixty-seven nations. The IGY's octagonal emblem features a globe cross-hatched by lines of latitude and longitude. The world's contrasting light and dark sides indicate its rotation relative to the Sun. A circular object inscribes a circumnavigator's track in orbit, something that had yet to happen.[7]

No one in the West knew it, but the Soviets, having agreed to the U.S. proposal to orbit a satellite, were determined to beat them to it. The USSR's rocket program had been initiated in order to launch warheads. The identity of its head, aerospace engineer Sergei Korolev, was a state secret—he was publicly known only as "Chief Designer." Korolev, who had survived a Stalinist purge and incarceration in a Siberian labor camp, was ambitious, yet shrewd. He persuaded Premier Nikita Khrushchev that Soviet participation in the IGY by launching a satellite would not compromise any military goal. A "Simple Satellite" would show that an object could be put into orbit and maintain its pressurization, while gathering data about the upper atmosphere, including its density and ability to carry radio waves.[8]

On October 4, 1957, the Simple Satellite, or Sputnik, became the first thing of human origin to orbit the Earth. The 185-pound aluminum sphere, 23 inches in diameter, had four aerial antennae between 7.8 and 9.5 feet long, back-leaning whiskery appendages that accentuated a sense of swooshing dynamism. Every ninety-six minutes, it ghosted around the world in an elliptical circuit inclined 65 degrees to the equator, its closest point to Earth 145 miles and its farthest 560.[9]

Sputnik means "Companion," a reminder of the original meaning of *satellite* as "escort," and it was an excellent name because Sputnik may have been a thing, but it was not dumb. It spoke the relatively new language of *telemetry*, in which data are generated in one place and transferred to another. The telegraph had made telemetry possible in the late nineteenth century, a power expanded by discovery of how to manipulate radio waves. Sputnik's duties included keeping up a companionable chatter to the peoples of Earth by emitting a steady radio beep. Korolev requested a signal audible even to "the most dilapidated receiver" and the Soviets organized twenty-six amateur radio clubs, in a line from the Baltic to Siberia, to report on the satellite's progress. The English-speaking journal *Radio* likewise announced frequencies for tracking Sputnik.[10]

At a time when people were accustomed to fiddling with radio dials and when "ham" radio operators were creating a geeky democracy of communication, the first satellite's beep was a public relations coup. Even when pollsters asked leading questions about how the Soviet object must surely frighten them, many Americans responded that they thought the world's first satellite, which would transmit for twenty-two days (and orbit for another seventy days), was a marvelous testament to human ingenuity. Opinion changed once some U.S. politicians claimed that Sputnik was nothing but a test of Communist rocket power. Still, on the whole, and certainly outside the United States, the context of the IGY ensured that science trumped politics. Earthlings welcomed their new companion.[11]

Objects had of course circled the Earth before, starting with Magellan's *Victoria*. Some of them had been described as circumnavigating independently, as human travelers did. In 1939, for example, the Kellogg's breakfast cereal company had sent two suitcases around the world—one east, one west—to race each other as a promotional gimmick. Information also circled the Earth. Telegraph and airmail, augmented by radio and telephone networks, had put so many girdles around Gaia that she was by 1957 trussed as tight as a spider's prey. But mute objects carry stories that humans supply. The same is true of information systems—an around-the-world message is an echo. Sputnik's beep announced a new possibility: information generated as the object circled the world, the full story of a circumnavigation *in real time*.[12]

It was a triumph of telemetry, and a hint about the future of *cybernetics*. In 1948, the American theorist Norbert Wiener had redefined that term (Greek for "human self-governance") as the study of the self-regulation of mechanisms. He compared such mechanical systems to living creatures, both animals and humans. Not everyone liked the comparison. Were organic creatures essentially robots? It did not help that cybernetics had first proved useful to military technology—during World War II, Wiener worked on antiaircraft guns. Theorists in the Soviet Union initially rejected cybernetics as a capitalist theory, even as Soviet engineers were encouraged to create feedback systems, including computer programs. After Stalin's death, cybernetics became vastly popular in the Soviet Union. It would be indispensable to a great deal of modern technology, especially computing, but also the self-guiding systems necessary for rocketry and satellites.[13]

Sputnik was such a success that Khrushchev demanded a follow-up. Korolev suggested the launch of a living test subject. The publicity stunt was ordered to commemorate the fortieth anniversary of the Russian Revolution in early November 1957 (also the hundreth anniversary of Tsiolkovskii's birth). The tight schedule did not permit the development of a capsule that could be recovered. Re-entry and landing were impossible; the test subject would die in space. Obviously, killing a human for science would kill any positive publicity. Something else would have to be sacrificed.[14]

A patient dog would embody the USSR. The Soviet scientists and engineers decided that primates, another obvious choice, were too excitable and unpredictable, the shrieking, sulking opera divas of the animal world. Dogs, in contrast, were soldiers who followed orders. Small female dogs were selected for their compactness, which included their ability to urinate without standing up; those with light coats were preferred because they showed up best on camera. To adapt them to the space capsule, the dogs were placed in ever smaller cages; those unable to cope were released from the program. Some of the subjects were, rather forebodingly, recruited from the Pavlov Institute in Leningrad (now St. Petersburg). Others were pets volunteered by people who worked on the project. Captured strays were prized for their resilience—they would gladly eat the strange gelatinized food developed for canine space travel.[15]

Because the best candidate was a favorite of the scientific staff, who were disinclined to condemn her to death, they pushed forward, instead, a stray from Moscow. They had named her *Kudryavka,* "Little Curly" (for her spiral tail), and nicknamed her "Little Bug" and "Little Lemon," Russian endearments that betrayed some affection for her, too. (Shortly before the launch, one man brought her home to play with his children, a final treat.) Kudryavka became *Laika* in all official press releases because her other names were too difficult for Westerners to spell or pronounce. (*Laika* describes a type of dog, a "Barker" with Husky ancestry, in Kudryavka's case, Husky mixed with terrier.) On October 31, she was placed in the capsule of Sputnik 2. For over two days, she waited in her tiny compartment at the tip of the rocket, far above the launch pad. She received heat, food, and water, and was attached to a sanitary waste system, but was unable to turn around or roll over, let alone scratch an ear with a hind paw, or run in easy loping circles, or do any of the thousand doglike things she had been conditioned to resist. At last, she was fitted with her flight harness and medical sensors. The capsule was sealed and the rocket uttered its monstrous roar.[16]

The payload of Sputnik 2 was 508.3 kilograms or 1,118.3 pounds, six times the weight of Sputnik 1. Yet that made it only about twice the weight of Baron Koenig-Warthausen's plane and less than a quarter of the full payload of Wiley Post's *Winnie Mae.* Laika had access to water and food, and was supposed to consume a ration of poisoned food before her oxygen gave out. That part of the plan offered mercy, but the fatal ration was time-released and unavailable to the dog as the capsule's heat insulation failed at a rate guaranteed to prolong her suffering. Those details remained secret—they revealed too much about the flaws in Sputnik 2.[17]

Before she died, Laika communicated through telemetry's new dialect, biotelemetry, in which an organism's body generates the data. The sensors had immediately shown that liftoff made Laika panic. Although her training may have helped her calm down once in orbit, the rising heat within the capsule caused more distress. Like Sputnik, she sent a story back in real time, in a language humans could not understand perfectly, but whose likeliest translation was "Get me out of here."

The Soviets simply announced that they had launched Laika. "Muttnik," American journalists dubbed the stray mongrel who was Earth's first creature to go into orbit. When news that she would not be recovered came out inadvertently, some people were outraged. Britons in the National Canine Defence League asked dog owners everywhere to observe a minute of silence on each day that Laika circled the globe.[18]

For those critics, it seemed a step backward to embed death into a planetary expedition. When the Soviets introduced Laika to the world—she was photographed for newspapers and her bark recorded for radio broadcasts—she resembled the animal mascots that had flown around the world not that long before. Yet Tanim and Tail Wind, who like Laika were small, light-colored, and photogenic, had been appealing because they risked only as much as the people they accompanied. Their presence on an airplane signaled human confidence that they could maintain circumnavigation's zero mortality rate, even for nonhuman passengers. None of that was true of Sputnik 2.

Members of the Soviet bloc insisted that Laika's sacrifice was science's gain. Postage stamps and postcards commemorated her. And yet the most common image on those commemorations had been carefully selected. A series of publicity photos of her show her in various prelaunch poses. In one shot, she poses in a mock-up of her capsule. Her concentration indicates that she was watching someone off-camera, probably a trainer, for her next cue. To the human eye, her close attention gives her a look of anxiety. In another photograph, her eyes are not so anxiously attentive and her muzzle is in sharper focus, which makes her jaw look set, more determined—more suited to the expeditionary harness she wears. That photograph was turned into a colored drawing (see illustration 19) and used on biscuit tins, cigarette packages, and the many postal issues, with the same posture and sometimes details of the harness. The image of the brave soldier dog does not match all the original photographs, let alone the biotelemetry.[19]

After the fall of the Soviet Union, the people involved in Sputnik 2 expressed regret about Laika. In 1998, her main trainer, Oleg Gazenko, said, "the more time passes, the more I'm sorry about it. We did not learn enough from the mission to justify the death of the dog." At a meeting of the World Space Congress in 2002, scientist Dimitrii Malashenkov gave details about Sputnik 2's faulty heat shielding. These were not just sentimental confessions. If the goal had been to learn how to send humans into space, recovery would also be necessary. Laika's ordeal solved only half of the problem—her launch served USSR propaganda purposes more than it did scientific ones.[20]

In her complex role in geodrama, as victim yet hero, Laika is the one captive circumnavigator who enjoys international cult status. She has done so continuously, within and beyond the Soviet bloc. During Christmas 1957, for example, a Sputnik-shaped decoration containing a toy dog was hung above the Ginza shopping district in Tokyo. Probably because 1958 was the Year of the Dog in Asia, at least two Japanese planes were renamed *Laika*. Statues went up to Laika in France and, eventually, Russia. Several novels and films have featured her as a character, sometimes a central one. She is history's most famous dog. Compare that to all the humans who were forced to help circumnavigators. Even the best-known fall short of pop-culture celebrity. Laika is more famous than Enrique de Malacca, for example.[21]

Laika's cult status is the result of the history of round-the-world travel, which has required, over time, that it display confidence by minimizing mortality. That Enrique de Malacca and the other captives are forgotten is the product of the same history, which has, since the early nineteenth century, rebranded circumnavigation as a non-fatal adventure, with heroically active participants who can testify about their voyages. The use of passive animal subjects, an everyday occurrence in laboratories, diminishes the glamour.

Although animals would continue to be used as test subjects in space, solicitous recovery plans for them have become standard procedure. The plans may fail, or the animals may be recovered only to be euthanized, but the solicitude seems to matter. The many other dogs that the Soviets would rocket into space were intended to return. (In one case, an emergency team raced to rescue a dog from a downed capsule timed to explode.) The U.S. space program's test primates also were supposed to come home. When the chimpanzee Enos survived two orbits in NASA's Mercury spacecraft on November 29, 1961, he was retired from space travel. That is now the protocol for all the named animals that have been sent into space, by any nation.[22]

In contrast, human astronauts and cosmonauts have volunteered for repeat launches. The disparity is an admission that one set of creatures can agree to be sent into space, while the others cannot. For the non-humans, one forced mission apiece is plenty. For that reason, the pool of human "testers" in the Soviet Union, who were pushed even further than cosmonaut trainees, would be kept secret; they resembled animal test subjects too much.[23]

The furor over Laika represented the first signs of unease over the mechanics of space travel. The technology that carried objects and information over a staggering number of miles and through several mediating links was impressive, but intimidating—and it was danger-ous. If a human were to explain how it felt to be inside it all, the story would fit into the history of round-the-world travel; if a machine told the story, that might represent a brave new era of technology. But any-thing in between—an animal bleeding a distress signal through her carotid artery, for instance—well, that was unsettling. Post-Laika study of animals sent into space has for that reason tended to focus on their bodily responses, not anything that would resemble human emotions.

Aside from its failure as a life-support system, Sputnik 2 was suc-cessful. It completed 2,370 orbits before it re-entered the atmosphere on April 14, 1958. As it came back to Earth, it ignited and burned up, Laika's funeral pyre.

Subsequent satellites would be equipped to tell more elaborate stories about the Earth and the physical layers around it. Like the first Sputnik—and unlike animals—these orbiting robots have been uncomplaining and disposable, though also capable of being anthropo-morphized as observers who tell stories about their travels.

The first satellite to relay a meaningful message was Explorer 1 (see illustration 20), the U.S. response to Sputnik. Launched on January 31, 1958, Explorer 1 carried multiple sensors, the most important of which turned out to be a Geiger counter. The *New York Times* described the counter as transmitting, in a "singing tone," a "combination of four 'voices.' " For eleven days, Explorer sang an oscillating tune about radiation, mostly to itself. Only when it was above one of the twenty receiving stations on Earth was its message recorded on audiotape. The data confirmed James Van Allen's hypothesis that radiation was trapped within the Earth's magnetosphere. Subsequent satellites would corroborate the existence of these bands of charged particles, now called the Van Allen radiation belts. It was the IGY's most important discovery about the composition of the cosmos.[24]

Explorer 1's data were disseminated among scientists everywhere, including at the annual International Cosmic Ray Conference, held in Moscow in 1959. The taped data, stored at Iowa State University, are still available to researchers. All this means that, despite the Cold War, Explorer 1 spoke the most international language of any round-the-world traveler to date: numerically rendered data. To be sure, the audience for the data belonged to a tiny global cross-section, meaning scientists. For that reason, the data tapes resemble Maximilian of Transylvania's first report on the Magellan expedition. Written in Latin, that publication also had an international audience, though one restricted to the learned population within Europe. Data have replaced Latin (or other human languages) as the broadest means for communicating an around-the-world journey. Robotics and telemetry seemed to be perfectly good ways for humanity to explore the cosmos.[25]

But given the cultural and historical significance of *human* planetary exploration, it was inevitable that flesh-and-blood people would boldly go where only dogs and robots had gone before. The first unearthly realm to be completely circumnavigated by human beings was not space but, rather fittingly, the sea over which so many around-the-world expeditions had taken place. That mission matched the outer-limits agenda of the IGY, merging polar research and space exploration, which undersea oceanography resembled in several compelling ways.[26]

Over eighty-four days in 1960, the submarine USS *Triton* completed the first fully submerged circumnavigation, as secret a project as any missile or satellite launch, and only publicized when completed. The *Triton*'s commander, Edward L. Beach, was a highly decorated naval submarine expert and author of the best-selling novel *Run Silent, Run Deep* (1955). Beach wrote up the *Triton* expedition for *National Geographic* and published a book on the voyage in 1962. (Because submarines were strategic Cold War instruments, technical information about the *Triton* was kept secret.)[27]

Everything about the voyage was experimental. *Triton* was the first nuclear submarine to have two reactors. It displaced 8,000 tons, twice as much as any other submarine at the time. Its orders were "to collect oceanographic and gravitational data in one continuous circuit around the world," which included a "world-girdling recording of the bottom contour" of the ocean. (One discovery was a new undersea peak in the mid-Atlantic.) As medical test subjects, the crew generated data about the human response to closed (and cramped) physical environments and to radioactivity, both of which were assumed to have application for space travel. "Our first space pioneers" would face similar risks— sealed from an unbreathable atmosphere and exposed to radiation. It was a remarkable experiment in a closed system, a recreation of an Earth-like atmosphere for an entire maritime circumnavigation, the achievement that early modern sailors had long dreamed of. *Triton* fairly burst with 77,613 pounds of food—16,487 pounds in frozen form, with another 6,631 pounds of canned meat and 12,130 pounds of canned vegetables. Air was recirculated, fresh water distilled.[28]

And then one sailor threatened to kill the experiment, poor fellow, by developing a kidney stone he could not pass. As his agony increased, he had to be sedated, so he would groan occasionally rather than scream all the time. Beach knew he had to resurface to get the man to a hospital. Would the *Triton*'s record of submerged circumnavigation still be valid? Could its secrecy be maintained? What if "the newspapers of the Soviet bloc were to get hold of the [news] item?" The submarine's scientific work would not be affected, but national morale, "particularly in view of the Russian successes with their Sputniks," might suffer. The *Triton* went 2,000 miles off course to deliver the crewman to a small band of rescuers, sworn to secrecy. To maintain independence from land, Commander Beach declined to take on any new supplies. Back down he went, and the brief time above water is not held against the submarine's eighty-four-day record of submersion.[29]

Beach could have used the near-fatal episode to bolster his claim to a genealogy that went back to Magellan. He read a biography of Magellan en route. He told the crew that they would be bound together in their "knowledge of having recorded one of history's great voyages." The submarine made a point of following the route of the *Victoria*, though it had to round the Horn rather than thread the strait named for Magellan. It delivered a plaque to install at Magellan's place of embarkation and surfaced to photograph his place of death in the Philippines, to mark the journey's halfway point. This was the only time the expedition's secrecy was compromised, when the periscope surprised a man in a canoe. Aside from that, and the offloaded patient, the experiment worked: humans could go around the world in vessels sealed away from normal Earth conditions for months at at a time.[30]

Survival in orbit was harder. To carry humans into space, the Soviets developed a post-Sputnik craft, *Vostok*, meaning "East" or "Dawn," that could re-enter the atmosphere. The first working Vostok was launched with two dogs on July 28, 1960, but exploded before achieving orbit.[31] A subsequent mission on August 19 successfully orbited two dogs, Belka and Strelka, plus mice, insects, plants, fungi, and other specimens, including strips of skin from some of the scientists. The total payload was 4,600 kilos, and it included 241 vacuum tubes, over 6,000 transistors, 56 electrical motors, and 15 kilometers of electrical cables. Cameras and biosensors monitored the dogs. When Belka vomited, space adaptation syndrome—space sickness—was born. Yet all the animals and materials were recovered in good condition and the strips of skin regrafted onto their donors. (Belka and Strelka were retired from the service.) A "human-rated" Vostok was successfully tested on March 9, 1961, with another dog, a wealth of rodents, some reptiles, many other biological samples, and a mannequin named "Ivan Ivanovich," a dummy human to represent the human space travelers to come.[32]

Those pioneers bulk larger in history than they did in real life. The earliest space capsules were as self-sufficient as submarines, but had less room than early airplanes, let alone the old sailing ships. Instead of Pigafetta's reams of paper, or Dampier's bamboo-and-wax tube for his journal, a Soviet cosmonaut had a mere pencil and plastic slate. Every ounce added enormously to the cost of a launch. Passengers had to fit, and be fit. Soviet cosmonaut recruits had to be between twenty-five and thirty years old, no taller than 1.75 meters (5 feet 9 inches), and weigh no more than 71 kilos (156.2 pounds). One of the candidates, a test pilot, was so short he boosted himself with a cushion when he sat in a cockpit. NASA specified men under age forty, later reduced to thirty-five, no more than 5 feet 11 inches in height, and under 180 pounds, less than Sputnik 1 itself. The lightest American recruit was only 150 pounds and the shortest 5 foot 7 inches.[33]

As important as a trim physique was a high panic threshold, whether natural or acquired. The first cosmonaut and astronaut trainees were experimental subjects. They knew they would don one pressurized life-support system, their space suits, squeeze into another, their space capsules, and then enter the void. Once there, they had to cope with isolation, sensory deprivation, and technical failures. They had to withstand the tremendous pressure of takeoff and re-entry, with the sensation of weightlessness in between, plus centripetal forces that induced nausea. To prepare, they had to be in top physical condition—they lived in fear that a twisted ankle or a case of sniffles would reduce their eligibility. "Hard in training, easy in battle," Korolev assured his weary charges. And yet they were to remain composed and cheerful at all times, delighted to be puking and sweating for science. "One absolutely essential trait of a cosmonaut is coolness and calm in any possible situation," a Soviet trainee observed. Their American counterparts said that "confidence" was their key trait.[34]

Whatever it was called—cool, calmness, the "right stuff"—it was a severe test of proprietary confidence. In both countries, the men trained to an irregular beat of unintended rocket explosions. The worst occurred in Kazakhstan when a military rocket at the main Soviet launch site detonated on October 24, 1960; more than one hundred people were killed. A Soviet trainee cosmonaut died on the ground when an alcohol-saturated pad, used to remove a medical sensor, caused a spark within an oxygen-rich training unit. The inferno claimed the first human casualty of biotelemetry. The other candidates were probably not told about the accident for several weeks, and the public not until 1986.[35]

The most promising Soviet candidates were Yuri Gagarin (primary cosmonaut), Gherman Titov (backup), and Alexei Leonov (the backup's backup). They were chosen for their political backgrounds as much as for their physical fitness. All had the right rural or proletarian heritage, military training, and ability to articulate the larger patriotism of the project. Gagarin was the most winning. He was the small fellow who needed the booster seat when he flew, but he possessed one of the twentieth century's biggest smiles, a constant signal of delight at being alive—at all—let alone at the dawn of space travel. Gagarin was the perfect, and perfectly photographable, spokesman for manned rocketry.

Biotelemetry would generate a parallel soundtrack to whatever he chose to say; Gagarin said that "the doctors uncovered the entire life story of each of us from the heart." The cosmonauts' training, however, meant they could manipulate that alternative source of information. On the night before the launch, for example, Gagarin and Titov retired to mattresses that had been secretly equipped with pressure sensors to detect whether they tossed and turned. They declined sleeping pills, and the sensors showed no signs of stress. Did they sleep? Their training, including awareness that they were probably being monitored, could have masked any pre-flight insomnia. It was no secret that Korolev slept badly. He worried most about the Vostok's ascent. If that went wrong, the capsule would plummet into the ocean near Cape Horn, in the Roaring Forties that had shredded Elcano's *Victoria* back in 1522. Retrieval would be difficult, if not impossible.[36]

But as Korolev had promised, it was easy in battle: *Vostok 1*'s launch from the Baikonur cosmodrome in Kazakhstan on April 12, 1961, went perfectly. Gagarin did his orbit and, despite a technical glitch on re-entry, returned safely at Saratov in southern Russia. Throughout, telemetry returned streams of data on the equipment and the atmosphere through which it traveled. Gagarin generated still more information through biotelemetry, and sent reports via radio and television signals. (Twenty minutes after launch, as Vostok passed over Alaska, American observers picked up Gagarin's TV link.)[37]

The press releases, vetted by the Soviet news agency TASS, were as selective as anything Charles V had released about the Magellan/Elcano voyage. Mostly, the news items stressed the novelty of the Soviet achievement and the heroism of the pilot. As Vostok had shuddered into its ascent, Gagarin had the presence of mind to shout a quotable phrase, *Poyekhali!* usually translated as "Let's go!" It continued the restless, leave-this-minute confidence of other round-the-world utterances, including Phileas Fogg's "I'm always ready" and the 1924 U.S. airplane mechanic's "I'm ready now"—"Let's go." All of Gagarin's publicized quotations emphasized his good physical condition and excellent spirits. "I feel fine," he kept assuring his contacts. Weightlessness did not impair his ability to eat, drink, and stay alert, though it did make him lose his pencil. Biotelemetry backed up his story: his pulse was reported as having stayed between 70 and 75 beats per minute, even during launch.[38]

Is there a perfect duration for a circumnavigation? Gagarin's flight took 108 minutes, with 89.01 minutes in orbit. It was the first around-the-world journey measured in minutes—"these minutes shook the world," TASS proclaimed. A slow circumnavigation on the Earth's surface makes clear the globe's immensity—the journey has too many details to be put into the final account. Record-setting aerial transits, like Wiley Post's solo seven-day, nineteen-hour ordeal, don't allow record keeping. Orbit offers the temptation of real-time narration, but it goes by too quickly. Gagarin observed how, when leaving and re-entering the Earth's shadow, the change in light "comes instantly," completely unlike any dawn or dusk seen on the planet's face. Just as well his pencil drifted out of reach. Even at half the speed of Puck, how could anyone take notes? Instead, Gagarin summarized what the Earth looked like from space: "It is indescribably beautiful."[39]

In one notable instance, the official Soviet story about the orbit was false and Gagarin had to repeat the falsehood. He had not returned to Earth in Vostok's capsule, but by parachute. The International Aeronautical Federation stipulated that, to set a new aerospace record, a pilot had to take off and land in the same vessel. The Soviet authorities wanted to claim a new altitude record, so were stingy with the truth. The first press release said "the spaceship 'Vostok' landed safely" at 10:55 a.m., Moscow time, without specifying where Gagarin was. At a press conference on April 16, Gagarin admitted there was "a parachute variant" but claimed (in the third person) that "the pilot remained inside the cabin." His memoirs include a diagram showing the flight path of the capsule, but not him. The truth about the parachute came out in 1978.[40]

And only in 1991 did a full transcript of Gagarin's mission appear, in Russian. Among other technical details about the flight, it gives a running commentary based on his biotelemetry. "I hear you loud and clear," one of the ground personnel told him before the launch—"your pulse is 64, respiration is 24." "That means my heart's beating," Gagarin playfully shot back. At this point, his banter matched his heart rate. Promised some recorded music to kill time, the songs proved hard to pipe into the capsule and the delays and back-and-forth about them became an unintended test of Gagarin's sang-froid. His pulse accelerated during the launch and into orbit, yet never exceeded that admirable 70 to 75 beats per minute. Meanwhile, Gagarin gave descriptions of the Earth that began with detached observation, then slipped into rhapsody:

> I'm observing clouds. . . . The landing site. . . . It's pretty, what beauty! . . . Attention. I see the horizon of the Earth. A very pretty halo. At first there's a rainbow from the very top of the Earth's surface and down. A rainbow like this is transitioning. Very beautiful! Everything's going through the rightmost porthole. I'm seeing stars through the "gaze" [porthole], seeing the stars passing by. A very beautiful spectacle.[41]

More details about the Soviet space program continue to emerge, as documents become public and old memories are aired. But it is unlikely that they will ever bridge the abyss between what Gagarin said and what biotelemetry recorded about the first human orbit around the world. The achievements in telemetry were crucial to the mission. An official statement from the USSR Academy of Medical Sciences lauded the "new branch of science" that had, on April 12 of that year, "converted" the human body's signals into electrical impulses and zoomed them over planetary distances. The signals were accompanied by verbal reports that a "Muttnik" could never supply.[42]

Yet the two narratives don't match. Even as Gagarin rhapsodized about seeing the Earth, his heart never raced. He claimed an exhilaration he had been trained to damp down. Emotional coolness may be necessary to negotiate the bizarre experience of orbit—the techno-swaddling, the monstrous units of distance, the dizzying speed—but it renders the achievement, and the telling of it, anticlimactic.

Whatever tale he told, for whom did Yuri Gagarin speak? Obviously, he was a hero in the USSR, and his reputation has survived the fall of the Soviet Union. But he was an international celebrity too, fêted in Cuba, kissed by Gina Lollobrigida, lunched by Elizabeth II of England. The Soviets called Gagarin's overseas tours an "earthly 'orbit of friendship.' " Yet Charles de Gaulle, president of France, diminished the internationalism by claiming Gagarin's orbit as a triumph for European culture. The Martinique-born cultural critic Frantz Fanon, speaking as part of an emerging assault on the legacy of empire, rejected the claim: "Colonel Gagarin's exploit, whatever General de Gaulle thinks, is not a feat which 'does credit to Europe.' " The Socialist International did not exist to be appropriated by the capitalist leaders of the old imperial centers. Formerly colonized people, including Fanon, could now claim science and world expeditions as part of their history, too.[43]

Humanity's common dependence on Earth was excellent proof of that global citizenship, as well. The first orbiting spokesman for that position was Gagarin's backup, Gherman Titov. On August 6–7, 1961, Titov did seventeen loops around the planet in just over twenty-five hours, a long day. He was at least as ebullient as Gagarin, and even more poetic, especially about the multiple sunrises and sunsets he witnesssed in succession:

> Marching across the planet in a circle of deep red-orange, the vivid sunset yielded unwillingly to the dark surrounds that advance to envelope the light of day, flashing colour to the horizon where the blue halo increased its richness of hues until, by some magic of transformation, night reigned supreme.

Again, the verbal emoting did not match the biotelemetry. Titov found it odd to keep telling ground personnel how he felt, given that they could see his heart rate and blood pressure for themselves. (U.S. tracking stations picked up a signal from *Vostok 2* nearly as steady as Sputnik's beep—Titov's carefully monitored heartbeat.) He sent greetings to all major cities over which he passed, including Washington, D.C. In orbit, his and the U.S. capitals lay only eighteen minutes apart; "Cosmic space brought our cities together."[44]

In his memoirs, published in Russian in 1961 and translated into English in 1962, Titov updated two circumnavigators' traditions. The Earth had rotated beneath him as he passed over it, of course, so his circumnavigator's tracks were never the same. A map at the start of his book shows the oscillating pattern of his seventeen orbits, together weaving a net around the planet (see illustration 21). When Titov celebrated his twenty-sixth birthday, about five weeks after his day in orbit (he is still the youngest person to have gone into space), he joked that he should subtract seventeen days from his age, given the "seventeen cosmic sunrises" he saw within twenty-five hours, a truly cosmic Circumnavigator's Paradox.[45]

Above all, Titov stressed that seeing the planet from space made him long for it, not for the stars. When a journalist had asked him why he wanted to go into space, he had responded as if telling a riddle: "I suppose, because I love everything on earth so truly and deeply." From space, he saw his beloved from a new angle. In another set of interviews, he recalled that the "sphere" looked "like something right out of a fairy tale." His memoir told how, in flight, he had concluded, "It is very handsome, our earth!" He emphasized his authority on that matter, as someone whose vision of the planet gave him a unique perspective: "I saw it, our earth, I saw it all. It is lovely, but it is really small, when you look at it from space . . . I suddenly realized with all my being how careful we must be with it, how we have to work and think so as to have peace reign eternally on all six continents."[46]

That was quite a statement. Titov spoke just as U.S.-USSR negotiations over a moratorium on testing nuclear devices were reaching a climax; the Limited Nuclear Test Ban Treaty would be signed in 1963. His planetary vision revived the internationalist hopes for round-the-world travel that had taken a beating during World War II. It also radically redefined human effort on a global scale. For a long time, geodrama had urged humans to master the Earth, all of it. Titov warned that the concern should be to protect the planet, all of it.[47]

The American space program, whatever its antagonism toward its Soviet counterparts, would nonetheless endorse their view of humanity's shared status as terrestrial beings. The parallel with earlier eras of circumnavigation is instructive. European maritime circumnavigators may have been rivals, but they had shared a goal of making the planet useful to them; Khrushchev and President John F. Kennedy may have been as coldly calculating as Philip of Spain and Elizabeth of England, but their nations' respective space travelers shared a vision of preserving the Earth.

The National Aeronautics and Space Administration was created in 1958, in response to Sputnik. Its leading rocket scientist, the German-born Werner von Braun (who had surrendered to the Americans at the end of World War II), defined the program's goals in stages. Orbit was to lead to exploration of the Moon and Mars. (That was the Soviet plan, too.) The "von Braun paradigm" envisioned an orbiting space station as a stage halfway between satellite observation of Earth and manned landings on the Moon, and as a place where Mars-bound spaceships could be assembled and fueled. But after Gagarin's orbit, Kennedy rearranged the stages. In a May 25, 1961, speech to the U.S. Congress, he stated that the United States should commit its resources to placing a man on the Moon by the end of the decade. The Mercury program, the U.S. equivalent of Vostok, deferred the creation of an orbiting space station.[48]

NASA guarded its technical information at least as closely as the Soviets did theirs, but to emphasize that the Mercury astronauts were free to speak about parts of their experience, journalists were granted a chaperoned access to the seven men. The result brought space travel into line with the media coverage that had become standard for round-the-world travel. The Mercury Seven sold their story to *Life* magazine, for example, which circulated well beyond the United States. (The Soviet cosmonauts read parts of it.) The collated final product, *We Seven, by the Astronauts Themselves* (1962), was very like Lowell Thomas's *First World Flight* (1925). In it, an engaging composite storyline (by journalist John Dille) weaves together the astronauts' first-person accounts, culminating with John Glenn's testimony about his orbit in *Friendship 7*, an American first, on February 20, 1962.[49]

The main point of *We Seven* was that all the telemetry and robotics in the world could not replace the human perspective. It was an arguable but not obvious point, given how NASA training habituated its astronauts to the techno-swaddling and inhuman dimensions intrinsic to space travel. (The men teased each other that NASA had wanted to send up a dog, instead of them, "but they thought that would be too cruel.") The candidates stressed the comedy of human adaptation to tiny spaces. "We sometimes joked," Glenn said, " 'You don't climb into the Mercury spacecraft, you put it on.' " Indeed, the astronauts' pressure suits and flight capsule's contour couches were as sharply tailored as any suit from Savile Row. Walter M. Schirra, Jr., said that "sliding into the couch is a little like slipping a gingerbread man back into the cooky cutter."[50]

Although the astronauts maintained that they were not "just one more piece of equipment," and much more than "Spam in a can," their perception of the material world around them was highly mediated. "We could not have done it without telemetry," admitted Alan B. Shepard, Jr. *Friendship* 7 depended on global tracking stations that extended over 25,000 miles, connected with an estimated 140,900 miles of wiring. The data "started streaming" from the vessel "long before it went into orbit and it kept flooding into the computers until after the spacecraft landed in the ocean." Shepard protested that "even though the electronic machines were clever, we did not let them run the show." But at a traveling velocity of 17,500 mph or 5 miles per second, a person was hard-pressed to keep track of time and events. The capsule had several clocks and the astronauts wore watches. Glenn also described the monitor that "shows you—on a small scale model of the earth that revolves as the spacecraft moves along in orbit—exactly what section of the earth you are passing over." (Vostok had a revolving globe, too.) A day passed in forty minutes, and a night melted away as quickly.[51]

Whatever the bewildering thrill of those minutes, NASA training was designed to condition any response to it. When Alan Shepard made a non-orbital launch, his pulse was only slightly elevated. (Other bodily responses were harder to control: his flow of norepinephine was over twice its normal rate.) When Glenn looked at the Earth during his orbit, his joy was measured: "Oh," he said, "that view is tremendous!" The verbal cool was the result of good training. Glenn's blood pressure had actually become slightly elevated, and he sweated off five and a half pounds during the flight. Scott Carpenter turned out to be the poet of the Mercury mission. He regretted the brevity of orbit. "I wanted to be weightless again," he said, "and see the sunsets and sunrises, and watch the stars drop through the luminous layer." But he also craved more time in the Mercury capsule in order to "learn to master that machine a little better."[52]

It was a way of saying that the Mercury Seven were not *cyborgs*, living organisms fully integrated into cybernetics because enhanced by electromechanical technology. That term was defined in 1960, precisely in relation to astronautics. Carpenter was not alone, however, in insisting that he mastered the machine, not the other way round, that he, unlike it, could see beauty in sunrise and sunset.[53]

The pioneer generation of orbital travel ended with the first space walk. From one of the post-Vostok ships, a *Voskhod* ("Sunrise"), Alexei Leonov ventured out into space on March 18, 1965. His first words upon embarking on this unprecedented cosmic experience? "I can see the Caucasus." Somehow, even as humans soared above Gaia, she was still the center of attention, and the distant stars her spotlights.[54]

But surely space travel meant leaving Earth? The answer continued to be yes, though also no. Or else: maybe?

There was still a sense that orbit was on the threshold of something. Glenn said the Mercury astronauts were "helping to break the bonds that have kept the human race pinned to the earth." The TASS communiqués on Gagarin's orbit announced that he had "made his exit into cosmos." Noting that the flight took place in the morning, and in a spaceship named for the dawn, no less, TASS claimed that April 12, 1961, was "the morning of a new era." It was especially bold of the Soviets to announce a break from their pre-revolutionary history of maritime exploration. An official statement about Gagarin's orbit insisted that the Revolution of 1917 had initiated true human progress. There was no science under the tsars; all those earlier around-the-world voyages had been imperialist nonsense.[55]

And yet the foundational imagery of space travel was modeled on seafaring. The names *cosmonaut* and *astronaut* both called upon humanity's nautical heritage. The word *astronaut* had first been used, in English, in a piece of science fiction published in 1880, to describe a space vessel: "In shape my Astronaut somewhat resembled the form of an antique Dutch East-Indiaman." By the 1920s, the word referred to the men who used spacecraft, not the craft itself. The Soviets coined *kosmonavt* in the 1950s as a parallel term; the Chinese would create *tai-konaut* in the 1990s. *Spaceship* has a similar history, appearing first in science fiction in the late nineteenth century, later adopted by the science fiction–reading engineers of the mid-twentieth century. Korolev, for instance, who called Earth "the shore of the universe," in 1960 suggested the term *spaceship* (*kosmicheskii korabl'*) for what his engineers were building. TASS picked up the word in the same year. The name *Vostok* had been a common one for Russian ships. (Bellingshausen had commanded a *Vostok* when he went around the world from 1819 to 1821.)[56]

NASA's splashdowns were obvious reminders of space travel's nautical heritage. While the Soviets preferred to bring their craft back to land, often without publicity, the Mercury and succeeding Apollo space capsules made televisually dramatic "landings" in the sea. (The astronauts' equipment included shark repellant.) Because of the splashdowns, the U.S. Navy and Coast Guard had starring roles in space launches. When Glenn was fished from the Atlantic and taken aboard the USS *Noa*, the sailors made him an honorary crew member and voted him "Sailor-of-the-month." They marked his footprints, where he had first stepped on deck, in white paint. From the *Noa*, Glenn enjoyed his fourth sunset of the day.[57] During the splashdown that concluded *Gemini 3*, Gus Grissom and John W. Young became seasick. "John managed to hang on to his meal," Grissom confessed, "but I lost mine in short order." Maritime knowledge came in handy when Jim Lovell accidentally erased the navigational data on *Apollo 8*'s computer. Lovell restored the calculations by making star sightings with a sextant.[58]

President Kennedy, a former navy man, gave space its saltiest tang in his "We choose to go to the moon" speech at Rice University in Houston on September 12, 1962. "We set sail on this new sea," he said, "because there is new knowledge to be gained, and new rights to be won, and they must be won and used for the progress of all people." That was the latest claim that scientific exploration benefited all of humanity. And yet Kennedy also evoked the more guarded internationalism of the postwar period: "only if the United States occupies a position of pre-eminence can we help decide whether this new ocean will be a sea of peace or a new terrifying theater of war . . . space can be explored and mastered without feeding the fires of war, without repeating the mistakes that man has made in extending his writ around this globe of ours."[59]

As that cautious optimism hinted, aviation supplied another precedent for space travel, especially among the first generation of cosmonauts and astronauts, who were predominantly pilots. Gagarin revered Saint-Exupéry, especially the novel *Night Flight*, about the dangerous glories of French postal service on the Line. His own memoir, *Road to the Stars,* was evidently named for a favorite Saint-Exupéry quotation: "Your road is paved with stars." Mercury astronaut Gordon Cooper remembered his father, who had also been a pilot, telling stories about Wiley Post and Amelia Earhart. Another of the Mercury Seven, "Deke" Slayton, said of aviation, "I never assumed that switching over to the space age would be much different."[60]

Orbit in particular seemed close to Earth, and similar to other kinds of circumnavigation. In 1961, the magazine *New Scientist* described Gagarin as having made a "circumnavigation of the Earth," and the terms *orbit* and *circumnavigation* have become, for space travel, synonyms. "The radio connected me like a navel cord with the earth," Gagarin said, so he never felt "bored and lonely" in space. Titov was blunter: "No man ever truly leaves the earth. This is impossible. To survive in that bitterly alien environment beyond our world, he must take with him the most precious and essential items of life that earth provides."[61]

Gagarin and Titov differed, however, over what their connections back to Earth represented. Gagarin said that he had, by 1959, "heard something about cybernetics," and particularly its claim that "electronic machines would replace the human brain." But he retained the old Stalinist suspicion of cybernetics. "The human brain is nature's most perfect work," he said; "there is nothing to replace it and never will be," and the Mercury Seven would have agreed. Titov, like NASA's engineers, was more enthusiastic about technology. Space travel entailed "a strange and wonderful blending" of human and machine.[62]

The globe, and geopolitics, determined the nature of the blend. To maximize the helpful little eastward shove that Earth's rotation gives to objects launched into orbit, each superpower constructed its main rocket launch site as close to the equator as its territory allowed: Bio-konur in Kazakhstan and Cape Canaveral in Florida. But what then? Glenn recalled that "the world did not happen to be set up as an ideal tracking range for us"—shades of the earliest circumnavigators' puzzlement over the globe's odd configuration. The United States negotiated treaties with its allies to set up tracking stations on foreign soil, either islands or "friendly continents around the world." When Glenn flew over Australia during its night, he liked how the residents of Perth and other cities had kept their lights on for him in a big, glowing "g'day!" In contrast, the Soviets used self-contained naval vessels for their tracking stations. Nonetheless, the USSR made requests that, should any cosmonaut land in foreign territory, he be assisted and returned, not treated as a spy or enemy.[63]

If the sea and the air each supplied precedents for space travel, they nevertheless implied different legal possibilities. Open ocean was international territory; airspace over land belonged to distinct nations. But outer space is not exactly an extension of airspace. Because the Earth rotates, "above" perpetually shifts—who can't claim the Moon? It also mattered what spacecraft might be doing. This "functional approach" betrayed the continuing suspicions of the earliest years of aviation and rocketry, and their associations with war. The context of the IGY had soothed some doubts. Since the United States had announced plans in 1955 to launch a satellite, it, and other nations, had tacitly approved such flights. But when the USSR landed a probe on the Moon in 1959, debate over the sovereignty of celestial bodies became critical. UN Secretary-General Dag Hammarskjold expressed his belief that individual nations had no right to claim celestial bodies.[64]

The solution would draw on the past centuries of round-the-world travel, as the United States and the USSR's different ground-support systems had already indicated. By brokering access to the sovereign territory of others, the United States had revisited the terms of amity under the Hundred Years' Peace, somewhat modified by aviation's shifting negotiations; by using ships stationed on open ocean, the Soviets evoked the Grotian ideal of the free seas.

The Congress of Vienna was most powerfully revisited in a resolution adopted unanimously by the UN General Assembly in December

1961, when "outer space and celestial bodies" were declared to be "free for exploration and use by all States." The decision matched the concurrent Antarctic Treaty (June 1961), which declared the southern continent to be the object of scientific exploration for "all mankind," as well as the principles of the upcoming Convention on the High Seas (September 1962), which would reiterate that individual nations could not possess the high seas; these agreements reflected, as well, the scientific cooperation of the IGY, which had included international investigation of polar regions and the oceans. The Outer Space Treaty (January 1967) would make the Grotian ideal into international law: outer space, including all celestial bodies, would not be "subject to national appropriation by claim of sovereignty." Almost all of the world's nations signed the treaty, including the United States and the USSR.[65]

But were the new international decisions built on the old imperial order? A sense of shared planetary space had been compromised by several Western trespasses. These included the voyage of USS *Triton*, which had passed, submerged, through the Surigao Strait of the Philippines and the Macassar Strait of Indonesia, waters that those nations considered proprietary. Non-Western nations with penetrable gulfs and archipelagos feared similar incursions. It didn't help that rocket launches were often done from colonized areas. Kazakhstan was part of Russia's old imperial territory. France would make its initial launches from Algeria, then, after the Algerian War, from Guiana. Was rocket science truly international?[66]

It was all somewhat speculative, given that the human body might be unfit for prolonged space travel anyway. People could survive and even thrive at sea—children had been conceived and born there—as well as *under* the sea and in polar environments. *Triton* was at sea four days longer than it had taken Phileas Fogg to circle the globe. And there has been a continuously occupied station, Mawson, south of the Antarctic Circle since 1954. In contrast, humans had spent mere hours in orbit.

The hours were hard-won—the goal of zero mortality proved impossible to maintain, as Vostok-era disasters had already shown. On January 27, 1967, the first flight-ready Apollo capsule ignited during a routine checkout procedure, killing three astronauts and dimming

NASA's luster. On April 24 of the same year, the first manned Soyuz spaceship, having experienced multiple equipment failures, landed without its parachute deploying; its pilot died.[67]

Even if the equipment got better, the humans might not. As Titov had made his seventeen loops around the Earth, his condition had deteriorated. He became nauseated and lethargic, unable to complete many of the tests on his roster. On the other hand, his condition assisted an auto-experiment, proving that a person could sleep and awaken in orbit. Toward the end of the twelfth orbit, he began to feel better. Perhaps the longer the orbit, the greater the capacity for recovery? And then a brave new life in space? [68]

But life would end in space if it somehow impaired the human ability to reproduce. Gagarin, Titov, and Glenn were all married men and fathers, so the first test of human generation following space travel fell to a woman. While NASA was initially reluctant to send women into space, the Soviets recruited female cosmonauts to prove the social equality women enjoyed under communism. They put Valentina V. Tereshkova into orbit on June 16, 1963. The exercise did not quite prove gender equality. Soviet space officials were obsessed with Tereshkova's femaleness to the point of misogyny. Titov's struggle to control his (male) body in space had been forgiven, for instance, but not Tereshkova's similar effort. And everyone worried whether the unmarried female cosmonaut would be able to have children. When she became pregnant the year after she returned, she was carefully monitored. Baby and mother were fine.[69]

There was related concern over animal test subjects. Strelka's postflight litter of puppies was good news for science. (One of the space dog's progeny, Pushinka, was presented to the Kennedy family in the White House.) Less encouraging data came from two female dogs that, in 1966, spent a record three weeks in orbit. One dog received drugs to counteract radiation from the Van Allen belts, the other was the control. But both dogs returned to Earth extremely weak. Their blood and urine showed elevated levels of calcium—under conditions of weightlessness, their bones had been leaching away.[70]

It was the first evidence that prolonged release from Earth's gravitational force, a necessary consequence of travel in outer space, might have medical consequences. An object circling Earth is still affected by gravity, but its orbit makes it perpetually fall toward Earth, and humans perceive this as the same kind of bodily soaring that they might feel, more briefly, on a roller coaster. Without the resistance felt on Earth, muscle and bone atrophy. This was spaceflight's most unwelcome resemblance to the scurvy-plagued around-the-world maritime voyages. The bad news was confirmed by the two-man crew of a Soviet spaceflight in 1970. Despite using pressure suits and undertaking special exercises to counteract weightlessness, the cosmonauts became listless, puffy, and irritable. On landing, they had to be carried from their capsule, and it took them almost two weeks to recover. Even if the effects of orbit were temporary, it was sobering not to see space travelers bound away from spacecraft under their own steam. Yet again, it seemed that robotic emissaries might be physically sturdier than their human creators.[71]

Satellites were indeed the easiest way for nations to enter the space age, in a big cosmic *us too*. In fairly quick succession, France (November 1965), Japan (February 1970), and China (April 1970) became the third, fourth, and fifth nations to put satellites in orbit. The leap forward in participation was rapid—as if the earlier history of circumnavigation, from the 1500s to early 1800s, had been compressed into a decade.

Data-gathering satellites sent back unprecedented reports on the state of the planet. The first TIROS (Television Infrared Observation Satellite) went up from Cape Canaveral on April 1, 1960, and into a polar orbit where it photographed weather patterns and the world's surface beneath them. Polar-orbiting satellites are excellent Earth observers. They survey all latitudes in each pass, and the Earth's rotation beneath them continuously renews their longitudinal orientation. When Landsat 1 was put into orbit in 1972, for example, its near-polar orbit permitted unprecedented photographic and radiometric images of Earth. Its images verified that some of the world's deserts were spreading, diminishing land for farms and pastures.[72]

Geostationary satellites offer a different option. By performing geosynchronic circuits at 22,000 miles above the equator, orbiting in time with the planet, they are not only girdles round about the Earth but also hovering belt buckles over it. They work best as communications satellites that gather signals from one place below, then relay them back or to other satellites. Such devices do not generate information. Like the telegraph and airmail, their messages are echoes:

> *Love love love*
> *Love love love*
> *Love love love . . .*

The Beatles sang "All You Need Is Love" for the first time to an audience gathered together by three telecommunications satellites. For two hours on June 25, 1967, at the start of what would be called "the Summer of Love," an estimated 400 to 600 million viewers in twenty-six nations tuned in to *Our World,* the first global television broadcast. Organized by the European Broadcasting Union and the British Broadcasting Corporation, *Our World* included contributions from seventeen different countries. The Beatles represented the United Kingdom. Their earnest anthem was perfect for the occasion: old-fashioned in its yearning that round-the-world communications would unite humankind; space-age in its use of the latest round-the-world technology. In the broadcast, balloons printed to look like world globes float above the Fab Four, who had decided to embed into their song bits of "international" music—that's why the *Marseillaise* is its lead-in.[73]

Our World meant the developed world. All of the Beatles' song snatches were from Europe, except for a Glenn Miller fragment from the United States. And the broadcast only reached nations in the industrialized world. Canada's segment included an interview with media doyen Marshall McLuhan, who had to be prodded multiple times to admit that most of the world lacked the technology to tune in to the historic program.

That would change. For many people, satellite communications represent freedom from empire. To be sure, the first satellites were launched by the old imperial powers, often from colonized zones. But compared to telegraph offices and postal systems, satellite links have been less burdened with the paraphernalia of occupying nations. As satellites multiplied, they offered the worldwide network that Arthur

C. Clarke had predicted—what McLuhan called "a global embrace, abolishing both space and time as far as our planet is concerned," and "reducing its scope to the extent of an evening stroll."[74]

As the cost of rocket launches declined, and as formerly colonized nations generated their own wealth, they sent up their own satellites. India did it first. On April 19, 1975, Aryabhata (named for the classical Indian mathematician-astronomer) was launched from a Soviet rocket platform. Designed to perform scientific experiments, it did so only for five days before it malfunctioned. But the brief transmissions were significant achievements for a part of the world that had previously depended on the global communications networks that had emerged with Western empires. Other post-colonial places have since then sent up their own satellites, Indonesia (1976), the Arab League (1985), and Pakistan (1990), on a growing list of places that have emancipated themselves from Western-dominated communications networks.[75]

But orbit was not emancipation from the planet itself, as several science fiction landmarks began to acknowledge. At first, science writers had emphasized the otherworldliness of space. The pioneer American SF magazine *Astounding* (1930–60) had predicted that space travel would be, well, astounding. And at the start of the space age, *Star Science Fiction Stories* (1953–59) had celebrated the starry path presumed to lie ahead. But a different vision appeared in an anthology published from 1966 to 1980: *Orbit*. That title would probably not have occurred to earlier generations of fantasy writers.[76]

Two pearls of science fiction likewise revealed the pull of the planet. The first is the novel *Solaris* (1961), by the Polish writer Stanislaw Lem, subsequently made into two films and a television program. The book describes events on a space station that orbits a planet covered by a sentient ocean. The ocean has begun to invade the consciousnesses of the station's scientists—each man is haunted by a visitor culled from his memories of Earth. One of them argues that, in their clumsy search of the cosmos, they have asked for this fate, to circle one planet while tormented by memories of another: "We have no need of other worlds. We need mirrors. We don't know what to do with other worlds. A single world, our own, suffices us; but we can't accept it for what it is."[77]

That was one message, as well, of *Star Trek*, the remarkably popular and influential television series, in its three seasons from 1966 to 1969. The series title suggests that the starship USS *Enterprise* (NCC-1701) spends its time hurtling around far-flung solar systems. But less than a quarter of the episodes take place in space. Slightly over half of them plant the *Enterprise* in orbit around a planet, while the drama unfolds below. (The remainder of the shows are hybrids, with deep-space action and terrestrial adventures.) In a recurring shot, familiar to all within the *Star Trek* community, the starship curves (left to right) around a planet that has a conveniently Gaia-like atmosphere. As in *Solaris*, the Earthlings never really leave Earth. In two of the *Star Trek* episodes set in orbit, the planet at the center *is* Earth.

If there is a threshold humans must cross to get to the rest of the cosmos, it is the Moon, Earth's original around-the-world traveler. The Moon has made its circuit (every 27.3 days, at a speed of 1.022 kilometers per second, or 2,286 miles per hour) ever since it was created, roughly 4 billion years ago, long before human beings existed, let alone thought about going around the world.

In a story he wrote over Christmas 1948, Arthur C. Clarke defined the Moon's liminal position by imagining a permanent human outpost on it. "The Sentinel" is set in 1996, as an expedition explores Mare Crisium. The team's geologist ("selenologist if you want to be pedantic") sees something glint in the mountains high above. It doesn't look natural. He and a companion hike up and discover an artificial pyramid protected by a force field. An atomic blast opens the field, but disintegrates the pyramid into inscrutable fragments. What was it? Probably a sentinel that another intelligent race had installed to monitor Earth, silently circling the planet for centuries. Its destruction may be a galactic signal to check back on the Earthlings, who now have the capacity for space travel, at least as far as their faithful satellite.[78]

Clarke's target year for a manned lunar outpost, 1996, seemed plausible. The USSR's Luna program sent twenty robotic devices to the Moon, three in the progam's first year, 1959. Luna 1 missed its mark and became the first artificial object to orbit the Sun; Luna 2 landed on the Moon; and Luna 3 went around it, photographing its far side for the first time. NASA responded with the manned Gemini and Apollo programs, to fulfill Kennedy's goal to put an American on the Moon by 1969.

The technology that made it possible also told the story of it in real time, and to the public. NASA routinely broadcast its launches and splashdowns, and the Soviets would eventually do so as well. But however exciting it was at the start, space news lost its appeal due to launch delays, soupy audio, and fuzzy television footage. After some botched attempts to bring spaceflight to television, NASA staged a live broadcast from the capsule of *Apollo 7*, complete with goofy cue cards.[79]

The connection between camera and planet was most apparent during the *Apollo 8* mission, the first manned flight to make it to the Moon. (The capsule did not land—it verified that an Apollo spacecraft could get to the Moon and back.) The mission produced dramatic black-and-white photographs of the Earth emerging from behind the Moon. On Christmas Day 1968, the astronauts broadcast that image of "earthrise" while reading the first ten verses of the King James Version of the Bible: "In the beginning God created the heaven and the earth . . ."

Apollo 8's Yuletide gift may have been intended, in part, to rebuke the Soviets' secular view of space, though it followed the Soviet tradition of gazing back at Earth in wonder. *Apollo 10*'s grainy color photographs of Earth increased the sense of rhapsody, as did *Apollo 17*'s hyperreally clear "blue marble" photographs of the planet taken in 1972.[80]

Public fascination with the images showed that, rather than a blow-by-blow circumnavigator's narrative, space may require haiku. A photograph is the visual equivalent: a crisp image, quickly absorbed, yet with haunting resonance. The swiftness of orbit made any narration difficult, and the conditioned coolness of astronauts and cosmonauts, plus their preoccupation with tightly scheduled technical tasks, had made it dull. The snapshot quickness of the earliest Soviet observations of the Earth—"it's beautiful!"—had been better. The camera kept the Earth at center stage, though without the human voice that had been an essential part of geodrama. If robotic cameras could send back the visual haiku, the humans themselves would be superfluous. A central element of geodrama would end.

Much has been written on how public reaction to the photographs represented an entirely new conception of the Earth, a break with the human past, when the whole planet could only be imagined, never seen. It's true that the whole Earth had never before been seen in one glance. But it's not true that humans lacked any meaningful sense of the entire planet before they saw it from space. So what did that view of it actually represent?

It would have been surprising had a protective wonder at seeing the planet not been the reaction, given rising public concern over the physical safety of a world threatened by nuclear weapons. Titov had first stated that worry in 1961, in a text available in the West since 1962. In a speech to the World Women's Congress of 1963 in Moscow, Tereshkova also remembered of her time in orbit that "every 1½ hours the dawn met us. The sun rose in an inimitable gamut of colors. Can we ever allow the mushroom clouds of atomic explosions to eclipse the sun?"[81]

That question repeated the long-standing concern among many round-the-world travelers to harmonize the world's peoples, to be socially responsible. However carefully phrased, Titov's and Tereshkova's pleas to end the arms race were extraordinary—overt criticism of the Soviet military was dangerous. A variety of political and social pressures eventually brought détente, a thaw in relations between the United States and USSR that culminated in the Strategic Arms Limitation Talks (SALT) Agreement, which concluded in 1972, the same year the "blue marble" photos were taken.

But Titov and Tereshkova had added something to the desire to be responsible to humanity: to be responsible to the physical planet. That was new, and it was reiterated in many environmentalist reactions to images of the "Whole Earth." The post-Apollo idea that the planet was a spaceship, a closed system with finite resources, was perhaps best expressed by the American utopian Buckminster Fuller, whose *Operating Manual for Spaceship Earth* (1968) used the concept of prolonged spaceflight to discuss the state of the planet. Spaceship Earth became the model of environmental conservation.[82]

This was fundamentally different from earlier conceptions of the Earth. People had been defining themselves and their actions in relation to the planet since the 1520s. But the earlier definitions had been antagonistic: fear of the globe's unforgiving dimensions gave way, after three centuries, to confidence that, despite its bulk, the planet could be made useful, even subservient, a sprawling beast finally tamed. The conquest seemed a dubious goal by 1968, and this represented a momentous doubt over the human place on the planet. That epiphany was not instantaneous. It had been seventeen years in the making, from Gagarin's initial rhapsody to the *Apollo 8* photographs. Seventeen years is nothing, however, compared to four-plus centuries, the two first acts of geodrama, during which humans thought they had to fight the planet, not protect it. That explains why any Whole Earth epiphany has had such limited impact—the planet's condition has improved in some ways since 1968, but deteriorated in too many others.

If anything, the lunar expeditions gave human beings a new sense of confidence by defining another measure of their reach: to the Moon and back. That metric had existed as a hypothetical unit of travel, whereas "around-the-world" had measured a human activity since 1522. In 1935, the total miles that two United Airlines pilots had flown was calculated as "a distance equal to 2,000 trips around the world at the equator," which actual people had done, or "as equivalent to two round trips to the moon with enough miles left over for two or more flights around the world," which was at that point just dreaming. By 1967, when Pan American Airways gave a "Space Age View of 10,000 Round-the-World Flights," it compared two distances that humans had actually traveled: 10,000 times around the world amounted to forty trips to the Moon and back.[83]

Beginning with *Apollo 11*, U.S. astronauts spent precious hours on the Moon. *Apollo 17* delivered Eugene Cernan, the last person, so far, to have strolled the lunar surface. This expedition of 1972, the longest Moon stay, lasted 75 hours, about 11 percent of the 655.73 hours that it takes for the Moon to circle the planet.

By that measure, we have not yet passed from the shores of the planet to the ocean of space—we've been on a merry-go-round. Starting in 1522, circumnavigators rode the horses near the center, the revolving Earth itself. Aviators went toward the edge of the carousel; submarine sailors moved back toward the center. Astronauts in orbit rode an outer ring of horses. But we have never ridden the outermost horse, the Moon, because no one has done what Clarke predicted in "The Sentinel." There has not yet been a permanent lunar outpost, so no one has gone around the entire world while on the Moon.

In our defense, we've been rather busy down on the face of the planet.

Chapter 11

Army and Navy Surplus

As Fogg and *Passepartout skim the Alps in their balloon, the resource-ful valet scoops up snow to chill his master's wine . . .* On October 17, 1956, the film *Around the World in 80 Days* had a gala premiere at New York's Rivoli Theatre. Producer Michael Todd's two-hour-and-fifty-five-minute epic, shot in his trademark Todd-AO widescreen process, was filmed in over ten countries and featured an international cast. "A smasheroo from start to finish," *Variety* proclaimed. The *New York Times* declared it "a sprawling conglomeration of refined English comedy, giant-screen travel panoramics and slam-bang Keystone burlesque . . . undeniably, quite a show." The film made money and won five of the eight Academy Awards for which it was nominated.[1]

The film's success—even beyond the United States—matched a major transformation of round-the-world travels. State-sponsored expeditions had been the original prototype for such journeys, though gumption and greater affordability had, by the late nineteenth century, allowed some private travelers to circle the globe. But unprecedented numbers of civilians could, by the middle of the twentieth century, use the surplus bounty of the industrialized world to embark on world circuits that had once required the support of an army or a navy. Fogg the private traveler was often their *me-too* model. His cinematic popularity signified an incomplete transition. Were the imperial politics that had ushered him around the world defunct? Would a circumnavigation ever be able to rid itself entirely of the proprietary confidence—including the use of armed force—that had favored Westerners against others?

• • •

There have been many adaptations of Jules Verne's novel. Two silent films, American (1913) and German (1919), had tackled it early on; in an American serial (1922–23), the characters use airplanes and submarines to get *Around the World in 18 Days*. Orson Welles, a great fan of Verne's novel, did three radio adaptations, and there would be a French radio play in 1948. Welles wrote a screenplay in 1941. When Hollywood producers were uninterested, he managed to convince Cole Porter to write songs for a staged version and Michael Todd to produce it on Broadway.[2]

Welles modeled his *Around the World* on Verne's stage play, which he had seen as a child. He rivaled the stupendous expense of that production with elaborate costumes, a thirty-four-person orchestra, and sets that required seventy-five stage hands. In early spring of 1946, reviews of the out-of-town premieres dwelled on the play's everything-but-the-kitchen-sink dimensions: "sensational—in every sense of that repugnant word"; "lavishly miscellaneous"; "et cetera, et cetera." One notable enthusiast declared, "This is wonderful. This is what theatre should be": Bertolt Brecht.[3]

A slightly streamlined version opened on Broadway in June 1946. The premiere had fourteen curtain calls and the play ran for seventy-five performances. Many people conceded that the show was amazing, sometimes in a good way. Welles had made old-fashioned silent films to cover scene changes—these were very popular. He played several roles himself, naturally, and did magic tricks, which everyone loved. But the miscellany of music, film, circus acts, card tricks, et cetera, struck many as odd. *The New Yorker* critic voiced a concern that "if God will forgive me, Cole Porter's music and lyrics are hardly memorable." (Porter's talents shone in the following year's *Kiss Me, Kate*, which is why people go around humming "Too Darn Hot" instead of "There He Goes, Mr. Phileas Fogg.") Everyone lost money, Welles especially. His multimedia variety act might have worked a generation earlier, in music halls and vaudeville, or it could have wowed a later avant-garde (as it did Brecht) as a kind of performance art. But in 1946, it seemed odd to complicate Verne's story with layers of extra fun.[4]

Welles was onto something, however, in his kitchen-sink approach, and in his idea to film the results. Alexander Korda had bought the film rights to Welles's play, and Welles did some location shots in Morocco, but nothing came of it. Michael Todd, who had abandoned Welles's stage project, bought out Korda and beat the project into existence, then into shape. (One of his weary scriptwriters, the humorist S. J. Perelman, described Todd to T. S. Eliot as "a combination of Quasimodo and P. T. Barnum.")[5]

Todd prodded viewers to compare Fogg's quest to state-sponsored expeditions, though not those of the past, which he may have considered too musty. A prologue instead features contemporary footage of rocket launches and clips from Georges Méliès's silent film version of Verne's *From the Earth to the Moon* (1902). Newsman Edward R. Murrow, who hosts the prologue, compares the age of steam to the era of rocket power, adding the Cold War homily that "speed is good only when wisdom leads the way" to use it on "this shrinking planet."

After the prologue, the film is faithful to Verne's novel, except when it isn't. Trapped in Paris, Fogg and Passepartout escape by balloon to the Mediterranean. *No,* yelp the purists; *not in the book!* But, after all, Verne had written several adventures involving balloons. A more interesting anomaly comes when Fogg and Passepartout transit the Suez Canal, the bit of Anglo-French engineering that had provided the swift passage to India essential to Verne's imaginary journey in 1872. And yet, by the autumn of 1956, when the glitterati turned out for the film's New York premiere, the canal was no longer that passage.

On July 26, 1956, President Gamal Abdel Nasser of Egypt had announced his intention to nationalize the Suez Canal. By September, Egypt fully controlled it. Beginning in late October, a joint French-British-Israeli force invaded Egypt to preserve Western authority over the canal. Although the three aggressors won the war, they lost the battle of international opinion. The United States and United Nations insisted on their withdrawal; France and Britain ceded any remaining claim to the canal.[6]

It was a local victory, and part of a global story. In social and political terms, the most significant development of the late twentieth century was decolonization. From the end of World War II onward, many parts of the old European empires steadily regained their freedom. The largest empire, the British, sustained the largest losses. (In the form of the Commonwealth of Nations, Great Britain retained an alliance of some of its formerly colonized zones.) It should be said that decolonization had British advocates. But for most white Britons, the receding empire represented an ebbing national confidence. Britain's lack of a manned space program, the new form of planetary domination, also hinted that it was no longer in the game.[7]

As empires withered, other international organizations bloomed. The three main variants have been political, corporate, and not-for-profit. They are not mutually exclusive. An around-the-world traveler who visited New York City in the 1950s and 1960s, for instance, could visit the United Nations, refresh herself with a Coca-Cola, and check into a YWCA; she was a citizen of the world, a participant in a global consumer's democracy, and a beneficiary of the old internationalism that had fostered Scout troops and Esperanto clubs.[8]

As a result, more people could go around the world, even though fewer possessed Fogg's personal wealth and sense of entitlement. Cheap, patched-together circumnavigations were popular, not because of a free-floating beatnik or hippie zeitgeist, but because the world was a patchwork. A new generation of vagabonds used any means necessary, including ships and boats, railways, and airplanes, but increasingly paved roads, which had proliferated beyond belief after World War II.[9]

Whatever their private politics, the English vagabonds showed that a British sense of global command died hard. John Lennox Cook planned a rather desperate getaway from a dead-end job in 1951 by going around the world by motorcycle, a conveyance he had never ridden—he didn't even have a driver's license. Nevertheless, Cook advertised for a companion, selected Tim Hamilton-Fletcher from the applicants, and got himself a motorcycle and driver's license. He began his story with lines from Robert Louis Stevenson:

> *We travelled in the print of olden wars;*
> *Yet all the land was green;*
> *And love we found, and peace,*
> *Where fire and war had been.*[10]

Wobbling between expeditionary and private modes, Cook and Hamilton-Fletcher traced Britain's fading global presence. Rather than stop at toll bridges, they simply "wore determined looks and shouted 'military' " and got through. They relied on actual army surplus in the form of British War Office maps. The map for Afghanistan included an area labeled "unexplored." In that mysterious zone, the two men prosaically lunched on tinned chicken. "Aftewards we drank beer and played table-tennis. For adventurers it was depressing." Substantial help came not from their nation but from corporate sponsors: the Dunlop tire company, and the Anglo-Iraqi and Anglo-Iranian oil companies, later folded into British Petroleum.[11]

In a 1955 speech, Prince Philip, the Duke of Edinburgh, invited more British subjects to prove their ability—still—to tackle the globe. He suggested that all British university students try to work their way around the world with initial capital of no more than £5, more like £100 today. When Alistair Boyd finished exams at Cambridge in 1956, he thought *why not?* (Prince Philip would do his own round-the-world tour in 1957; he probably would have liked to do it on £5, though royal protocol made that impossible.) Boyd quickly learned that cargo shippers had little interest in casual workers; airlines had none whatsoever. Tramp steamers were his only (and colorful) option. He got up the Amazon on a boat smuggling booze and tobacco. The smugglers' "hush money" doubled his capital. Irregular work continued to provide transport and cash. He did very well out of media appearances. On the Art Linkletter show in Los Angeles, he was a prize specimen: a new-style eccentric British globe-trotter.[12]

Boyd was consistent in preferring non-governmental actors to officialdom. He resented the border and labor regulations that frequently derailed him. And he felt the burden of British citizenship because he traveled during the Suez Crisis—"the enormous traffic block of war." He gave paid lectures arranged by two organizations founded in the heyday of internationalism, the English-Speaking Union (1918) and the British Council (1934), and he sometimes bunked with the YMCA and the Salvation Army, which didn't care what passport or work permit he carried. He concluded that the world was "wound round with red tape but filled with good people."[13]

That was Wendy Myers's conclusion as well. On her book's dust-jacket, our wandering heroine leans on a battered khaki rucksack. A Union Jack sewn on the sack's lower pocket is worn halfway off—perfect allegory. On the rucksack's top flap, the green triangular emblem of the Youth Hostels Association is fresh and whole, signifying an alternative: non-governmental, international associations, including the (prewar) YHA.

Myers represented her travels as her "seven-year hitch-hike around the world" (1960—67), though she moved by many means—ferry, bus, automobile, even airplane. She paid her way much of the time, extracting cash sewn into her brassiere, and taking odd jobs. She wrote scripts for Radio Ceylon, nursed in a Sydney hospital, taught English in Senegal, and sold shoes in Punta Arenas near the Strait of Magellan. As a British subject, she could use the Commonwealth Employment Agency to find jobs in Commonwealth countries, which is how she rebuilt her travel funds in Australia.[14]

Charm was as good as money. Photographs throughout Myers's account display her extraordinary adaptability. She dons a variety of national costumes, boards a range of transport, samples multiple cuisines, and always with new friends: Japanese pals in China, a U.S. serviceman in South Vietnam, a woman on a ferry crossing the Volta, journalists in Laos—on and on. Wearing a pleased-to-be-here smile, her arm is linked through her companion's, or hugged around a shoulder. Her openness guaranteed help. An Iranian family living in Japan paid for her train ticket to Tokyo. One man provided an airline ticket between Rangoon and Bangkok; another gave her judo lessons, in case of assault. She was only threatened with that twice, and only needed to fight once. "Kind people—all over," she concludes about the world. It's too naïve to be true, but its sense of postwar internationalism beats the imperialist alternatives, and indicates that people other than Myers also believed in it.[15]

Certainly, some vagabonds perceived the fall of empire as their opportunity to move about the world. 'Adnān Ḥusnī Tallū had served in the Kuwaiti armed forces against the 1956 invasion of Egypt during the Suez Crisis. That experience made him want to represent the Arab world to the West. He decided to do a circumnavigation by motorcycle, on a one-man goodwill mission. His sense of internationalism was apparent in his decision to become a Rover Scout. After creating the Boy Scout movement in 1908, Sir Robert Baden-Powell had founded Rover Scouting in 1918 for young men to perform public service and enjoy outdoor activities. As a Rover, Tallū gained support from the international Boy Scout Association during his seven-year journey—he represented Arab peoples and participated in an international brotherhood. His travels had their difficulties, but he was mostly impressed by the ease of going around the world: "Now we see travel across continents as if it was an excursion that we are invited to." (Writing long after his return, he further reflected that "we have even reached the moon, and descended upon it.")[16]

Aviation was uniquely suited to service the new internationalism, especially as airlines lost government subsidies and insignia. Qantas was a case in point. The airline had been an imperial mainstay—its three founders had served in World War I's Australia and New Zealand Army Corps, Anzac, and modeled their airline's crisp acronym on that. In its junior partnership with Imperial Airways, Qantas had dutifully moved mail and people along the Kangaroo Route, the world's "Longest Hop." But the airline's postwar leaders knew the empire was in decline and began to shed their military and imperial heritage.[17]

In 1957, Qantas's directors brokered rights to U.S. airspace and refueling points, creating a North American segment they could link to the Kangaroo Route in order to make a circle around the planet. World-round service, east and west, was inaugurated on January 14, 1958, with two piston-powered super Constellations named *Southern Aurora* and *Southern Zephyr*, which converged back on Sydney on January 20. Special first-cover postal memorabilia were issued for the flights (see illustration 22). The journalists and travel agents herded onto the planes provided priceless publicity. Qantas was able to maintain two weekly round-the-world services, one from Melbourne, one from Sydney. The Constellations did the circuit between five and six days, with seventeen stops for refueling and to release passengers who did not care to be circumnavigators. Round-trip tickets in first class ranged from 776.5 to 798.15 Australian pounds; tourist class cost 585. In 2011 U.S. dollars, these sums would range from $12,600 to $17,200.[18]

Qantas's achievement was unique. Congressional legislation still prevented Pan American Airways from offering world-round service; passengers had to transfer to a domestic airline, typically United, to cross the United States. Pan Am, as it was known after 1958, continued to petition the U.S. Civil Aeronautics Board for permission to offer domestic routes, and lobbied to counteract U.S. concessions to Qantas. Meanwhile, BOAC, the corporate entity that succeeded Imperial Airways (and would become British Airways), focused on the old imperial routes that stretched from Britain down through Africa and across to the eastern Pacific, leaving service over the Atlantic, Pacific, and North America to others.[19]

Although Qantas launched its new service to make money, its proprietary magazine announced a nobler goal: to unite the peoples of the world. If state-sponsored and private aviators had dropped that burden, civil airlines would take it up. Qantas's chairman, Sir Wilmot Hudson-Fysh, reflected that the "aviation industry has telescoped time and distance," and claimed that "rapid air communication between nations is the surest way to promote world peace and understanding." The journalists on the inaugural world flights had interviewed world leaders along the way, and Qantas reported that "the common theme of these views was confidence and hope for the future of mankind." Pope Pius XII said that Qantas "will indeed help to bring about a harmonious union among nations indispensable for the peace man craves."[20]

Having abandoned the British Empire, Hudson-Fysh did not wish to be tied to either of its successors, the United States or the USSR. To balance his U.S. commercial alliance, he visited Moscow in 1958 to get rights to Soviet airspace and stops for refueling. His contact with Aeroflot and Soviet authorities, made as Khrushchev was opening Russia to tourists, promised that Qantas's round-the-world passengers could go where they liked, free of empires old or new, members of a global consumer's democracy.[21]

Then, in October 1959, the airline announced the first around-the-world jet service. BOAC and Aeroflot had already introduced jets to civil aviation, another example of postwar army surplus; the predictable claim, as one U.S. headline suggested, was that jets would "Give Globe a Tighter Girdle." Qantas's jet service made it possible to circle the globe in 51 hours and 45 minutes, a historic acceleration (rounding down a bit) "from round the world in eighty days to round the world in 50 hours." The airline stressed jet service's unique blend of comfort and speed—worth the money—and gave its jet aircraft special decor: murals with Australian wildflowers, dome lights representing the stars of the southern hemisphere, Australian wool and leather on the seats. It was "like floating in your own lounge chair all the way from Sydney to San Francisco," one passenger reported.[22]

Although jets were not more fuel-efficient than pistons, it was thought that jet travel would be more cost-effective. "Speed doesn't cost, it pays," one commentator claimed, in a bid for business flyers. Leisure travelers, including female passengers and families, offered an even bigger market. One Qantas advertisement shows a woman and her son testing the stability, quiet, and speed of their jet aircraft with three objects: a coin could be balanced on edge, the tick of a watch was audible, and the posy on the lady's lapel would be fresh on arrival. Jet travel made it "twice as easy," Qantas claimed, "for the well-dressed woman to step from the aircraft as confident in her smart appearance as when she started her trip."[23]

At the end of 1959, Qantas celebrated a year of round-the-world service, nearly a year of jet service, and the twenty-fifth anniversary of the Kangaroo Route. On that route alone, it had flown 121,916,000 miles. That was 5,000 times around the world "or, in this age of space travel, 250 return trips to the moon."[24]

An around-the-world jet plane ticket was a luxury but, for consumers in the developed world, in the midst of a postwar economic boom, an affordable one. Qantas' timing was perfect. In 1957, passengers were evenly divided between sea and air services. By the next year, airlines carried 1,292,000 passengers over and around the Atlantic while shipping lines accounted for only 964,000. The balance would keep tipping. Commercial aircraft had carried 9 million passengers globally in 1945, but 1.17 billion by 1993.[25]

Other airlines began to offer world-round service, though usually in alliances. By 1957, Trans World Airlines (TWA) and Northwest Orient, as it was then known, jointly offered round-the-world service. When Pan Am laid on jet service, the *New York Times* explained that it would "girdle the earth except for the trans-continental United States." In 1962, Pan Am estimated that 50,000 people were circling the world in its aircraft each year and the airline, much more successfully than Qantas, marketed itself as a global icon, the Coca-Cola of the skies. Five years later, in July 1967, Pam Am announced what it called its 10,000th scheduled round-the-world flight. The airline presented its accelerated world-round service as the culmination of a historic speedup, from Magellan to Nellie Bly to Pan Am's gleaming silver jets.[26]

Airplanes would have seemed swiftly convenient in any case, and especially for an around-the-world journey, but they gained a critical boost because of repeated crises over the Suez Canal. Western shippers had to pay more to use the canal after its nationalization. When the conduit closed entirely during the Six-Day War in 1967 between Israel and its Arab neighbors, ships traveling between Europe and Asia had to round the Cape of Good Hope, as in the old days, and that was expensive. Air cargo had been costly, too, but now not so much, not in comparison.[27]

There were two niggling doubts. The first was time-zone fatigue, by 1965 renamed "jet lag." It initially puzzled civilians. When the American journalist Beatrice Cobb flew around the world in 1948, she couldn't understand her listlessness. A PAA stewardess confided that "she always has to go to bed for several days to recover from the effects of changes in food, water and climate." When he flew around the world in 1956, the Spanish writer Enrique Aguilar also complained that air travel prevented acclimatization. Steaming to the Philippines had taken forty days: "you would gradually get used to the heat. But now, with aviation, you pass from a cold climate to a tropical climate in a matter of hours, without transition . . . and this is what is unbearable." Qantas air hostesses were trained to offer "a meal," not breakfast, lunch, or dinner, because a jet passenger was on "Tummy Time." "Forget your watch," the airline urged its guests; "jet time is merely relative to enjoyment."[28]

"Enjoyment" was unlikely. To zip through time zones faster than the Earth turned was contrary to nature. When the Indian writer Anant Gopal Sheorey took an around-the-world jet tour in 1968, he concluded that "the jet age had made all travel so fantastically fast but, after all, the earth is an earth and the risks and imponderables involved in circling round it are inherent in the situation." As he crossed the IDL, Sheorey lost a day—"a very interesting phenomenon, isn't it?"—and confessed that he "felt absolutely tired." More pitiable than medically serious, jet lag was nonetheless a bad surprise for anyone who thought that a jet tour around the world should be the nicest circumnavigation in history.[29]

The other drawback was an immensely increased dependence on petroleum fuel. In the 1950s, the back of every Qantas in-flight magazine featured an advertisement for Shell Oil, a reminder of what made round-the-world travel so easy and glamourous, with all the dirty work of oil extraction well offstage. Aguilar noted that one energy-hungry technology seemed to require another: "in the century of the airplane, air conditioning is an almost indispensable and complementary invention" in order to compensate for lack of acclimatization between cold and hot places. Like no other material, petroleum seemed to clear the world of brute labor and fill it with conveniences. But, as recurring political instability in the oil-rich Arab states was beginning to show, universal dependence on petroleum meant global slowdown whenever oil failed to flow.[30]

At the very least, the new consumer democracy of the air allowed a more diverse set of travelers to become circumnavigators and to claim a place in the tradition of telling the circumnavigator's story. The Hungarian poet and priest Hieronymus Fenyvesi (or Fenyvessy) had long dreamed of going around the world. PAA's "great silver bird" granted his wish. Fenyvesi also flew the "white bird of Air France" and United's planes across North America, but only the PAA logo adorns the book's final illustration, a montage of the world's great sites. In 1962, Egyptian writer Anīs Manṣūr used jet service to follow a series of journalism assignments around a world that he concluded was a "big, rich book," though the swift flights through multiple global zones meant he spent his whole tour in summertime, never escaping the heat.[31]

Jet travel generated speedy "world tours" for many other political and cultural figures. Outgoing Philippine president Diosdado Macapagal did an around-the-world tour in early 1966, using a United Nations of airlines. Richard Nixon made his first round-the-world tour, as president, in 1969. Jet service let the Israel Philharmonic circle the world in 1960, Quincy Jones and some of his star recording artists in 1961, and Japanese Judo master Kazuo Ito in 1969.[32]

And yet there are surprisingly few narratives about jet circumnavigations. As with around-the-world cruises, the glossy advertisements had already told the story. When the British historian and government adviser Arnold J. Toynbee circled the world in 1956–57, he deplored the commercial airlines that made it possible. He learned that, on a world tour, "one may hope to see a fraction of the surface of the globe," and would have to fight "two deadly enemies" of global comprehension, "the capital city and the aeroplane," which kept travelers from what Toynbee considered to be the real world of country folk and countryside. Anīs Manṣūr disagreed. "I saw the world and circled the planet," he said; "I saw more of the world than pioneers of space who were like captives in metal barrels that race along at 28 thousand miles per hour at an elevation of 200 miles from the earth. Indeed, they saw the world from above, and I walked through it; they saw the forests and oceans and I saw the cities, the villages, and the people."[33]

But with 50,000 people or more going around the world in a year, the novelty soon wore off, the price of democratization. Enrique Aguilar admitted that while "a trip around the world, only a few years ago, was a novel thing," it was by the 1950s "almost child's play" and "one of the most vulgar things that can exist." A news item could quickly tell the story. For an issue of *Travel* magazine in 1967, Ira Wolfert quipped that flying around the world required 82,388 gallons of fuel, 60 of coffee, 36 of liquor, and "a ground staff of 286, stretched in a narrow belt around the world." Local newspapers sometimes ran stories on someone's once-in-a-lifetime circumnavigation, though rarely in any detail.[34]

It was too easy to parody. Enter S. J. Perelman, the humorist who sweated over the filmscript for *Around the World in 80 Days*. With the comic artist Al Hirschfeld, "who had done the Giant Swing once before," Perelman circled the world on airplanes and cruise ships and wrote up the result as *Westward Ha!, or, Around the World in Eighty Clichés* (1948). He and Hirschfeld returned, Perelman bragged, with "none of that rich harvest of serenity and wisdom" supposedly gained from a world tour. When Perelman made a second Giant Swing, recorded for posterity in *Eastward Ha!* (1977), he mocks the airplane, the "big silver bird," that cheapened the planetary experience of yore. When a fly alights his nose on the 707 taking him from New York to London, he incongruously remembers Slocum's spider aboard the *Spray*. Told that the morning flight had switched to London time and is serving drinks, he orders a double. At journey's end, he spoofs the Aouda-Fogg romance: he sinks "down on one knee—no mean achievement in economy class"—and pops the question to his object of desire. There is still neither serenity nor wisdom; a world circled quickly is not worth knowing.[35]

Yet again, the planet had been made invisible, just as Toynbee had complained. For most paying customers, a circumnavigation at jet speed represented a vague triumph over the short time it took to get around something glimpsed occasionally from a pressurized cabin. If it took little effort to get around the world by airplane, it took a lot to do something unusual with the experience.

Behold two circumnavigators who embarked from the San Francisco airport in 1963. A pair of young men in dark suits and skinny ties, they look like Peace Corps workers, or Mormon missionaries, but for their excellent tans and the surfboards they carry. It was November and the waters off northern California were cold. The duo's buddy, the surfer-filmmaker Bruce Brown, decided to send them, by scheduled airline flights, to a succession of warmer places. "Filmed in Actual Locations Around the World," *The Endless Summer* (1966) documented an unlikely planetary quest: Dakar, Accra, Lagos, Cape Town and Durban, Melbourne and Sydney, New Zealand, Tahiti, Hawai'i, and then a return to California. The movie's geopolitics have not aged well: a prologue features Hawai'i, with no hint that surfing was not originally the sport of young white Americans; the perfect wave exists in Cape St. Francis, South Africa, where apartheid goes unmentioned. The movie's planetary politics are harder to dismiss. Brown insists that the world exists for physical pleasure. To circle the planet is to maximize that pleasure. "With enough time and enough money, you can spend the rest of your life following the summer around the world."[36]

Ah, that seductive, misleading "you." Only a tiny fraction of the world's population could enjoy an endless summer of surfing, though it sure looks like fun. That cannot always be said about the sport that has redefined naval circumnavigations, competitive sailing.

Single-handed circumnavigations had continued in Joshua Slocum's wake and for Slocum's reasons, as antitheses to commercial travel. Small boats were well suited to athletic young men with a little money and a lot more time. Harry Pidgeon did an around-the-world cruise on *Islander*, westabout from Los Angeles, between 1921 and 1925, the first since Slocum. (When he toured a tourist steamship, Pidgeon declared it fit for "invalids and elderly ladies.") Frenchman Alain Gerbault made a westward circle in *Firecrest*, from 1923 to 1929. He passed and mocked Cunard's luxury *Franconia*, hauling Champagne and tourists around the world. *Firecrest* was a minnow among Leviathans—spectators gawked as she went through the looming Panama Canal. Another Frenchman, Louis Bernicot, circumnavigated in the *Anahita*, also westward, from 1936 to 1938. Each of the three sailors took his time, saw the sights, and benefited from the imperial order that had placed consuls and yacht clubs around the world. They knew their predecessors, including Darwin, Slocum, and London. In the Torres Straits, Gerbault chatted with a man who had met Slocum and seen lantern slides of his voyage.[37]

After World War II, the yacht was no longer the antithesis of the cruise ship but of the airplane. Small boats were sporty alternatives for circumnavigators who were willing to take their time and pay a bit more money. Even before the war, some married couples had sailed around the world and, after 1945, entire families would give it a try, fulfilling Lady Brassey's prediction that yachts would give children a global adventure under parental supervision. Eric and Susan Hiscock became, in 1965, the first couple to have made two circumnavigations (they went on to do a third). Suttie Adams set out from San Francisco with her four children in 1961, completing a circumnavigation by 1965.[38]

That sailing around the world had become so easy may have tempted certain sailors to make it harder, more like the bad old days. If kids could be made into crew, for example, why not into captains? On his departure around the world in 1965, a sixteen-year-old Californian described his intended solo journey in the first-person singular. "I want to see the world," Robin Lee Graham declared, "and not on a tourist's itinerary with a passport stamped full of one-day visas." His statement was a sharp (and perhaps intended) rebuke to what had just become another common way for young Americans to see the world: through military service in the Vietnam War.[39]

Graham had learned to sail on his father's boat, the prophetically named *Golden Hind*. He had an appreciation for circumnavigators, including Magellan, Cook, and Bligh, but especially Slocum. To do his own voyage, Graham acquired a 24-foot fiberglass sloop named *Dove*. *National Geographic* gave him an advance and he set off in the fall of 1965. After the shakedown leg from California, Graham left Hawai'i on September 14 with $75 in cash and five hundred pieces of secondhand clothing to barter for stuff he might need on the way. He had a small battery-powered radio to follow news and weather, and acquired a "Gibson Girl," a surplus World War II emergency transmitter, crank-driven. He stocked canned goods and other preserved food, plus fishing gear and a .22-caliber pistol. Two kittens were a parting gift from an uncle.[40]

The cats were essential. Graham had underestimated "the awful loneliness and exhaustion a singlehander has to face," which went on and on and on. He had only intermittent radio contact at sea and, because airplanes were replacing ships, he encountered few other sailors and had to be self-sufficient. He became expert at using sextant, chronometer, and nautical tables to calculate his position. When a shark nipped off the taffrail's spinner (part of the log that calculates the distance sailed), he shot it. When he needed money for some new gear, he worked as an electrical fitter's assistant for two months in Darwin, Australia. There, he came closest to encountering someone else sailing solo around the world with a new ambition: speed.[41]

"Long, long ago, as the best fairy stories begin," Francis Chichester explained, "I wanted to fly around the world alone." When flying was the very latest thing, in the 1930s, Chichester, a London businessman, had owned a de Havilland Gipsy Moth, the type of plane that the de Sibours had flown around the world. He converted his into a seaplane and set several records. In an attempted around-the-world flight, he made it to Japan, but was severely injured when his plane hit a cable. He took up long-distance sailing after another brush with death, a serious lung condition.[42]

He decided to make another round-the-world attempt, this time by sea and on a long-neglected route, "the old clipper way" (about which he had just written a book). He admired the clipper captains who had "experimented with" the winds for over one hundred years. He also knew that yachts had been circumnavigating the globe along the trade wind belts, though rarely via Cape Horn.[43]

Putting it all together, Chichester decided to make "a voyage round the world faster than any small boat had made before." He aimed to beat an Argentine, Vito Dumas, who had set a speed record while sailing 7,400 miles around the bottom of globe—technically not a circumnavigation. Chichester wanted to go even faster on a full circumnavigation, through two diametrically opposite points on the globe, which at sea would require a journey of about 30,000 miles. He also wanted to beat the clippers on their own route. The record for the run from Australia to England had been 123 days; he planned to do it in 100. To maximize his speed, he would go west to east, with the prevailing wind. (Slocum and Graham did it the hard way, against the wind.)[44]

Chichester, like Graham, was obviously disenchanted with the technologies that had made global travel faster and safer, yet blander. He asked an interesting historical question: why had people abandoned sailing for steam before determining how fast the wind could take them? The editors of the *Sunday Times* and the *Manchester Guardian* gambled that Chichester's experiment would boost circulation and each bought weekly radio dispatches from him. Chichester left Plymouth on August 27, 1966, a year and a month after Graham, sailing the 53-foot *Gipsy Moth IV,* purpose-built for the journey.[45]

Often, a circumnavigation begins well and then goes badly, but Chichester's went badly from the start. A leg injury, sustained before departure, never really healed. He was slow to get his sea legs anyway. "Seasickness is very anti-romantic," he confessed. On September 17, he celebrated his sixty-fifth birthday; four days later, he spotted the last ship he would see for two months. Once across the equator, the "spiritual loneliness of this empty quarter of the world" was indescribable.

His health deteriorated. He broke a tooth and couldn't fix it; eating enough to maintain his strength was hard. By November 5, he said, "I could not stand on my legs without support, just as if I had emerged from hospital after three months in bed." It was almost a relief when the self-steering mechanism broke: now he had to get help. Unfortunately, he was still 2,300 miles short of Sydney. Jury-rigged repairs let him keep up the speed necessary to set a new record, 107 days, England to Australia.[46]

A nice quiet collapse in Sydney would have been super, but within ten minutes of his arrival on December 11, 1966, Chichester was fielding questions from representatives of ninety-four different press outlets, the biggest press conference in the city to date. (Graham, who would arrive on the opposite side of Australia three months later, was untroubled by fanfare.) Then Chichester was free to convalesce, put back some of the 40 pounds he had lost, and get *Gipsy Moth* ready for the run home. Bad luck struck shortly after departure from Sydney. A rogue wave tipped the yacht over until her masts touched water. Everything inside was soaked or broken, though, miraculously, the boat was intact.[47]

The isolation was almost worse than the physical danger. To maximize speed on his inbound run, Chichester did not plan to stop anywhere (see Map 4). He had a "vast stretch of the Pacific ahead of me, 5,000 miles of the lonely rim of the world." (Even the single-handers who made stops dreaded solitude—Gerbault had taken over two hundred books.) At least the yawning expanse did not present medical hazards. Because he had read so many early narratives of circumnavigation (Drake and Anson, among others), Chichester feared scurvy and appreciated "the knowledge painfully acquired by humanity from the sufferings of those old seamen." His study of the clipper route showed that scurvy had continued to plague modern sailors, especially in cold climates. He ate strategically and steadily.[48]

His spirits were harder to manage. Unlike Graham, he carried no pets, but began to feed the albatrosses that followed him. He amused himself by racing the Sun as he rounded the Horn. Above all, he adapted to solitude by believing, on some level, that he didn't need anyone else in the world. That made subsequent encounters with people almost unbearable. Chichester grew to dislike the "girl reporter work-ing for the *Sunday Times,*" who, by interviewing him over the radio, was just doing her job. But the frequent radio interruption "poisons the romantic attraction of this voyage." He cringed when other ships came close and was horrified to find a flotilla of well-wishers waiting to ac-company him on the final stretch to Plymouth. He wished they hadn't bothered.[49]

But he was famous—of course they bothered. By the time he crossed a point on one side of the globe directly opposite another on his out-bound route, he had beat Dumas. And when he put in to Plymouth on May 28, 1967, he had made the fastest voyage around the world done by any small vessel. His average speed had been 130 miles a day, over 29,630 miles, with a total of three technical circumnavigations, those antipodean cross-points. It had required 200,000 words in eight log books to record the journey, plus all the radio dispatches. In all, he set seven new records.[50]

Francis Chichester's fairy tale, his romantic quest, was intensely appealing to the British public. The conservative *Times* and the left-leaning *Guardian* do not ordinarily share heroes, but their editors agreed that Chichester had revived Britain's flagging national confi-dence. His very provisions underlined his Englishness: beer on tap, 10 pounds of marmalade, and toothbreaking Kendal Mint Cake, the energy-rich stuff of Everest ascents. He had already been knighted while in Australia, but on July 7, 1967, he was dubbed with his title at Greenwich by Elizabeth II, who used the same sword with which Elizabeth I had enobled Drake, an earlier Sir Francis and prior circum-navigator. In the published account of the voyage, an epilogue thunders away at the convergence of myth and history. Just like "Drake, Anson, Cook and Nelson," Chichester exemplified "that incredible endurance that the people of England have shown when it was needed of them."[51]

But Drake's and Anson's voyages, unlike Chichester's (or Graham's), had been state-sponsored; the modern-day equivalent was not a lone yachtsman but an astronaut. Sailing alone around the world is derring-do, but it is DIY derring-do. The postwar rise in extreme adventure—off-road cycling, climbing Everest, and so on—showed how rising numbers of civilians could, for fun, flirt with dangers that imperial or national expeditions had once monopolized. Graham's "Gibson Girl" and Chichester's Kendal Mint Cake were good examples of that army-surplus repackaging, in which Western consumers repurposed militarism to express, if not empire over people, then over the planet.[52]

The excitement over DIY derring-do inspired a unique contest, the *Sunday Times* Golden Globe Race. Several sailors had clocked that Chichester went around the world with only the one stop in Sydney. Why stop at all? As one man concluded about competitive sailing, "That's about all there's left to do now." Aware that several sailors were preparing to do it, the *Sunday Times* offered a trophy to whoever was first to sail around the world single-handed and non-stop, and £5,000 to the fastest. Contestants had to leave England between June 1 and October 31, 1968, had to go around both southerly capes, and could neither anchor nor receive any material assistance. A well-wisher's foot on the deck or a spare part tossed aboard would mean disqualification. Words were the only acceptable form of outside contact. Because sea traffic was declining, and because it could take eight hours for a ship's radio signal to be heard on shore, contact was guaranteed to be infrequent.[53]

The three best-known contestants had different responses to the challenge. The man tipped as the winner, Frenchman Bernard Moitessier, already famous as a yachtsman and author, went to sea to escape humanity. He declined a free radio, preferring a signal mirror for Morse code and a slingshot to hurl written messages onto passing vessels. He shuddered at the race's frank publicity, disgusted that sailing was becoming commercialized. At the other extreme, British engineer Donald Crowhurst, a weekend sailor, entered the race to advertise his flagging electronics business. In between those opposites was Robin Knox-Johnston, a sociable fellow, Navy man, and patriotic Briton—he went to sea for queen, country, and Commonwealth. Of the nine entrants, these are the three to watch. They generated the most interesting accounts of sailing alone around the world.[54]

Alone at sea, Moitessier withdrew into himself, his boat, and the planet he circled, two microcosms within a macrocosm. *Joshua,* named for Slocum, is his "little red and white planet made of space, pure air, stars, clouds and freedom in its deepest, most natural sense." He measures himself against "the curve on the little globe" upon his chart table and marvels at the "fullness of body and mind" he can achieve "after five months in a closed system." He is startled to see a clean, sawed plank, a drifting reminder of the crafted world of humanity. He finds companionship in other creatures, especially the birds that follow him, which he feeds and wishes he could stroke. (Lucky birds, to have found a sailor free with the cheese, butter, and pâté.) However much Moitessier wanted to escape ordinary sociability, he longed for news of his brother competitors. By definition, they were some of the hardest people on Earth to reach. Circling the globe in the same race, they were each "alone facing infinity." [55]

He rather liked infinity. Upon completing his circumnavigation, recrossing a longitude in February 1969 he had passed the previous September, he was reluctant to finish—and probably win—the race. He and *Joshua* had "sailed round the world . . . but what does that mean, since the horizon is eternal?" It was ridiculous to measure the planet for a race, for sport: "round the world goes further than the ends of the earth, as far as life itself." To accept the terms of the contest would be to accept "the modern world—that's the Monster. It is destroying our earth, and trampling the soul of men." He decided to forfeit and keep going until he felt like stopping, which was in Tahiti, two months after the race had ended.[56]

Knox-Johnston stayed in the race. He would publish his account of the experience as *A World of My Own*. That "world" could refer, equally, to his boat, *Suhaili*, or to the globe he circled alone. One night, when *Suhaili* tips nearly sideways in the water, Knox-Johnston is "jerked awake by a combination of a mass of heavy objects falling on me and the knowledge that my world had turned on its side." Unlike Moitessier, he missed human contact and raged when told that the rules had been altered and he was no longer allowed physical mail, even family correspondence. He felt consigned to a terrible silence. As he rounded the Horn, he tuned in to the British Antarctic Survey radio stations and was solaced to hear English voices. He was a rapt solo audience for their technical updates. "It was fascinating," like listening to a radio play.[57]

On Christmas Day 1968, when Knox-Johnston picked up some

U.S. commercial radio stations, he heard a broadcast from the crew of *Apollo 8*, the famous message with readings from Genesis. He felt humbled:

> The contrasts between their magnificent effort and my own trip were appalling. I was doing absolutely nothing to advance scientific knowledge. . . . Nothing could be learned of human endurance from my experiences that could not be learned more quickly and accurately from tests under controlled conditions.[58]

He was too modest.

The non-stop boats, not unlike spaceships, were experiments in life support within a "closed system," as Moitessier had pointed out. To be independent from land, each vessel required a crucial balance among boat, sailor, and supplies. Food stores consisted of heavy cans, the old military technology—freeze-dried foods, the new government issue, were not yet available to consumers. (Moitessier's fatty snacks for albatrosses all came from tins.) Damp sea air is bad for cans. Paper labels slip off, so that meals become nutritionally unbalanced surprises, and rust weakens seams. For long voyages, each tin's label must be removed, the contents noted in paint, and the can varnished against rust. It had taken Knox-Johnston and five assistants "four boisterous evenings in the garage" to process his 1,500 cans of food. Carrying surplus added safety, but slowed down a racer. Moitessier jettisoned 375 excess pounds, including twenty-five bottles of wine, 30 pounds of a jam he didn't care for, extra rope, and four cans of kerosene. Water was too bulky to pack; sailors captured rain when they could.[59]

The intricate calculations resembled those that went into spaceflight and submarine travel, both of which descended from the Cook plan, the historic effort to wean sailing ships from land. Again, the boundary between military and civilian efforts was blurring: the Golden Globe contestants were not state-sponsored, yet they successfully operated within the isolated, closed systems historically associated with military expeditions. Moitessier and Knox-Johnston were at sea for ten months apiece, scurvy-free, teeth firm in their heads.

True, unlike astronauts, sailors did not have to ration air, along with food and water. On the other hand, they outdid the spacemen in their extended solitude. Knox-Johnston wondered "how the crime rate would be affected if people were sentenced to sail round the world alone, instead of going to prison. It's ten months solitary confinenent with hard labour." In contrast, the first real test of orbital duration would come in 1970, when two Soviet cosmonauts stayed in space for two weeks. Even as space flights grew longer, they were made in groups and with frequent contact with ground support—crowded and chatty, compared to sailing alone around the world.[60]

Deprived of humanity, Knox-Johnston, like Moitessier, became keenly aware of his place within the natural world. He hated sharks, and shot any that came too close, as when one visitor began to scrape the anti-fouling paint off *Suhaili*'s hull. But he welcomed albatrosses and dolphins. He was afraid of whales—a head-butt or tail-smack could be dangerous—but confessed, "I always felt a little lonely when the whales left." In the Sargasso Sea, he improvised an aquarium for captured crabs, and tenderly fed them tinned sardines.[61]

The tenderness showed the greening of the circumnavigator's con-science, another parallel with spaceflight. If astronauts' worries over the planet had been largely inspired by fear of atomic holocaust, mariners were more concerned to conserve natural resources and maintain eco-logical balance. Knox-Johnston's library on the *Suhaili* included two books on the sea by Rachel Carson, famous for her *Silent Spring* (1962), with its criticism of the human impulse to control the natural world through pesticides. Moitessier's fear of "the Monster," the modern civi-lization represented by "the bulldozer and the concrete mixer," showed a similar turn.[62]

Donald Crowhurst, in contrast, gambled on new technology to win the Golden Globe. He had mortgaged his home and business to build an experimental trimaran, a boat with one main hull and two outrig-gers; he needed to finish the race to fulfill obligations to the press and creditors, and hoped to win the prize money. But unlike Moitessier and Knox-Johnston, Crowhurst had no publication agreement in place be-fore he set out. Telling a coherent story about himself was not part of his mission, and the omission may have been fatal.[63]

Under pressure to make the contest's deadline, Crowhurst set out in his trimaran, which began to fall apart almost immediately. He hesitated to return, but feared to continue. He thought he might be able to bluff his way home by giving false positions in his radio reports, then falling silent when he was presumed to be beyond the Cape of Good Hope. He would lurk in the South Atlantic and fall in behind the winner or winners once they had rounded the Horn. (Finishing without winning would, he thought, prevent any close scrutiny of his log.) For a while it worked. The *Sunday Times* printed his self-reported positions, a fake course plotted behind the front-runner, Nigel Tetley. But when Tetley's trimaran sank in the North Atlantic, on its return, Crowhurst knew he would be exposed. He vanished at sea, a presumed suicide. Knox-Johnston was the only contestant to finish and therefore win the race. When a fund was organized to help Crowhurst's bankrupt family, Knox-Johnston contributed his £5,000 prize.[64]

It is meaningful when humans conquer the planet, though no less so when they cannot, and it might have helped had Crowhurst realized that. There is an honorable tradition of failed circumnavigations. Yachtsmen respect Jack London's attempt—Moitessier had named one of his earlier boats the *Snark*. Crowhurst could have written about defeat, and could have emphasized the experimental value of his and Tetley's trimarans, which have been vindicated in subsequent races. Instead, his surviving log, found on the abandoned trimaran, cribs from published accounts of successful round-the-world voyages, especially Chichester's. If he had wanted to convince people that he had gone around the world, he must also have prepared a false log, though it vanished with him. The writing that most consumed him, however, was a bizarre fantasy about physics and the hidden nature of the cosmos. It was an escape from the real world of nature that absorbed and sustained the other yachtsmen in the race.[65]

Given the tragedy that marked the Golden Globe Race, Robin Lee Graham—the teenager who was still making his way around the world—seemed wise to go slow, to use global distance to grow into himself. The three stories Graham filed with *National Geographic* emphasize his youth. He kept losing his cats, but always replaced them, warm patches of furry insulation against a windswept loneliness. He tape-recorded a diary and read whenever he could. "I never could have stayed sane without books." His reading list—Tolkien and Ayn Rand, among others—is another reminder of his age; kids on land read these same books. He named some of his cats for favorite characters, including Piglet and Pooh. Still, he could kill sharks and handle a boat with the best of them. When a storm in the Indian Ocean broke his mast, he jury-rigged another and sailed 2,300 miles to Mauritius in twenty-three days, his original goal. Here *National Geographic* came to the rescue by flying him a new mast—via Qantas.[66]

And yet in its domestic cosiness, Graham's *Dove* resembled the Brasseys' *Sunbeam* as much as Slocum's *Spray*. At Suva in the Fiji Islands, he met a fellow Californian, Patricia Ratterree, who was also going around the world. They reconnected and married in South Africa. Despite his father's concern that he might abandon the solo circumnavigation, Graham carried on, meeting his wife at anchor. He stepped up his search for animal companions, with special care for any that needed it. When the newlywed Grahams encountered a pelican too injured to feed, for example, they caught it, sutured its ripped pouch, and wired together the eroded bones of its bill before releasing it. Graham's final cats were two ailing strays. He got them to a vet and named them for a pair of Tolkein's dwarves, Fili and Kili. Fili developed rather good sea legs, considering she was blind. The sightless hussy managed to sneak ashore in the Canal Zone, locate a tomcat, and sashay back full of kittens for the *Dove*'s lonely master.[67]

After acquiring a new boat, *Big Dove*, in the Virgin Islands, Graham returned to California on April 30, 1970. During his world tour, he had aged five years, grown over an inch, married, and was expecting a child. The timing of his return was legally significant—he had passed his twenty-first birthday.[68]

His five-year odyssey redefined circumnavigation for young people, another way that expeditionary ordeals were transformed into sports. Graham knew that young men had always gone to sea. (In the Galápagos, he realized that Charles Darwin had been only six years older than he was during his scientific visit to the islands.) For centuries, boys had served rough apprenticeships on ships. But Graham was his own master for a whole round-the-world voyage, an achievement chronicled not only in *National Geographic* but also a subsequent book and film. His example has proved liberating for other young people, who can now sail around the world as a leisure activity, not in servitude.[69]

But it has not been a liberating choice for everyone, as seen in the official lists of single-handed sailors kept by two organizations. The Joshua Slocum Society International (JSSI, 1955–2011) was founded "to record, encourage, and support long distance passages in small boats"; the World Sailing Speed Record Council (WSSRC), was set up in 1972 by the International Yacht Racing Union (now the International Sailing Federation) to determine new sailing records. Since 1972, no official maritime record can be claimed unless a voyage has been registered with the WSSRC and monitored en route.[70]

The two organizations' lists of sailors do not represent a random cross-section of humanity. Of the sixty-seven round-the-world single-handers recorded by the JSSI, the majority, thirty-seven (55 percent), are from the English-speaking world: twenty-two from the United States, eight from Britain, three from anglophone Canada, and four from Australia or New Zealand. (This does not include an Irishman.) Only one, Vito Dumas, is from Latin America. Some are European (ten Frenchmen, four Germans, four Poles, a scattering of others), even one Russian. Another circumnavigator is Israeli, and three are Japanese. The WSSRC's list is similar. It documents all nations that have had record-breaking solo circumnavigators, a total of 8, a small minority out of the 192 member nations of the United Nations. Most single-handers come from the northern hemisphere and all from industrialized places in the temperate zones. The overwhelming majority are male.

Around-the-world sailing is a useful rebuke to the easy loops that airplanes put around the planet, but the expeditionary heritage of circumnavigations survives. Single-handed circuits mostly interest people, mainly men, from nations that had significant—and voluntary—involvement in around-the-world travel during the age of empires. An imbalance in material resources and proprietary confidence has survived decolonization, whatever the redistribution of army surplus.

"Of course, he had no way of knowing that Fogg was traveling to save the world, not just to girdle it. . . ." Philip José Farmer's *The Other Log of Phileas Fogg* (1973) offers "the Cosmic Truth Behind Jules Verne's Fiction!" It is a seminal text in steampunk, the trans-historical mash-up of brass-bound Victoriana and sci-fi fantasy. Farmer reveals that Fogg cultivated his reputation for dullness to hide that he was in reality a superhuman extraterrestrial battling rival extraterrestrials. For some people, including Farmer, going around the world was itself no longer interesting. It had to be jazzed up with spaceships. The augmentation matched a continuing admiration for space travel as the ultimate in human endeavor, perhaps best exemplified in Tom Wolfe's *The Right Stuff* (1979), a semi-comic valentine to the U.S. manned space program.[71]

By 1968, it was thought that commercial travelers might venture onto spaceships, too. Pan Am's waitlist for lunar trips eventually tallied 93,005 hopefuls. In Stanley Kubrick's *2001: A Space Odyssey* (1968), an official leaves Earth on a commercial aircraft to investigate some strange events on the Moon. (That part of the movie is based on Arthur C. Clarke's "The Sentinel.") The investigator takes a Pan Am shuttle to a way station in orbit, then another shuttle to the Moon. People who see the film today snicker: 2001 came and went without commercial space traffic.[72]

Other fantastic planetary circuits were possible, however. In 1971, two scientists used scheduled commercial jet flights to test Einstein's theory of relativity. The special theory of relativity hypothesized that a moving clock would record time more slowly than one in a fixed place; the general theory of relativity predicted that gravity would have a similar effect, proportioned to the difference in gravitational potential affecting a flying clock versus one on the ground. J. C. Hafele and Richard E. Keating arranged for four atomic clocks to jet around the world, first eastward, then westward. They predicted that the flying clocks would not only differ from a ground reference clock, but also "depending on the direction of the circumnavigation": going with the Earth's rotation would produce a "time loss" while flying against it would represent a gain. They were right. Relative to the atomic time scale at the U.S. Naval Observatory, the eastbound clocks lost 59 nanoseconds (give or take 10 nanoseconds) while the westward timepieces gained 273 nanoseconds (plus or minus 7). Hafele and Keating thanked the agencies that had handled the expedition: Pan Am, TWA, and American Airlines.[73]

In October 1977, Pan Am organized its own unworldly expedition in the form of a polar circumnavigation, a special Flight 50 to commemorate the company's fiftieth anniversary. (With the sponsorship of two aviation companies, two TWA pilots had done the first circumpolar flight in 1965; their Boeing 707, *Pole Cat*, flew the 26,275 miles in fifty-one hours, with four stops for refueling.) Pan Am's 747 *New Horizons* completed the loop in just over forty-eight hours, with first-class tickets priced at $3,333 and economy at $2,222. (Commemorative stationery showed Pan Am's proprietary globe with a polar circuit indicated.) Passengers were astonished that, as they passed over the pole, the Sun was first on one side of the plane and then the other. This was a one-time event, but another aircraft offered an out-of-this-world experience on a daily basis.[74]

Oh, to be among the lucky few who know the dreadful truth about the ultimate in luxury travel: there was no legroom on Concorde. Cruising at around 1,350 mph—over twice the speed of sound—that aircraft, jointly operated by British Airways and Air France, offered the only supersonic ride money could buy, a stupendously fast adventure that was otherwise the preserve of military test pilots. During its commercial run from 1976 to 2003, Concorde was Europe's answer to the Vostok and Apollo programs. It outran the course of the Sun, covering the five time zones between London and New York in less than three and a half hours. Under those conditions, weight and space were at a premium, and the passengers a bit squashed together.[75]

Although most Concorde flights bolted between New York and either Paris or London, BA and Air France offered occasional treats, including round-the-world tours. The first circumnavigatory test flight, made in late 1986, took Concorde just a minute under thirty hours of flying time to cover 28,238 miles. Stops for passengers to enjoy de luxe locations brought the world tour organized for that year up to a span of eighteen days. A second tour in 1987 cost £15,000. The first circumnavigation to be offered from the United States, also 1987, was made "in the spirit of Jules Verne's Phileas Fogg, but in the ultimate of luxury." Indeed, indeed: tickets for the 1988 U.S.-marketed tour started at $32,995. On another global sweep, passengers could sip wines from the parts of the world they visited, plus four different kinds of Champagne. "Mrs P. L." gushed her approval: "maybe one day I will come back to earth again."[76]

And every once in a while, British Airways or Air France would use a Concorde to boom around the planet just to see how much faster they could go. An Air France Concorde circled the world westward, including stops to refuel, in 32 hours, 49 minutes and 13 seconds, setting off from Lisbon on October 12, 1992. On this voyage, faster than any commercial aircraft had ever circled the globe, the Sun never set. Air France reduced the record three years later, down to just over 31 hours in an eastbound flight.[77]

And yet, in a widely repeated pun, supersonic transport never quite took off, not commercially. It failed to expand because airline travel was no longer a luxury. The revolution in transport was not sleek Concorde, the glamorous almost-spaceship for the few, but the 747, introduced in 1968 as the sturdy omnibus of the skies, progenitor of all other jumbo jets. Economy class expanded faster than Concorde could fly. The trend accelerated several times. U.S. deregulation of airlines let upstarts compete with the older airlines that had once enjoyed special national status. And the repeated oil crises of the 1970s made economy class the preferred option.[78]

The future of commercial aviation has not been futuristic. Instead, it has been "no-frills," the apt designation for ever cheaper flights, with their vanishing services and amenities. More people can fly, but the future of consumer democracy, in the air, might be having to pay to use a toilet, the wretchedness of which would be absolutely apparent if done all the way around the world. Even worse, civil aviation achieves no greater social good. Repeated studies have failed to find any correlation between world peace and long-distance airline travel.[79]

Slow Food gets all the attention but slow travel predates it. The rebellion against fast food began in 1986. Decades earlier, an assortment of slow neo-Foggs had been determined to dawdle round the world, a great reversal of the global acceleration that Verne had blessed.

To retain some of their shrinking market, shipping lines presented slowness as an advantage. In 1955, making a virtue of necessity, British-owned Cunard offered a world-round tour that avoided the newly troubled Suez Canal and went via Cape Town, the old "tavern between two seas." Norwegian-America Line also organized world tours that omitted the Suez Canal. By the 1970s, American President Lines sold around-the-world cruises that used neither the Suez nor Panama canals; the Magellan Suite went for $20,580. (The cheapest fare was $3,235.) Airplanes may have been faster and less expensive but, as one cruise organizer pointed out in 1961, ships were more luxurious. For passengers with more time than money, students and retired people especially, cargo shippers began to offer no-frills accommodation; in the 1960s, around-the-world cargo ship fares ran between $1,200 and $1,400. (Accounts of this kind of tour stress the places visited, not the featureless accommodations.) And, it had to happen: shippers organized cruises within Fogg's eighty-day schedule. Holland-Amerika offered that standard tour annually in the 1960s, an almost-circumnavigation of 15,967 miles, with twenty ports-of-call for sightseeing. In the 1970s, their cruise expanded to eighty-nine days— "Don't let anyone tell you the world is getting smaller." [80]

As the planet regained its girth, an array of civilians tested themselves against it, sometimes prompted by a new authority, the *Guinness Book of Records*. Originally conceived after a disagreement among British sportsmen over which gamebird flew fastest, the book became a surprise hit. (By its fiftieth anniversary in 2005, it was the best-selling copyright book in the world, with sales that topped 100 million.) The *Guinness Book* states that any record-setting action must have two witnesses and press coverage; different kinds of feats have specifications to prevent cheating, cruelty to animals, and exploitation of other humans, especially children. [81]

From the *Guinness Book*, Minnesota native David Kunst discovered that there was no record for walking around the world. "I could hardly believe it," he said—"Man had been to the moon, but he hadn't yet walked around his own planet." Kunst set out with his brother, John, who suggested that they raise awareness of UNICEF on their way. The brothers carried a World Citizenship Declaration from Minnesota and "Planetary Passports" from an NGO, documents that did not have the authority of the old Nansen passport, but indicated confidence that the world's future would be transnational. The two men distributed

UNICEF pamphlets along their route. That part of the plan went awry when bandits in Afghanistan, thinking the brothers were collecting money for UNICEF, waylaid and shot them. John died.[82]

David Kunst went home to bury his brother and convalesce, but he itched to complete his walk. In an interview in his hometown of Waseca, a Mrs. James Sybilrud criticized him for leaving a wife and three children in the first place, then continuing to pursue the world record after his brother's death. "Thank you, Mrs. Sybilrud," Kunst remarked. "It's you I'm leaving behind. It's you I'm walking away from." He returned to Afghanistan and carried on. Another brother joined him for part of the trek through Asia and Australia. Kunst also went through four mules, each of which pulled a small wagon of supplies. The mules suffered so badly in hot climates they had to be flogged to stagger on. When his Australian mule died, David Kunst convinced his girlfriend (later his second wife) to pull the wagon alongside him with her car. It was all properly miserable, a reminder of the size of the world and of the surface obstacles that airline passengers could avoid.[83]

And yet the English journalist Ted Simon was anything but miserable during his slow motorcycle circumnavigation from 1973 to 1977. "In spite of wars and tourism and pictures by satellite, the world is just the same size it ever was," he happily reported. Simon derided airplanes for making travel bland and the places it connected indistinguishable, as if travelers were sucked into one end of a silver tube and spat out the other. He quoted Henry David Thoreau's criticism of the rush to connect Maine and Texas by telegraph, even though Maine and Texas might "have nothing important to communicate." Information wasn't knowledge. (The *Sunday Times*, which bankrolled Simon, might have disagreed.) To *know* the planet, "you have to stay on the ground and swallow the bugs as you go. Then the world is immense."[84]

Simon appreciates the comic mismatch between his huge global goal and his small, often hapless self. When he drops by the *Sunday Times* office to collect some last-minute "oddments," he sees the story of his departure announced in the proof of next day's paper: "Ted Simon left England yesterday," as if he no longer existed where he stood. Later, in India, he thinks a soothsayer has declared him to be Jupiter. "Why not?" Simon preens, pleased to be planet-sized. Later, he is deflated—the man had been referring to Jupiter's influence in Simon's personal zodiac.[85]

Whatever his outsized ego, he knew his limits. Counseled to carry firearms, he refused. "Merely to carry a gun invites attack," he concluded, and, truly, he was never in a situation where it would have helped. As it turned out, his best weapon was a clipping from the *Sunday Times* that had announced his departure and bore his picture: "it opened more doors than my passport did."[86]

A motorcycle turned out to be a good way to savor the size of the world. Simon patriotically chose a British-made Triumph 500-cc Tiger Hundred; he thought of it "as constituting a sort of space capsule that could travel at will." When he had cycled down through Europe, taken the ferry to Tunis, and then done the length of Africa, he had an "extraordinary flash" he was never able to achieve again. At rest in a café, his brain registers "the whole of Africa in one single vision, as though illuminated by lightning." He can also imagine "the real circle" he intends to put around the whole Earth. "I would have laid my tracks around the surface of this globe," he predicts, "and at the end it would belong to me, in a way that it could never belong to anyone else."[87]

He loves it when others appreciate the size of the world, and therefore his effort at encircling it. An Irish doctor, working in a hospital in Kenya and following Simon's progress in the *Sunday Times*, had invited the motorcyclist to visit, if possible. And one incredible day, Simon is there. In South Africa, Simon tells someone at a petrol station that he has come from England overland: "Oh! It is too big," the man demurs. And as Simon had vaguely promised, he shows up with his battered bike at the Triumph showroom in Los Angeles, though without warning the staff that he had reached the United States. " 'Holy cow,' they say. 'It's Ted Simon.' "[88]

Simon gains, as well as a sense of the planet's immensity, an appreciation of the resources necessary to keep him zooming around it. He became a connoisseur of road surfaces, and expert at classifying the falls he took: merely annoying, actually painful, or medically serious. After he had estimated his fuel costs, the Yom Kippur War sent oil prices into orbit—there went his budget. He packed Michelin maps, which bore little gas-pump symbols to show where he might slake his steed. Sometimes, he had to buy small amounts of petrol siphoned out of people's cars, one time out of a lawn mower. In Mexico, where buses and trucks spewed diesel fuel, his "face and clothes became black with oily droplets."[89]

It all gives Simon a wary appreciation of the parts of the world that are clean, comfortable, *easy*. Comfortless and begrimed in South America, he fantasizes about the developed world's material plenty. It is as if a "television set" switches on in his head and flashes relentless images—culled from the consumer democracy absent in rural Latin America—of cold beer, hot food, real coffee, and "a perfect and most loving woman in a large, clean bed." The United States is the cleanest and most comfortable stretch on the route. And yet, once he's there, Simon misses all the parts of the world where "children came running" to see Jupiter on his Triumph.[90]

What may be most remarkable about Simon's story is its refusal of nostalgia, either for empire or for pre-modern technology. He likes how decolonization has rendered him a traveler rather than invader. He may hate planes, but a motorcycle, after all, suited Marinetti the Futurist. Not for Simon the pith helmet and the penny-farthing. All things in moderation—only the Earth is allowed to maintain its outrageous proportions.

Most of the doubt over of a world swiftly circled by jets has been nostalgic. That was certainly true of the last great expeditionary circumnavigation. From 1979 to 1982, the British explorer Ranulph Fiennes led a team around the world, through both poles, and without use of aircraft (see Map 4). Yes, the Transglobe Expedition did everything the slow hard way: over the Sahara (sand, sandflies, and vipers); through the Roaring Forties (seasickness); across Antarctica in sixty-four miserable days (frostbite all round and, because of the need to cook and generate heat, a constant risk of fire); by sea through the Pacific (rather pleasingly uneventful); four weeks threading the Northwest Passage (never done as part of a marathon global expedition); then over the north pole (see above: frostbite and fires). Fiennes made the agonizing circuit at the suggestion of his wife, Virginia, who noted that no one ever had done it before. Nor since. Humans have traveled to the Moon multiple times, but Fiennes's transglobal feat has never been duplicated.[91]

It is all the more remarkable, given that the Transglobe was a semi-DIY expedition. True, Fiennes did have royal patronage and support from the Royal Geographical Society, though this fell short of state sponsorship. And he and his two teammates had military training and used military and state connections to do much of the planning and coordination for the exercise. (Delivery of ground support and the distribution of supply caches meant that at least one other circle was inscribed around the planet, along with the official one.) But sponsorship came from over six hundred agencies, most of them private. And the expedition had to do without support from the armed forces while in Antarctica, where the Antarctic Treaty forbade anything resembling a military exercise.[92]

As with Chichester's solo circumnavigation, Fiennes's expedition warmed the hearts of Britons who thought the loss of empire should not reduce Britain's place in the world. Fiennes in fact compared himself to Chichester. He also sought Robin Knox-Johnston's advice for his expedition's sea segments, and noted that a transpolar expedition, unlike a voyage to the Moon, had never been done. Prince Charles, the expedition's royal patron, called it "a mad and splendidly British enterprise." The *New York Times* concluded that "the British aren't so weary as they're sometimes said to be" and that the expedition "makes one wonder how the sun ever set on the empire."[93]

The English writer Nicholas Coleridge agreed and, in a smaller way, set out in 1984 to show the lasting beauty of a transit around the Earth on its surface, in homage to Phileas Fogg. Coleridge knew that Jean Cocteau and S. J. Perelman had already emulated Fogg, but without historic authenticity. He decided to go around the world as close to Fogg's route as possible, from the Reform Club and back, without the use of airplanes.[94]

It must be said that the plan would have puzzled Verne. His novel had been about maximizing velocity on a global scale; his extraordinary voyages were studded with equally extraordinary technology—he had claimed Fogg was in orbit. But Coleridge was interested in the age of steam's slowness rather than its speed.

The ebbing of Victorian technology and imperial geography made the journey harder than Coleridge might have expected. By 1984, the route Fogg had taken went through nineteen countries, only one of which, Hong Kong, still had a British governor-general. A British passport, Coleridge complained, no longer eased transit. He learned that people in the former British colony of Singapore still read *Around the World in Eighty Days*, though they considered a re-enactment of it "wilfully reactionary." They had a point, given that air travel had steadily eroded ground transport everywhere. In two instances, between Cairo and Port Said, and then between Suakin (the Sudan) and Djibouti (Ethiopia), Coleridge was stuck. With no elephants to purchase, he hired taxis. He took a total of ten ships and, unlike whist-addicted Fogg, was unprepared for the tedium of ocean crossing. Had he read the narratives of the single-handers, he would have known to pack books. Instead, on the dhow he took across the Arabian Sea, he lingered over every word of an old copy of *Newsweek International,* even the masthead. He was "hoarding a paragraph about zip codes, printed in microscopic type, for a rainy day." As it happened, he got back in seventy-eight days.[95]

But here's a crucial question: stretched end to end, how many pythons would it take to go around the world? In 1989, the answer was *two*—if they were Monty Pythons. That year, the BBC did its bit for the ongoing Verne revival by broadcasting two versions of *Around the World in 80 Days*, each starring a former Python.

One program was a dramatization of Verne's novel, the last (so far) to be pitched to adults. Eric Idle played Passepartout to Pierce Brosnan's Phileas Fogg. A great many French actors must have deplored the use of an Englishman with a fake French accent. But there is an intriguing tradition of mismatch between the actors and the characters they play in Verne's story.

For his movie, Michael Todd had cast the Mexican actor Cantinflas (Mario Moreno) as Passepartout. A brilliant choice, it guaranteed that the film played very successfully throughout Latin America. On the posters for the Latin American premiere in Caracas, Cantinflas is billed above the English David Niven, who played Fogg. But Todd did not dare cast a South Asian actress against Niven's Fogg—Shirley McLaine played Mrs. Aouda. Staging a romance between a white actor and an Asian actress would have been risky for a major feature in 1956. On the other hand, McLaine's Aouda authentically proposes marriage to the repressed English circumnavigator, and her action, as in the novel, reveals the gift of an extra day. Neither the casting, nor the lighthearted look at the British Empire, elicited overt criticism. *The Times of India* recommended the film on the eve of its Bombay premiere in 1960. Over time, however, the story's imperialism has become less attractive.[96]

By replicating yet criticizing the original story's imperialism, the 1989 BBC series tried to have it both ways. The writers added a segment, for instance, in which Fogg and party are toted in sedan chairs. The series did have an Asian actress for Mrs. Aouda, who gives anti-imperial speeches to Fogg. When "Indians" attack their American train, she refuses to shoot back. But this outspoken Aouda somehow cannot propose to Fogg. Instead, Pierce Brosnan sinks down on one knee, just like S. J. Perelman in economy class. The female character can now be Asian, or bold, but not both. (In Philip José Farmer's steam-punk extravaganza—which has spaceships, for crying out loud—Aouda demurely waits for Fogg to pop the question.) The planetary topsy-turvy that Edgar Allan Poe and Jules Verne had imagined, subversive in the nineteenth century, has begun to look clumsily compensatory.[97]

In the other 1989 BBC program, another former Python took a real eighty-day tour. Given that Michael Palin has many of Fogg's attributes—polite, uncomplaining, good posture—it is surprising that he was the BBC's fourth choice to host a travel series based on the Verne novel. The company sent him around the world in 1988 with five assistants, each known by her or his first name plus "Passepartout." The journey was to start and end at the Reform Club within eighty days, and be performed only by surface travel. Occasionally, a Passepartout flew ahead to do reconnaissance, though everyone regarded "going on aeroplanes as a cheat."[98]

Palin's contract specified six fifty-minute television episodes, plus a book, though no one knew whether the voyage would work. Scheduled surface travel had continued to disappear since Coleridge's circumnavigation. To have the plan fail on camera would be awkward. The first two episodes are mostly safe travelogue—interesting, not fascinating.

Things pick up with a railway strike in Italy, and then the original plans for getting to India are scumbled. All of a sudden, it's more like Fogg's ordeal than anyone had expected.

To cross the Arabian Sea from Dubai to Bombay, Palin and the crew end up on a dhow, seemingly a smaller one than Coleridge had used, which Palin describes, correctly, as "a medieval sailing ship." He seems uneasy at the start, but as the historical re-enactment becomes real, he relaxes into a different life. He sleeps on deck, admires the sail-handling, eats the food prepared by the Gujarati crew, and develops diarrhea—Enrique de Malacca's revenge. Details of the dhow's much-frequented toilet, an open-ended bucket in a rickety frame, perched over the water, make one wonder whether the BBC considered their star expendable.

The slow dhow was the best part of the series; the program was increased to a total of seven episodes to accommodate it. By the time it aired, Palin had wound his way back to the Reform Club. In a final photograph, taken with a London newspaper on the day of his return, he appears thin and tired. He reaped the benefit of a journey that turned out to be unexpectedly difficult. The first episode in the series had an estimated 7.8 million viewers, but demand for Palin's book surged after the broadcast of the dhow episode, and the series concluded with 11.5 million viewers.[99]

In Palin's book, slow is good. He knows "aeroplanes" could have whipped him around the world in thirty-six hours. "But air travel shrink-wraps the world leaving it small, odourless, tidy and usually out of sight," which is exactly what proprietary confidence had done to the planet. Somewhat like Chichester, Palin tries to work out whether there might have been an optimal historical moment, when speed did not alienate travelers from the world. "The reason why Phileas Fogg's 80-day journey retains its appeal," Palin decided, "is that it is still the minimum time needed to go round the world and notice it. To see it, smell it and touch it at the same time"—to taste its bugs.[100]

He concludes that speed diminishes the satisfaction of human effort while requiring too many resources. He watches ships passing through the Suez Canal, not a passenger vessel among them. (He is four days behind Fogg's schedule and can see why.) He is delighted to discover steam locomotives in China; "what makes steam engines so good to watch is that you can see them working." Conversely, he is dismayed that there are no regular passenger services across the Pacific. "When Fogg arrived anywhere," Palin grumbles, "there was an outward-bound passenger service leaving that same day." On the Amtrak trains he takes across North America, he cites fellow eccentrics who, he claims, are "not average Americans. They're people who care about their environment, who despise and fear what big business is doing to it."[101]

Palin was convinced that slower travel could repair relations between human beings and their planet. Perhaps eighty-day circumnavigations might become popular, even an Olympic sport?[102]

Would everyone benefit from a slower, labor-intensive world? Or is slow travel, like Slow Food, an amenity for the well-off, a discretionary holiday from the jet travel that the global poor might never afford?

Palin ponders the trade-offs. One of the men on the dhow had said he preferred that vessel to a big ship, though the latter surely paid better. The *Neptune Garnet,* which took Palin across the Pacific, circled the world every sixty-three days, six times a year. He considers this a good thing. Tropical produce, once flown to Europe at great expense for the rich, was by 1989 likely to be transported on ships, which reduced the cost and spread the pleasure around. Palin knows that fossil fuel kept the ships moving, yet notes that a slower pace saved fuel. The *Neptune Garnet* used 10 tons of fuel to keep at 20 knots, but 20 tons for 23 knots.[103]

Perhaps some global travel could be done more slowly. But speed may appeal to people who are unlikely to be nostalgic for the imperial order that had cushioned Phileas Fogg.

In 1989, Saloo and Neena Choudhury set out from Delhi in a Hindustan Motors "Contessa Classic," a luxury sedan, noted for its comfortable seats. Over the world's continents, the Choudhurys covered 24,901 road miles (40,075 kilometers), more than an equator's worth of driving, and they did it in 69 days, 19 hours, and 5 minutes. When the *Guinness Book* downgraded their achievement, citing a subsequent record by the British army, the Choudhurys protested that an expeditionary force, unlike civilians, did not lose time by fussing with passports and such. They determinedly set off again in 1991. By the time they crossed the finish line in Delhi, they had set a new record of 39 days, 20 hours, 15 minutes—3 hours and 45 minutes less than the British team. "It gave me great pleasure to have defeated the British Army," said Mr. Choudhury, who somehow implied a greater victory than the driving record.[104]

Chapter 12

Magellan redivivus

"*Show Mr. Ferdinand Magellan in,* what does that pest want now?" Mr. Magellan is a persistently undead character in António Lobo Antunes's novel *The Return of the Caravels* (1988), translated from Portuguese into English in 2002. It is a historical novel, though one in which past and present tangle into each other. The Portuguese overseas empire is being created even as it is collapsing. Unemployed Renaissance sailors wander into cocktail bars, clutching their astrolabes. Vasco da Gama keeps soul and body together by playing cards and mending shoes. The empire surges outward—the caravels depart— even as Portugal relinquishes its Asian and African colonies: the caravels return. Among those who trail back home to Lisbon is the nation's black sheep, the Portuguese captain who went to the Indies the wrong way (westward) and for the wrong nation (Spain), and who never sailed around the world although everyone thinks he did. "That pest" Magellan is brought back to life, *redivivus*. Show him in, by all means—he has a lot to answer for.[1]

He's made a kind of comeback. Certainly, reminders of his existence now vie with those of Phileas Fogg. He lends his name to a commercial global positioning system. Magellan the GPS is supposed to prevent anyone from getting as lost as Magellan the Iberian explorer. In development since the 1960s, a rudimentary United States GPS network was operating by 1978, with four orbiting satellites. By 1993, a full complement of twenty-four satellites circled the world. (There are now thirty-one, plus residuals that can be activated should any primary satellite fail.) Each device orbits the globe twice a day, with at least four satellites in view from any given spot on Earth. The busy robots constantly report their locations and times of transmission. In effect, the GPS is a chronometer-plus-map of the entire world generated by constant round-the-world surveillance. It is an excellent example of army surplus. The precision bombing that the GPS enabled during the 1991 Persian Gulf War was an unintended demonstration of its commercial viability. The GPS now has widespread civilian applications, from counseling lost motorists to irrigating parched crops; sales of GPS devices and services were worth $15 billion by 2006.[2]

We are rich in swoops around the planet. But to what end? Just as the past never dies within Antunes's novel, so it survives within geodrama. If every generation gets the circumnavigation it deserves, then our generation somehow deserves every possible kind, and all of them at once. At any given moment, around-the-world travel is taking place in nearly every version that took place in the past. All of geodrama's three acts play out simultaneously, as if in a variety show with too many headliners, each trying to upstage the others.

The sheer volume of the globe-circling traffic is evidence of continuing confidence, a never-ending human capacity to find fulfillment and even joy in planetary adventure. But most people don't go around the world, because they can't. The Earth's circumference still measures the distance between haves and have-nots; confidence is unevenly distributed, and it is doubtful that around-the-world travel could ever become a universal human experience, even as the *us-too* and *me-too* enthusiasms for it continue. The circumnavigator's green conscience asks whether everyone should be included in the experience, given the toll the accumulated circuits would take on the planet. That protective attitude toward Gaia has emerged in the last fifty years, only a tenth of the history of circumnavigation. Most of geodrama represented a longer, deeper enmity toward the globe, a fear that every circumnavigation must be a kind of war of attrition against a forebodingly big Earth. If the battle seemed to have been won in the nineteenth century, the victory seemed dubious by the second half of the twentieth century.

To see those multiple thoughts and possibilities in constant interplay, come with me. We are about to take a few final turns around the planet. Our virtual circumnavigator's track will follow notable points on the routes of real and imagined circumnavigations from the past thirty-odd years. The travelers we'll encounter will be using very different modes of transport, and they're not all going in the same direction, but no matter. They trace an excellent, representative spiral round about the Earth, on its surface, then in the air, next in orbit, and then, at last, up and out into space. What's interesting about the folks we'll meet en route is whether they are fearful, confident, or doubtful. Sometimes they helpfully embellish their preferences with historical references, sometimes not, though their occasional forgetting of the past is itself intriguing.

Luckily for us, during our final circuits round about the world, we don't need to prepare ship's biscuit, portable soup, or malted anything. Like *Punch*'s reviewer on his tour of the Great Globe in Leicester Square, we need neither toothbrush nor carpetbag. Just as Fogg was, we're always ready.

First stop: Lunenburg, Nova Scotia. The shock of the historically authentic was apparent when the tall ship *Picton Castle* returned to Canada in 1998. She had set out in 1997 with seventy-seven sailors but, over the eighteen-month voyage, they drifted away. Only twenty-three of the original hands returned, just under 30 percent, an attrition rate that was probably the most historically accurate part of the journey. The ship went through thirteen cooks alone. Obviously, few people are willing to do pre-industrial labor, the kind that coal and petroleum made obsolete, if they have other options, often petroleum-fueled options. The *Picton Castle*'s crew jumped ship anywhere there was an airstrip.[3]

Compare that vote of no confidence to pleasure cruises. "Wish you were here?" Cunard's Web site asks, over a montage of alluring destination snapshots taken from the classic Grand Tour itinerary. At prices that, for 2013, range from $19,994 to $72,995, Cunard's *Queen Victoria* will take passengers on an eastabout circumnavigation, with Gilded Age luxuries, such as white-glove tea service, never found aboard the old sailing ships. Between the *Picton Castle* and the *Queen Victoria*, the historically different states of dread and confidence are still available, sometimes at the same time on the same ocean.[4]

But now across the Atlantic. By the time she sailed back into the *Bay of Biscay* that nestles between France and Spain, on her final approach to Les Sables d'Olonne, Ellen MacArthur knew she had lost the race. For days, e-mail and satellite telephone had confirmed that Michel Desjoyeaux held the lead in the 2000–01 Vendée Globe. As MacArthur "ranted down the phone" to a supporter at the finish line about what she should have done better, she "heard the fireworks going off for Mich's arrival." Anyone who sails alone around the world is tough and must anticipate failure. But when a distant geostationary satellite uplinked the percussive proof of her competitor's victory, then downlinked it to MacArthur, she wept.[5]

The Vendée Globe Race is one of several created since the Golden Globe Race. Their participants rarely mention any dangerous maritime heritage, though they could. In the first year of the Whitbread Round the World Race, 1973, three people died. There were calls for the race to be discontinued, but it still occurs every four years, and in 2001–02 it gained a new corporate sponsor, Volvo. The Volvo Ocean Race's organizers do not dwell on its hazards. Instead, they claim ancestry with two successful and recent sport circumnavigators, Francis Chichester and Robin Knox-Johnston. The Vendée Globe, first raced in 1989, also sails every four years. The race requires a single-handed, unassisted, non-stop circumnavigation of the world, with a circumnavigation of Antarctica thrown in just to keep things lively. A successful "campaign" to enter and pursue the race costs between $7.7 to $15.4 million. About 40 percent of entrants don't finish; there have been two deaths so far. And yet entrants have increased in number since the race started. That's confidence.[6]

Finally, there is the Jules Verne Trophy, awarded for the fastest circumnavigation that departs and returns to a point midway between Britain and France. Sailors must pass around the three main southern capes, Horn, Good Hope, and Leeuwin (in Australia). Warnings about once-in-a-lifetime stress and strain are offset by a claimed nineteenth-century ancestry which, in this case (as the race's prize implies), goes back to Jules Verne's fictional story, in which no one dies, rather than to any historical maritime circumnavigation, especially the ones that killed nearly everyone. The goal of the first race, in 1993, was to beat Phileas Fogg's time of eighty days, which Bruno Peylon and his crew just managed to do. Since then, the time has been whittled down to 48 days, 7 hours, 44 minutes, and 52 seconds, "the time to beat."[7]

The speed records have been shrinking due to better boat designs, but also because yachtsmen have weather reports and communications systems that pre-satellite circumnavigators never dreamed of. The old-timers put one circle around the planet apiece; today's round-the-world sailor benefits, like Milton's Satan, from multiple passes. With the help of satellites, sailors can avoid weather that would slow them down and get into wind patterns that speed them up. And they use satellite telephones to contact people ashore, either for technical advice or emotional support.

This means that even the non-stop sailors, who never anchor or receive material assistance, are electronically moored to land, as nearly all ocean traffic is. Michael Palin regretted that big container ships, unlike his beloved Indian Ocean dhow, used satellites rather than sextant and charts to navigate. In the case of the Yugoslavian freighter that toted him across the Bay of Bengal, "eight satellites orbiting the earth send down signals against which the ship's position can be checked. The vagaries of the weather—clouds or fog—have been defeated, the days of the sextant superseded."[8]

No competitive around-the-world sailor forgoes satellite assistance, whatever the cost. In the late 1980s, Kay Cottee (the first woman to do a non-stop solo circumnavigation) had to arrange a special corporate donation of $8,000 for her satellite phone bill; about a decade later, Jesse Martin, who set a record as youngest circumnavigator, had e-mail and satellite phone bills that ran to $50,000, for a voyage that cost $300,000 in total. (Martin spent only $7,400 on food.) For the big round-the-world races, with budgets in the millions of dollars, big chunks of those budgets cover communications, to gather information for the skippers and crews, and to broadcast press releases to the media. The onboard systems are cybernetic, in the sense that they monitor situations they have created. "On modern race boats you cannot just lash the helm,"

Ellen MacArthur points out; "they need computers to control them, especially since we sail at such high speeds," which the electronic systems have helped to foster.[9]

At the very least, the satellite systems help racers keep track of each other, something that Bernard Moitessier had wished for during the Golden Globe Race. During the 1996–97 Vendée Globe Race, Englishman Pete Goss, sailing *Aqua Quorum* and fighting his way through the Roaring Forties, heard a mayday call over his systems. It came from a fellow racer, Frenchman Raphael Dinelli, whose boat was 160 miles away and downwind, but sinking fast. "I had to go," Goss decided, and forfeited any chance of winning. The Royal Australian Air Force had managed to locate Dinelli and drop life rafts, but could not extract Dinelli amid the ocean's chop and grind. "No, sir, we're pinning our hopes on you," they told Goss. When Goss got close enough to pull Dinelli aboard, the Frenchman insisted on passing over some distress beacons, a box of provisions and—*bien sûr*—a bottle of Champagne. Goss played to stereotype, too. He "bunged on the kettle" and gave Dinelli a cup of tea, plus continuing physiotherapy, and got him back to Hobart in Australia.[10]

The Vendée Globe officials carefully calculated out the time, provisions, and repairs that Goss had used in his rescue of Dinelli, and allowed him to return to his original position and carry on. He didn't win the race; it didn't matter. Goss was one of only six sailors to finish at all (out of an original sixteen), and the first Briton to do so. For his heroism at sea, he was awarded the French Légion d'Honneur and a British MBE. He was best man at Raphael Dinelli's wedding. The two men co-skippered the 1997 trans-Atlantic Jacques Vabre Race, and won in their class. Goss said of his Vendée Globe triumph that he felt like "a dog with two dicks." Goodness! Only two?[11]

A quick stop in Utrecht shows that a Dutch court (in 2009) was more fearful than confident about sailing around the world, whatever the improvements in boats and navigation. The court forbade a thirteen-year-old sailor, Laura Dekker, to depart on a solo circumnavigation, even though she had her parents' permission. Dekker had planned her voyage in defiance of authority from the start. No official organization would have recognized her accomplishment, not even the *Guinness Book of Records*, which requires record-setters to be fourteen at the outset.[12]

The desire to set an around-the-world record is so powerful, it has taken on a life of its own. People want to do it because others are doing it, because it seems within reach to do so. But in fact most of the low-hanging fruit has long since been picked, if not cooked down into jam. To reach up higher requires access to expensive equipment that can go faster, over longer stretches of territory, and into ever more punishing climates. The armed forces and national budgets may no longer have a monopoly on such technology, but corporate sponsorship—the other option—may be no less exclusive.

Another option is to claim a record based on identity: first woman, youngest person, first citizen of a certain nation. As the old expedition-ary heritage of maritime circumnavigations continues to fade, more around-the-worlders come from places that had not used circum-navigations to create empires; the 2011–12 Volvo Ocean Race went through Abu Dhabi and the United Arab Emirates fielded a team. More women have joined the masculine ranks of single-handers. That means that many "firsts" have been claimed. Strict guidelines are en-forced by entities that determine whether new records have been set: the *Guinness Book of Records* for using the Earth's terrestrial surface; the Fédération Aéronautique Internationale (FAI) for feats of aerospace; and the World Sailing Speed Record Council (WSSRC) for maritime accomplishments.

A glance at the WSSRC shows the odds against setting a new re-cord, and the consequent narrowing of focus on the present day, not the longer history of round-the-world travel. Qualifying voyages must follow WSSRC guidelines and be announced to the council in a formal letter of intent before departure. For the category of round-the-world non-stop speed records, the official list begins with Knox-Johnston and includes Ellen MacArthur, who went around the world in 2005 in sev-enty-one days, single-handed, to win the Vendée Globe—at last! The WSSRC's designation of eastabout and westabout circumnavigations that cover a minimum of 21,760 nautical miles includes racers from 1969 onward. "Other kinds" of voyages, the most historically inclusive group, begins with Slocum, not, for instance, the clipper captains who set earlier speed records. It acknowledges the Australian Kay Cottee as the first woman to sail around the world, non-stop, single-handed, and unassisted (1987–88), and recognizes Jesse Martin as the youngest ever non-stop, solo, unassisted circumnavigator, aged eighteen when he returned. But Martin's record bears a small warning: the WSSRC no longer recognizes records based on age.[13]

That has meant trouble for several young sailors. In 2009, Mike Perham, a Briton, became the youngest person, so far, indisputably to sail around the world—he finished when aged seventeen—though the WSSRC recognizes only his deed, not his age. An Australian, Jessica Watson, and an American, Zac Sunderland, have also claimed to set new age records while sailing around the world, though the WSSRC cannot verify that either sailed its minimum number of miles (21,760) or complied with other regulations. Watson and Sunderland have countered that, because the WSSRC no longer recognizes records set according to the age of the sailor, it should mind its own business.[14]

If you care to check the blogosphere, opinion runs hot in defense of Watson and Sunderland, though just as hot against them. Far more interesting than the snickety details of qualification and disqualification, and strong opinions about them, is the conviction among many sportsmen and women that adjudication about who has gone around the world in some memorable fashion needs to be done in the first place. The clear desire is to keep the numbers on the low side, to maintain a sense of achievement—to keep the planet safe from day-trippers, for heaven's sake.

The lower age limit does seem to be sinking fast. In January 2011, at age fifteen, Laura Dekker sent out on the circumnavigation that Dutch authorities had thwarted earlier. She made a solo circumnavigation, returning when she was sixteen years and four months old. Her circuit was not non-stop. Even if it had been, no official body would have recognized her achievement based on her age. That is unlikely to prevent still younger sailors from establishing new "records" through news headlines and Internet commentary.[15]

At what point would sailing around the world become child's play? Perhaps yachts could be rigged so that older children could handle them. Maybe grown-ups could use satellite systems to monitor a child's progress and coach her or his decisions. Would a voyage made under such conditions still count? Would an eleven-year-old solo circumnavigator be taken seriously? Or would that age mark the point at which sailing alone around the world was no longer a test of the sailor, but of the equipment?

Recall Wiley Post's planetary sweet spot. Post managed to fly around the world alone, in less than eight days, with an autopilot that let him catch enough sleep to keep going. Had he rounded the Earth faster, he would have proved less about his stamina; slower, less about his flying ability. As Howard Hughes had graciously conceded when he overtook Post's record, it was the aircraft that got better, not the pilot. Post's ordeal marked a unique measure of human, plane, and planet. So too might there be a uniquely interconnected measure of age, yacht, and planet.

An alternative would be around-the-world sailing races that require old forms of navigation. Non-commercial sailors in small craft are not required to have GPS. They must use radio and, at a minimum, radar to communicate and to avoid traffic. Satellite navigation is discretionary. For historically authentic races, one category might include those sailors who opt to use the LORAN (*LO*ng *RA*nge *N*avigation) system, a land-based radio navigation system, immediate precursor to GPS. And there could be another category for those brave enough to use celestial navigation. Those races would solve the problem of ever-shrinking opportunities to achieve new round-the-world milestones. They would be dangerous, especially for young or inexperienced sailors. That would reduce anyone's confidence.

Bilbao, Spain. When Robin Knox-Johnston, in 2007, finished one of the many round-the-world sailing achievements he had done since winning the Golden Globe Race in 1969, he used the media attention to denounce the human-engineered environment that dominated the planet. He complained about sailing's reliance on satellite phones, which didn't always work—"It was all so much easier 38 years ago when none of these gadgets had been invented." He deplored the profusion of container ships, many of which nearly ran him down. The boxy metal Leviathans had replaced the whales he had seen during his first circumnavigation, forty years earlier, now glimpsed rarely. The world's oceans were "ruined," he said, and he could not bear to sail around them again. The end of the race in Bilbao was, he declared, the end for him. For a man so clearly in love with the sea, it was quite an indictment.[16]

Paris, France. Let's wobble along and try to get our land legs back, here in the City of Light. This is a fine place to consider that, every year, for almost two decades now, there has been a race in which people circle the world for a prize of $1 million, with surprising challenges along the way, as well as friendship- and love-straining obstacles, and constant media attention. Despite these obvious similarities to Phileas Fogg's steam-era adventure, the executives at CBS consider *The Amazing Race*, twenty seasons and counting (in 2011–12), to be without precedent. The first season's race went straight through Paris, but did anyone whisper the name "Jules Verne"? *Non,* at least not on camera.[17]

It's not an unusual decision. The trend has been for leisure travelers, especially those trying to beat each other around the world, to have a highly selective interest in the past, or none whatsoever. Their confidence is perhaps all the stronger because they don't know where it comes from. Verne is not completely forgotten. Nicole and Yvon Borie-Chalard cited him, their fellow Frenchman, when they did a *Tour du monde en 44 jours* without use of aircraft. But the long nineteenth century's proprietary confidence works best when it is simply there, shorn of any description of how it came into being, lest that indicate how it might unravel.[18]

Syros, Greece. I hope you have exact change, or else the correct transit card, because we've arrived at an entrance to *METRO-Net*. From 1993 until his death in 1997, Martin Kippenberger installed seven portions of this planetary underground, with entrances in Syros, Greece, and Dawson City, Canada, and ventilation grates and tubes in other places, including Los Angeles. Kippenberger was an artist, not an engineer. *METRO-Net*'s entrances led nowhere—nowhere physical, that is. They were designed to make people think about how they were and weren't connected to other parts of the world. The Internet is a virtual network that can rapidly connect the world's peoples on a daily basis. Could it ever have a physical counterpart? Could a subway ride from Greece to Canada be an everyday affair, or would it be a Concorde-like adventure for the few? [19]

In his novel *Ghazalnus w Baghakani Khayal* (*Ghazalnus and the Gardens of Imagination*, 2008), the Kurdish writer Bakhtyar Ali registers the continuing disparity between globalization and immobility by describing two "Magellans." One is "the real Magellan," an immigrant who, after many wandering adventures, comes home to Kurdistan. The other is "the imaginary Magellan," who dreams of travel but has never left his hometown. [20]

Rising fuel prices and climate change may be questioning the sustainability of a consumer democracy of travel opportunities, but the idea remains appealing. Those who have enjoyed full-blown consumerism are reluctant to give it up, and ought to be reluctant to say that others cannot aspire to it. Because the consumerism was formed after, and sometimes in implied liberationist opposition to Western imperialism, it has been difficult to criticize, though with its now obvious toll on the planet, it may prove at least as dangerous to humanity as empires were.

We are near *Gallabat, Sudan,* having traveled down Africa by motorcycle, again, in the company of Ted Simon, who is seriously lost. Simon has inspired many "RTW" motorcyclists, including the actors Charley Boorman and Ewan McGregor, who filmed their journey as a television series. In 2001, at the age of seventy, Simon decided to revisit the route he had traveled about twenty years earlier. "It's for significance," Simon explained; "I want to tell people something important, something valuable about the world, before I leave it." [21]

The second journey was in some ways easier than the first, though Simon found that disconcerting. He used a friend's air pass for first-class travel over oceans; he took several *Lonely Planet* guides, and loaded up on new-fanglements in the forms of a laptop, Web site, CD burner, and GPS. "Thanks to the Cold War," he recognized, "I had become a creature of communication . . . incapable of sinking peacefully into my surroundings." More than that, the surfeit of information was misleading. Simon had tried to find Gallabat with the help of his GPS device, which gave him the wrong directions. Humans set him straight. "After that I never gave the GPS another chance." [22]

The worst part, for Simon, was seeing the accumulating physical damage to the world. In Libya, he said, buildings surrounded roads that once were rural, and rubbish lay everywhere. Farther into Africa, Simon missed seeing trees, which were being felled and burned for charcoal. Without the forests, much of the wildlife had vanished. He knew he should not condemn anyone's desire for a less wild and more developed life, the kind of life he enjoyed. But he worried that the desire would "lead the human race to disaster" if combined with a growing population. "So much of what had fascinated me on my first voyage through the world was disappearing—cultures, customs, animals, whole ecologies, all diluted, muddied or driven to extinction." He didn't think the dilution or extinction was the fault of the people he met along the way. Rather, he said, "I regret that one culture had become so powerful that it has made all the others slaves or tributaries to it, even though it is my culture." [23]

The middle of the Indian Ocean. Over nine months in late 2003 and early 2004, Nicole covered 12,400 miles, a speed record for her category of "contestant," marine animals. Toothy Nicole was a great white shark named for shark-loving actress Nicole Kidman. Marine biologists had briefly detained her and (very carefully) fitted her with a tag that recorded her locations, then popped off and transmitted its data via satellite back to the waiting scientists. It seems possible that Nicole was a celestial navigator—she swam in a remarkably straight line and spent 60 percent of her journey near the surface, where she could see the Sun and stars. Her journey went across the Indian Ocean and back, not around the world, but she was still young, after all, not yet in search of a mate. Do animals self-circumnavigate, or do they only go around the world in captivity, as the Goat and Laika did? Who knows what a mature great white shark might be capable of, given the proper incentive? [24]

Sharks and albatrosses, the fast-traveling companions of maritime circumnavigators, may not make Great-Circle tours of the planet. But if one of them is eventually tracked as crossing all lines of longitude in a single direction, at whatever latitude, she or he would qualify for membership in the Circumnavigators' Club in New York City—the club's application does not specify that potential members must be human.[25]

If circumnavigators were the first people to demonstrate that the planet is a stage for human action, the relationship between stage and actor has changed in geodrama's third act. In the past few decades, circumnavigators have begun to ask whether the Earth is not merely a physical platform, but an actor that can react. The apprehension resembles the fear of early modern circumanvigators that the globe might simply shrug them off. But the newer fear represents an even deeper doubt about the consequences of human impact on the Earth, with a growing apprehension that a planet girdled and bound into position, for the convenience of humanity, might bite back, not just at the circumnavigators but at every living being on the globe, great white sharks and all.

It is a rare circumnavigator these days who does not have some green in his or her conscience. Most racing yachts use some kind of sustainable power, and most sailors feel a sympathy with marine life. When a large shark chewed off his yacht's rudder and massaged itself against the small boat, Teddy Seymour, rather amazingly, blamed himself. (Without sponsorship, in the 1980s, Seymour did what he called a "No Frills Circumnavigation"—he is the first black person to have completed a single-handed circumnavigation using both the Panama and Suez canals.) Having failed to paint his rudder a darker color, Seymour decided his vessel looked like a shark and invited fond attention. Vistors should be shooed away, not shot. Kay Cottee installed a wind generator and solar panels on her boat, *First Lady*, and whereas earlier circumnavigators had eaten the flying fish that landed on their decks, Cottee returned them to the sea. Ellen MacArthur also hated to have dead flying fish on her conscience. So did Jesse Martin. He rigged *Lionheart* to generate solar and wind power, and used no fossil fuels for his journey of 328 days. But he noted that he had generated waste and other kinds of pollution—more could be done.[26]

A visit to the archipelago of *New Caledonia* took a month out of the August 2004–January 2006 circumnavigation of *Sorcerer II*. The 95-foot yacht belonged to scientist J. Craig Venter, who circled the world while taking samples of surface seawater. That may sound familiar, and was meant to. Venter was inspired by the nineteenth-century voyages of HMS *Beagle* and HMS *Challenger*, which had made scientific observations of the world at regular intervals. He said those journeys focused on descriptions of the physical appearance of natural specimens, most of them visible to the naked eye. He sought smaller quarry, the microbes scooped up in the seawater, and wanted to isolate their genetic material, not describe their physical appearance. (In that way, his venture resembled Captain FitzRoy's chain of chronometrical readings around the world, which also identified planetwide patterns within nature.) The expedition identified 1,800 new species and 1.2 million new genes: microbial life on Earth, only beginning to be understood, will provide "information that could be used to address some of the world's environmental problems."[27]

Just south of the Canal Zone is where the crew of the *Tûranor* discovered a sea turtle caught in a loop of plastic line, in a trash-littered section of the Pacific Ocean. The humans freed the turtle and blogged about the garbage. Their catamaran *Tûranor* (a name from Tolkein's *Lord of the Rings*) was entirely solar-powered. The Planet-Solar group launched her on a circumavigation in 2011, as a "project for the benefit of our planet." The journey, successfully completed in May 2012, was intended to show "that we already have the technology required for sustainability" but must nurture the resolve to use it.[28]

Circumnavigations that vindicate renewable energy combine doubt and hope—doubt that the old ways of mastering the planet are sustainable, but hope that new methods might be. The point of these journeys is to take less out of the Earth than an ordinary circumnavigation would do, both as a goal in itself, and as a new test of human endurance and ingenuity on a planetary scale. Of course some of the older journeys had been sustainable, if unconsciously. The eccentrics who had cycled or walked around the world had not necessarily had the good of the planet in mind. Even as recently as the 1990s, for instance, Ffyona Campbell, the first woman credited with walking around the world, was primarily interested in getting into the *Guinness Book of Rec-*

ords. But she and others who had done painfully self-powered circuits over the Earth's land surface had shown that the whole damn thing could be done with one's own body.[29]

That was why Colin Angus and Julie Wafaei did a circumnavigation "under their own steam." They used nothing with a motor, not even boats with sails, just their own strength to cycle, row, and walk around the world. Nor were they isolated eccentrics—they beat two other teams trying to do the same thing. Angus and Wafaei intended their journey as "a loud and clear statement about the urgency of climate change" and as publicity for the viability of alternative means of transport, even on a global scale. The National Geographic Society named them "Adventurers of the Year" in 2007, also extolling them as "the New Magellans." Because Angus and Wafaei married on their return, they were also a new Phileas Fogg and a new Mrs. Aouda (though who was which?).[30]

Oakland, California. That slight bump means we're airborne, on our way around the world for a second time, this time with a little more altitude. Meet our pilot, Linda Finch, who completed Amelia Earhart's intended round-the-world route in 1997, with a eastbound takeoff from California. Finch's reconstructed Lockheed Model 10 Electra was historically accurate, except for its modern navigation systems.[31]

The same was true of Robert Ragozonni's vintage open-cockpit biplane, which he flew around the world in 2000, the first such attempt since the U.S. Army Air Service did it with similar planes in 1924. Finch and Ragozonni had no choice: aircraft cannot be legally operated without up-to-date navigation and communication systems. But that makes them historical approximates. This was especially true for Finch. Earhart went down very probably because she lacked good radio contact with Howland Island, never a problem for Finch. And yet despite his state-of-the-art navigation systems, Ragozonni shaved only five days off the 1924 record.[32]

Château-d'Oex. The historical paradoxes continue in the Swiss Alps, where Bertrand Piccard and Brian Jones set out in 1999 to circumnavigate the world in a balloon. It may seem odd that a balloon, the very first successful aircraft, would be the last to make it around the world. But, even more than sailing ships, balloons are at the mercy of the winds. They need help from that newest round-the-world technology, weather satellites. Only with near-constant weather updates did global balloon travel begin to be conceivable. Piccard, from a family of famous explorers, had made two earlier balloon attempts but each time the weather had defeated him. Third time lucky: waving to supporters from *Breitling Orbiter 3*, Piccard and Jones rose up and over the Alps, like Fogg and Passepartout in the 1956 movie.

For three weeks, Piccard and Jones hot-bunked, one man sinking into the capsule's single berth while the other rose to stand watch over all the information streaming into the balloon. The men had to avoid a patchwork of no-fly zones over Egypt and Yemen (another product of decolonization), and adhere to China's requirement to stay south of the 26th parallel. Coming down on land would have been bad; landing in the sea disastrous. The Pacific was for that reason the most terrifying segment, a reminder that nature could be more formidable than any human power. "For me this flight is a unique opportunity to establish a friendlier relationship with our planet," Piccard noted in his expedition narrative, published in *National Geographic*. "Human beings always want to control nature, but to fly around the world in a balloon, even using our most advanced technology, we must harmonize with nature." When *Breitling Orbiter* thumped down into the Egyptian desert on March 21, 1999, Piccard and Jones had to wait three days to be retrieved, but it was worth it; they had done it. [33]

To some extent, the revival of old technologies and the airborne historical re-enactments have been done because, as with sailing, many other aerial records have already been set. Fixed-wing aircraft have made non-stop circumnavigations with midair refueling (1949), then non-stop without refueling (1986), and, finally, solo and non-stop (2005). The Earth has been circled by helicopter going west to east by a team (1983), by helicopter east to west flown by a single-hander (also 1983), and then by helicopter over the poles by a team (2007). Stay tuned for the inevitable solo version of that. After Piccard and Jones completed their balloon circuit in 1999, another adventurer did it solo in 2002. All of these milestones have turned (or will turn) into ever-shrinking time records, harder and harder to beat.[34]

The record-setters may have little interest in the extended history of circumnavigations, much like their maritime counterparts. That was true of the stupendous all-rounder Steve Fossett. Before his fatal airplane accident in 2007, Fossett had set 115 new sporting records. He was the one who achieved the first solo balloon flight around the world (2002), plus the (then) fastest sail around the world—a total of 58 days and 9 hours (2004), and the first solo non-stop circumnavigation done in a fixed-wing aircraft (2005). And yet he was only interested in his generation of record-setters, his competitors, with one quick glance back at round-the-world American aviators, including Wiley Post.[35]

Other aerial circuits are historical in the sense that they are intended to show a way forward. In 2014, *Solar Impulse* HB-SIA, a solar-powered aircraft, is supposed to fly around the world. Piccard is one of the founders of the initiative to develop the airplane, as are Buzz Aldrin, the astronaut, and Jean Verne, great-grandson of the author. If it works, it will be the first aerial circumnavigation to be entirely solar-powered. The project's Web site features a quotation from Jules Verne: "all that is impossible remains to be achieved."[36]

Cape Canaveral, Florida. Ten, nine, eight . . . shall we head into orbit? The countdown for the first U.S. space shuttle took place on April 12, 1981 (. . . three, two one). Walter M. Schirra, Jr., one of the legendary Mercury Seven, took questions from the press. NASA officials might have hoped for statements as sunny as the clear day of the launch (zero: liftoff!), but Schirra did things his way. "Mostly it's lousy out there," Schirra warned about space. "It's a hostile environment, and it's trying to kill you. The outside temperature goes from a minus 450 degrees to a plus 300 degrees. You sit in a flying thermos bottle."[37]

He may not have picked the best occasion for these foreboding images, but Schirra was right in 1981, and he still is. By 2012, over five hundred people had gone into space. But repeat experience has not erased doubts that the human body is suited to the outer limits. Before we blast off, let's continue to hover cautiously just above the planet's surface. From there, we can take stock of some revealing anxieties about the cost of leaving Earth.

The space shuttle launch in 1981 represented a post-Apollo commitment to experimentation with orbit which, except for those brief periods on the Moon, still represents the uncrossed threshold to the cosmos. Independently, the United States and USSR decided to use orbit as a test for longer voyages, most likely to Mars. "We simply have no data on man's ability to live and work in space for long periods of time," NASA's acting administrator concluded in 1971, and that data is what space stations would generate.[38]

The experiment is crucial. Unlike planetary explorers, even circumnavigators, interplanetary explorers cannot ever enjoy shore leave—until they come back to Earth, there is no removal, however brief, from the microcosms of spaceship and space suit. Beyond our solar system, might space travelers find a Gaia-like planet, with real shorelines on real oceans, where they could unlatch their helmets and gasp fresh air? That is the real goal of manned spaceflight. Predictions that we should and must seek out other habitable planets beyond our solar system have never ceased. If anything, they have gained traction because of fear that an overstressed or too-crowded Earth might cease to support life.[39]

At forty years and counting, the age of orbiting space stations has now outlasted any other period of human space exploration. The USSR established nine orbital outposts between 1971 and 1982, all named *Salyut*, meaning "Salute" or "Fireworks." America's Skylab orbited from 1973 to 1979, and three successive crews of three men occupied the station for a total of 171 days during 1973 and 1974, with the longest single stretch lasting 84 days. A subsequent Russian station, Mir, orbited for fifteen years, a total of 86,000 times around the world (2.2 billion miles), from 1986 to 2001. *Mir* is often translated as "peace," but can also mean "village," "community," or "new world," each of which conveys hope that human life could be re-created in orbit.[40]

Finally, there is the world's current extraterrestrial outpost, the International Space Station (ISS), first orbited in 1998, after the United States and Russia agreed to cooperate on the project and open it to international scientific research. Much of the design and work on the ISS was ceded to the Russians, after President Richard Nixon made it clear that he would support space shuttles or a space station, but not both. The shuttles (135 missions, 1981–2011) were designed to ferry materials and people to the ISS. Unlike other functioning space capsules, they were reusable. The rockets that shot them into space from Cape Canaveral were not; the bodies that traveled in them were supposed to be.

London, England. In a 2006 speech to the Royal Society of London, NASA head Michael D. Griffin lauded Captain James Cook, who had won the society's Copley Medal in 1776 for efforts at defeating sea scurvy. While conjecturing that Cook's expeditions had faced greater dangers than any mission to Mars, Griffin compared scurvy to the effects of weightlessness:

> This rate of bone loss for astronauts is 10 times worse than for those who suffer from osteoporosis here on Earth. Thus, like sailors of the 18th century, our astronauts on the space station or in future missions to Mars face significant medical hazards in the form of bone fractures and kidney stones that could jeopardize their health and their mission. The equivalent of [antiscorbutic] sauerkraut for modern-day astronauts is the unpleasant but necessary nutrition and exercise regimen to create muscle tension and mitigate bone loss. But these are stopgaps, incomplete and unsatisfactory at best.[41]

When they use the age of sail as a guide to the new ocean of space, NASA officials tend to focus on the expeditions with higher survival rates, from the late eighteenth century onward. The first shuttle to go into space was called the *Columbia*, after the American ship that fur trader Robert Gray had sailed around the world in the 1790s, an American first. A subsequent name, the *Endeavour,* for the ship Cook commanded on his first circumnavigation, was selected from suggestions submitted by schoolchildren. The *Challenger* was named for the British scientific vessel that had dredged sea samples around the world in the 1870s.[42]

The space agency's Web site, like Griffin's Royal Society speech, claims that the original "Endeavour and her crew reportedly made the first long-distance voyage on which no crewman died from scurvy, the dietary disease caused by lack of ascorbic acids." Moreover, "Cook is credited with being the first captain to use diet as a cure for scurvy, when he made his crew eat cress, sauerkraut and an orange extract." (You of course know better: John "Foul-Weather Jack" Byron was the first circumnavigator recorded as having maintained a scurvy-free ship.)[43]

A sense of the human body's innate terrestriality may be maritime circumnavigation's most powerful legacy. It turned out to be relatively easy to engineer the closed systems of spacecraft to support human life: air and water can be recycled, waste removed, and humidity controlled. Those successful life-support systems do nothing, however, to counteract weightlessness. Because of mounting evidence that the human body deteriorates in orbit, Salyut's cosmonauts were carefully monitored and required to do two hours of exercise each Earth day. Success with that regimen has made exercise a standard element of space-station life—it is considered the best therapy against bone loss and muscle and tissue deterioration. On Mir, Valerii Poliakov, who holds the record of longest stint in space, 438 days (1994 to 1995), exercised as frenziedly as a caged hamster running in a wheel. Poliakov returned in no worse shape than cosmonauts who weathered shorter space assignments. That set the bar low. After more than a week in orbit, most returning astronauts and cosmonauts don't have the strength to pull themselves from their capsules, let alone walk away from them. A voyage to Mars and back would take 520 days, ample time for deteriorating bone mass to form kidney stones. Remember that when a sailor aboard the USS *Triton* had developed his kidney stone, the commander resurfaced and offloaded him. Beyond Earth, out of orbit, that's not an option.[44]

Experiments with plants and animals have been equally discouraging, even though life in space might require cultivation of food, rather than the extraterrestrial equivalents of hardtack and salt beef, which in any case might have the nutritional content pummeled out of them by radiation. In 1971, Russians grew the first plants in space. Four years later, another Russian crew produced the first edible vegetables, onion sprouts. By 1982, a space-grown plant (*Arabidiopsis*, the lab rat of the plant world) produced seeds that germinated on Earth. By 1997, Mir produced only the second success (in twenty-six years) at growing plants with viable seeds. Animals fared even worse. Quail eggs hatched on Salyut produced headless chicks. In a later attempt on Mir, the chicks seemed healthy, but soon died.[45]

The successful breeding of mammals in space is the gold standard—it would provide the greatest reassurance that prolonged weightlessness and exposure to elevated radiation would not medically compromise humans. So far, the full mammalian cycle, from conception to conception, hasn't happened in space. For that reason, it remains unclear whether humans, as mammals that live and reproduce on Earth, could survive a long physical separation from Earth, at least not in the form they have today.[46]

In short, it is not yet apparent that the metal microcosms could fully protect their human occupants. To the public, NASA presents a magic-bullet scenario: just as a few food items solved the problem of scurvy, so good nutrition and exercise will combat weightlessness in orbit or absence of gravity in space, and perhaps future remedies will be even less intrusive and more portable. The equivalent of vitamin C tablets against scurvy might be a space station that rotates just enough to generate a medically meaningful level of artificial gravity.[47]

But remember that it took about four centuries to beat scurvy. Over two hundred years divide Magellan's first circumnavigation from the emergence of the "Cook plan." And that regimen was not really vindicated until the twentieth century, because the Congress of Vienna opened unprecedented territory (and fresh provisions) to circumnavigators after 1815—frequent shore leave introduced variables that negated many of the Cook plan's experimental results. Even if modern science could reduce that four-hundred-year period by a quarter, to one hundred years, we would still be at the halfway mark. That might be about right, given the twenty-six years it took to replicate results in growing reproductive plant seeds in space, if we assume that human beings are only four times as complicated as plants.

Is it worth it? A group of idealists in *Moscow, Russia,* thought it should be. The first working meeting of the Association of Space Explorers took place in Moscow, in 1983. The association's membership criterion is merely having been in Earth orbit. When the group's first international congress met in Paris, in 1985, it stated as its main goal "to protect and conserve the Earth's environment." Cosmonaut-artist Alexei Leonov, the first person to perform a space walk, designed the congress's emblem, a space suit helmet with the Earth reflected on its visor. According to the association's logic, the greater the number of people who go into space, the larger the percentage of humanity who will achieve the environmentalist epiphany of seeing the big Blue Marble from space: line 'em up.[48]

Orbit represents another example of army surplus, as a more diverse set of travelers blast off and commercial space services replace the neo-imperialism of the space race. Diversity was initially a Soviet goal, in order to show socialism's international embrace. The first person of African ancestry to go into space, Cuban scientist Arnaldo Tamayo Mendez, did so on a visit to Salyut in 1980. NASA and the Soviets then launched several Muslim men into space. The first was Sultan Salman Abdulaziz al-Saud, from Saudi Arabia, dispatched by NASA in 1985. The fourth Muslim, Abdul Ahad Mohmand, read the Quran in space for the first time in 1988, after the Soviets sent him to Mir.[49]

As the USSR devolved into Russia, the Russians had another reason to diversify space travel: for profit, by charging for travel on Soyuz rockets. In 1990, to win a ratings war, the Tokyo Broadcasting System paid $12 million to send television reporter Toyohiro Akiyama to Mir. Akiyama was the first Japanese and the first civilian to orbit the world; Helen Sharman, a chemist, would be the first female civilian and the first Briton. Because Akiyama and Sharman were aboard Mir to work, they were not tourists. That honor fell to businessman Dennis Tito, who in 2001 paid a reported $20 million to soar to the ISS.[50]

Space tourism will require the continued commercialization of rocket launches. But most national governments are backing away from manned space ventures. Only three nations have sent their citizens into orbit from their own territory. The USSR (later Russia) and the United States have done the most launches. China joined the club in 2003, with the successful launch of Yang Liwei in *Shenzhou 5*, described as "a good means to stimulating the national self-esteem" of the Chinese. China has successfully launched three more manned missions, including one with the first woman in June 2012. The European Union abandoned attempts to develop a manned space program, and Japan has yet to develop one. Great Britain has always declined to do so—it is the first conspicuous example of a nation having the capacity for an around-the-world activity, yet renouncing it. With the end of its space shuttle program in 2011, the United States gave up a manned space program. Although an American corporation, SpaceX, was able to send supplies to the ISS in May 2012, human transport to the station must now be done aboard Russian craft. The ISS itself is slated to be de-orbited in 2016, with a splashdown in the Pacific Ocean.[51]

Several companies are banking on space tourism becoming the successor to Concorde, an out-of-this-world experience for any aspirant Buck Rogers with enough bucks. Among other things, an uptick in the number of people who see Earth while in orbit would be a crucial test: does seeing the planet that way automatically generate a planetary green conscience? Or had that Whole Earth epiphany blossomed in the unusual conditions of Cold War concern over nuclear annihilation? Without that worry, has the green conscience withered?

In Kazakhstan, site of Soviet and now Russian rocket launches, it has at times been painfully apparent that going into orbit is not necessarily easier than going to the Moon. The first crew sent to Salyut in 1971 (on *Soyuz 10*) was unable to dock and came right back to Kazakhstan. *Soyuz 11* did dock and its three crewmen spent twenty-three days aboard. Their expedition had its glitches, but they made tremendously popular television broadcasts, light and funny dispatches from the outer limits. That made it all the sadder when, after a seemingly successful re-entry and landing, the men were found dead within the return capsule, which had depressurized too quickly during descent. Equally unsettling were two U.S. space shuttle disasters. In January 1986, *Challenger* broke apart after a seal on a rocket booster failed during takeoff. If any of the seven-person crew survived the disintegration, they perished when the broken capsule hit the surface of the Atlantic Ocean. In February 2003, *Columbia* disintegrated during re-entry; all seven astronauts aboard died.[52]

The Soyuz and space shuttle disasters represent 100 percent mortality on three highly visible expeditions spaced about fifteen years apart. As horrified public responses indicated, the incidents dashed the expectation of zero mortality essential to the confidence of the long nineteenth century.

• • •

Seoul, South Korea. The engineer-artist Hojun Song proposes another way to democratize orbit. Song runs an Open Source Satellite Initiative (OSSI) and plans to launch a global orbiting device. This impiously named GOD is a small box, 60-cubic inches packed with four LED lights, a solar cell, and a battery. The device costs only $500; the launch, to be done by the commercial rocket company NovaNano, will cost $100,000. Once in orbit, GOD will send messages in Morse code. Song considers his device to be just the beginning. Perhaps satellites will become as cheap and common as personal computers. Sputnik might be the equivalent of one of those room-sized, institutional mainframe computers of the 1950s, and GOD an early PC. Plus, having a personal satellite is safer than going into orbit yourself. "Realizing [a] fantasy and dreaming of another fantasy is as valuable as [the] practical and scientific missions [of space travel] and probably even more to some people," Song has said.[53]

He may be onto something. For over fifty years, amateur radio enthusiasts, hams, have had fun with free satellite radio frequencies. The first Orbiting Satellite Carrying Amateur Radio (OSCAR) was launched in 1961. It said "hi" in Morse code, four dots ("H") followed by two dots ("I"). In 1969, Australians formed AMSAT (Radio Amateur Satellite Corporation) to encourage and track future such devices. An even smaller CubeSat (a satellite with a volume of 1 liter, or 10 cm cubed) can be launched as a cheap piggyback on a commercial rocket payload. Project OSCAR and AMSAT have launched over thirty satellites; a total of twenty-three nations have launched OSCARs. And to do low-cost space science, several colleges and universities have launched CubeSats. Why not let them represent us in space?[54]

Maybe because they are not the same as unmediated interaction among human beings or between humans and the natural world. Certainly, they vindicate the prediction that the robots will take over space: almost all human experience of space has been mediated by orbiting satellites, and almost all civilian applications of space technology have been robotic. Alistair Boyd, who went around the world on five quid in 1956, spent the rest of his life doing development work. He appreciates how "the internet and mobile phone have shrunk the world. They do not, however, make it any easier to get around on a fiver." On at least some level, the physical planet requires physical engagement; humans need human interaction; robotic surrogates are no substitute.[55]

In Haruki Murakami's *Spuutoniku no koibito,* or *Sputnik Sweetheart* (1999), the character Miu is the elusive sweetheart of the title. Her nickname beeps several reminders of existential solitude: Laika, lost "in the infinite loneliness of space"; "a little lost Sputnik"; "a poor little lump of metal, spinning around the earth." Even two humans who are in love, "traveling companions" or Sputniks, who, from a distance "look like beautiful shooting stars," are locked into solitary prisons that only occasionally cross paths, then burn into nothingness. "Was the earth put here just to nourish human loneliness?" the novel's narrator wonders. He scans the sky and imagines "all the man-made satellites spinning around the earth":

> I closed my eyes and listened carefully for the descendants of Sputnik, even now circling the earth, gravity their only tie to the planet. Lonely metal souls in the unimpeded darkness of space, they meet, pass each other, and part, never to meet again. No words passing between them. No promises to keep.[56]

International Date Line. If crossing the IDL has reduced the Circumnavigator's Paradox to an everyday affair on Earth, the weary ritual of the jet-lagged globe-trotter, the temporal wonders of geodrama persist in orbit. Aboard Skylab in 1973, Alan Bean decided that "it's not so much that we are 270 miles up in space that isolates us from the rest of the world[,] it's that we are going so fast." He calculated that the station crossed the IDL "every 93 minutes so we gain a day but of course as we go east we lose [in] it an hour every time." "It all averages out," he breezily concluded. Bean and his colleagues amused themselves by using the weird time lapses, and their required exercise on a stationary bicycle, to set records for "pedaling non-stop around the world." As Skylab's orbit decayed, they had to pedal faster.[57]

Orbit's inhuman speed was one reason Sergei Krikalev hesitated when a journalist invited him to consider the changes that had occurred while he was on Mir in 1991—the Soviet Union had fallen; Krikalev's hometown was no longer Leningrad but St. Petersburg. The cosmonaut finally spoke, though cryptically: "Just now it was night; now it's light and the seasons rush past." His statement may have been a way to avoid expressing a political opinion in a volatile situation. But it also cited the Earth's indifferent enormity and velocity as reasons to ignore humanity's small historical blips.[58]

And, just as the Circumnavigator's Paradox had originally confounded the religiously faithful (meaning Christians confused by the relativism of Earth time), so do its orbital equivalents. Malaysia's first astronaut, a Muslim, Sheikh Muszaphar Shukor, was scheduled to be on the ISS during Ramadan in 2007. Foremost among Ramadan's obligations are fasting from dawn until sundown and bowing in prayer toward Mecca five times a day. How could Shukor observe either duty while in space? The Malaysian government convened 150 scientists and clerics to write "Guidelines for Performing Islamic Rites at the International Space Station." The experts stipulated that the time at Mecca determined night and day. They were sympathetic to the difficulty of locating Mecca while rapidly circling the planet—when in doubt, they said, pray toward Earth.[59]

Even the GPS succumbs to the Circumnavigator's Paradox. As round-the-world jet airplanes had demonstrated in 1971, to anyone standing still, a fast-moving clock would appear to lose time. Plus, a clock less affected by gravity runs slower than one under greater gravitational pressure. Each GPS satellite carries four atomic clocks—as precise a technology as is currently available. But GPS signals must be adjusted, not only for the time it takes them to downlink their position to Earth, but also for the apparently slow count they keep of the passing seconds, a constant reminder that, on a planetary scale, there is no common time scale.[60]

We press eastward and back to Canada, to hover above *Montreal*. There, in a 1987 meeting organized by the United Nations, forty-six nations adopted a protocol to reduce substances in the atmosphere that deplete the Earth's protective layer of ozone, our screen against solar radiation. The signing nations assessed many kinds of evidence of ozone depletion, including satellite data from Nimbus 7. Launched in 1978, the device was at that point the largest metereological monitor put into space. Its 2,176 pounds included eight sensors to detect natural and artificial pollutants in Earth's atmosphere. By the late 1980s, Nimbus 7's data corroborated other indications that atmospheric ozone had been rapidly thinning over the past decade, with an alarming hole over Antarctica. Human-driven emission of chlorofluorocarbons (CFCs) was eventually deemed the likeliest culprit. In parallel with the Montreal Protocol, the Vienna Convention of 1985 established a framework for proscribing use of CFCs. The Vienna Convention is the only UN treaty ever to achieve universal ratification by all member states. If any set of international laws is the successor to the Congress of Vienna, as an arbiter of planetary behavior, it may be the Vienna Convention.[61]

Nimbus 7 is only one of several satellites that keep an orbital eye on the globe. In 1990, NASA announced its Mission to Planet Earth program for an Earth Observing System. The most successful part of the system was Landsat 7, which from 1999 onward has photographed the entire surface of the Earth every sixteen days. Other satellite applications have been predicted to solve environmental problems. Solar-powered satellites, for instance, might provide energy that does not require carbon emissions, a major cause of climate change. The Sun has long since been the standard power source for orbiting objects. Space solar power (SSP) would build upon that success: large solar panels in orbit would collect energy from the Sun and transmit it down to Earth.[62]

Orbital technology is not foolproof, however. Notable failures include an extravagant CIA-Boeing contract for a spy satellite program, Future Imagery Architecture, at a loss of $4 billion. Twice, NASA has tried to launch an orbiting carbon observatory (OCO). That satellite might have generated a better understanding of which carbon dioxide emissions stay in the atmosphere and which somehow dissipate. In some years, all excess carbon dioxide disappears. Knowledge of why that happens might generate ways to minimize the impact of the emissions. But two rocket launches for the OCO failed. NASA has suspended the project.[63]

Low Earth Orbit. We've put it off long enough. Into orbit, *poyekhali*, let's go! Once we're there, notice how remarkably busy it is, depopulated void no more. Different craft make their independent circuits at different levels, which we'll visit in the order that they rise above the Earth, continuing our spiral around, up, and out.

Most human space travel has occurred between 200 and 2,000 kilometers above the surface of the planet. The ISS, for instance, oscillates between 320 and 380 kilometers above Earth. You can tell you're close to the world because you cannot see the Whole Earth. Instead, you witness sections of its surface moving constantly and quickly beneath you.

Watching the planet go by has been a very popular pastime aboard the various space stations, so much so that different crew members have requested more time to do it. The globe mesmerizes them, draws them back to it, somehow or another.

Russian cosmonauts have been most expressive about their longing for Earth. On Salyut, the cosmonauts discussed how they missed the planet that they saw slipping below them at a distance. When they slept, they dreamed vivid dreams of woods and of rivers. On Mir, new arrivals were welcomed with bread and salt, the traditional Russian offerings—sustenance harvested from the earth—that are given in hospitality down on Earth. Anatolii Artsebarskii, asked when it was that a cosmonaut was "home," said that it happened "as soon as he breathes the earth's air through the [landing capsule's] open hatch. The smell of the earth. That's the feeling of being home again."[64]

The stifling opposite possibility forms a bleak backdrop in Jonathan Lethem's novel *Chronic City* (2009), which unfolds during a U.S.-Chinese war. A doomed U.S.-Russian space station, *Northern Lights,* supplies one of the book's many subcomponent stories. The cosmo-astronauts avoid looking at Gaia's blue face because they must see it "through a coy lace veil of mines" planted by the Chinese. The mines prevent their return, even as their metal microcosm fails. The "cramped homely craft" is smeared with what seems like earthly familiarity, "farts and halitosis filling the chambers with odor." A shipboard garden yields some food, and photosynthesizes precious oxygen, but it fills the cosmo-astronauts' quarters with hot, damp rot. When an anti-freeze line explodes, the near-zombie crew, emaciated, exhausted, sponge up the mess, discard their saturated clothing, and wring the last, poisonous drops from their already filthy hair.[65]

The novel represents a creeping disillusion about space travel: too difficult for humans, too easy to weaponize with obliging, war-like robots. Don Delillo's story, "Human Moments in World War III," documents a two-man crew on an orbital mission to gather intelligence about the planet and about "possibly hostile satellites." Their drab daily schedule is a parody of the housekeeping and medical regimens necessary to spacecraft. Constant surveillance of the Earth, the shared task of the two men, has pulled them apart. "He doesn't see it anymore," one character realizes about his sense of the planet, whereas his partner is consumed by it, but cannot explain why:

> Lotus-eater, smoker of grasses and herbs, blue-eyed gazer into space—all these are satisfied, all collected and massed in that living body, the sight he sees from the window.
>
> "It is just so interesting," he says at last. "The colors and all."
>
> The colors and all.

In his story, first published in 1983, Delillo confines a Whole Earth hippie and a "right stuff" soldier within the same, tiny, orbiting microcosm, in order to show them circling the planet in mutual incomprehension. A parallel disillusion with Soviet space exploration was apparent in Viktor Pelevin's *Omon Ra* (1992). The (fictional) manned Moon launch that is the novel's centerpiece is a hoax, as many of the historic Soviet expeditions turn out to be, in whole or in part. In secret, Laika has survived her ordeal, if only to represent the clapped-out state of Russia. The aging, red-eyed soldier-dog wears a uniform jacket "with the epaulettes of a major-general and two orders of Lenin on the chest." She is an alcoholic. "A small flask of brandy appeared in the Flight Leader's hands, and he poured some into the bowl. Laika made a feeble snap at his wrist." Aboard the *Northern Lights*, in Lethem's novel, something in the plant funk and chemical spew gives the crew's lone woman cancer. As she dies, she sends "pale, stinky stories" to her fiancé, Chase. She questions whether humans whirling around the planet really exist: "We're soaring atoms, Chase, that's what orbit consists of, the in-human hastening of infinitesimal speck-like bodies through an awesome indifferent void."[66]

So far, the only space travelers to die in the void have been the crew of *Soyuz 11*; all others have perished on the ground or within Earth's atmosphere. But it is true, as the bleak fictions note, that a truly effective international agreement on the peaceful use of space does not exist. Even before the last space shuttle flew into the sunset, the United States had, in spring 2010, launched an unmanned surveillance space plane. The X-37B's orbit tracks over "many global trouble spots," as the *New York Times* put it, including Iraq, Iran, Afghanistan, Pakistan, and North Korea. Meanwhile, Russia, China, and the European Union have or are developing their own global positioning systems and would like to break the U.S. monopoly, each citing security reasons for not wishing to rely on the U.S. military for satellite navigation.[67]

Without a definitive international accord, peaceful and civilian use of orbit could be jeopardized if nations with the capacity to go into space ever went to war. The Outer Space Treaty of 1967, though a beginning, has yet to be substantially updated or clarified. Because U.S. trade laws currently categorize satellite technology as weapons, for example, it has been impossible for Americans to purchase space on Chinese rockets in order to launch their devices. When, in 2007, Chinese engineers targeted an aging communications satellite to test a ground-based missile system, their action questioned whether orbital technology was inherently peaceful. A sense of shared international purpose might change the situation, if Washington and Beijing could agree over their respective diplomatic, military, and commercial goals.[68]

Even if Lethem was pessimistic, in *Chronic City*, about international law on space, he was optimistic about perpetuating human life in space. His details of the constant housework and imminent bodily deterioration are spot-on; but that shipboard garden? A fantasy. Perhaps the future of space travel will someday resemble what was predicted at the start: hearty humans who connect planets, maybe even stars, finally able to recreate whatever keeps them hearty on Earth aboard their space vessels. But in the meantime, a doomed Ancient Mariner haunts each spaceship or space station. Just when things seem to be going well, up he floats, with his non-stop yammering about death. He's hard to silence, let alone evict.

● ● ●

Slightly higher, 569 kilometers above Earth. Robotic emissaries are a different expression of confidence, a prediction that human technology has an ever-expanding planetary capacity that the human body might lack. The confidence is not unwarranted. By the end of 2007, close to nine hundred satellites were in orbit, operated by authorities in over forty different countries. Unlike humans, satellites can be programmed to ignore the world they orbit. They include a famous telescope that goes round and round, about 200 kilometers above the ISS, with never a glance back to its place of origin.[69]

A telescope launched by rocket was first suggested at least as early as 1923, in order to capture images of space unclouded by Earth's atmosphere, the interference that gives stars their charming twinkle. NASA initiated a space telescope project in 1969, with support, after 1975, by the European Space Agency. The launch of the final version, the Hubble Space Telescope, depended entirely on U.S. space shuttles. After many delays, *Discovery* took the five instruments that constituted Hubble into space on April 24, 1990. It was immediately clear that the telescope's images weren't clear. The main mirror had a major flaw. In 1993, *Endeavour*'s crew worked for five days to fit Hubble with corrective lenses, robotic spectacles called corrective optics space telescope axial replacement (COSTAR).[70]

And then the stupendous cosmic snaps came down to Earth, along with a steady rain of data. As it orbits every ninety-seven minutes, Hubble reorients to focus on different, distant phenomena. Scientists submit requests to reserve time for the orientations necessary for their research. A total of 20,000 individual observations are made each year; these have been the basis for over 8,000 scientific articles. So far, the biggest discoveries include the identification of dark matter and a more precise dating of the age of the universe (13 to 14 billion years).[71]

Another finding, on how galaxies evolve, was unexpectedly connected to the first circumnavigation. The Magellanic Clouds, the first widely republished description of which came from Magellan's voyage, are examples of fragmentary galaxies, primitive ancestors of full star systems. They constitute Earth's closest fragmentary systems, a mere 210,000 light-years away, with the greatest potential for showing how galaxies form. When astronomer Antonella Nota trained Hubble on the Small Magellanic Cloud in 2005, she discovered three populations of infant stars, a total of 70,000 stellar babies. The youngest of the populations emerged only 5 million years ago, about the time our ancestors began to walk on their hind feet.[72]

Up where most of the satellites roam. Far above the level of space stations and of Hubble, GPS satellites rotate at around 20,200 kilometers above Earth. And above them, at around 36,000 kilometers, geostationary communications satellites revolve more slowly, once per Earth day. Each is positioned to avoid the others, needless to say, but anything in orbit cannot stay there forever. What goes up really and truly must come down, however slowly, and the smaller it is, the less stable its position.

The Outer Space Treaty of 1967 made only a weak statement about the need to protect space as a natural environment, and little protective effort has followed. In 1978, two scientists, Donald J. Kessler and Burton G. Cour-Palais, predicted that a "debris belt" would form as more satellites went into orbit. Once orbital patterns began to decay, two satellites were bound to collide—Kessler and Cour-Palais said it would happen sometime between 1985 and 2005—and then their detritus would collide, a succession of ever smaller smashups, with each soundless crash generating debris at an exponential rate, much the way asteroid belts are presumed to be created. The resulting debris would spin along in unpredictable patterns. In an encounter with an orbiting spacecraft, any object larger than two inches could be dangerous, given that both objects would be traveling at around 17,500 miles per hour.[73]

When satellites die, they become exhibits in a huge, orbiting museum of obsolete technology. Items in low orbit tend to re-enter the atmosphere and burn up, but anything higher up sweeps around and around for much longer. Of the estimated half-million objects in orbit, about 15,000 have been cataloged and 13,000 of them are two inches or larger. The oldest artificial orbiter is Vanguard 1, launched in 1958. It keeps company with Syncom 3 (1964) and Astérix 1, France's first satellite (1965). In February 2009, four years later than they'd predicted, Kessler and Cour-Palais were proven correct when a non-functioning Russian communications satellite smacked into an operating American satellite, a belated and unintended clash between the two Cold War rivals. The collision created yet more trash. That same year, an oncoming object forced the crew of the ISS to prepare for an emergency evacuation. In June 2011, the ISS crew again manned the "lifeboats" (two escape capsules) when an unidentified orbiting object approached at 29,000 mph. It missed the ISS by only 1,100 feet. A third emergency drill, in early 2012, resulted from the debris from the satellite that China had blown up in 2007.[74]

Solutions have been proposed, grand plans to collect the trash, de-orbit and incinerate it, and so forth. These are yet more of those predictions that are half-confident, half-fearful. They posit technological solutions to technological problems, not unlike those plans for big solar panels to beam power down to Earth.

But any prediction for the future would benefit from a look at the past: no globe-girdling technology has by itself ever saved the world. Circumnavigators lacked an effective anti-scorbutic regimen, for centuries. The telegraph was supposed to usher in world peace, but didn't. Aircraft represented another way to unite the peoples of the world, until they delivered bombs. Satellite communications promise another glorious future, but tyrants can use them just as easily as freedom fighters.

In all the history of round-the-world travel, technology has never been effective in the absence of international agreements. The biggest boon to circumnavigators was not citrus, steampower, canals, or airplanes, but the Hundred Years' Peace. That accord was effective, though selfish. It favored the West's imperialist powers and their imitators. Before it could be reconfigured to become truly international, it vanished, and has only begun to be replaced with a small number of United Nations treaties. The confidence of the nineteenth century, the effortless sense of commanding the world, depended on a unique coincidence of technology and politics, historically unprecedented and not yet replicated.

And why must we look only to what new technology could do? It might be equally useful to wonder, as Francis Chichester and Michael Palin did, whether technologies from the past could be worth revisiting, if not reviving. The physical ability to take on the entire planet has a long history—we are not unique in being able to do it. Do we focus on present or future technologies because we are truly confident in them? Or do we just lack imagination about possible ways to blend present and future with the past?

Let's end with a turn around other planets, or, more realistically, watch others do it for us. The heritage of circumnavigation has finally been applied to the otherworldly endeavors of robotic explorers. Some of those automata are named for explorers famous in the history of round-the-world travel. A European Space Agency project, called Darwin, was designed to investigate the possibility of inhabitable planets. At a position 1.5 million kilometers from Earth, the Darwin probe would look for telltale patterns of infrared light around star-orbiting planets, indications of the possibility of life. This Darwin never got off the ground, but Magellan lent his name to a NASA space probe launched on May 4, 1989. That robotic *Magellan* went to Venus, which it circumnavigated for four years. Even more appropriately, given its namesake, its mission ended with a suicidal plunge toward the planet. Until it failed, the descending device collected data on Venus' hot, dense atmosphere. As with project Darwin, the probe Magellan honored an earlier earthly explorer by projecting his discoveries into unearthly realms.[75]

Maybe that's appropriate. Those missions represent truly new goals in exploration. Rather than dance around Gaia yet again, Magellan *redivivus* and Darwin reconceived were designed to circumnavigate other planets and search for still more. The ventures might be parts of a deferred plunge into the cosmos, a necessary abandonment of Earth, sooner or later, depending on how long our home planet or Sun holds out.

Is the third act of geodrama also the last act? Have the most recent circumnavigators been, not actors on the Earth, but stage hands who tug a final curtain around the planet, which we bid goodbye as we launch into the beyond? That would be the ultimate test of what makes us essentially human, and whether that essence is linked to Earth or not.

So far, the terrestrial link has been strong. Awe before the planet's dimensions, confidence that humans could tame the vast globe, and then doubt that the domestication has been entirely beneficial—each stage in the geodrama identified humans in relation to the planet, not away from it. The earliest centuries of round-the-world travel generated copious evidence of humanity's longing for the Earth. The scorbutic sailor's desire for the smell of fresh-cut turf, and his cheered spirits at seeing a waterfall hanging like a bright earring from a green island peak, revealed his sense of his innate terrestriality, a sensation represented not just to the eye or the mind, but felt deep within the body— the gut, the flesh, the bone.

The longing may have gone into abeyance in the long nineteenth century, but that era may be the least representative in human history; its proprietary confidence would be tremendously difficult to replicate, let alone extend to all. Instead, doubt that the human relationship to the planet should be a matter of controlling it, and worry that we might need the Earth more than it needs us, was revived as cosmonauts and astronauts ventured out into the "new ocean" and went around the world in an entirely new way, still unable to stop gazing and thinking about it. Recall those Salyut cosmonauts who dreamed of the planet's woods and rivers, hungry to see and to smell them again.

The circumnavigator's near-physical cravings for the planet, the unwillingness to turn away from it, the eagerness to encircle and embrace it, again and again, are all powerful reminders that, over the past five hundred years, the physical globe has been as central to our collective history as it was to each individual around-the-world traveler. Even if we might somehow survive elsewhere in the cosmos, we should probably not draw a final curtain around the Earth, but instead take very good care of it. We might miss it.

Acknowledgments

For their financial support of this project I am extremely grateful to the Huntington Library of San Marino, California, especially its former director, Robert C. Ritchie, and to the Harvard University Center for the Environment, particularly its director, Daniel Schrag.

For access to research materials I would like to thank the Mitchell Library of the State Library of New South Wales, Sydney, Australia; the Huntington Library; the British Library, London; the Library of Congress, Washington, D. C.; Qantas Heritage Collection, Mascot, New South Wales (Des Sullivan in particular); the Pan Am Collection, University of Miami; the Museum of Modern Art, New York City; the Victoria and Albert Museum of Childhood, London; the British Airways Heritage Centre, Harmondsworth, England; the Smithsonian Institution (Air and Space Museum Library and Postal Museum Library), Washington, D. C.; the John Carter Brown Library, Providence, Rhode Island; the Beinecke Library, Yale University; the Rare Books and Special Collections Department, Princeton University Library; and the many collections and services of Harvard Library.

For their feedback on different portions of this book I greatly benefited from audiences at the following venues (and the people who invited me to give seminars or lectures there): the Rice University Department of History (Rebecca Goetz), the Huntington Library (Robert Ritchie), the UCLA Department of History (Sanjay Subrahmanyam), the Northeastern University Global History Seminar (Laura Frader), the National Maritime Museum, Greenwich, and Royal Society of London (Simon Schaffer), the National University of Ireland-Galway (Nicholas Canny), the Harvard University Center for the Environment (Daniel Schrag and James Clem), the Georgetown University Early Modern Global History Seminar (Alison Games), the Massachusetts Historical Society Environmental History Seminar (Conrad Wright), the American Historical Association 2010 Annual Meeting (Laurel Thatcher Ulrich), the Florida International University Department of History (Kenneth Lipartito), the John Carter Brown Library (Ted Widmer), the University of Edinburgh and National Library of Scotland (Charles W. J. Withers, Innes M. Keighren, Bill Bell, and David McClay), the University of Maryland Department of History (Richard Bell), the University of Sydney Department of History (Andrew Fitzmaurice), the Center for Pacific and American Studies, University of Tokyo (Yasuo Endo), the Institute for American and Canadian Studies at Sophia University, Tokyo (Shitsuyo Masui), the University of Utrecht Centre for the Humanities

(Rosi Braidotti), the Princeton Institute for International and Regional Studies (Sarah Rivett and Nigel Smith), the Columbia University Center for International History (Deborah Coen), the Massachusetts Institute of Technology Seminar on Environmental and Agricultural History (Harriet Ritvo), the British Association of American Studies (Scott Lucas and Ian Scott), CERI-Institut d'études politiques de Paris (Romain Bertrand and Stephane Van Damme), and the Stanford University Program in History and Philosophy of Science and Technology (Londa Schiebinger).

David Armitage, Michael D. Bess, Clint Chaplin, and Chris McAuliffe read and critiqued full drafts of the book in manuscript. In each case, this was an extremely generous gift of time, for which I am very grateful. I received advice and assistance on specific parts of the project from Peder Anker, Alison Bashford, Graham Burnett, Edward Collins, Nick Crawford, Volker Dehs, Lila Dlaboha, Yasuo Endo, Peter Eli Gordon, Jeremy Greene, Shawn Hill, John Huffman, Maya Jasanoff, Nicholas Jose, Daniel Jütte, Lauren Kerr, Mary Lewis, Anna Pritt, Harriet Ritvo, Claire Roberts, Helen Rozwadowski, Glenda Sluga, Brian P. Taves, Shuichi Wanibuchi, and Max D. Weiss. I was also very fortunate to have had the help of several translators: Jennifer van der Grinten (Japanese), Gregory Afinogenov (Russian), Dzavid Dzanic (Arabic), Ilsoo Cho (Korean), and Asher Orkaby (Hebrew).

Scott Moyers championed this book even before it existed and Andrew Wylie has been cheerful and forceful in seeing it through to the end. I am forever grateful to Alice Mayhew for "getting" the project immediately and to Alice and Jonathan Karp for supporting the book so enthusiastically. I thank the many helpful hands at Simon & Schuster (Jonathan Cox, Karyn Marcus, Emily Remes, Lisa Healy, Ruth Lee-Mui, Michael Accordino, John Wahler, Julia Prosser, Rachelle Andujar) for their patience at steering me through editing, production, and beyond. Ditto to Adam Eaglin and James Pullen at the Wylie Agency. John Huffman ably proofread the proofs at great speed. Jeffrey L. Ward did the characterful maps.

I am extremely lucky to live in a city, Cambridge, Massachusetts, which has a first-class foreign language bookstore. Thanks to that bookstore, Schoenhof's, I was able, at the drop of a hat, to select a short, readable paperback in French from shelves of options. It was an amazing privilege to take myself and the book that I chose, Jules Verne's *Le Tour du monde en quatre-vingts jours,* aboard the SSV *Corwith Cramer* on her cruise from Bermuda to Nova Scotia to Woods Hole in spring of 2005-I can never thank the Sea Education Association enough for taking me away to teach at sea.

Finally, I wish to note that, whatever the glories of going around the world, many were forced to do it unwillingly. This book is dedicated to two captive circumnavigators, who stand for all the rest.

Notes*

Prologue

1. Jules Verne, *Around the World in Eighty Days*, trans. Michael Glencross (London, 2004), 21. For a more scholarly edition of the novel, see Verne, *Around the World in Eighty Days*, trans. William Butcher (New York, 2008).

2. Robert Silverberg, *The Longest Voyage: Circumnavigators in the Age of Discovery* (Athens, OH, 1972); Donald Holm, *The Circumnavigators: Small Boat Voyagers of Modern Times* (London, 1975); Derek Wilson, *The Circumnavigators, a History: The Pioneer Voyagers Who Set off around the Globe* (New York, 2003); Jacques Brosse, *Great Voyages of Discovery: Circumnavigators and Scientists, 1764–1843*, trans. Stanley Hochman (New York, 1983), finely illustrated but factually unreliable; Raymond John Howgego, ed., *Encyclopedia of Exploration to 1800*, 4 vols. (Sydney, 2003–06). In a book that reports on recent round-the-world travelers, especially sailors, Israeli journalist Moshe Gilad considers some historic predecessors. See his *Mo'adon meḳife ha-'olam: be-'iḳvot ha-nos'im ha-gedolim* (*The Circumnavigators Club: In the Footsteps of the Great Travelers*) (Modan, Israel, 2003).

3. For claims that pre-modern humans focused on small environments (islands, gardens, regions) rather than planetary ones, and that they had short-term, slow-moving senses of historical time: Fernand Braudel, *The Mediterranean and the Mediterranean World in the Age of Philip II*, trans. Siân Reynolds (Berkeley, CA, 1995), I, 17–23; Mary Louise Pratt, *Imperial Eyes: Travel Writing and Transculturation* (London, 1992); Richard H. Grove, *Green Imperialism: Colonial Expansion, Tropical Island Edens, and the Orgins of Environmentalism, 1600–1860* (New York, 1995); Martin W. Lewis and Kären E. Wigen, *The Myth of Continents: A Critique of Metageography* (Berkeley, CA, 1997); Martin W. Lewis, "Dividing the Ocean Sea," *Geographical Review*, 89 (1999), 188–214; Richard Drayton, *Nature's Government: Science, Imperial Britain, and the "Improvement" of the World* (New Haven, CT, 2000); Paul Warde, "The Invention of Sustainability," *Modern Intellectual History*, 8 (2011), 153–70; Stephen Kern, *The Culture of Time and Space, 1880–1918* (Cambridge, MA, 1983). On space travel and the whole Earth: Pamela Mack, *Viewing the Earth: The Social Construction of the Landsat Satellite System* (Cambridge, MA, 1990); Denis Cosgrove, "Contested Global Visions: One-World, Whole-Earth, and the Apollo Space Photographs," *Annals of the*

Association of American Geographers, 84 (1994), 270–94; Denis Cosgrove, *Apollo's Eye: A Cartographic Genealogy of the Earth in the Western Imagination* (Baltimore, 2001), 257–67; Neil Maher, "Shooting the Moon," *Environmental History*, 9 (2004), 526–31; Robert Poole, *Earthrise: How Man First Saw the Earth* (New Haven, CT, 2008).

4. On whether the social networks intrinsic to globalism can be traced very far back into time, see Bruce Mazlish and Ralph Buultjens, eds., *Conceptualizing Global History* (Boulder, CO, 1993); Robert W. Cox, "A Perspective on Globalization," in *Globalization: Critical Reflections*, ed. James H. Mittelman (London, 1996), 20–22; Bruce Mazlish, "Comparing Global History to World History," *Journal of Interdisciplinary History*, 28 (1998), 385–95; Emma Rothschild, "Globalization and the Return of History," *Foreign Policy*, 115 (1999), 106–16; A. G. Hopkins, "The History of Globalization—and the Globalization of History?" in *Globalization in World History*, ed. Hopkins (London, 2002), 11–46; David Armitage, "Is There a Pre-History of Globalization?" *Comparison and History: Europe in Cross-National Perspective*, ed. Deborah Cohen and Maura O'Connor (New York, 2004), 165–76; Michael Lang, "Globalization and Its History," *Journal of Modern History*, 78 (2006), 899–931.

On the rereading of sources to discover human-to-non-human connections, see *The Ecocriticism Reader: Landmarks in Literary Ecology*, ed. Harold Fromm and Cheryll Glotfelty (Athens, GA, 1996); see also Lawrence Buell, *The Environmental Imagination: Thoreau, Nature Writing, and the Formation of American Culture* (Cambridge, MA, 1995). In essence, my book is an ecocritical reading of circumnavigators' narratives.

ACT ONE: FEAR

1. Magellan agonistes

1. Stephen Gosch and Peter N. Stearns, *Premodern Travel in World History* (New York, 2007).

2. Ovid, *Metamorphoses*, trans. Mary M. Innes (Harmondsworth, UK, 1955), 50–59.

3. On spices and the spice trade, see Paul Freedman, *Out of the East: Spices and the Medieval Imagination* (New Haven, CT, 2008).

4. Christine Garwood, *Flat Earth: The History of an Infamous Idea* (London, 2007); David Woodward, "The Image of the Spherical Earth," *Perspecta*, 25 (1989), 3–15.

5. Denis Cosgrove, "Mapping the World," in *Maps: Finding Our Place in the World*, ed. James R. Akerman and Robert W. Karrow, Jr. (Chicago, 2007), 78–90; Joan-Pau Rubiés, ed., *Medieval Ethnographies: European Perceptions of the World Beyond* (Burlington, VT, 2009).

6. Freedman, *Out of the East*, 96, 114–15.

7. E. G. R. Taylor, *The Haven-Finding Art: A History of Navigation from Odysseus to Cook* (London, 1971), 3–20, 89–171.

8. George E. Nunn, *The Geographical Conceptions of Columbus: A Critical Consideration of Four Problems* (Milwaukee, 1992), 1–30.

9. Felipe Fernández-Armesto, *Pathfinders: A Global History of Exploration* (New York, 2006), 174.

10. Ibid., 161–71; William D. Phillips, Jr., and Carla Rahn Phillips, *The Worlds of Christopher Columbus* (Cambridge, 1992), 220.

11. Carl Schmitt, *The Nomos of the Earth in the International Law of the Jus Publicum Europaeum*, trans. and ed. G. L. Ulmen (New York, 2003), 87–92.

12. E. G. Ravenstein, *Martin Behaim, His Life and His Globe* (London, 1908), 84–90; Freedman, *Out of the East,* 141–42; Elly Dekker and Peter van der Krogt, *Globes from the Western World* (London, 1993) 18, 26.

13. Fernández-Armesto, *Pathfinders*, 174–80; Sanjay Subrahmanyam, *The Career and Legend of Vasco da Gama* (New York, 1997), 76–145.

14. Fernández-Armesto, *Pathfinders*, 180–82; Subrahmanyam, *Vasco da Gama*, 145–54.

15. Charles E. Nowell, ed., *Magellan's Voyage around the World: Three Contemporary Accounts* (Evanston, IL, 1962), 40, 46; Janet L. Abu-Lughod, *Before European Hegemony: The World System,* A.D. 1250–1350 (Oxford, 1989), 260–86.

16. John W. Hessler, trans. and ed., *The Naming of America: Martin Waldseemüller's 1507 World Map and the Cosmographiae introductio* (London, 2008).

17. Nowell, *Magellan's Voyage*, 39–43, 46.

18. Ibid., 43–45; T. F. Earle and John Villiers, eds., *Albuquerque: Caesar of the East, Selected Texts by Afonso de Albuquerque and His Son* (Warminster, UK, 1990), 69–89.

19. Nowell, *Magellan's Voyage*, 50; Antonio Pigafetta, *Magellan's Voyage: A Narrative Account of the First Circumnavigation*, trans. and ed. R. A. Skelton (New York, 1994), 67.

20. Earle and Villiers, *Albuquerque*, 139, 141, 149.

21. Nowell, *Magellan's Voyage*, 13–15; Glyndwr Williams, *The Great South Sea: English Voyages and Encounters, 1570–1750* (New Haven, CT, 1997), 1–2.

22. Nowell, *Magellan's Voyage*, 15–17; nutmeg and mace come from the same plant, in any case.

23. Ibid., 47–49.

24. Ibid., 18–19.

25. Ibid., 18–39, 49–55; Marcel Destombes, "The Chart of Magellan," *Imago Mundi,* 12 (1955), 65–88.

26. Nowell, *Magellan's Voyage*, 55–56; [Henry Edward John] Stanley, ed., *The First Voyage round the World by Magellan* (London, 1874), xxviii–xliv.

27. Pablo E. Pérez-Mallaína, *Spain's Men of the Sea: Daily Life on the Indies Fleets in the Sixteenth Century,* trans. Carla Rahn Phillips (Baltimore, 1998), 129–31; Nowell, *Magellan's Voyage*, 59–60; Phillips and Phillips, *Worlds of Christopher Columbus*, 144–45.

28. Pérez-Mallaína, *Spain's Men of the Sea*, 134–40.

29. Michael Williams, *Deforesting the Earth: From Prehistory to Global Crisis* (Chicago, 2003), 193–96.

30. Pérez-Mallaína, *Spain's Men of the Sea*, 141–45.

31. Ibid., 63–83.

32. Ibid., 23, 35–39, 45–49, 115, 126–28.

33. Ibid., 99, 101, 191, 238.

34. Nowell, *Magellan's Voyage*, 57–59.

35. Margaret S. Creighton and Lisa Norling, eds., *Iron Men, Wooden Women: Gender and Seafaring in the Atlantic World, 1700–1920* (Baltimore, 1996), ix–xi; Pérez-Mallaína, *Spain's Men of the Sea*, 164–72.

36. Nowell, *Magellan's Voyage*, 64, 67; Alvaro da Costa to King Dom Manuel, c. 1518, in Stanley, *First Voyage round the World*, xxxv–xxxvii; Sebastian Alvarez to King Dom Manuel, July 18, 1519, ibid., xxxviii, xlii–xliii; Pérez-Mallaína, *Spain's Men of the Sea*, 55; Pigafetta, *First Circumnavigation*, 149 n. 10, 150 n. 2.

37. Pigafetta, *First Circumnavigation*, 2–8, 10–11.

38. Ibid., 38–41, 151 n. 3; Pérez-Mallaína, *Spain's Men of the Sea*, 2.

39. Genoese Pilot, "Account," in Stanley, *First Voyage round the World,* 1.

40. Pigafetta, *First Circumnavigation*, 42–45.

41. Genoese Pilot, "Account," 2–3; Pigafetta, *First Circumnavigation*, 46, 153 n. 12.

42. Francisco Alvo (or Alvaro), "Log-Book," in Stanley, *First Voyage round the World*, 218; Pigafetta, *First Circumnavigation*, 45, 46–47.

43. Pigafetta, *First Circumnavigation*, 48–49.

44. Ibid., 54–55.

45. Genoese Pilot, "Account," 5; Pigafetta, *First Circumnavigation*, 47, 50, 154 n. 26.

46. Genoese Pilot, "Account," 3–4; Pigafetta, *First Circumnavigation*, 50, 154 n. 2, 154 n.11; Pérez-Mallaína, *Spain's Men of the Sea*, 211–12.

47. Pigafetta, *First Circumnavigation*, 50–51.

48. Ibid., 51–52, 155 n. 4.

49. Ibid., 52.

50. Magellan to Duarte Barbosa, Nov. 21, 1520, in Stanley, *First Voyage round the World*, 175–76; Pigafetta, *First Circumnavigation*, 52–53, 155 n. 2, 156 n. 4.

51. Pigafetta, *First Circumnavigation*, 53.

52. Genoese Pilot, "Account," 8–9; Pigafetta, *First Circumnavigation*, 53–54, 57; Maximilian of Transylvania, "Discourse," in Nowell, *Magellan's Voyage*, 291.

53. Pigafetta, *First Circumnavigation*, 56, 57.

54. George E. Nunn, "Magellan's Route in the Pacific," *Geographical Review*, 24 (1943), 615–33.

55. Pigafetta, *First Circumnavigation*, 58.

56. Ibid., 57.

57. Ibid.

58. P. D. A. Harvey, "Local and Regional Cartography in Medieval Europe," in *The History of Cartography.* Vol. I: *Cartography in Preshistoric, Ancient, and Medieval Europe and the Mediterranean,* ed. John Brian Harley and David Woodward (Chicago, 1987), 482–84.

59. Pigafetta, *First Circumnavigation*, 43, 58, 157–58; Maria Portuondo, *Secret Science: Spanish Cosmography and the New World* (Chicago, 2009).

60. Alvo, "Log-Book," 223; Genoese Pilot, "Account," 9; Pigafetta, *First Circumnavigation*, 60.

61. Genoese Pilot, "Account," 9–10.

62. Pigafetta, *First Circumnavigation*, 60–61, 63.

63. Genoese Pilot, "Account," 10–11; Pigafetta, *First Circumnavigation*, 64, 65.

64. Pigafetta, *First Circumnavigation*, 67, 68.

65. Ibid., 68, 70–84, 87.

66. Genoese Pilot, "Account," 12; Pigafetta, *First Circumnavigation*, 87–88.

67. Pigafetta, *First Circumnavigation*, 89–90.

68. Ibid., 90; Mairin Mitchell, *Elcano: The First Circumnavigator* (London, 1958), 63.

69. Pigafetta, *First Circumnavigation*, 90.

70. Ibid., 94; Genoese Pilot, "Account," 14.

71. Pigafetta, *First Circumnavigation*, 90–91, 94, 100.

72. Ibid., 105.

73. Alvo, "Log-Book," 227; Genoese Pilot, "Account," 19–20; Mitchell, *Elcano*, 67; Pigafetta, *First Circumnavigation*, 102, 152 n. 8.

74. Alvo, "Log-Book," 227–29; Genoese Pilot, "Account," 16–22; Pigafetta, *First Circumnavigation*, 106, 108, 110.

75. Genoese Pilot, "Account," 22–23; Pigafetta, *First Circumnavigation*, 113, 132.

76. Pigafetta, *First Circumnavigation*, 113–18. On the region's politics, see C. O. Blagden, ed., "Two Malay Letters from Ternate in the Moluccas, Written in 1521 and 1522," *Bulletin of the School of Oriental Studies,* 6 (1930), 87–92.

77. Genoese Pilot, "Account," 24–25; Pigafetta, *First Circumnavigation*, 117–19, 121.

78. Genoese Pilot, "Account," 24; Pigafetta, *First Circumnavigation*, 119–20.

79. Genoese Pilot, "Account," 20, 25; Pigafetta, *First Circumnavigation*, 123–28, 135–36, 139.

80. Pigafetta, *First Circumnavigation*, 139–141.

81. Ibid., 143, 147; Alvo, "Log-Book," 235.

82. Pigafetta, *First Circumnavigation*, 147–48.

83. Ibid., 148.

2. *A World Encompassed*

1. Antonio Pigafetta, *Magellan's Voyage: A Narrative Account of the First Circumnavigation*, trans. and ed. R. A. Skelton (New York, 1994), 100 (gifts of paper totaling 11 quires, about a ream).

2. Ibid., 181 n. 21 and 23; Mairin Mitchell, *Elcano: The First Circumnavigator* (London, 1958), 105.

3. Pigafetta, *First Circumnavigation,* 157 n. 4; Mitchell, *Elcano*, 100, 102.

4. Joanne Snow-Smith, *The Salvator Mundi of Leonardo da Vinci* (Seattle, 1982), 57–58; Catherine Hofmann, et al., *Le Globe et son image* (Paris, 1995), 11–29, 31–45. Geographers were an exception; they were depicted making world maps and globes, though this showed their connection to representations of the world, not symbolic worlds. See Hofmann, *Le Globe et son image*, 33, 54, 55.

5. Clements R. Markham, trans. and ed., *Early Spanish Voyages to the Strait of Magellan* (London, 1911), 15–18, 44–50; Mairin Mitchell, *Friar Andrés de Urdaneta, O.S.A.* (London, 1964), 8–28.

6. Markham, *Voyages to the Strait of Magellan*, 21, 50–53, 55, 59–67; Mitchell, *Friar Urdaneta*, 28–36.

7. Genoese Pilot, "Account," in [Henry Edward John] Stanley, ed., *The First Voyage round the World by Magellan* (London, 1874), 26–29; "Account of the Ship *Trinity*," in ibid., 237–42; Charles E. Nowell, ed., *Magellan's Voyage around the World: Three Contemporary Accounts* (Evanston, IL, 1962), 263–65; Mitchell, *Friar Urdaneta*, 37–50.

8. Pigafetta, *First Circumnavigation,* 4–5; Nowell, *Magellan's Voyage*, 269–70.

9. Maximilian of Transylvania, in Nowell, *Magellan's Voyage*, 309.

10. Pigafetta, *First Circumnavigation,* 4–5.

11. Edward Luther Stevenson, *Early Spanish Cartography of the New World, with Special Reference to the Wolfenbüttel-Spanish Map and the Work of Diego Ribero* (Worcester, MA, 1909), 11–14.

12. Stevenson, *Early Spanish Cartography*, 15, 44–47; Pablo E. Pérez-Mallaína, *Spain's Men of the Sea: Daily Life on the Indies Fleets in the Sixteenth Century*, trans. Carla Rahn Phillips (Baltimore, 1998), 261 n. 126–28.

13. Markham, *Voyages to the Strait of Magellan*, 21–22, 74–89; Mitchell, *Friar Urdaneta*, 63–76.

14. Pigafetta, *First Circumnavigation,* 30–31; "Ambassadors' Globe," Nuremberg, c. 1525, Beinecke Library, Yale University; but see Richard Burleigh, "Note on Radiocarbon of the Ambassadors' Globe," *Imago Mundi*, 33 (1981), 20.

15. A. D. Baynes-Cope, "The Investigation of a Group of Globes," *Imago Mundi*, 33 (1981), 9–19; John North, *The Ambassadors' Secret: Holbein and the World of the Renaissance* (London, 2002).

16. Pigafetta, *First Circumnavigation,* 148.

17. Ibid., 13–18, 21–22; the best surviving copy is Antonio Pigafetta, "Navigation et descouvrement de la Inde Superieure et isles de Malueque . . . ," Beinecke Library, Yale.

18. Lucien Febvre and Henri-Jean Martin, *The Coming of the Book: The Impact of Printing, 1450–1800*, trans. David Gerard (London, 1976), 15–44, 112–14.

19. Pigafetta, *First Circumnavigation,* 15, 18; Petition of Pigafetta, August 1524, in Guglielmo Berchet, comp., *Fonti italiane per la storia della scoperta del nuovo mondo,* 2 vols. (Rome, 1892–93), I, 183n; George B. Parks, comp., *The Contents and Sources of Ramusio's Navigationi* (New York, 1955), 7.

20. Parks, *Ramusio's Navigationi*, 19–20; William Shakespeare, *The Tempest*, I, ii, l. 372; Pigafetta, *First Circumnavigation,* 18, 152 n. 3, 152 n. 5.

21. "Discourse of M. Giovanni Battista Ramusio upon the Voyage Made by the Spaniards Round the World," in Nowell, *Magellan's Voyage*, 270, 271.

22. C. Colin Smith, *Spanish Ballads* (Oxford, 1969), 13, 15–16, 19; J. Antonio Cid, et al., eds., *El Romancero Pan-Hispanico Catálogo General Descriptivo* (Madrid, 1983), III, 75.

23. David Quint, *Epic and Empire: Politics and Generic Form from Virgil to Milton* (Princeton, NJ, 1993).

24. Luís Vaz de Camões, *The Lusíads* [1572], trans. and ed. Landeg White (Oxford, 1997), canto 1, verse 8, ll. 1–4, and canto 10, verse 138, ll. 5–8.

25. Theodore J. Cachey, Jr., "Tasso's Navigazione del Mondo Nuovo and the Origins of the Columbus Encomium," *Italica*, 69 (1992), 326–44.

26. Ramusio, "Discourse," 273.

27. C. O. Blagden, ed., "Two Malay Letters from Ternate in the Moluccas, Written in 1521 and 1522," *Bulletin of the School of Oriental Studies*, 6 (1930), 96; Marcel Destombes, "The Chart of Magellan," *Imago Mundi*, 12 (1955), 79; Gregory C. McIntosh, *The Piri Reis Map of 1513* (Athens, GA, 2000); Giancarlo Casale, *The Ottoman Age of Exploration* (New York, 2010), 22–25, 37–39, 192–96 (oddly, Casale claims that Pigafetta did not complete the circumnavigation).

28. Pigafetta, *First Circumnavigation,* 148; Ramusio "Discourse," 272, 274. See also Ian R. Bartkey, *One Time Fits All: The Campaigns for Global Uniformity* (Stanford, CA, 2007), 10; George Alter, "David Gans: A Renaissance Jewish Astronomer," *Aleph,* 11 (2011), 102–03.

29. Nicholaus Copernicus, *On the Revolutions*, trans. Edward Rosen, ed. Jerzy Dobrzycki (Baltimore, 1992), 9–10, 345–47; Francis R. Johnson, *Astronomical Thought in Renaissance England: A Study of the English Scientific Writings from 1500 to 1646* (Baltimore, 1937), 93–210; Jerzy Dobrzycki, ed., *The Reception of Copernicus' Heliocentric Theory* (Dordrecht, NL, 1972); Owen Gingerich, *The Book Nobody Read: Chasing the Revolutions of Nicolaus Copernicus* (New York, 2005).

30. Theodor de Bry (after an engraving of Hans Galle), *America pars quarta* (Frankfurt, 1594).

31. "Narrative of a Portuguese," in Stanley, *First Voyage round the World,* 31; Battista Agnese, portolan atlas, Venice, c. 1642, John Carter Brown Library, Brown University, Providence, RI; Thomas R. Adams and Jeannette D. Black, "Oval World Map by Battista Agnese," in *Portfolio Honoring Harold Hugo* (Meriden, CT, 1978), [1–3].

32. Markham, *Voyages to the Strait of Magellan*, 23–24; Mitchell, *Friar Urdaneta*, 105–39.

33. Zelia Nuttall, trans. and ed., *New Light on Drake: A Collection of Documents Relating to His Voyage of Circumnavigation, 1577–1580* (London, 1914), 301 (citing Nuño da Silva); Harry Kelsey, *Sir Francis Drake: The Queen's Pirate* (New Haven, CT, 1998).

34. Nuttall, *New Light on Drake*, 7 (John Butler), 209 (Don Francisco de Zarate), 386 (John Winter).

35. "The famous voyage of Sir Francis Drake," in Richard Hakluyt, comp., *The Principall Navigations Voiages and Discoveries of the English Nation,* ed. David Beers Quinn and Raleigh Ashlin Skelton, 2 vols. (Cambridge, 1965), II [643A]; Nuttall, *New Light on Drake,* 24 (John Drake), 172–73 (San Juan de Anton), 186 (Giusepe de Parraces), 207 (Zarate), 288–89, 299 (da Silva).

36. "The famous voyage" [643D]; Nuttall, *New Light on Drake*, 162 (Anton), 207–08 (Zarate), 270, 303, 310 (da Silva), 405 (Don Antonio de Padilla).

37. Nuttall, *New Light on Drake*, 24 (John Drake), 387 (Winter).

38. Ibid., 25–27, 39, 41 (John Drake), 246 (da Silva); "The famous voyage," [643A-643D].

39. "The famous voyage" [643B, 643F–643G]; Pérez-Mallaína, *Spain's Men of the*

Sea, 84–85, 100, 122–23, 229–37; Nuttall, *New Light on Drake*, 45 (John Drake), 114 (Juan Solano), 353–55 (Francisco Gomez Rengifo), 393–94 (da Silva).

40. Helen Wallis, *The Voyage of Sir Francis Drake Mapped in Silver and Gold* (Berkeley, CA, 1979), 4–5; "The famous voyage," [643D]; Francis Fletcher, manuscript map, Sloane MS 61, f. 35 recto, British Library.

41. Nuttall, *New Light on Drake*, 27 (John Drake); "The famous voyage" [643E].

42. "The famous voyage," [643E]; Nuttall, *New Light on Drake*, 331 (Diego de Alarcon), 341 (Gaspar de Vargas).

43. Nuttall, *New Light on Drake*, 28–31 (John Drake), 243 (Martin Enriquez); "The famous voyage," [643F–643G].

44. "The famous voyage," [643G].

45. Ibid., [643E, 643G]; Nuttall, *New Light on Drake*, 66 (Pedro Sarmiento), 105 (Guatemala documents), 242 (Enriquez).

46. Nuttall, *New Light on Drake*, 81–82 (Sarmiento), 101 (Guatemala documents), 110–11 (Costa Rica documents), 162–63 (Anton).

47. Nuttall, *New Light on Drake*, 31 (John Drake), 113 (Costa Rica documents), 187 (Parraces), 251 (da Silva), 255 (Enriquez).

48. "The famous voyage," [643G, 643I].

49. Nuttall, *New Light on Drake*, 34 (John Drake); "The famous voyage," [643I–643J, 643K].

50. Nuttall, *New Light on Drake*, 32–33 (John Drake); W. S. W. Vaux, ed., *The World Encompassed by Sir Francis Drake* (London, 1854), 183–84.

51. Nuttall, *New Light on Drake*, 424–28 (Drake's company); "The famous voyage," [643L]; Vaux, *World Encompassed*, 162.

52. N[icholas] Breton, *A Discourse in Commendation of . . . Frauncis Drake* (London, 1581), Aii verso (preface), Aii recto to verso (new pagination), Avi recto; Wallis, *Voyage of Sir Francis Drake*, 1–2, 10–12.

53. Arthur H. Dean, "The Second Geneva Conference on the Law of the Sea: The Fight for Freedom of the Seas," *American Journal of International Law*, 54 (1960), 757; Nuttall, *New Light on Drake*, 245 (Enriquez).

54. Nuttall, *New Light on Drake*, 402 (Padilla); Geoffrey Parker, "David or Goliath? Philip II and His World in the 1580s," in *Spain, Europe, and the Atlantic: Essays in Honour of John H. Elliott*, ed. Richard L. Kagan and Geoffrey Parker (New York, 1995), 254; J. H. Elliott, *Spain, Europe, and the Wider World, 1500–1800* (New Haven, CT, 2009), frontis.

55. Stephen Parmenius, *De navigatione . . .* (1582), in *The New Found Land of Stephen Parmenius*, ed. David B. Quinn and Neil M. Cheshire (Toronto, 1972), p. 99, ll. 331–34.

56. William Camden, *Annales* (1580), Engl. trans. (London, 1625), Nuttall, *New Light on Drake*, 55 (John Drake); Kelsey, *Sir Francis Drake*, 222–24; Geoffrey Whitney, *A Choice of Emblemes* (Leiden, 1586), 203.

57. José Luis Salas, *Fray Martín Ignacio de Loyola: "gran obispo de esta tierra, eje del desarrollo humano y cristiano del Paraguay y regiones vecinas"* (Asunción, Paraguay, 2003), 43, 48–60.

58. Loyola cited in González de Mendoza, comp., *The History of the Great and*

Mighty Kingdom of China and the Situation Thereof (1586), ed. George T. Staunton, 2 vols. (London, 1853–54), II, 252–53.

59. Ibid., II, 338, 339.

60. "A Letter of Master Thomas Candish [Cavendish]," in Hakluyt, *Principall Navigations,* ed. Quinn and Skelton, II, 808, 813; Francis Pretty, "Admirable and Prosperous Voyage of Master Thomas Candish," in Hakluyt, *Principall Navigations . . .* (London, 1599–1600), III, 803, 804, 805, 806, 807.

61. Robert Hues, *Tractatus de Globis et Eorum Usu. A Treatise Descriptive of the Globes Constructed by Emery Molyneux . . .* [1594], ed. Clements R. Markham (London, 1889), xxxiv–xxxv.

62. Pretty, "Admirable and Prosperous Voyage," 804, 806, 808; "Candish Letter," 809, 810.

63. "Candish Letter," 810; Pretty, "Admirable and Prosperous Voyage," 811, 814.

64. "Candish Letter," 810, 811; Pretty, "Admirable and Prosperous Voyage," 806, 808, 810, 811, 814, 816.

65. "Candish Letter," 812; Pretty, "Admirable and Prosperous Voyage," 816–17.

66. "Candish Letter," 813–15; Pretty, "Admirable and Prosperous Voyage," 817.

67. "Candish Letter," 812–13; Pretty, "Admirable and Prosperous Voyage," 817–18, 822.

68. Edward Arber, ed., *A Transcript of the Registers of the Company of Stationers of London, 1554–1640,* 5 vols. (London & Birmingham, 1875–94), II, 505, 506, 509; David Beers Quinn, ed., *The Last Voyage of Thomas Cavendish, 1591–1592* (Chicago, 1975), 17.

69. *Franciscus Drack. Noblissimus Eques Angliae* (Germany?, c. 1588).

70. Quinn and Skelton, "Introduction," Hakluyt, *Principall Navigations,* I, xxiii, xxxiii, xliv, xlix; "Candish Letter," 813–15. Drake did feature in an epic, by the great Spanish playwright and poet Félix Lope de Vega Carpio, *Dragontea* (1598). Lope de Vega touches on the circumnavigation, but the epic is essentially about Spain's long-deferred defeat of the pirate.

71. "The famous voyage" [643B, 643C]; Pretty, "Admirable and Prosperous Voyage," 805, 815, 823–24.

72. Wallis, *Voyage of Sir Francis Drake,* 13–15.

73. Ibid., 20.

74. Camden, *Annales,* 429; Nuttall, *New Light on Drake,* xxvii–viii; Wallis, *Voyage of Sir Francis Drake,* 7–10, 19–21.

75. Helen Wallis, "The First English Globe: A Recent Discovery," *Geographical Journal,* 117 (1951), 277–90; Wallis, *Voyage of Sir Francis Drake,* 15–17; William Sanderson, *An answer to a scurrilous pamphlet* (London, 1656), A33 verso.

76. Wallis, *Voyage of Sir Francis Drake,* 16, 18.

77. Hues, *Tractatus de Globis,* 5–15; Francis R. Johnson, *Astronomical Thought in Renaissance England: A Study of the English Scientific Writings from 1500 to 1645* (Baltimore, 1937), 202–03.

78. Sir Francis Drake, engraving attributed to Jodocus Hondius, completed by George Vertue, c. 1583, National Portrait Gallery (NPG), London, 3905; Sir Francis Drake, unknown artist, oil on panel, c. 1580, NPG, 4032; Sir Francis

Drake, after unknown artist, line engraving, late 16th–early 17th century, NPG, D2283; Philip Nichols, *Sir Francis Drake Revived . . .* (London, 1626), title page; Thomas Cavendish, engraving by Jodocus Hondius, in *Franciscus Dracus redivivus* (Amsterdam, 1596); Thomas Cavendish, engraving by Robert Boissard, Bodleian Library, Oxford, Rawl. 170, fol. 121.

79. Andrew Gurr, *The Shakespearean Stage, 1574–1642*, 2nd ed. (Cambridge, 1982), 120–21; William Shakespeare, *The Comedy of Errors*, in *William Shakespeare: The Complete Works*, ed. Stanley Wells (Oxford, 2005), III, ii, ll. 116–17 (see also Wallis, *Voyage of Sir Francis Drake*, 21).

80. Shakespeare, *Henry V*, in *Complete Works*, prologue, l. 13; Gurr, *Shakespearean Stage*, 129–30, 196; Jon Greenfield, "Timber Framing, the Two Bays and After," in *Shakespeare's Globe Rebuilt*, ed. J. R. Mulryne and Margaret Shewring (Cambridge, 1997), 97–120; Pérez-Mallaína, *Spain's Men of the Sea*, 261 n. 126–28. (A ducat was worth about one English pound.)

81. Shakespeare, *A Midsummer Night's Dream*, in *Complete Works*, II, i, ll. 169–76; Gurr, *Shakespearean Stage*, 226.

82. Alastair Fowler, *Spenser and the Numbers of Time* (London, 1964), 24–33, 156, 161; William Empson, *Essays on Shakespeare*, ed. David B. Pirie (Cambridge, 1986), 224–29.

83. Shakespeare, *Midsummer Night's Dream*, in *Complete Works*, II, i, ll. 6–7; IV, i, ll. 96–97.

84. Quinn, *Cavendish's Last Voyage*, 52.

85. Clements R. Markham, ed., *The Hawkins' Voyages During the Reigns of Henry VIII, Queen Elizabeth, and James I* (London, 1878), 141–42.

3. Traffic

1. Francesco Carletti, *My Voyage around the World*, trans. Herbert Weinstock (New York, 1964), xiii, 3, 13.

2. Ibid., 42, 79, 102–03.

3. Ibid., 103–05, 201.

4. Ibid., 98, 129, 132, 152, 162.

5. Ibid., 50–51, 69, 82, 91, 192–94.

6. Ibid., 142.

7. Ibid., 220–21.

8. Ibid., 227–34.

9. Ibid., 234–43.

10. Ibid., 249, 258; *Ragionamenti di Francesco Carletti Fiorentino* (Florence, 1701), 368 ("*circondato*").

11. "The Voyage of Oliver Noort round about the Globe," in Samuel Purchas, comp., *Hakluytus Posthumus or Purchas His Pilgrimes, Contaying a History of the World in Sea Voyages and Lande Travells by Englishmen and Others* [1625], 20 vols. (Glasgow, 1905), II, 187–206; J. C. Mollema, *De Reis om de Wereld van Olivier van Noort, 1598–1601* (Amsterdam, 1937), 31; Raymond John Howgego, *Encyclopedia of Exploration to 1800*, 4 vols. (Sydney, 2003–6), I, 760.

12. "Voyage of Oliver Noort round about the Globe," 206; Howgego, *Encyclopedia of Exploration to 1800*, I, 760.

13. Arthur H. Dean, "The Second Geneva Conference on the Law of the Sea: The Fight for Freedom of the Seas," *American Journal of International Law*, 54 (1960), 757.

14. Ibid., 757–61; Hugo Grotius, *The Free Sea*, ed. David Armitage (Indianapolis, 2004), 48.

15. Joris van Spielbergen, *The East and West Indian Mirror . . . and the Australian Navigations of Jacob le Maire*, ed. J. A. J. De Villiers (London, 1906), 2–45; Howgego, *Encyclopedia of Exploration to 1800*, I, 988.

16. Villiers, *East and West Indian Mirror/Australian Navigations*, 166–67.

17. Ibid., 176–77, 184, 187, 188–89.

18. Ibid., 192, 229, 231.

19. Ibid., 148, 151–52, 162, 231.

20. World map showing track of Schouten's voyage in 1615–17 ([Amsterdam?, 1618?]), http://catalogue.nla.gov.au/Record/2576580.

21. "Voyage of Oliver Noort round about the Globe," 189–90; Villiers, *East and West Indian Mirror/Australian Navigations*, 64, 106, 108, 112, 176–77, 186–87, 191, 196, 223.

22. "Voyage of Oliver Noort round about the Globe," 189, 198; Villiers, *East and West Indian Mirror/Australian Navigations*, 17, 108, 114–15, 191, 196.

23. Antonio Pigafetta, *Magellan's Voyage: A Narrative Account of the First Circumnavigation*, trans. and ed. R. A. Skelton (New York, 1994), 63; "Voyage of Oliver Noort round about the Globe," 189; Villiers, *East and West Indian Mirror/Australian Navigations*, 17.

24. W. S. W. Vaux, ed., *The World Encompassed by Sir Francis Drake* (London, 1854), 76; Villiers, *East and West Indian Mirror/Australian Navigations*, 65.

25. "Voyage of Oliver Noort round about the Globe," 197, 199–200, 204; Villiers, *East and West Indian Mirror/Australian Navigations*, 30, 63, 85, 118, 121, 123–24, 130, 216, 218.

26. Verse cited in HelenWallis, *The Voyage of Sir Francis Drake Mapped in Silver and Gold* (London, 1979), 24; Peter Heylyn, *Mikrokosmos. A little description of the great world* (London, 1625), 470.

27. Purchas, *Hakluytus Posthumus*, I, 45, 56.

28. Ibid., I, frontis; ibid. II, 84–284; Mollema, *Olivier van Noort*, plate 3.

29. "Well-willer of the Common-wealth," in *A true relation of the fleete which went vnder the Admirall Iaquis Le Hermite through the Straights of Magellane towards the coasts of Peru . . .* (London, 1625), 4, 6; Benjamin Schmidt, "Exotic Allies: The Dutch-Chilean Encounter and the (Failed) Conquest of America," *Renaissance Quarterly*, 52 (1999), 452–56.

30. Schmidt, "Exotic Allies," 455n.

31. Vaux, *World Encompassed*, xv, 6, 22, 35, 90.

32. Ibid., 162.

33. Villiers, *East and West Indian Mirror/Australian Navigations*, 232; Purchas, *Hakluytus Posthumus*, I, 48.

34. Jerzy Dobrzycki, ed., *The Reception of Copernicus' Heliocentric Theory* (Dordrecht, NL, 1972).

35. John Milton, *Paradise Lost*, ed. Christopher Ricks (London, 1989), Book VIII, ll. 32, 70–71, 122–25, 168.

36. Ibid., Book IX, ll. 63–66, 89.

37. William Dampier, *Dampier's Voyages*, ed. John Masefield, 2 vols. (London, 1906), II, 108. For an overview of Dampier's eventful life, see Anton Gill, *The Devil's Mariner: A Life of William Dampier, Pirate and Explorer, 1651–1715* (London, 1997).

38. Marcus Rediker, " 'Under the Banner of King Death': The Social World of Anglo-American Pirates, 1716 to 1726," *William & Mary Quarterly*, 3d ser., 38 (1981), 205.

39. William Hacke, ed., *A Collection of Original Voyages: Containing I. Captain Cowley's Voyage round the Globe* (London, 1699), 6, 7.

40. Ibid., 6, 10.

41. Dampier, *Voyages*, I, 365.

42. Hacke, *Original Voyages*, 12, 29, 45.

43. Jack Beeching, Introduction to A. O. Exquemelin, *The Buccaneers of America*, trans. Alexis Brown (London, 1972), 16–18.

44. Pedro Cubero Sebastián, *Breve relacion de la peregrinacion que ha hecho a la mayor parte del mundo* (Madrid, 1680), "Aprobacion" by Juan Cortés Ossorio (unpaginated), 1–5, 108.

45. Dampier, *Voyages*, I, 496–500; II, 100; Thomas Hyde, *An Account of the Famous Prince Giolo . . .* (London, 1692).

46. Dampier, *Voyages*, I, 17; II, 312.

47. Ibid., I, 20; Harry Sieber, *The Picaresque* (London, 1977); Ellen Turner Gutiérrez, *The Reception of the Picaresque in the French, English, and German Traditions* (New York, 1995), 67–71.

48. Dampier, *Voyages*, I, 42, 98, 489.

49. Ibid., I, 292–93, 296; II, 31, 73, 284.

50. Ibid., I, 43.

51. Ibid., I, 44, 45, 48, 51, 54.

52. Ibid., I, 150, 160, 184–85, 264–66.

53. Ibid., II, 120–21.

54. Ibid., I, 42, 232.

55. Ibid., I, 46, 51–52; II, 161–64, 173, 426.

56. Ibid., I, 46; II, 174, 176–77, 183–84.

57. Ibid., I, 48.

58. Ibid., I, 89–90, 93, 128, 241, 248, 289, 303–08, 327–28, 579–80.

59. Ibid., I, 223, 306–07; II, 195.

60. Ibid., I, 47.

61. Jonathan Swift, *Gulliver's Travels*, ed. Robert DeMaria (London, 2003), 208; Daniel Defoe, *Robinson Crusoe*, ed. Angus Ross (Harmondsworth, UK, 1981), 27; *Benjamin Franklin's Autobiography*, ed. Joyce E. Chaplin (New York, 2012), 14–15.

62. *The Diary of John Evelyn*, ed. E. S. de Beer, 6 vols. (Oxford, 1955), V, 295.

63. Dampier, *Voyages*, II, 294–304 (seasons), 305–21 (tides and currents).

64. Ibid., I, 552; Herman Moll, "A Map of the World. Shewing the Course of Mr. Dampiers Voyage Round It: From 1679 to 1691" (London, 1697).

65. Dampier, *Voyages*, I, 287, 300.

66. Ibid., I, 378.

67. Howgego, *Encyclopedia of Exploration*, I, 297.

68. William Funnell, *A Voyage round the World* . . . (London, 1707), preface.

69. Ibid., 52, 68, 102, 103–04; Dampier, *Voyages*, II, 579.

70. Woodes Rogers, *A Cruising Voyage round the World* . . . (London, 1712), 6.

71. Ibid., xii, 11, 23, 106, 109–13, 125–31, 209.

72. Ibid., 273, 292–305, 309, 425–26; Glyndwr Williams, *The Great South Sea: English Voyages and Encounters, 1570–1750* (New Haven, CT, 1997), 154–60.

73. Rogers, *Cruising Voyage*, 162, 168, 194, 268, 278–79, 360, 387.

74. Samuel Taylor Coleridge, *The Rime of the Ancient Mariner* (1798), in *Coleridge: Poetical Works*, ed. Ernest Hartley Coleridge (Oxford, 1967), pp. 190–91 (ll. 115–18); Robert Fowke, *The Real Ancient Mariner: Pirates and Poesy on the South Sea* (Bishop's Castle, UK, 2010).

75. George Shelvocke, *A Voyage round the World by the Way of the Great South Sea* (London, 1726), 4, 6, 16–17, 27, 31–37, 72–73, 181–82, 188.

76. Ibid., 206–60, 285.

77. Ibid., 60–64, 292, 294, 332, 378–79, 392, 405–07, 424, 434–35, 439, 440–41.

78. Ibid., viii; Dampier, *Voyages*, I, 135; Rogers, *Cruising Voyage*, 131; William Betagh, *A Voyage round the World* . . . (London, 1728), 146; *The Journal of Jacob Roggeveen*, ed. Andrew Sharp (Oxford, 1970), 40–41, 143.

79. Pigafetta, *First Circumnavigation*, 149 n. 10; Clements R. Markham, ed., *Early Spanish Voyages to the Strait of Magellan* (London, 1911), 25–27, 30; Harry Kelsey, *Sir Francis Drake: The Queen's Pirate* (New Haven, CT, 1998), 84; Williams, *Great South Sea*, 31, 140; Mollema, *Olivier van Noort*, 31. Numbers for the men on these voyages are less certain than those represented in the next chapter. The attrition rates include a number of hazards: losses due mostly to deaths (Loaysa), to men being parted from the ships (Le Maire/Schouten), or to both factors (Dampier).

80. Daniel Defoe, *A New Voyage round the World, by a Course Never Sailed Before* . . . (London, 1725), 2, 3, 8.

81. Ibid., 92, 102, 112; "Voyage of Oliver Noort round about the Globe," 198.

4. Terrestriality

1. Richard Walter and Benjamin Robins, *A Voyage round the World* . . . *by George Anson*, ed. Glyndwr Williams (London, 1974; cited hereafter as *Anson's Voyage*), xv–vi.

2. Glyndwr Williams, ed., *Documents Relating to Anson's Voyage round the World, 1740–1744* (London, 1967; cited hereafter as *Anson Documents*), 236–37.

3. Mary Terrall, *The Man Who Flattened the Earth: Maupertuis and the Sciences in the Enlightenment* (Chicago, 2002), 88–129; Neil Safier, *Measuring the New World: Enlightenment Science and South America* (Chicago, 2008), 6.

4. *Anson Documents*, 5, 35, 37, 38, 40.

5. Leo Heaps, ed., *Log of the Centurion. Based on the Original Papers of Captain Philip Saumarez* (London, 1973), 26–30; *Anson's Voyage*, 23.

6. *Anson Documents*, 34.

7. Heaps, *Log of the Centurion*, 88, 99; *Anson Documents*, 70.

8. Heaps, *Log of the Centurion*, 165–66, 174, 177.

9. Heaps, *Log of the Centurion*, 163, 174, 175, 177; *Anson Documents*, 77, 78, 128, 130, 164.

10. *Anson Documents*, 137, 138, 145, 152–54, 163.

11. Heaps, *Log of the Centurion*, 222–27; *Anson Documents*, 185.

12. *Anson Documents*, 239–49.

13. Michael Williams, *Deforesting the Earth: From Prehistory to Global Crisis* (Chicago, 2003), 193.

14. *Anson's Voyage*, xxi–v; *Anson Documents*, 283; Washington Collection, Boston Athenaeum, Wa.84; title entry in http://realbiblioteca.patrimonionacional.es/.

15. *Anson's Voyage*, 9, 217.

16. Heaps, *Log of the Centurion*, 115, 137–38, 163; *Anson Documents*, 118, 123, 136, 160, 162, 169, 178, 184, 203; *Anson's Voyage*, 42, 139, 169, 211, 253, 343.

17. *Anson's Voyage*, 49.

18. Ibid., 268.

19. Ibid., 85–86, 104–07; *Anson Documents*, 77, 78, 93, 112, 118, 123, 124, 127–32, 160, 162, 164, 169, 178, 203; Heaps, *Log of the Centurion*, 163, 174, 175, 177.

20. *Anson's Voyage*, 112, 113; Heaps, *Log of the Centurion*, 182.

21. *Anson Documents*, 79, 131, 166; *Anson's Voyage*, 112, 114; Heaps, *Log of the Centurion*, 161.

22. *Anson Documents*, 124, 128; *Anson's Voyage*, 266, 267, 287.

23. *Anson Documents*, 166; *Anson's Voyage*, 267–68.

24. *Anson Documents*, 80, 135, 166, 167; *Anson's Voyage*, 119. On shifting European preferences in landscape, see Andrew Ashfield and Peter de Bolla, eds., *The Sublime: A Reader in British Eighteenth-Century Aesthetic Theory* (Cambridge, 1996), and Susan Stewart, *On Longing: Narratives of the Miniature, the Gigantic, the Souvenir, the Collection* (Durham, NC, 1993).

25. Johannes Hofer, "Dissertatio medica de nostalgia, oder Heimwehe" (trans. Carolyn Kiser Anspach), *Bulletin of the History of Medicine*, 7 (1934), 379–91; Jean Starobinski, "The Idea of Nostalgia" (trans. William S. Kemp), *Diogenes: An International Journal of Philosophy and Humanistic Studies*, 14 (1966), 81–103; J. D. Lyons, *Before Imagination: Embodied Thought from Montaigne to Rousseau* (Stanford, CA, 2005); *Anson's Voyage*, 112, 267.

26. Samuel Sutton, *An Historical Account of a New Method for Extracting the Foul Air out of Ships . . . to which are Annexed Two Relations Given thereof to the Royal Society . . . and a Discourse on the Scurvy by Dr. Mead* (London, 1749), iv–v, 72–75, 93–94, 96–97, 100, 109; James Lind, *A Treatise of the Scurvy . . .* (Edinburgh, 1753), 89–91, 100–03, 107–16, 119–20, 196, 213–15; Nathaniel Hulme, *Libellus de natura: causa, curationeque scorbuti . . . to which is annexed, A Proposal for Preventing the Scurvy in the British Navy* (London, 1768), 3–4, 6–7, 8–12, 49, 21–22, 83–88, 91–209. On scurvy and the Royal Navy, see William A. McBride, " 'Normal' Medical Science and British Treatment of the Sea Scurvy, 1753–75," *Journal of the History of Medicine and Allied Sciences* 46 (1991), 158–77; Christopher Lawrence, "Disciplining Disease: Scurvy, the Navy, and Imperial

Expansion, 1750–1825," in *Visions of Empire: Voyages, Botany, and Representations of Nature*, ed. David Philip Miller and Peter Hanns Reill (New York, 1996), 80–96. On British debates over air, see A. E. Clark-Kennedy, *Stephen Hales, D.D., F.R.S.: An Eighteenth-Century Biography* (Cambridge, 1929), 151–69; D. G. C. Allan and R. E. Schofeld, *Stephen Hales, Scientist and Philanthropist* (London, 1980), 84–87.

27. Richard Mead, "Discourse on the Scurvy," in Sutton, *New Method*, 118, 119; James Lind, *Treatise on the Scurvy*, 3d ed. (London, 1772), 534–35.

28. John Dunmore, *French Explorers in the Pacific: The Eighteenth Century* (Oxford, 1965), 57–113.

29. Raymond John Howgego, ed., *Encyclopedia of Exploration to 1800*, 4 vols. (Sydney, 2003–06), I, 140, 165, 194; George Robertson, *The Discovery of Tahiti: A Journal of the Second Voyage of H. M. S. Dolphin . . .* , ed. Hugh Carrington (London, 1948), xxiii, 245; J. C. Beaglehole, ed., *The Journals of Captain James Cook on His Voyages of Discovery*, 4 vols. (Cambridge, 1955–74), I, cxxvi–vii, 588–600; II, 874–87.

30. George Forster, *A Voyage round the World*, ed. Nicholas Thomas and Oliver Berghof, 2 vols. (Honolulu, 2000), II, 684.

31. On the presence of Anson in subsequent circumnavigators' accounts, and absence of the medical experts, see the keyword-searchable Eighteenth Century Collections Online (Detroit, 2003–) editions of Louis-Antoine de Bougainville, *A Voyage round the World*, trans. John Reinhold Forster (London, 1772), and John Hawkesworth, ed., *An Account of the Voyages undertaken by the Order of His Present Majesty . . .* (London, 1773).

32. Robert E. Gallagher, ed., *Byron's Journal of His Circumnavigation, 1764–1766* (Cambridge, 1964), xx.

33. Ibid., xxiii, xxix, xliv.

34. Ibid., xxx, 16, 18, 60, 83, 86, 156.

35. Robertson, *Discovery of Tahiti*, 6, 6n., 38–39, 101; James Cook, "The Method Taken for Preserving the Health of the Crew of His Majesty's Ship the *Resolution* during Her Late Voyage Round the World," *Philosophical Transactions of the Royal Society of London*, 66 (1776), 404–05.

36. Helen Wallis, ed., *Carteret's Voyage round the World, 1766–1769,* 2 vols. (Cambridge, 1965), I, 151–52; Bougainville, *Voyage*, 206, 213.

37. Wallis, *Carteret's Voyage*, I, 128–29, 144; Bougainville, *Voyage*, 1, 14, 123.

38. Robertson, *Discovery of Tahiti*, 42; Wallis, *Carteret's Voyage*, I, 113; Bougainville, *Voyage*, 74, 89, 125; Beaglehole, *Cook's Journals*, II, 45–46; Forster, *Voyage round the World*, II, 693.

39. Robertson, *Discovery of Tahiti*, 7n.; Beaglehole, *Cook's Journals*, I, 613; II, 899–900, 906–07, 910–11, 954–57; Forster, *Voyage round the World*, I, 33; II, 695–96; John Pringle, *Six Discourses . . .* (London, 1783), 163–64, 180.

40. Robertson, *Discovery of Tahiti*, 64, 65, 81; Beaglehole, *Cook's Journals*, II, 185n; Lawrence, "Disciplining Disease," 80–96.

41. Robertson, *Discovery of Tahiti*, 38; Bougainville, *Voyage*, 359; "The Journal of Nassau-Siegen," in John Dunmore, ed., *The Pacific Journal of Louis-Antoine de*

Bougainville, 1767–1768 (London, 2002), 281; Beaglehole, *Cook's Journals*, II, 647.

42. J. C. Beaglehole, ed., *The Endeavour Journal of Joseph Banks, 1768–1771*, 2 vols. (Sydney, 1962), II, 704.

43. Gallagher, *Byron's Journal*, xxx–xxxi, 15, 24, 60, 83, 110, 118, 156; Robertson, *Discovery of Tahiti*, 37; Bougainville, *Voyage*, 125–26, 199–200, 227, 328, 367–68; "The Journal of Caro," in Dunmore, *Journal of Bougainville*, 213.

44. Charles E. Rosenberg, "Medical Text and Social Context: Explaining William Buchan's *Domestic Medicine*," *Bulletin of the History of Medicine*, 57 (1983), 22–42; William Buchan, *Domestic Medicine* (London, 1769), xii (foreign medicines), 61, 439, 444 (raw fruits and vegetables), 55, 126, 306, 313, 374 (night air). Buchan did recommend the medical benefit of smelling herbage and freshly opened earth (143), but not at night. Beaglehole, *Cook's Journals*, I, 629.

45. Gallagher, *Byron's Journal*, 36, 37; Beaglehole, *Endeavour Journal*, I, 175.

46. Bougainville, *Voyage*, 313.

47. *The Poems of Samuel Johnson*, ed. David Nichol Smith and Edward L. McAdam, 2nd ed. (Oxford, 1974), 183–84; J. C. Beaglehole, *The Life of Captain James Cook* (London, 1974), 291 (which corrects some of the former).

48. Robertson, *Discovery of Tahiti*, 75; Bougainville, *Voyage*, 207, 231, 244.

49. Beaglehole, *Endeavour Journal*, II, 145; Forster, *Voyage round the World*, II, 601; Beaglehole, *Cook's Journals*, I, 418.

50. Gallagher, *Byron's Journal*, 122; Hawkesworth, *Account of the Voyages Undertaken*, I, 469, 476, 488; Bougainville, *Voyage*, 231, 244. On island ecologies, see Richard Grove, *Green Imperialism: Colonial Expansion, Tropical Island Edens, and the Orgins of Environmentalism, 1600–1860* (Cambridge, 1995).

51. Robertson, *Discovery of Tahiti*, 38, 64, 65, 99–100, 109, 112; Beaglehole, *Cook's Journals*, I, 7, 74.

52. Gallagher, *Byron's Journal*, 121; Robertson, *Discovery of Tahiti*, 94–95.

53. Bougainville, *Voyage*, 303–05; Londa Schiebinger, "Jeanne Baret: The First Woman to Circumnavigate the Globe," *Endeavour*, 27 (2003), 22–25; Glynis Ridley, *The Discovery of Jeanne Baret: A Story of Science, the High Seas, and the First Woman to Circumnavigate the Globe* (New York, 2010).

54. Gallagher, *Byron's Journal*, 75; Robertson, *Discovery of Tahiti*, 93; Wallis, *Carteret's Voyage*, I, 194–95, 201–02.

55. Glyndwr Williams, "Tupaia: Polynesian Warrior, Navigator, High Priest—and Artist," in *The Global Eighteenth Century*, ed. Felicity Nussbaum (Baltimore, 2003), 38–51.

56. Beaglehole, *Endeavour Journal*, I, 74, 82, 83, II, 251.

57. Robertson, *Discovery of Tahiti*, 112; Anders Sparrman, *A Voyage to the Cape of Good Hope . . .*, 2 vols. (Cape Town, SA, 1975), I, 113; Beaglehole, *Cook's Journals*, II, 112.

58. Schiebinger, "Jeanne Baret," 22–25; *Poems of Samuel Johnson*, 183–84; Beaglehole, *Life of Captain Cook*, 291.

59. Bougainville, *Voyage*, 226, 249.

60. Ibid., 217, 225; Robertson, *Discovery of Tahiti*, 216–21, 226.

61. Elly Dekker and Peter van der Krogt, *Globes from the Western World* (London, 1993), 92–93, 95, 109, 111; "A correct GLOBE with the new Discoveries," (London, c. 1780), British Library, Maps Division, C4.a.3.(5).

62. Bougainville, *Voyage*, xxix, xxx.

63. Ibid., vii, 223; Fanny Burney, *Memoirs of Dr. Burney*, 3 vols. (London, 1832), I, 270–71.

64. For the whole unhappy affair, see John Lawrence Abbott, *John Hawkesworth: Eighteenth-Century Man of Letters* (Madison, WI, 1982), xiii–xv, 137–86.

65. *Horace Walpole's Correspondence*, ed. W. S. Lewis, et al., 48 vols. (New Haven, CT, 1955), XXVIII, 96.

66. Wallis, *Carteret's Voyage*, II, 464–76.

67. *Baldwin's London Weekly Journal*, May 22, 1773; *Private Papers of James Boswell from Malahide Castle*, ed. Geoffrey Scott and F. A. Pottle, 18 vols. (Mt. Vernon, NY, 1928–34), XI, 218.

68. Jocelyn Hackforth-Jones, "Mai," in *Between Worlds: Voyagers to Britain, 1700–1850*, ed. Hackforth-Jones (London, 2007), 44–55; Wallis, *Carteret's Voyage*, I, 201; Williams, "Tupaia," 38–51; Dunmore, *Journal of Bougainville*, lix.

69. Charles Coulston Gillispie, *Science and Polity in France at the End of the Old Regime* (Princeton, NJ, 1980), 337–56; John Lough, *The Encyclopédie* (London, 1971), 1–60, 85–91.

70. David Henry, *An Historical Account of all the Voyages round the World, Performed by English Navigators*, 4 vols. (London, 1774–75); John Hogg, *A New, Authentic, and Complete Collection of Voyages round the World*, 6 vols. (London, [1785?]); Jean Pierre Berenger, *Collection de tous le voyages faits autour du monde, par les différentes nations de l'Europe*, 10 vols. (Lausanne & Geneva, 1788–89).

71. Denis Diderot, *Rameau's Nephew and Other Works*, trans. Jacques Barzun and Ralph H. Bowen (New York, 1964), 180; James Boswell, *The Life of Samuel Johnson*, ed. R. W. Chapman (Oxford, 1980), 722–23.

72. Beaglehole, *Endeavour Journal*, II, 329.

73. Forster, *Voyage round the World*, I, 7, 8.

74. Cook, "Preserving the Health of the Crew of the *Resolution*," 402–06. Away from Cook's direct supervision, the *Adventure* did have cases of scurvy.

75. Alexander Home, "An Account of Kamschatka," in Beaglehole, *Cook's Journals*, III, pt. 2, 1455–56.

76. Anthony R. Wagner, *Historic Heraldry of Britain: An Illustrated Series of British Historical Arms* (London, 1939), 87.

First Entr'acte

1. *The Journal of Jean-François de Galaup de La Pérouse, 1785–1788*, 2 vols., trans. and ed. John Dunmore (London, 1994), I, xx, cxxiv; II, 431, 525. See also John Dunmore, *French Explorers in the Pacific: The Eighteenth Century* (Oxford, 1965), 250–82.

2. *Journal of La Pérouse*, I, xxi, c, civ–cv, cxi, cxvii, cxlviii, cxlix, 24–25, 92; II, 246, 369, 432, 486, 519, 536–40.

3. Ibid., I, clxxxvi–vii, ccxix, ccxxviii, 177; II, 331, 365, 367, 446, 481, 484.

4. Ibid., I, xxx, ccviii, ccxix–ccxxviii; Dunmore, *French Explorers in the Pacific*, 283–341.

5. *Journal of the United States in Congress Assembled . . .* (New York, [1787]), XII, 145; James R. Fichter, *So Great a Proffit: How the East Indies Trade Transformed Anglo-American Capitalism* (Cambridge, MA, 2010), 48–49.

6. Frederic William Howay, *Voyages of the "Columbia" to the Northwest Coast, 1787–1790 and 1790–1793* (Boston, 1941), 112; John Torpey, *The Invention of the Passport: Surveillance, Citizenship and the State* (Cambridge, 2000), 4–56; Benjamin Franklin "To All Captains and Commanders" [Mar. 10, 1779], *The Papers of Benjamin Franklin*, ed. Leonard R. Labaree, et al., 40 vols. to date (New Haven, CT, 1959–), XXIX, 86–87.

7. Fichter, *So Great a Proffit*, 49–52; Howay, *Voyages of the "Columbia,"* 23, 24, 50, 85.

8. *Pennsylvania Mercury and Universal Advertiser*, Aug. 19, 1790; Howay, *Voyages of the "Columbia,"* 37–38, 51–52, 122, 124.

9. Fichter, *So Great a Proffit*, 52–54; Howay, *Voyages of the "Columbia,"* 181, 183, 185, 368, 369, 419.

10. *The Journal of James Morrison, Boatswain's Mate of the Bounty* (London, 1935), 18–22, 24, 27, 28, 29; William Bligh, *The Log of the Bounty*, 2 vols. (London, [1937]), I, 47, 49, 59, 66, 74, 107, 127, 129, 143, 155, 167, 169, 171, 173, 185, 206, 219–20, 235, 355, 359, 360, 361, 363, 367.

11. *Journal of James Morrison*, 40–41; *The Voyage of the Bounty's Launch as Related in William Bligh's Despatch to the Admiralty and the Journal of John Fryer* (London, 1934), 30, 53, 55–56, 61; Bligh, *Log of the Bounty*, I, 373, 393, 418; II, 16, 30, 65, 66, 82.

12. *Bounty's Launch/Journal of John Fryer*, 36, 60.

13. Andrew David, et al., eds., *The Malaspina Expedition, 1789–1794: Journal of the Voyage by Alejandro Malaspina*, 3 vols. (London & Madrid, 2001), I, xxxvii.

14. Ibid., I, xxxvi, l, lv, xci.

15. Ibid., I, xv.

16. C. P. Claret Fleurieu, *A Voyage round the World . . . by Etienne Marchand*, 3 vols. (London, 1801), I, cxx, 21, 175, 187–88; II, 126–27, 164–68. See also Dunmore, *French Explorers in the Pacific*, 342–53.

17. Fleurieu, *Voyage by Marchand*, I, xcix, cxi, 43–44, 88–89, 162, 171, 187; II, 8, 47, 117, 149.

18. Ken Alder, *The Measure of All Things: The Seven-Year Odyssey and Hidden Error That Transformed the World* (New York, 2002).

19. George Vancouver, *A Voyage of Discovery to the North Pacific Ocean and round the World, 1791–1795*, 4 vols., ed. W. Kaye Lamb (London, 1984), I, 2–5, 6, 15, 18, 22–23, 33–34.

20. Ibid., I, 271.

21. Ibid., I, 272–73, 282, 312; IV, 1470–71, 1538, 1542–43. The most serious complaint about Vancouver was also a backhanded compliment. At several times during the journey, he had punished a misbehaving midshipman by having him caned in a cabin. This discipline was much milder than the lashing done to ordinary sailors on deck, and Vancouver had the right to order it,

but the boy in question was Thomas Pitt, second Baron Camelford, cousin of the prime minister. Vancouver's political misjudgment ruined his career. Yet the episode, far from proving his brutality, indicates instead the new conviction that service on a maritime circumnavigation should be hazardous to no one. See ibid., I, 231–40.

22. Ibid., II, 463n, 698, 705, 734; III, 1018, 1019, 1083, 1101; IV, 1285n, 1314, 1497, 1628; Howay, *Voyages of the Columbia*, 336.

23. Vancouver, *Voyage of Discovery*, IV, 1501–02, 1533.

24. Paul W. Schroeder, *The Transformation of European Politics, 1763–1848* (Oxford, 1994), 574–82, 628–36.

25. Thomas Trotter, *Observations on the Scurvy* (London, 1792), 44–46, 174; Vancouver, *Voyage of Discovery*, II, 698; IV, 1422, 1481.

ACT TWO: CONFIDENCE
5. A Tolerable Risk

1. *Charles Darwin's Beagle Diary*, ed. Richard Darwin Keynes (Cambridge, 1988), 17, 46.

2. Xavier de Maistre, *A Journey round My Room* (1795), trans. H[enry] A[ttwell] (New York, 1871), 1, 11, 14, 78–79, 117, 118, 136.

3. *Monthly Magazine* (London), Feb. 1, 1807, 38; Michael M. Smith, "The 'Real Expedición Marítima de la Vacuna' in New Spain and Guatemala," *Transactions of the American Philosophical Society,* n.s., 64 (1974), 1–74.

4. Adelbert von Chamisso, *A Voyage around the World with the Romanzov Exploring Expedition in the Years 1815–1818 in the Brig Rurik, Captain Otto von Kotzebue*, trans. and ed. Henry Kratz (Honolulu, 1986), 185; John Dunmore, *French Explorers in the Pacific* (Oxford, 1969); N. A. Ivashintsov, *Russian round-the-World Voyages, 1803–1849* (Kingston, ON, 1980); Ilya Vinkovetsky, "Circumnavigation, Empire, Modernity, Race: The Impact of Round-the-World Voyages on Russia's Imperial Consciousness," *Ab Imperio*, 1–2 (2001), 191–209.

5. Jules-Sebastien-César Dumont d'Urville, *Voyage pittoresque autour du monde*, 2 vols. (Paris, 1834), I, i.

6. René Primevère Lesson, *Voyage médical autour du monde: Exécuté sur la corvette du roi La Coquille* (Paris, 1829), [iii] 3; Chamisso, *Voyage around the World*, 87; Otto von Kotzebue, *A Voyage of Discovery . . .* , 3 vols., trans. H. E. Lloyd (London, 1821), I, 178–80; Dunmore, *French Explorers in the Pacific*, 65, 138, 343.

7. Lesson, *Voyage médical autour du monde*, [iii]; Georg Heinrich von Langsdorff, *Remarks and Observations on a Voyage around the World from 1803 to 1807*, trans. Victoria Joan Moessner, ed. Richard A. Pierce, 2 vols. (Kingston, ON, 1993), I, 16; *The First Russian Voyage around the World: The Journal of Hermann Ludwig von Löwenstern, 1803–1806*, trans. Victoria Joan Moessner (Fairbanks, AL, 2003), 26, 87–88, 91, 115, 437; Kotzebue, *Voyage of Discovery*, I, 180; J[acques] Arago, *Narrative of a Voyage round the World, in the Uranie and Physicienne Corvettes* (London, 1823), 217–18, 236; *The Voyage of Captain Bellingshausen to the Antarctic Seas, 1819–1821*, ed. Frank Debenham, 2 vols. (London, 1945), I, 11, 15, 19–21, 40, 42, 121, 161, 190, 193; II, 405, 465.

8. Amasa Delano, *A Narrative of Voyages and Travels . . . Comprising Three Voyages round the World* (Boston, 1817), 253, 254–55; Auguste Bernard Duhaut-Cilly, *A Voyage to California, the Sandwich Islands, and around the World in the Years 1826–1829,* trans. and ed. August Frugé and Neal Harlow (San Francisco, 1997), 242; Ivashintsov, *Russian Round-the-World Voyages,* 103; Lesson, *Voyage médical autour du monde,* 1; Dunmore, *French Explorers in the Pacific,* 152.

9. Chamisso, *Voyage around the World,* 15; Arago, *Narrative of a Voyage,* pt. 2, 153; Langsdorff, *Remarks and Observations,* II, 1–2, 51–54.

10. Ivashintsov, *Russian Round-the-World Voyages,* 42, 50; Louis-Claude de Saulces de Freycinet, *Voyage autour du monde . . . pendant les années 1817, 1818, 1819 et 1820,* 2 vols. (Paris, 1824[–39]), I, 5–6, 11, 492; II, 69.

11. Arago, *Narrative of a Voyage,* pt. 2, 231; *Voyage of Bellingshausen,* I, 15; II, 461.

12. John Torpey, *The Invention of the Passport: Surveillance, Citizenship, and the State* (Cambridge, 2000), 4–56; Jane Caplan, " 'This or That Particular Person': Protocols of Identification in Nineteenth-Century Europe," in *Documenting Individual Identity: The Development of State Practices in the Modern World,* ed. Jane Caplan and John Torpey (Princeton, NJ, 2001), 49–66.

13. Chamisso, *Voyage around the World,* 227, 232; Delano, *Narrative of Voyages and Travels,* 26, 309–11, 374, 559–60.

14. Chamisso, *Voyage around the World,* 127; James Holman, *A Voyage round the World . . . ,* 4 vols. (London, 1827–32), IV, 495.

15. Lesson, *Voyage médical autour du monde,* 4–5.

16. *First Russian Voyage around the World,* 78, 157; Roger Collins, *Charles Meryon, A Life* (Devizes, UK, 1999), 44; Chamisso, *Voyage around the World,* 20, 29.

17. *First Russian Voyage around the World,* 32, 354, 434; Chamisso, *Voyage around the World,* 106, 109–10; Duhaut-Cilly, *Voyage to California,* 84.

18. Collins, *Charles Meryon,* 56, 82.

19. Chamisso, *Voyage around the World,* 21; Kotzebue, *Voyage of Discovery,* I, 95; *Voyage of Bellingshausen,* I, 161; *First Russian Voyage around the World,* 177. See also Katherine Plummer, *The Shogun's Reluctant Ambassadors: Japanese Sea Drifters in the North Pacific* (Portland, OR, 1991), 53–62.

20. Duhaut-Cilly, *Voyage to California,* xvi–xvii; Chamisso, *Voyage around the World,* 8.

21. Langsdorff, *Remarks and Observations,* I, 3; Arago, *Narrative of a Voyage,* 165; *First Russian Voyage around the World,* 73, 80–81, 333, 427; Chamisso, *Voyage around the World,* 19.

22. Nicholas Lane, terrestrial pocket globe (London, c. 1780), GLB0028, National Maritime Museum, Greenwich, UK. There are multiple other examples of pocket globes that feature Cook's death: C.4.a.5.(1); C.4.a5.2; C.4.a5.8; C.4.a4 (4); C.4.a4 (6); C.4.a4 (10); C.4.a4 (13), British Library, Maps Division.

23. British Library, Maps Division, British examples with Anson's and/or Cook's tracks: C.4.a.5.(9); C.4.a5.5; C.4.a5.2; C.4.a5 (4); C.4.a5.8; C.4.a.3.(5), C.4.a4(4); British examples with Cook plus another explorer (not Anson): C.4.a.5.(1); C.4.a4(6); C.4.a4(10); C.4.a4(12); C.4.a4(13); German example with Cook: C.4.a.5.(10).

24. Chamisso, *Voyage around the World*, 41, 52; *Voyage of Bellingshausen*, II, 359, 438–39; Collins, *Charles Meryon*, 85.

25. *Voyages of the "Columbia,"* 368; *First Russian Voyage around the World*, 20.

26. *First Russian Voyage around the World*, 65, 92; Chamisso, *Voyage around the World*, 60.

27. Y. Laissus, "Les voyageurs naturalistes du Jardin du Roi et du Muséum d'Histoire Naturelle," *Revue d'Histoire des Sciences*, 34 (1981), 259–317; *A Woman of Courage: The Journal of Rose de Freycinet on Her Voyage around the World, 1817–1820*, trans. Marc Serge Rivière (Canberra, 1996), xiii, xix; Marnie Bassett, *Realms and Islands: The World Voyage of Rose de Freycinet in the Corvette Uranie, 1817–1820* (London, 1962).

28. *Woman of Courage*, 19, 24, 61, 63, 67, 72, 77, 100, 109, 110, 120; Freycinet, *Voyage*, I, 3.

29. *Woman of Courage*, 122, 132; Ivashintsov, *Russian round-the-World Voyages*, 109.

30. Bassett, *Realms and Islands*, x, 10–11, pl. 19.

31. Holman, *Voyage round the World*, I, 2.

32. Ibid., II, 66–67, 105, 201, 297–99, 413; III, 311, 409; IV, 39, 389, 395.

33. James Zug, *American Traveler: The Life and Adventures of John Ledyard, the Man Who Dreamed of Walking the World* (New York, 2005), 139–207; Jason Roberts, *A Sense of the World: How a Blind Man Became History's Greatest Traveler* (New York, 2006), 169–220.

34. Holman, *Voyage round the World*, I, 474; II, 178–79, 257, 291–92; IV, 508; Frederick George Hilton Price, *A Handbook of London Bankers* (London, 1890–91), 81.

35. Holman, *Voyage round the World*, I, 459; III, 194, 384–85, 413.

36. See the reviews in the *Literary Gazette*, June 8, 1833, 363, and Nov. 14, 1835, 724–26; *Metropolitan Magazine*, July 1834, 81–83, Dec. 1835, 110–11; Roberts, *A Sense of the World*, 295.

37. Duhaut-Cilly, *Voyage to California*, 147–48.

38. Ivashintsov, *Russian round-the-World Voyages*, iii; Michael Williams, *Deforesting the Earth: From Prehistory to Global Crisis* (Chicago, 2003), 193–96.

39. *An Accompaniment to Mr. G. Pocock's Patent Terrestrial Globe* (Bristol, 1830), and "G. Pocock's Patent Terrestrial Globe," British Library, Maps Division, G.49 (cataloged under "E. Pocock").

40. Charles Darwin, *Journal and Remarks, 1832–1836* (London, 1839), 243, 496, 607; Charles Darwin, *The Beagle Letters*, ed. Frederick Burkhardt (Cambridge, 2008), 181.

41. Robert FitzRoy, *Narrative of the Surveying Voyages of His Majesty's Ships Adventure and Beagle . . .*, 3 vols. (London, 1839), I, 26, 144, 145, 149–53, 586–87.

42. Ibid., II, xii, xiv, 19–20, 107.

43. Ibid., II, 21, 23, 58, 78, 261–62, 316, 316n, 639.

44. Ibid., IV (appendix to II), 318–26, 331–52.

45. Darwin, *Beagle Diary*, 66, 121, 147, 376, 425, 447; *Beagle Letters*, 23, 30, 52, 66.

46. FitzRoy, *Narrative*, II, 361, 638–39.

47. Darwin, *Journal and Remarks*, 83, 139; *Beagle Letters*, 111.

48. John Beck, *A History of the Falmouth Packet Service, 1689–1850: The British Overseas Postal Service* (Exeter, 2009), 1–26, 33–46.

49. Darwin, *Beagle Letters*, 113, 120, 156, 175, 180, 254, 255, 285; FitzRoy, *Narrative*, II, 499. On the letter drop at Port Famine, see Jules Sebastien-Cesar Dumont d'Urville, *Voyage au pole sud et dans l'Océanie . . . Histoire du voyage,* 10 vols. (Paris, 1841–46), I, 95.

50. Darwin, *Journal and Remarks*, 571, 602, 603; *Beagle Letters*, 122, 139, 249–50.

51. Darwin, *Journal and Remarks*, 46–47, 171, 191, 480, 483; FitzRoy, *Narrative*, II, 191; Christine Garwood, *Flat Earth: The History of an Infamous Idea* (London, 2007), 36–78.

52. Darwin, *Beagle Letters*, 396n.

53. "Three Sundays in a Week," *The Complete Works of Edgar Allan Poe*, ed. James A. Harrison, 17 vols. (New York, 1965), IV, 227–35.

54. Francis Warriner, *Cruise of the United States Frigate Potomac round the World* (New York, 1835), 11–12, 75–85, 126, 299, 311.

55. "Three Thursdays in One Week," *Philadelphia Public Ledger*, Oct. 29, 1841; letter from "NAVV.," ibid., Nov. 17, 1841; T. O. Mabbott, "Poe and Dr. Lardner," *American Notes & Queries*, 3 (November 1943), 115–17; Archer Taylor, Answer to "Poe, Dr. Lardner, and 'Three Sundays in a Week,' " ibid., 3 (January 1944), 153–55.

56. Isaac Howe, "An Abstract of a Cruise in the U.S. Frigate Constellation in the Years of 1840, 41, 42, 43 & 44," Huntington Library, San Marino, CA, 37, 70, 85, 132.

57. See Alan Gurney, *The Race to the White Continent* (New York, 2000).

58. Charles Wilkes, *Narrative of the United States Exploring Expedition*, 5 vols. (Philadelphia, 1849), I, xxv–xxvi, xxvii–xxix; William H. Goetzmann, *Exploration and Empire: The Explorer and the Scientist in the Winning of the American West* (New York, 1972), 233–36; Nathaniel Philbrick, *Sea of Glory: America's Voyage of Discovery, the U.S. Exploring Expedition, 1838–1842* (New York, 2003); Herman J. Viola, "The Story of the U.S. Exploring Expedition," in *Magnificent Voyagers: The U.S. Exploring Expedition, 1838–1842,* ed. Herman J. Viola and Carolyn Margolis (Washington, DC, 1985), 9–23.

59. Wilkes, *Narrative*, I, 137; II, 319–20; *Saturday Evening Post* (Philadelphia), Dec. 7, 1839; Philbrick, *Sea of Glory*, 237, 282, 291.

60. Wilkes, *Narrative*, III, 265; Viola, "Story of the U.S. Ex. Ex.," 20, 21; E. Jeffrey Stann, "Charles Wilkes as Diplomat," in *Magnificent Voyagers*, 205–25.

61. Douglas E. Evelyn, "The National Gallery at the Patent Office," in *Magnificent Voyagers*, 227–54.

62. D'Urville, *Voyage*, I, 38, 138; II, 336; IV, 193; VIII, 13–14, 123–25, 129–31, 137, 150; X, 179–80; Viola, "Story of the U.S. Ex. Ex.," 18.

63. D'Urville, *Voyage*, I, 1, 166; X, 55–57.

64. Mary Louise Pratt, *Imperial Eyes: Travel Writing and Transculturation* (New York, 1992); http://www.hakluyt.com.

65. "A Journey round the Globe," *Punch*, 21 (July–Dec. 1851), 4.

66. Richard D. Altick, *The Shows of London* (Cambridge, MA, 1978), 464–67.

67. Ibid., 464; Charles Louis-Lesur, *Annuaire historique universal . . .* (Paris, 1826), 236–37.

68. Robert L. Carothers and John L. Marsh, "The Whale and the Panorama," *Nineteenth-Century Fiction,* 26 (1971), 319–28; Kevin J. Avery, " 'Whaling Voyage round the World': Russell and Purrington's Moving Panorama and Herman Melville's 'Mighty Book,' " *American Art Journal,* 22 (1990), 50–78; Altick, *Shows of London,* chaps. 10–15.

69. Ida Pfeiffer, *A Lady's Voyage round the World* [1851], intro. Maria Aitken (London, 1988), 32; Charles E. Lee, *"The Blue Riband": The Romance of the Atlantic Ferry* (London, [1930]), 3–64; Robert A. Margo, *Wages and Labor Markets in the United States, 1820–1860* (Chicago, 2000), table 5B4, p.117.

70. Pfeiffer, *Lady's Voyage,* 26, 28–29, 34–35, 42, 192, 203, 227.

71. Ibid., 41, 143–44, 161, 185, 213, 236, 242, 243, 248; Charles Steinwedel, "Making Social Groups, One Person at a Time: The Identification of Individuals by Estate, Religious Confession, and Ethnicity in Late Imperial Russia," in *Documenting Individual Identity,* 67–82.

72. Pfeiffer, *Lady's Voyage,* 13, 54, 73, 87, 138, 143.

73. Ibid., 3, 39, 67–69, 80, 82, 103, 169, 176, 270–72.

74. Haskell Springer, "The Captain's Wife at Sea," in *Iron Men, Wooden Women: Gender and Seafaring in the Atlantic World, 1700–1920,* ed. Margaret S. Creighton and Lisa Norling (Baltimore, 1996), 92–117; Lisa Norling, *Captain Ahab Had a Wife: New England Women and the Whalefishery, 1720–1870* (Chapel Hill, NC, 2000), 214–61.

75. *She Went A-Whaling: The Journal of Martha Smith Brewer Brown from Orient, Long Island, New York, around the World on the Whaling Ship Lucy Ann, 1847–1849,* ed. Anne MacKay (Orient, NY, 1993), 23, 24, 33, 35, 36, 37–38, 43, 45, 51, 65; Springer, "Captain's Wife at Sea," 92–117. For an example of an interesting yet never published circumnavigation account by a woman, see Charlotte A. Babcock, "Life on the Ocean Wave: Reminiscences of my Voyages Around the World, 1851–8," MS, c. 1886, Huntington Library.

76. Christian Diener and Graham Fulton-Smith, *Franz Hanfstaengl: Album der Zeitgenossen* (Munich, [1975]), unpaginated.

77. George Coffin, *A Pioneer Voyage to California and round the World, 1849 to 1852* (Chicago, 1908), 197.

78. Ibid., 10, 14, 16, 32, 52.

79. Ibid., 30, 34, 35, 41, 45.

80. Ibid., 60, 134, 184, 185.

81. Jacques Arago, *Curieux Voyage autour du monde* (Paris, 1853).

82. Herman Melville, *Moby-Dick,* ed. Harrison Hayford and Hershel Parker (New York, 1967), 143–44, 203, 204, 470.

6. Fast—faster

1. Jules Verne, *Around the World in Eighty Days* (1872), trans. Michael Glencross (New York, 2004), 33, 49.

2. Ibid., 1, 2, 3, 6.

3. J. R. Planché, *Mr. Buckstone's Voyage round the Globe* (IN LEICESTER SQUARE). *A Cosmographical, Visionary Extravaganza, and Dramatic Review, In One Act and Four Quarters* (London, 1879), 15, 17, 19, 25.

4. V. Askechninskii, *Iakov Amfiteatrov, ordinarnyi professor Kievskoi D. Akademii (Biograficheskii ocherk) (Iakov Amfiteatrov, Professor Ordinary at the Kiev Spiritual Academy [Biographical Sketch]*) (Kiev, 1857), 55; see also Pavel Nebol'sin, *Razskazy proezzhago (Tales of a Traveler)* (St. Petersburg, 1854), 77–78. Translations by Gregory Afinogenov.

5. Thomas E. Beaumont, *Pencillings by the Way: A Constitutional Voyage round the World, 1870 and 1871* (London 1971); Alexander Graf von Hübner, *A Ramble around the World, 1871*, trans. Lady Herbert (New York, 1874); Raymond Cazallis Davis, *Reminiscences of a Voyage around the World* (Ann Arbor, MI, 1869), 3, 4.

6. Jeremy Black, *The British Abroad: The Grand Tour in the Eighteenth Century* (New York, 1992); James Brooks, *A Seven Months' Run, up, and down, and around the World* (New York, 1872).

7. J. M. Peebles, *Around the World: or, Travels in . . . "Heathen" Countries*, 2nd ed. (Boston, 1875); Ian Tyrrell, *Woman's World/Woman's Empire: The Women's Christian Temperance Union in International Perspective, 1880–1930* (Chapel Hill, NC, 1991), 50, 62–80.

8. Verne, *Around the World in Eighty Days*, 12–14.

9. Robert Evans, Jr., " 'Without Regard for Cost': The Returns on Clipper Ships," *Journal of Political Economy*, 72 (1964), 32–43; Francis Chichester, *Along the Clipper Way* (New York, 1966), 7–8, 13–15.

10. John Jennings, *Clipper Ship Days: The Golden Age of American Sailing Ships* (New York, 1952); Octavius T. Howe and Frederick C. Matthews, *American Clipper Ships* (Salem, MA, 1927); J. Y. Millar, *Brief Account of Voyage around the World of the Clipper Sailing Ship "Maulesden" in 1883* [Oakland, CA, 1927], 19 (time charter).

11. Verne, *Around the World in Eighty Days*, 14; John Adams, *Ocean Steamers: A History of Ocean-going Passenger Steamships, 1820–1970* (London, 1993), 21.

12. Ida Pfeiffer, *A Lady's Second Journey round the World* (New York, 1856), 39, 72, 111, 117, 191, 325, 457, 461–62, 489.

13. Ibid., 134, 326.

14. George Francis Train, *My Life in Many States and in Foreign Lands* (New York, 1902) 338; [Ludovic de Beauvoir], *Voyage autour du monde par le comte de Beauvoir* (Paris, 1874), 6, 878.

15. Clive Ponting, *A New Green History of the World: The Environment and the Collapse of Great Civilizations* (New York, 2007), 283; William Perry Fogg, *"Round the World": Letters from Japan, China, India, and Egypt* (Cleveland, OH, 1872), 23, 26.

16. Ponting, *New Green History of the World*, 281; *The Times* (London), June 5, 1871.

17. Lewis Mumford, *Technics and Civilization* (New York, 1934), 156–63; Asa Briggs, *Victorian Things* (London, 1988), 289–326; Etienne Barillier, *Steampunk!* (Lyon, 2010).

18. Jean Chesneaux, *The Political and Social Ideas of Jules Verne*, trans. Thomas Wikeley (London, 1972), 36, 69.

19. Verne, *Around the World in Eighty Days*, 8; *Jules Verne's Twenty Thousand Leagues Under the Sea: The Definitive, Unabridged Edition Based on the Original French Texts,* trans. Walter James Miller and Frederick Paul Walter (Annapolis, 1993), 164–82.

20. Verne, *Around the World in Eighty Days*, 16, 17, 19.

21. Ibid., 19; Bill Cormack, *A History of Holidays, 1812–1990* (London, 1998), 23–25.

22. Alan Sillitoe, *Leading the Blind: A Century of Guidebook Travel, 1815–1914* (London, 1995), 3, 5, 166; Planché, *Mr. Buckstone's Voyage round the Globe*, 12.

23. Verne, *Around the World in Eighty Days*, 23.

24. Ibid., 4; Sillitoe, *Leading the Blind*, 8, 213; John Torpey, *The Invention of the Passport: Surveillance, Citizenship, and the State* (Cambridge, 2000), 57–92.

25. Verne, *Around the World in Eighty Days*, 20, 31.

26. Roland Barthes, "The *Nautilus* and the Drunken Boat," *Mythologies*, trans. Annette Lavers (New York, 1972), 65–67; William Butcher, *Verne's Journey to the Centre of the Self: Space and Time in the Voyages extraordinaires* (Houndmills, UK, 1990).

27. Pamela Horn, *The Rise and Fall of the Victorian Servant* (Dublin, 1975), 30, 151–83.

28. Ibid., 13, 26, 72, 86–87. Jeeves tries to get Wooster to go around the world. See P. G. Wodehouse, *The Code of the Woosters* (New York, 2011), 2–3, 79.

29. Verne, *Around the World in Eighty Days*, 24, 37.

30. Samuel Irenaeus Prime, *The Life of Samuel F. B. Morse* (New York, 1875), 302.

31. Tom Standage, *The Victorian Internet: The Remarkable Story of the Telegraph and the Nineteenth Century's On-line Pioneers* (New York, 1998), 58.

32. Bern Dibner, *The Atlantic Cable* (Norwalk, CT, 1959).

33. *Nuovi Annali delle scienze naturali*, ser 3, vol. I (Bologna, 1850), 150; Charles F. Briggs and Augustus Maverick, *The Story of the Telegraph and a History of the Great Atlantic Cable* (New York, 1858), 12; *The Atlantic Telegraph Report of the Proceedings at a Banquet given to Mr. Cyrus W. Field . . .* (New York, 1866), 47; John B. Dwyer, *To Wire the World: Perry M. Collins and the North Pacific Telegraph Expedition* (Westport, CT, 2001).

34. C. T. McClenachan, *Detailed Report of Proceedings Had in Commemoration of the Successful Laying of the Atlantic Telegraph Cable* (New York, 1859), 50; Louis Figuier, *Les Merveilles de la science* (Paris, 1867), 283; Dibner, *Atlantic Cable*, 39; *Atlantic Telegraph Report*, 49. See also Salvador Manera, ed., *Historia de los progresos sociales . . .* , vol. I (Barcelona, 1870), 117.

35. Verne, *Around the World in Eighty Days*, 26–27.

36. C. P. Claret Fleurieu, *A Voyage round the World . . . by Etienne Marchand*, 3 vols. (London, 1801), II, 156–63.

37. D. A. Farnie, *East and West of Suez: The Suez Canal in History, 1854–1956* (Oxford, 1969), 3–86.

38. Ibid., 86–93.

39. Verne, *Around the World in Eighty Days*, 39–40, 41–42; Farnie, *East and West of Suez*, 120–37.

40. Verne, *Around the World in Eighty Days*, 51; Ian J. Kerr, *Building the Railways of the Raj, 1850–1900* (Delhi, 1995), 38, 211.

41. Ponting, *New Green History of the World*, 321.

42. Henry David Thoreau, *Walden, a Fully Annotated Edition*, ed. Jeffrey S. Kramer (New Haven, CT, 2004), 51; Jules Verne, *Claudius Bombarnac* (Paris, 1892), 19.

43. Verne, *Around the World in Eighty Days*, 45–46, 74.

44. Ibid., 54–57.

45. Ibid., 61–64.

46. Ibid., 15.

47. Pfeiffer, *Second Journey*, 263; Bernd Langensiepen and Ahmet Güleryüz, *The Ottoman Steam Navy, 1828–1923*, trans. James Cooper (London, 1995), 1–2; Daniel Panzac, *La Marine Ottomane: De l'apogée à la chute de l'Empire (1572–1923)* (Paris, 2009), 294–95.

48. Verne, *Around the World in Eighty Days*, 133.

49. Sheila Hones and Yasuo Endo, "History, Distance and Text: Narratives of the 1853–1854 Perry Expedition to Japan," *Journal of Historical Geography*, 32 (2006), 563–78; *Commodore Perry and the Opening of Japan: Narrative of the Expedition of an American Squadron to the China Seas and Japan, 1852–1854, the Official Report of the Expedition to Japan*, comp. Francis L. Hawks (Stroud, UK, 2005), 88, 481.

50. Kume Kunitake, *The Iwakura Embassy, 1871–73*, ed. Graham Healey and Chushichi Tsuzuki, 5 vols. (Chiba, 2002), I, xv, xvi; Ian Nish, ed., *The Iwakura Mission in America and Europe: A New Assessment* (Richmond, UK, 1998).

51. *San Francisco Daily Evening Bulletin*, Jan. 15, 1872; Kunitake, *Iwakura Embassy*, I, xvi, xxx, xxxiii, xxxvi, 30.

52. Kunitake, *Iwakura Embassy*, I, 13, 19, 33, 35; II, 100; III, 145; *Bangor Daily Whig & Courier*, Jan. 25, 1872; Shigeru Nakayama, "Diffusion of Copernicanism in Japan," in *The Reception of Copernicus' Heliocentric Theory*, ed. Jerzy Dobrizycki (Dordrecht, NL, 1971), 153–82.

53. Kunitake, *Iwakura Embassy*, I, 6; V, 272–84, 317.

54. Verne, *Around the World in Eighty Days*, 97, 112.

55. Ibid., 110–11, 114–26, 139–43.

56. Ibid., 154; Stephen E. Ambrose, *Nothing Like It in the World: The Men Who Built the Transcontinental Railroad, 1863–1869* (New York, 2000); *Entertainment Given to Mr. A. A. Low . . .* (New York, 1867), 41, 44.

57. R. M. Devens, *Our First Century* (Chicago, 1878), 914.

58. *Le Magasin pittoresque*, Apr. 1870, 108. See also Allen Foster, *Around the World with Citizen Train: The Sensational Adventures of the Real Phileas Fogg* (Dublin, 2002), 197.

59. Train, *My Life*, 331, 339; Foster, *Around the World with Citizen Train*, 175–93, 331, 332.

60. Cormack, *History of Holidays*, 26–27.

61. Verne, *Around the World in Eighty Days*, 184, 190–98, 208–16.

62. Ibid., 217–22, 227–29.

63. Bret Harte, *The Writings of Bret Harte: Poems and Two Men of Sandy Bar, a Drama* (Boston, 1896), 107–11, 315.

64. Verne, *Around the World in Eighty Days*, 35, 53, 144; Verne, *Le Tour du monde en quatre-vingts jours* (Paris, 1873), 305; John Sutherland, *Who Betrays Elizabeth Bennet? Further Puzzles in Classic Fiction* (Oxford, 1999), 211–14; Butcher, *Jules Verne*, 226.

65. Werner Sollors, *Neither Black nor White Yet Both: Thematic Explorations of Interracial Literature* (New York, 1997), 173–82, 189–219, 337–59 (on happy endings; the rest of the book discusses the more common tragic marriages). See also Shuchi Kapila, *Educating Seeta: The Anglo-Indian Family Romance and the Poetics of Indirect Rule* (Columbus, OH, 2010), and Damon Ieremia Salesa, *Racial Crossings: Race, Intermarriage, and the Victorian British Empire* (New York, 2011).

66. Jules Verne, "Edgard Pöe et ses oeuvres," *Le Musée des familles*, 31 (1864), 204; François Rabelais, *The Histories of Gargantua and Pantagruel*, trans. J. M. Cohen (Harmondsworth, UK, 1955), 171; Verne's copy of Rabelais is noted in Volker Dehs, "La bibliothèque de Jules et Michel Verne," *Verniana*, 3 (2011), 95.

67. Jules Verne, "Les Méridiens et le calendrier," *Bulletin de la Société de Géographie* (Paris), vol. 6, no. 6 (July 1873), 423–29. Butcher, *Jules Verne*, 226, claims that the second use of *inconsciemment* was in Verne's *Mysterious Island* (1876).

68. Ian Bartky, *One Time Fits All: The Campaigns for Global Uniformity* (Stanford, CA, 2007), 21–22, 26, 35–47, 59–99.

69. Verne, *Around the World in Eighty Days*, 36, 229; Horn, *Victorian Servant*, 129, 186; Verne, *Le Tour du monde en quatre-vingts jours*, frontis.

7. The Club of Eccentrics

1. A. d'Ennery and Jules Verne, *Le Tour du monde en 80 jours*, in *Les Voyages au théâtre* (Paris, [1881]), 48–49, 67–73, 87–92, 125; Marguerite Allotte de la Fuÿe, *Jules Verne* (New York, 1956), trans. Erik de Mauny, 143–46. The play differs from the novel in many respects, with added characters and events. See the editorial comments in Jules Verne, *Around the World in Eighty Days*, trans. and ed. William Butcher (Oxford, 1995), 207.

2. *Le Temps*, Nov. 7, 1872, and Dec. 15, 1872; "Jules Verne at Home," *McClure's Magazine*, Jan. 1894; "Jules Verne at Home," *Strand Magazine*, Feb. 1895; *The Pioneer*, Allahabad (begins Feb. 11, 1876); Brian Taves and Stephen Michaluk, Jr., *The Jules Verne Encyclopedia* (Lanham, MD, 1996), 134–35.

3. David Parlett, *The Oxford History of Board Games* (Oxford, 1999), 34–35, 88–89, 95–100.

4. Jules Verne and Gabriel Marcel, *Découverte de la Terre. Histoire générale des grands voyages et des grands voyageurs* (Paris, 1870–80).

5. Allotte de la Fuÿe, *Jules Verne*, 137; Piers Brendon, *Thomas Cook: 150 Years of Popular Tourism* (London, 1991), 141–50; Thomas Cook, *Letters from the Sea and from Foreign Lands Descriptive of a Tour round the World* (London, [1873]), vi.

6. Cook, *Letters from a Tour round the World*, 10, 15, 16–18, 20–22, 24–25, 32, 35.

7. Brendon, *Thomas Cook*, 150, 151.

8. Eugene Vetromile, *A Tour in Both Hemispheres; or, Travels around the World* (New York, 1880), 107; William Perry Fogg, *"Round the World": Letters from Japan, China, India, and Egypt* (Cleveland, OH, 1872), 30, 32, 34.

9. E. Hepple Hall, *The Picturesque Tourist. A Handy Guide round the World* (New York, 1877); Fogg, *"Round the World,"* 155.

10. Cook, *Tour round the World*, 9–10, 23, 67.

11. John Dunmore Lang, *Brief Notes of the New Steam Postal Route from Sydney and New Zealand to London* (Sydney, 1875), 48, 64.

12. William J. J. Spry, *The Cruise of Her Majesty's Ship "Challenger"* (London, 1876), 4. On the expedition in general, see Helen Rozwadowski, "Small World: Forging a Scientific Maritime Culture," *Isis*, 87 (1996), 409–29; Richard Corfield, *The Silent Landscape: Discovering the World of the Oceans in the Wake of HMS Challenger's Epic 1872 Mission to Explore the Sea Bed* (London, 2004).

13. Spry, *Cruise of HMS Challenger*, 1, 179; T. H. Tizard, et al., *Narrative of the Cruise of H.M.S. Challenger*, 2 vols. (London, 1885), I, pt. 1, xliv–xlvi, xlviii–xlix; C. Wyville Thomson, *The Voyage of the "Challenger": The Atlantic*, 2 vols. (London, 1877), I, 1–9.

14. Spry, *Cruise of HMS Challenger*, 4; Tizard, *H.M.S. Challenger*, I, xxxviii–xxxix; Thomson, *Voyage of the "Challenger,"* I, 11–59, 93–106.

15. Spry, *Cruise of HMS Challenger*, 4, 383.

16. Ibid., 61, 92, 176, 244–45, 327, 367; Tizard, *H.M.S. Challenger*, I, pt. 2, 769–71; Gerhard Kortum, "The German Challenger of Neptune: The Short Life and Tragic Death of Rudolph von Willemoes-Suhm," *Challenger Society for Marine Science*, 8 (1998), 1–5; Corfield, *Silent Landscape*, 74.

17. Spry, *Cruise of HMS Challenger*, 59, 93, 118, 146, 171, 226–27, 230, 252, 302–03, 326, 335, 345.

18. Ibid., iv, 374, 384; Thomson, *Voyage of the "Challenger,"* I, xvi, 93–106.

19. Earl of Suffolk and Berkshire, Hedley Peek, and F. G. Aflalo, eds., *The Encyclopedia of Sport*, 2 vols, (New York, 1898), II, 615; Mrs. [Anna] Brassey, *Around the World on the Yacht "Sunbeam": Our Home on the Ocean for Eleven Months* (New York, 1878), 467–68; Helen Rozwadowski, *Fathoming the Ocean: The Discovery and Exploration of the Deep Sea* (Cambridge, MA, 2008), 118–21, 128.

20. [Harden Sidney Melville], *A Boy's Travels round the World, or the Adventures of a Griffin . . . Related by Himself* (London, 1867), 3, 71, 139–41.

21. Samuel Smiles [Sr.], ed., *Round the World . . . by a Boy* (New York, 1871), v, 15, 21, 35, 62; Corfield, *Silent Landscape*, 86.

22. P.-J. Stahl, *Voyage de découvertes de Mlle Lili et de son cousin Lucien* (Paris, 1866–67). Jules Verne's publisher, Pierre-Jules Hetzel, produced this work.

23. Brassey, *Around the World on the "Sunbeam,"* 6; Lady Annie [Anna] Brassey Papers, Huntington Library, vol. 70, p. 13.

24. Brassey, *Around the World on the "Sunbeam,"* 2, 13, 59–60, 83, 91, 184–85, 200–01, 326.

25. Ibid., 119–20, 122, 130, 132, 207.

26. Ibid., 281, 284, 308, 333, 335, 337, 387–88.

27. Ibid., 106, 107, 111, 315.

28. Ibid., 391, 417–18, 428, 429, 445–46.

29. A. W. Brian Simpson, *Cannibalism and the Common Law: The Story of the Tragic Last Voyage of the Mignonette and the Strange Legal Proceedings to Which It Gave Rise* (Chicago, 1984); Brassey, *Around the World on the "Sunbeam,"* 315.

30. Rozwadowski, *Fathoming the Ocean*, 186.

31. Torcuato Tárrago y Mateos, *Gran viaje universal alrededor del mundo, descrito por una sociedad de viajeros modernos*, 2 vols. (Madrid, 1881–82), I, 5–7, 10–13, 27, 48, 492, 624–29.

32. Enrique Dupuy de Lôme, *De Madrid a Madrid, dando la vuelta al mondo* (Madrid, 1877), 18, 153, 183–84; Ugo Bedinello, *Diario del viaggio intorno al globo delle regia corvetta italian "Vettor Pisani"* (Trieste, 1876), 65, 84, 178, 179; Ludwig Salvator, *Um die Welt ohne zu wollen* (Prague, 1886).

33. Fogg, *"Round the World,"* 95, 132–33; Andrew Carnegie, *Notes of a Trip round the World* (New York, 1879), 1, 2, 133, 154, 211–12.

34. Dupuy de Lôme, *De Madrid a Madrid,* 80–81, 240; Bedinello, *Viaggio intorno al globo,* 35, 86; Salvator, *Um die Welt ohne zu wollen,* 56, 251.

35. Dupuy de Lôme, *De Madrid a Madrid,* 32, 68, 94, 246, 303; Fogg, *"Round the World,"* 122.

36. Bedinello, *Viaggio intorno al globo,* 34, 174, 188; Smiles, *Round the World,* 205–07; Vetromile, *Tour in Both Hemispheres,* 469.

37. Smiles, *Round the World,* 208; Cook, *Letters from a Tour round the World,* 31; Carnegie, *Trip round the World,* 97, 107; M. Le Baron de Hübner, *A Ramble round the World,* trans. Lady Herbert, 2 vols. (London, 1874), I, v; Thérèse Yelverton, *Teresina Peregrina; or, Fifty Thousand Miles of Travel round the World,* 2 vols. (London, 1874), I, 97–106, 126, 130–31, 135 (not all of these comments flatter the Chinese, but the mixed account is still exceptional). For another example of the visible equator game, see J. Y. Millar, *Brief Account of Voyage around the World of the Clipper Sailing Ship "Maulesden" in 1883* [Oakland, CA, 1927], 9.

38. Li Gui, *A Journey to the East: Li Gui's A New Account of a Trip around the Globe,* trans. and intro. Charles Desnoyers (Ann Arbor, MI, 2004), vii, 2, 269.

39. Nathan Sivin, "Copernicus in China," *Studia Copernica,* 6 (1973), 63–122; John B. Henderson, *The Development and Decline of Chinese Cosmology* (New York, 1984), 61–62, 68–71, 137–73, 214–15, 237–41; Betty Peh-T'i Wei, *Ruan Yuan, 1764–1849: The Life and Work of a Major Scholar-Official in Nineteenth-Century China Before the Opium War* (Hong Kong, 2006), 310–11.

40. Li, *Journey to the East,* 87, 167, 180, 245–48.

41. Ibid., 177, 192–93, 194, 195, 268–69, 271, 274, 277.

42. Ian R. Bartky, *One Time Fits All: The Campaigns for Global Uniformity* (Stanford, CA, 2007), 48–58.

43. A. M. W. Downing, "Where the Day Changes," *Journal of the British Astronomical Association,* 10 (1900), 176–78.

44. Michitsura Noau, *Ōbei junkai nisshi* (Tokyo, 1886); Kiyotaka Kuroda, *Kan'y ū nikki* (Tokyo, 1887); Takeshirō Nagayama, *Shūyū nikki* (Tokyo, 1889); Hōsei Yorimitsu, *Sekai shūyū jikki* (Tokyo, 1891); Enryō Inoue, *Inoue Enryō sekai ryokōki* (Tokyo, 2003). Synopses from translator Jennifer van der Grinten.

45. Yŏnghwan Min, *Haech'ŏn ch'ubŏm: 1896nyŏn Min Yŏnghwan ŭi segye ilju/(Min Yŏnghwan's Trip around the World in 1896),* ed. Chaegon Cho (Seoul, 2007), 40–41, 159–60. Translations by Ilsoo Cho; McCune-Reischauer romanization of Korean. Most of the account was probably composed by Min's secretary—cf. Michael Finch, *Min Yŏng-hwan: A Political Biography* (Honolulu, 2002), 78–114.

46. P. C. Mozoomdar, *Sketches of a Tour round the World* (1884), 2nd ed. (Calcutta, 1940), 120.

47. David V. Herlihy, *Bicycle: The History* (New Haven, CT, 2004), 75–221.

48. Nicholas Thomas, *Entangled Objects: Exchange, Material Culture, and Colonialism in the Pacific* (Cambridge, MA, 1991), 83–124; Peter N. Stearns, *Consumerism in World History: The Global Transformation of Desire* (New York, 2001).

49. Thomas Wentworth Higginson, preface, in Thomas Stevens, *Around the World on a Bicycle*, 2 vols. (New York, 1887), I, [1].

50. Stevens, *Around the World on a Bicycle*, I, 2, 3, 15; II, 114, 126, 284.

51. Ibid., I, 7, 12, 19, 124; II, 30, 285, 355, 461–62.

52. Ibid., I, 23, 29, 40, 43, 404–05, 452–53, 505.

53. Ibid., I, 208–11, 410–12; II, 85–86, 177–78, 211.

54. Ibid., II, 197–208, 223–25, 230, 246, 256–82.

55. Ibid., II, frontis., 161.

56. Thomas Gaskell Allen, Jr., and William Lewis Sachtleben, *Across Asia on a Bicycle: The Journey of Two American Students from Constantinople to Peking* (New York, 1894), xi, xii, 6, 7, 8, 203; David V. Herlihy, *The Lost Cyclist: The Epic Tale of an American Adventurer and His Mysterious Disappearance* (Boston, 2010), 25–48, 71–94.

57. Herlihy, *Lost Cyclist*, 59–70, 93–118, 125–39, 147–69, 173–76.

58. Ibid., 176–275.

59. *Nellie Bly's Book: Around the World in 72 Days* (New York, 1890), 3–5, 17, 56.

60. Ibid., 7–14, 256.

61. Ibid., 18–19, 21.

62. Ibid., 45–57, 185, 213, 282–83.

63. Ibid., 75, 114, 122, 217.

64. Ibid., 248, 260, 267, 280, 283.

65. Jason Marks, *Around the World in 72 Days: The Race Between Pulitzer's Nellie Bly and Cosmopolitan's Elizabeth Bisland* (New York, 1993); "She's Broken Every Record!" *New York World*, Jan. 26, 1890; "Ralph" [John MacGregor], *The Girdle of the Globe; or the Voyage of Mister Mucklemouth* (London, 1890), 336–37.

66. "Race around the World" (New York, c. 1891), Elliott Avedon, Virtual Museum of Games, University of Waterloo, Ontario, Canada; "Around the World with Nellie Bly" (New York, 1890), New-York Historical Society; Parlett, *Oxford History of Board Games*, 100–01.

67. George Francis Train, *My Life in Many States and in Foreign Lands* (New York, 1902), 335–39.

68. Peter Zheutlin, *Around the World on Two Wheels: Annie Londonderry's Extraordinary Ride* (New York, 2007), the best history of her adventure.

69. *El Paso Daily Times,* June 29, 1895; Nellie Bly Junior, "Around the World on a Bicycle," *New York World*, Oct. 20, 1895.

70. Zheutlin, *Annie Londonderry's Extraordinary Ride*, 23–24; *Los Angeles Times*, May 29, 1895.

71. H. Darwin McIlrath, *Around the World on Wheels for the Inter-Ocean* (Chicago, 1898), 61, 90–91, 110–11; John Foster Fraser, *Round the World on a Wheel* . . . (New York, 1899), v, 66–67, 80, 81, 114, 339–40, 356–57, 384, 410–11.

72. Joshua Slocum, *Sailing Alone around the World*, ed. Thomas Philbrick (New York, 1999), xxx–xxxi, xxxix; Walter Magnes Teller, *Joshua Slocum* (New Brunswick, NJ, 1971); Geoffrey Wolff, *The Hard Way Around: The Passages of Joshua Slocum* (New York, 2010).

73. Slocum, *Sailing Alone*, 14, 23, 25, 127.

74. Ibid., 15, 21, 182.

75. Ibid., 25, 127, 130, 132, 164, 207.

76. Ibid., 34–38.

77. Ibid., 45–47, 76–77, 79–81, 100.

78. Ibid., 19, 223–24, 225.

79. Ibid., 39, 133, 152–53, 160, 223, 242.

80. Ibid., 50, 104–05, 207–08, 214; Christine Garwood, *Flat Earth: The History of an Infamous Idea* (London, 2007), 79–218.

81. Slocum, *Sailing Alone*, 243.

8. Pure Pleasure

1. Clive Ponting, *A New Green History of the World: The Environment and the Collapse of Great Civilizations* (New York, 2007), 281; "British Empire, Showing the Commercial Routes of the World and Ocean Currents" *The Times*, [London, 1895], David Rumsey Map Collection: http://www.davidrumsey.com.

2. Henry Gaze & Sons, *Gaze's Tourist Gazette* (London and New York, 1895).

3. Ibid., 11, 18, 42–43, 103.

4. Adam M. McKeown, "Global Migration, 1846–1940," *Journal of World History*, 15 (2004), 155–89; John Torpey, *The Invention of the Passport: Surveillance, Citizenship, and the State* (Cambridge, 2000), 93–111; William Corry, *A Tour round the World* (London, 1879), 226; Frederick Treves, *The Other Side of the Lantern: An Account of a Commonplace Tour round the World* (London, 1922), 261; Andrew Carnegie, *Notes of a Trip round the World* (New York, 1879), 14.

5. http://www.circumnavigators.org/about; *Certificate of Membership in the Circumnavigators' Club* (New York, 1912), held by Francis B. Purdie (member no. 176), vii, xvii, xix, private collection.

6. *Voyage autour du monde par un petit Français* (Paris, c. 1905), http://douglasstewart.com.au/objects/wonderful-panorama (last accessed 11/15/11).

7. Collver Tours, *Round the World: Exceptional Journeys under Escort* (Boston, 1908), t.p., 6, 31, 35–75, 76; Whitney Coombs, *The Wages of Unskilled Labor in Manufacturing Industries in the United States, 1890–1924* (New York, 1926), table B, p. 136.

8. Collver Tours, *Round the World*, 5, 30; William G. Frizell and George H. Greenfield, *Around the World on the Cleveland* [Dayton, OH, 1910], 11, 33.

9. Ken Kobayashi, *Nihon hatsu no kaigai kankō ryokō : 96-nichikan sekai isshū* (Yokohama, 2009), 7, 19, 34; Kaneyuki Ishikawa, *Sekai isshū gahō* (Tokyo, 1908); Eiichi Shibusawa, *Ōbei kikō* (Tokyo, 1903; rpr. 1989), 39. Synopses and quotation from translator Jennifer van der Grinten.

10. Mark Twain, *Following the Equator: A Journey around the World* (Hartford, CT, 1897), 62, 508.

11. Ibid., 615, 616, 631.

12. Ibid., 75.

13. Ibid., 95, 352.

14. Ibid., 506, 623, 625.

15. Jean d'Albrey, *Du Tonkin au Havre . . .* (Paris, 1898), 99–105, 113.

16. Ibid., 106–08, 145–46.

17. Ibid., 168, 242, 290; Twain, *Following the Equator*, 200–05.

18. Twain, *Following the Equator*, 712.

19. Philip Dawson, *The Liner: Retrospective and Renaissance* (London, 2005), 39–45, 52–57; Annabelle Kent, *Round the World in Silence* (New York, 1911).

20. Frank C. Clark Co., *Clark's Cruise of the "Cleveland" . . . around the World "Eastward" Leaving New York, October 16, 1909* (New York, 1909); Frizell and Greenfield, *Around the World on the Cleveland*, 9–10.

21. R. H. Casey, *Notes Made During a Cruise around the World in 1913* (New York, 1914), 4–6, 121.

22. Ibid., 6; Frizell and Greenfield, *Around the World on the Cleveland*, 9–10; Paul S. Junkin, *A Cruise around the World* (Creston, IA, [1910?]), 2, 3, 4, 124; George Tome Bush, *40,000 Miles around the World* (Howard, PA, 1911), 6, 7–8.

23. Bush, *40,000 Miles around the World*, 8, 9, 10, 16, 17; Junkin, *Cruise around the World*, 61–62.

24. Frizell and Greenfield, *Around the World on the Cleveland*, 14; Bush, *40,000 Miles around the World*, 8–9, 10–11, 36.

25. Frizell and Greenfield, *Around the World on the Cleveland*, 91, 93; Bush, *40,000 Miles around the World*, 40.

26. Junkin, *Cruise around the World*, 35–36; Bush, *40,000 Miles around the World*, 40.

27. Li Gui, *Journey to the East: Li Gui's A New Account of a Trip around the Globe*, trans. and intro. Charles Desnoyers (Ann Arbor, MI, 2004), 292–93; John Dunmore Lang, *Brief Notes of the New Steam Postal Route from Sydney and New Zealand to London* (Sydney, 1875), 14–15, 55; *Nellie Bly's Book: Around the World in 72 Days* (New York, 1890), 97–98.

28. *Nellie Bly's Book*, 146, 152; Junkin, *Cruise around the World*, 15–17, 50–51, 78; Frizell and Greenfield, *Around the World on the Cleveland*, 102; Bush, *40,000 Miles around the World*, 21, 47; Casey, *Cruise around the World*, 26.

29. Dawson, *The Liner*, 79; Thomas Stevens, *Around the World on a Bicycle*, 2 vols. (New York, 1887). II, 258.

30. Treves, *Other Side of the Lantern*, 4, 11, 15, 16–17.

31. Frank C. Clark Co., *Clark's Fifth Cruise around the World* (New York, 1924).

32. Aldous Huxley, *Jesting Pilate: The Diary of a Journey* (London, 1926), 3.

33. John F. Anderson, *An Endeavorer's Working Journey around the World* (St. Louis, 1903), 3, 5–6, 12.

34. Ibid., 4, 5–6, 92, 123, 134, 137, 169–70, 179.

35. Ibid., 201–04, 220, 230, 233, 235, 243, 270, 278, 279, 290, 297, 309.

36. Jack London, *The Cruise of the Snark* (New York, 1911), 1, 7, 57–58; Alex Kershaw, *Jack London: A Life* (New York, 1997), 177; James L. Haley, *Wolf: The Lives of Jack London* (New York, 2010), 212–13, 227, 235, 240.

37. London, *Cruise of the Snark*, 3, 6, 10.

38. Ibid., 10, 15, 20; Kershaw, *Jack London*, 130, 199; Haley, *Wolf*, 239, 254.

39. London, *Cruise of the Snark*, 10, 36–46; Martin Elmer Johnson to Jack London, Nov. 5, 1906, Jack London Papers, Box 226, JL8466, Huntington Library.

40. Kershaw, *Jack London*, 190; Haley, *Wolf*, 244.

41. London, *Cruise of the Snark*, 47–57, 235–61 (quotations at 243, 248).

42. Ibid., 15; David Standish, *Hollow Earth: The Long and Curious History of Imagining Strange Lands, Fantastical Creatures, Advanced Civilizations, and Marvelous Machines Below the Earth's Surface* (New York, 2006), esp. 143–85.

43. London, *Cruise of the Snark*, 332; Kershaw, *Jack London*, 205, 233; Haley, *Wolf*, 256; *Nation*, 93 (Aug. 17, 1911). Charmian London and one of the crewmen also published accounts of the voyage, focusing on their South Seas passage, as if that had been the intention all along. None of the three accounts was a best seller; see Haley, *Wolf*, 248, 259.

44. Harry A. Franck, *A Vagabond Journey around the World* (New York, 1911), xiii, xiv, xv n.

45. Ibid., 5, 23, 59, 102–04, 129, 239–41, 278–88, 410–43, 489, 502.

46. Ibid., xiv, 106.

47. NYK postal stationery, Nihon Yusen Kabushiki Kaisha, folder 41, Kemble Maritime Ephemera, Huntington Library (cited hereafter as KME).

48. H. J. Eckenrode and Pocahontas Wight Edmunds, *E. H. Harriman: The Little Giant of Wall Street* (New York, 1933), 50–58, 93–98; Maury Klein, *The Life and Legend of E. H. Harriman* (Chapel Hill, NC, 2000), 283–87, 290–91.

49. P. C. Mozoomdar, *Sketches of a Tour round the World* (1884), 2nd ed. (Calcutta, 1940), 1–2, 3, 36, 38–39.

50. Piers Brendon, *Thomas Cook: 150 Years of Popular Tourism* (London, 1991), 152.

51. William D. Wray, *Mitsubishi and the N.Y.K., 1870–1914: Business Strategy in the Japanese Shipping Industry* (Cambridge, MA, 1984), 213–25, 290–91; *NYK Maritime Museum Guidebook* (Yokohama, 2005), 42, 124.

52. Donald Murray, "How Cables Unite the World," *The World's Work*, 4 (1902), 2298–299.

53. *Telephony*, 61 (1911), 484.

54. *Por eso mundos*, vol. 10, no. 120 (1905), 46; O. Moll, *Das Untersee-Kabel in Wort und Bild* (Cologne, 1904), 84–85; *Russkaia mysl'* (*Russian Thought*), vol. 16, no. 11 (Moscow, 1895), 171–72; Nikolai Kablukov, *Ob usloviiakh razvitiia krestianskogo khoziaistva v Rossii* (*On the Conditions of the Development of the Peasant Economy in Russia*) (Moscow, 1908), 116; Léon Say, *Les Finances de la France sous la troisième république*, vol. 3 (Paris, 1900), 1–2; *Izvestiia knizhnykh magazinov tovarishchestva M. O. Vol'f po literature, naukam, i bibliografii* (*Literary, Scientific, and Bibliographical Gazette of the M. O. Vol'f Bookstore Company*), vol. 11, no. 3 (March 1908), 53. Russian translations by Gregory Afinogenov.

55. Alfred, Viscount Northcliffe, *My Journey round the World (16 July 1921–26 Feb. 1922)*, ed. Cecil and St. John Harmsworth (Philadelphia, 1923), 161, 222.

56. Tom Chaffin, *Sea of Gray: The Around-the-World Odyssey of the Confederate Raider Shenandoah* (New York, 2006); Angus Curry, *The Officers of the CSS Shenandoah* (Gainesville, FL, 2006), esp. 294–317; Kenneth Wimmel, *Theodore Roosevelt and*

the Great White Fleet: American Sea Power Comes of Age (Washington, DC, 1998); James R. Reckner, *Teddy Roosevelt's Great White Fleet* (Annapolis, MD, 1988), xi; "Trying on Her New Necklace," *Literary Digest*, 35 (1908); Roman John Miller, *Around the World with the Battleships*, intro. by James B. Connolly (Chicago, 1909), 128, 361.

57. Reckner, *TR's Great White Fleet*, xi, 16–17, 30, 32, 103–05, 140.

58. For a narrative of the canal's building, see David McCullough, *The Path Between the Seas: The Creation of the Panama Canal, 1870–1914* (New York, 1977).

59. Ibid., 591, 610–11; full-length portrait of President Theodore Roosevelt, Rockwood Photo Co., c. 1903, Library of Congress.

60. McCullough, *Path Between the Seas*, 606–07, 609.

61. *NYK Maritime Museum Guidebook*, 124; *Ripples in Time: Collection of NYK History* (Yokohama, 2004), 82.

62. Hanna Khabbaz, *Ḥawla al-kurah al-arḍīyah/Around the Globe*, (New York, 1920), back of front cover, 17, 19–20, 26, 27, 30–31, 154, 157, translation by Dzavid Dzanic; Bill Cormack, *A History of Holidays, 1812–1990* (London, 1998), 81–82; Torpey, *Invention of the Passport*, 93–121; John Torpey, "The Great War and the Birth of the Modern Passport System," *Documenting Individual Identity: The Development of State Practices in the Modern World*, ed. Jane Caplan and John Torpey (Princeton, NJ, 2001), 256–70.

63. A representative example is Hendrik Willem Van Loon, *A Voyage of Re-discovery* (New York, 1934), for the 1934 tour of Cunard and Thomas Cook & Son on the *Franconia*.

64. "Round the World Cruises 1923," United American Lines, folder 9, KME; "Empress of Britain World Cruise, 1931–32," [1931], Canadian Pacific, folder 76, KME.

65. Thos. Cook & Son, *The Supreme Travel-Adventure: Around the World on the 'Franconia' 1929* (New York, 1928), 7; "Round the World by the Way of the Orient . . ." [1924–25], 1, Dollar Steamship Lines, folder 1, KME; Northcliffe, *My Journey round the World*, 9.

66. "Round the World with N.Y.K." (1937), 4, Nihon Yūsen Kaisha (NYK), folder 8, KME; Walter L. Haworth, *Ninety Days* (Los Angeles, [1939]), 16.

67. Brendon, *Thomas Cook*, 264; Thos. Cook & Son, *Around the World—A Cruise* (New York, 1925), 16, 67, 69.

68. See itineraries and prices for: "A Cruise Round the World," 1924, Canadian Pacific, folder 82, KME; "World Cruise Empress of Australia," 1927, Canadian Pacific, folder 69, KME; "Clark's Fifth Cruise around the World," 1925, Anchor Line, folder 18, KME; "To the Far East . . . and around the World," c. 1926, NYK, folder 7, KME; "Round the World Cruises 1923," 7, 72, 75, United American Lines, folder 9, KME; "The Raymond-Whitcomb Cruise Round the World 1924," 56, ibid.; "Two Cruises around the World" [1925], 59–60, United American Lines, folder 9, KME; "The Wonder Cruise of the S.S. Resolute around the World," 1926, United American Lines, folder 11, KME. See also Cormack, *History of Holidays*, 82.

69. Cook, *Around the World* (1925), 9–10; "Round the World Cruises 1923," 6, United American Lines; "The Raymond-Whitcomb Cruise Round the World

1924," 10, United American Lines; "World Cruise Empress of Australia" (1927), Canadian Pacific; "Fifth Annual Round the World Cruise" (1927), Canadian Pacific, folder 69, KME; "Around the World Cruises" (1938), American President Lines, folder 1, KME.

70. "New Oil-Burner 'California'" (c. 1925), Anchor Line, folder 18, KME; "Round the World Cruise S.S. Empress of Scotland" (1925), 4, Canadian Pacific, folder 102, KME; "Around the World—110 Days" (c. 1912), Hamburg-Amerika, folder 57, KME; "World Cruise Empress of Australia," Canadian Pacific; "A Cruise Round the World" (1924), 3, Canadian Pacific, folder 82, KME.

71. Cook, *Around the World* (1925), 19; Thos. Cook & Son, *Around the World via the Southern Hemisphere* (New York, 1927), 77–79; Cook, *Supreme Travel-Adventure*, 51; "A Cruise Round the World" (1924), Canadian Pacific; "World Cruise of the Red Star Line S.S. Belgenland" [1929], 127, International Mercantile Marine Co., KME; "Fifth Annual Round the World Cruise" (1927), 86–87, Canadian Pacific.

72. "Canadian Pacific Cruise Glimpses" (1929), Canadian Pacific, folder 25, KME; Cook, *Around the World* (1925), 67–70, 73–74; Cook, *Supreme Travel-Adventure*, 75, 76, 78.

73. "Two Cruises around the World" [1925], United American Lines; "Round the World by Way of the Orient . . ." [1924–25], 21; "Your Own Cruise—Round the World," (1930s), Dollar Steamship Lines, folder 1, KME; "The Marco Polo Log" (1933), 1, Dollar Steamship Lines, folder 37, KME; table of fares [c. 1928], Dollar Steamship Lines, folder 2, KME; "A First-Class Vacation Tour for College Folks" (1928), Dollar Steamship Lines, folder 11, KME; "Nippon Yusen Kaisha—Japan Mail S. S. Co." (1905), NYK, folder 7, KME.

74. *NYK Maritime Museum Guidebook*, 118, 119; *Ripples in Time*, 106; "Around the World Cruises" (December 1938), American President Lines, folder 1, KME; "A Ship within a Ship" [c. 1930–33], Dollar Steamship Lines, folder 3, KME; "Around the World $556" (1934), NYK, folder 8, KME; "Around the World in 'Second Class'" (1935), NYK, folder 8, KME; "Thrift Tour" [1930s], Dollar Steamship Lines, folder 12, KME.

75. "Round the World Cruises 1923," 31, 43, United American Lines; Cook, *Around the World* (1925), 58; "Equator Dinner" (1926), United American Lines, folder 13, KME; "World Cruise of the Empress of Britain" [1931]; *Ripples in Time*, 108; Haworth, *Ninety Days*, 22.

76. "Dancing Round the World," International Mercantile Marine Co., c. 1920s, KME.

77. "World Cruise Empress of Britain 1934," pages for "Dressing around the World," Canadian Pacific, folder 77, KME; "Empress of Britain World Cruise from New York" (1937), pages for "Dressing around the World" and "Shopping around the World," Canadian Pacific, folder 77, KME; "Shopping around the World" (1928), NYK, folder 43, KME; "World Cruise of the Red Star Line S.S. Belgenland" [1929], with blank pages for notes, International Mercantile Marine Co., KME; "World Cruise of the Empress of Britain: A 'Memogram'" [1931–32], with monthly calendars for memoranda, Canadian Pacific, KME.

78. "World Cruise of the Red Star Line S.S. Belgenland," notes by Ira Oscar

Kemble, 91, 154, 189; Esther G. Leggett, *Extracts from Letters of Esther G. Leggett to Caroline Hazard* (Peace Dale, RI, 1938); Frederick James Hill, *Lantern Slides for a World Tour* (New York, 1928), 1.

79. Ivan Bunin, *The Gentleman from San Francisco,* trans. Bernard Guilbert Guerney (New York, 1934), 281–82.

80. Ibid., 282–86, 300, 307.

81. D. J. Richards, "Comprehending the Beauty of the World: Bunin's Philosophy of Travel," *Slavonic and East European Review,* 52 (1974), 514–32.

82. Jean Cocteau, *Round the World Again in 80 Days (Mon Premier Voyage),* trans. Stuart Gilbert (London, 2000), 1, 5.

83. Ibid., 9, 62, 136, 200–01.

84. Ibid., 77, 105, 115.

85. Ibid., 16, 17, 78, 127, 197, 211–12, 222.

86. Ibid., 224.

Second Entr'acte

1. Richard P. Hallion, *Taking Flight: Inventing the Aerial Age from Antiquity Through the First World War* (New York, 2003), 3–60; Joyce E. Chaplin, *The First Scientific American: Benjamin Franklin and the Pursuit of Genius* (New York, 2006), 293–302.

2. Hallion, *Taking Flight,* 61–268; Guillaume de Syon, *Zeppelin!: Germany and the Airship, 1900–1939* (Baltimore, 2002), 1–39.

3. F. T. Marinetti, "The Founding and Manifesto of Futurism" (1909), in *Futurism: An Anthology,* ed. Lawrence Rainey, Christine Poggi, and Laura Wittman (New Haven, CT, 2009), 49–52.

4. Robert Wohl, *A Passion for Wings: Aviation and the Western Imagination, 1908–1918* (New Haven, CT, 1994), 97–123, 157–201; Robert Wohl, *The Spectacle of Flight: Aviation and the Western Imagination, 1920–1950* (New Haven, CT, 2005), 49–107.

5. Wohl, *Passion for Wings,* 57; Martin Staniland, *Government Birds: Air Transport and the State in Western Europe* (Lanham, MD, 2003), 1–10.

6. Akira Iriye, *Global Community: The Role of International Organizations in the Making of the Contemporary World* (Berkeley & Los Angeles, 2002), 9–36; Jay Winter, *Dreams of Peace and Freedom: Utopian Moments in the Twentieth Century* (New Haven, CT, 2006), 11–74; Glenda Sluga, "Was the Twentieth Century the Great Age of Internationalism?" *Australian Academy of the Humanities Proceedings,* 34 (2010), 155–74.

7. Quoted in Carroll V. Glines, *Around the World in 175 Days: The First Round-the-World Flight* (Washington, DC, 2001), 48.

8. Antoine de Saint-Exupéry, *Southern Mail/Night Flight,* trans. Curtis Cate (Harmondsworth, UK, 1976), 112, 130, 147, 158, 160–61.

9. Glines, *Around the World in 175 Days,* 2. European and Chinese delegates at an international conference on railroads claimed, in 1902, that the circuit could be made in forty days, but no one seems to have tried it. See "Le Tour du monde en quarante jours," *Revue scientifique,* 18 (1902), 602.

10. Harriet White Fisher, *A Woman's World Tour in a Motor* (Philadelphia, 1911), 7, 8, 15, 17–18, 21–22, 23, 40–41, 50, 54, 308, 353.

11. Ibid., 15, 23, 30, 61–62, 73, 81, 105, 159, 166–70, 209, 213, 217, 313, 347–51.

12. James A. Ward, *Three Men in a Hupp: Around the World by Automobile, 1910–1912* (Stanford, CA, 2003), 4, 111, 118.

13. Ibid., 2, 59–60, 116, 157–58, 160, 211–18; Whitney Coombs, *The Wages of Unskilled Labor in Manufacturing Industries in the United States, 1890–1924* (New York, 1926), table C, p. 138.

14. Clärenore Stinnes, *Im Auto durch zwei Welten: Die erste Autofahrt einer Frau um die Welt 1927 bis 1929,* ed. Gabriele Habinger (Vienna, 1996); "Around-the-World-Motor Oil" can, c. 1930s, private collection.

15. "Herbert Strang," *Round the World in Seven Days* (New York, 1910); *Aero and Hydro,* Feb. 21, 1914.

16. Wohl, *Passion for Wings,* 69–94, 203–51, 282–89; Hallion, *Taking Flight,* 296–315, 335–79; de Syon, *Zeppelin!,* 71–109.

17. Stuart Banner, *Who Owns the Sky?: The Struggle to Control Airspace from the Wright Brothers On* (Cambridge, MA, 2008), 16–68; Robert J Millichap, "Airline Markets and Regulation," in *Modern Air Transport: Worldwide Air Transport from 1945 to the Present,* ed. Philip Jarrett (London, 2000), 37–40.

18. W. T. Blake, *Flying round the World* (London, 1923), 11, 205; Glines, *Around the World in 175 Days,* 15, 22.

19. Glines, *Around the World in 175 Days,* 32, 65–66, 165; Dan Hagedorn, *Conquistadors of the Sky: A History of Aviation in Latin America* (Gainesville, FL, 2008), 197–99; Lowell Thomas, *The First World Flight: Being the Personal Narratives of Lowell Smith, Erik Nelson, Leigh Wade, Leslie Arnold, Henry Ogden, John Harding* (Boston, 1925), 6.

20. Glines, *Around the World in 175 Days,* 165.

ACT THREE: DOUBT
9. Flight

1. Lowell Thomas, *The First World Flight: Being the Personal Narratives of Lowell Smith, Erik Nelson, Leigh Wade, Leslie Arnold, Henry Ogden, John Harding* (Boston, 1925), 19; Carroll V. Glines, *Around the World in 175 Days: The First Round-the-World Flight* (Washington, DC, 2001), 25–26; Patrick M. Stinson, *Around-the-World Flights: A History* (Jefferson, NC, 2011), provides a full inventory.

2. Glines, *Around the World in 175 Days,* 26–30; Thomas, *First World Flight,* 21–46.

3. Glines, *Around the World in 175 Days,* 23; Thomas, *First World Flight,* 53.

4. Glines, *Around the World in 175 Days,* 14, 18–21; Thomas, *First World Flight,* 108.

5. Glines, *Around the World in 175 Days,* 24–25; Thomas, *First World Flight,* 17, 72, 87, 89, 90, 91.

6. Glines, *Around the World in 175 Days,* 23, 107, 156.

7. Thomas, *First World Flight,* 205, 207, 256.

8. Glines, Around *the World in 175 Days,* 81–82, 88, 132–38, 145; Thomas, *First World Flight,* 105, 110, 139–42, 271.

9. Glines, Around *the World in 175 Days,* 113; Thomas, *First World Flight,* 162–75.

10. Thomas, *First World Flight,* 236–37; Glines, *Around the World in 175 Days*, 55.

11. Thomas, *First World Flight,* 237.

12. Glines, *Around the World in 175 Days*, 151; Thomas, *First World Flight,* 197–98, 296.

13. Glines, *Around the World in 175 Days*, 148, 149, 154, 155; Thomas, *First World Flight,* 301, 311–12, 315.

14. Glines, *Around the World in 175 Days*, 159.

15. *The Times* (London), Sept. 30, 1924; Thomas, *First World Flight*, 312.

16. Thomas, *First World Flight,* title page, xxi, 2–3, and illus. between pp. 26 and 27 and facing pp. 34, 323.

17. Ibid., 150, 221.

18. Charles Stuart Dennison, *Around the World with Texaco* (Houston, 1925), vii, 14, 97, 115, 164.

19. "A Flight around the World" ([Nuremberg], Germany, 1928), Museum of Childhood, Victoria & Albert Museum, London.

20. Linton Wells, *Around the World in Twenty-Eight Days* (Boston, 1926), xix, 4–5, 19, 233.

21. Ibid., 118–19, 199, 213, 268.

22. Ibid., 1–89, 108–09.

23. Dan Hagedorn, *Conquistadors of the Sky: A History of Aviation in Latin America* (Gainesville, FL, 2008), 219–20; D. Costes and J. M. Le Brix, *Notre Tour de la terre* [Paris, 1928], 4, 5, 168–69, 180, 216–17.

24. Costes and Le Brix, *Notre Tour de la terre*, 62–64, 88–92, 126–28, 134, 157, 174, 178, illus. facing 192.

25. Thos. Cook & Son, *The Supreme Travel-Adventure: Around the World in the "Franconia" 1929* (New York, 1928), 83; Priya Satia, "The Defense of Inhumanity: Air Control in Iraq and the British Idea of Arabia," *American Historical Review*, 111 (2006), 16–51.

26. Violette de Sibour, *Flying Gypsies: The Chronicle of a 10,000 Mile Air Vagabondage* (New York, 1930), 4, 6, 8, 10, 60, 293.

27. F. K. Baron von Koenig-Warthausen, *Wings around the World* (New York, 1930), 4, 6–7, 9, 11, 97.

28. Ibid., 19, 23, 67, 117, 149, 175.

29. Ibid., 8–9, 91; de Sibour, *Flying Gypsies*, 85, 98–99, 127, 179–80, 226, 244.

30. De Sibour, *Flying Gypsies*, 246–47; Koenig-Warthausen, *Wings around the World*, 108, 113, 116, 120, 121, 131, 181.

31. De Sibour, *Flying Gypsies*, 19–20, 134, illus. facing p. 144, 161.

32. Ibid., 289.

33. Koenig-Warthausen, *Wings around the World,* 137, 146, 161, 168, 175.

34. Ibid., 146–47, 156–59; de Sibour, *Flying Gypsies*, illus. facing p. 168.

35. Palle Huld, *A Boy Scout around the World: A Boy Scout Adventure*, trans. Eleanor Hard (New York, 1929), 1–2, 15.

36. Ibid., 5, 7, 9–10, 12–14.

37. Ibid., 14–15, 22.

38. Ibid., v-vi, 190.

39. Ibid., 93, 173, 177, 188.

40. Ibid., 185–86, 192, 193, 195, 197. One year later, in 1929, the Belgian illustrator Georges Prosper Remi, aka Hergé, introduced another *Boy's Own* fantasy, *The Adventures of Tintin*. Both Huld and Tintin were freckled, red-haired boy reporters who wore plus fours, breeches buckled around the knee. (In similar fashion, "Smilin' Jack," a cartoon aviator introduced in the *Chicago Tribune* in 1933, shared a nickname with one of the mechanics of the 1924 Army Air Service circumnavigation.) Hergé denied that Huld was his model for Tintin, who never goes around the world.

41. *Japan & America*, Feb. 1903, 13; ibid. Apr. 1903, 26–27.

42. Adi B. Hakim, Jal P. Bapasola, and Rustom B. Bhumgara, *With Cyclists around the World* (New Delhi, 2008), vii, viii, ix, x, 262, 363, back cover.

43. Ibid., 31, 133, 154, 168, 202, 208, 224, 230, 246, 257, 273, 276–77, 294, 296, 310, 311, 323–29, 332, 337.

44. I. S. K. Soboleff, *Nansen Passport: Round the World on a Motor-Cycle* (London, 1936), viii, 71, 81, 86, 88, 89, 92; John Torpey, *The Invention of the Passport: Surveillance, Citizenship, and the State* (Cambridge, 2000), 127–29.

45. Soboleff, *Nansen Passport*, 94, 127, 133, 136, 138–39, 181, 228, 229, 235.

46. John Henry Mears, *Racing the Moon (and Winning): Being the Story of the Swiftest Journey Ever Made, a Circumnavigation of the Globe by Airplane and Steamship in 23 Days, 15 Hours, 21 Minutes and 3 Seconds by Two Men and a Dog* (New York, 1928), 11, 15–16, 33, illus. between pp. 48 and 49.

47. Mears, *Racing the Moon*, 32, 146, 314–15; *Pravda*, no. 171, July 25, 1928 (trans. Greg Afinogenov). *Pravda* had on July 8, 10, and 11, 1928, given updates on the Americans' passage across the Soviet Union.

48. Mears, *Racing the Moon*, 22, 235, 306.

49. Guillaume de Syon, *Zeppelin!: Germany and the Airship, 1900–1939* (Baltimore, 2002), 110–46.

50. "The First Airship Flight around the World," *National Geographic*, 57 (June 1930), 553–88; Léo Gerville-Réache, *Autour du Monde en Zeppelin* (Paris, 1929), 56–62; Douglas Botting, *Dr. Eckener's Dream Machine: The Great Zeppelin and the Dawn of Air Travel* (New York, 2001), 150, 157.

51. "First Airship Flight," 655, 680, 688; Koenig-Warthausen, *Wings around the World*, 161; Gerville-Réache, *Autour du Monde en Zeppelin*, 74, 81.

52. Max Geisenheyner, "Around the World with the 'Graf Zeppelin,' A Novel Cardgame" (Bavaria, Germany, & New York: [1929]), Smithsonian National Air and Space Museum; Botting, *Dr. Eckener's Dream Machine*, 188; Gerville-Réache, *Autour du Monde en Zeppelin*, 27, 50, 62, 65.

53. Wiley Post and Harold Gatty, *Around the World in Eight Days: The Flight of the Winnie Mae*, intro. by Will Rogers (New York, 1932), 32–34, 40–41, 77, 264, 282; Stanley R. Mohler and Bobby H. Johnson, *Wiley Post, His Winnie Mae, and the World's First Pressure Suit* (Washington, DC, 1971), 4, 9; Bruce Brown, *Gatty: Prince of Navigators* (Sandy Bay, Australia, 1997), 63–71.

54. Mohler and Johnson, *Wiley Post*, 19, 20, 21, 26.

55. Ibid., 19; Post and Gatty, *In Eight Days*, photographs opposite pp. 112, 257.

56. Mohler and Johnson, *Wiley Post*, 20, 23; Post and Gatty, *In Eight Days*, 126, 216–17.

57. Post and Gatty, *In Eight Days*, 54, 70, 90, 175, 193, 216, 232; Brown, *Gatty*, 78; John McCannon, *Red Arctic: Polar Exploration and the Myth of the North in the Soviet Union, 1932–1939* (New York, 1998), 68–80.

58. Mohler and Johnson, *Wiley Post*, 20, 24; Post and Gatty, *In Eight Days*, 27, 68, 162, 213.

59. Post and Gatty, *In Eight Days*, 11, 70, 174, 197.

60. Mohler and Johnson, *Wiley Post*, 39; Brown, *Gatty*, 149.

61. Moher and Johnson, *Wiley Post*, 44–46, 50–51, 57.

62. Ibid., 59, 65, 66–67, 121–22.

63. Anthony Sampson, *Empires of the Sky: The Politics, Contests, and Cartels of World Airlines* (New York, 1985), 36, 37.

64. Gordon Pirie, *Air Empire: British Imperial Civil Aviation, 1919–39* (Manchester, 2009), 112–59, 198–214; Sampson, *Empires of the Sky*, 25–33, 50; Robert Wohl, *The Spectacle of Flight: Aviation and the Western Imagination, 1920–1950* (New Haven, CT, 2005), 165–210.

65. Obituary for Sir Hudson Fysh, *The Times* (London), Apr. 9, 1974; John Gunn, *The Defeat of Distance: Qantas, 1919–1939* (St. Lucia, Australia, 1985), esp. 186–255; Sampson, *Empires of the Sky*, 44–47, 50–53.

66. Martin Staniland, *Government Birds: Air Transport and the State in Western Europe* (Lanham, MD, 2003), 11–52; Sampson, *Empires of the Sky*, 36; David R. Jones, "The Rise and Fall of Aeroflot: Civil Aviation in the Soviet Union, 1920–91," in *Russian Aviation and Air Power in the Twentieth Century*, ed. Robin Higham, John T. Greenwood and Von Hardesty (London, 1998), 236, 244.

67. Certificate for Celeste Briggs, Aug. 22/21, 1937, Box 627, folder 4, Accession I, Pan American World Airways Records, University of Miami (Pan Am Recs.).

68. Emil Hurja, *Westward Ho—Fare Paid* (Juneau, AK, 1936), 72–73.

69. H. R. Ekins, *Around the World in Eighteen Days, and How to Do It* (New York, 1936), 1, 12–13, 25, 160; Bolivar Lang Falconer, *Flying around the World* (Boston, 1937), i, 2–5.

70. Ekins, *Eighteen Days*, 21, 33, 40, 42, 44; Falconer, *Flying around the World*, 17–18, 55, 78–79; de Syon, *Zeppelin!*, 191–93.

71. Charles H. Holmes, *A Passport round the World* (London, 1937), 13, 14, 25, 26; *Courier-Mail* (Brisbane), July 26, 1935.

72. Holmes, *A Passport round the World*, 33, 99; Falconer, *Flying around the World*, 29, 33.

73. Ekins, *Eighteen Days*, 47, 60–61, 149.

74. Jean Cocteau, *Round the World Again in 80 Days (Mon Premier Voyage)*, trans. Stuart Gilbert (London, 2000), 224; Falconer, *Flying around the World*, 17; de Syon, *Zeppelin!*, 189, 195–202.

75. Cocteau, *Round the World Again*, 218, 222.

76. Ibid., 223–24.

77. Aldous Huxley, *Jesting Pilate: The Diary of a Journey* (London, 1926), 82–84; Cocteau, *Round the World Again*, 229.

78. Holmes, *Passport round the World,* 254; Ekins, *Eighteen Days*, 16.

79. Falconer, *Flying around the World*, map facing p. 4; Ekins, *Eighteen Days*, endpapers.

80. Susan Schulten, *The Geographical Imagination in America, 1880–1950* (Chicago, 2001), 137–39, 205–06, 214–26, 231–34; McCannon, *Red Arctic.*

81. Ekins, *Eighteen Days*, 3, 160; Ian Mackersey, *Smithy: The Life of Sir Charles Kingsford Smith* (London, 1998).

82. On Earhart's celebrity, see Susan Ware, *Still Missing: Amelia Earhart and the Search for Modern Feminism* (New York, 1993); on Earhart as an aviator, Kathleen C. Winters, *Amelia Earhart: The Turbulent Life of an American Icon* (New York, 2010), esp. 167–216; Amelia Earhart, *Last Flight*, arr. George Palmer Putnam (New York, 1937), 4, 50–52.

83. Winters, *Amelia Earhart*, 184–91; Earhart, *Last Flight*, 85.

84. Earhart, *Last Flight*, 46, 63, 100, 137, 187; Winters, *Amelia Earhart*, 172–73, 175, 191–92.

85. Winters, *Amelia Earhart*, 205; Earhart, *Last Flight*, 146.

86. Winters, *Amelia Earhart*, 205–10; Earhart, *Last Flight*, 222–23.

87. Earhart, *Last Flight*, title page.

88. R. E. G. Davies, *A History of the World's Airlines* (London, 1964), 434; 1942 Annual Report, Box 82, folder 16, Accession I, Pan Am Recs.; "Register of Airline Slogans," IATA memorandum, Dec. 15, 1953, p. 10, Box 496, folder 1, ibid.

89. *The Times* (London), Aug. 22, 1939.

90. "Good-Will Flight," *Evening Post* (Wellington, NZ), Sept. 22, 1939.

91. *Osaka Mainichi (Daily News)*: Aug. 26, Sept. 1, Sept. 2, Sept. 4 (extra), Sept. 6, Sept. 24, Oct. 21, 1939.

92. Wohl, *Spectacle of Flight*, 213–74.

93. Alva L. Harvey, *Memoirs of an Around-the-World Mechanic (1924) and Pilot (1941)* (Manhattan, KS, 1978), 21–28.

94. Timeline of Pan Am history, box 50, folder 17, Accession I, Pan Am Recs.; Ed Dover, *The Long Way Home: Captain Ford's Epic Journey* (McLean, VA, 1999), quotation at p. 53.

95. Donald H. Agnew and William A. Kinney, "American Wings Soar around the World," *National Geographic,* 84 (July 1943), 57, 60, 78; Andrew R. Boone, "U.S. and Britain Complete Round-the-World Air Lines," *Popular Science* (May 1945), 123–24.

96. Wendell L. Willkie, *One World* (New York, 1943), 1, 2.

97. "Airways to Peace: An Exhibition of Geography for the Future," *Bulletin of the Museum of Modern Art*, 11 (New York, 1943), 3, 5, 7, 8, 9, 23.

98. "Airways to Peace," 23–24; MAID Catalogue #IN23627A, Museum of Modern Art, New York.

99. Sampson, *Empires of the Sky*, 62–71.

100. *Ten Thousand Times around the World* (n.p., 1945), p. 2, Box 623, folder 78, Accession I, Pan Am Recs.; "Around the World Today and Yesterday," press release, c. 1945, Box 204, folder 20, ibid.

101. "Register of Airline Slogans," IATA memorandum, Dec. 15, 1953, p. 10, Box

496, folder 1, Accession I, Pan Am Recs.; Sampson, *Empires of the Sky*, 82; USSR airmail stamp (1949), Allen-Lee Collection, Album 18, item 15, National Postal Museum, Washington, DC; Marylin Bender and Selig Altschul, *Chosen Instrument: Pan Am, Juan Trippe, the Rise and Fall of an American Entrepreneur* (New York, 1982), 256–57, 423–24.

102. Allene Talmey and Irving Penn, "Round the World Flight: 30 Days," *Vogue* (November 1947), 119, 120, 134–35, 199.

10. The Outer Limits

1. Colin Burgess and Chris Dubbs, *Animals in Space: From Research Rockets to the Space Shuttle* (Chichester, UK, 2007), 160–61.

2. Isaac Newton, *A Treatise of the System of the World* (London, 1728), v, 5–8; Frank H. Winter, *Rockets into Space* (Cambridge, MA, 1990); Rex Hall and David J. Shayler, *The Rocket Men: Vostok and Voskhod, the First Soviet Manned Spaceflights* (London, 2001), 3; Frank Winter, *Prelude to the Space Age: The Rocket Societies, 1924–1940* (Washington, DC, 1983); Asif Siddiqi, *The Red Rockets' Glare: Spaceflight and the Soviet Imagination, 1857–1957* (New York, 2010), esp. 18–30, 371.

3. J. D. Bernal, *The World, the Flesh, and the Devil: An Enquiry into the Future of the Three Enemies of the Rational Soul,* 2nd ed. (Bloomington, IN, 1969), 18–28, 33–47, 61; Roger D. Launius and Howard E. McCurdy, *Robots in Space: Technology, Evolution, and Interplanetary Travel* (Baltimore, 2008), xvii, and passim.

4. Isaac P. Abramov, et al., *Russian Spacesuits* (London, 2003), 5–9; Stanley R. Mohler and Bobby H. Johnson, *Wiley Post, His Winnie Mae, and the World's First Pressure Suit* (Washington, DC, 1971), 71–106.

5. Asif A. Siddiqi, *Challenge to Apollo: The Soviet Union and the Space Race, 1945–1974* (Washington, DC, 2000), 24, 57.

6. Arthur C. Clarke, *Wireless World,* 51 (February 1945), 58; Arthur C. Clarke, "Extra-Terrestrial Relays: Can Rocket Stations Give World-wide Radio Coverage?" ibid. (October 1945), 305–08.

7. Fae L. Korsmo, "The Birth of the International Geophysical Year," *Physics Today,* 60 (July 2007), 38–43; Siddiqi, *Challenge to Apollo,* 145–46; Siddiqi, *Red Rockets' Glare,* 335.

8. Siddiqi, *Challenge to Apollo,* 150, 163.

9. Ibid., 168; Hall and Shayler, *Rocket Men,* 62–63; Siddiqi, *Red Rockets' Glare,* 359.

10. Siddiqi, *Challenge to Apollo,* 165; Siddiqi, *Red Rockets' Glare,* 337–38, 342–43.

11. Walter A. McDougall, . . . *the Heavens and the Earth: A Political History of the Space Age* (New York, 1985), 141–56; Gretchen J. Van Dyke, "Sputnik: A Political Symbol and Tool in 1960 Campaign Politics," in *Reconsidering Sputnik,* ed. Roger D. Launius, John M. Logsdon, and Robert W. Smith (Amsterdam, 2000), 365–400.

12. *Broadcasting,* 16 (1939), 54.

13. Steve J. Heims, *John von Neumann and Norbert Wiener: From Mathematics to the Technologies of Life and Death* (Cambridge, MA, 1980); Donna J. Haraway, "A Cyborg Manifesto: Science, Technology, and Socialist-Feminism in the Late Twentieth Century," in Donna Haraway, *Simians, Cyborgs, and Women* (New

York, 1991); Geoffrey Bowker, "How to Be Universal: Some Cybernetic Strategies, 1943–70," *Social Studies of Science*, 23 (1993), 107–27; Peter Galison, "The Ontology of the Enemy: Norbert Wiener and the Cybernetic Vision," *Critical Inquiry*, 21 (1994), 228–66; Paul Edwards, *The Closed World: Computers and the Politics of Discourse in Cold War America* (Cambridge, MA, 1996); Slava Gerovitch, *From Newspeak to Cyberspeak: A History of Soviet Cybernetics* (Cambridge, MA, 2002).

14. Siddiqi, *Challenge to Apollo*, 172.

15. Ibid., 92–97; Hall and Shalyer, *Rocket Men*, 24–26.

16. Siddiqi, *Challenge to Apollo*, 173.

17. Ibid., 173, 174.

18. Hall and Shayler, *Rocket Men*, 66.

19. Burgess and Dubbs, *Animals in Space*, jacket and p. 153.

20. Many historical monographs published before 2002 did not accurately describe Laika's experience. The fuller story was finally related in a monograph, Burgess and Dubbs, *Animals in Space*, 163–65, and a graphic novel: Nick Abadzis, *Laika* (London, 2007).

21. *Sydney Morning Herald*, Dec. 21, 1957 (Ginza decoration); "Japan Reveling 12 Days for 1958," *New York Times*, Jan. 1, 1958; Yuri Gagarin, *Road to the Stars: Notes by Soviet Cosmonaut No. 1*, trans. G. Hanna and D. Myshne (Moscow, [1962]), 108; Viktor Pelevin, *Omon Ra*, trans. Andrew Bromfield (London, 1994; Russian ed., 1992); Jeanette Winterson, *Weight* (New York, 2005), 121–27; Abadzis, *Laika;* Lasse Hallström, dir., *Mitt liv som hund/My Life as a Dog* (1985); David Hoffman, dir., *Sputnik Mania* (2007).

22. Burgess and Dubbs, *Animals in Space*, 259–69.

23. Jamie Doran and Piers Bizony, *Starman: The Truth Behind the Legend of Yuri Gagarin* (London, 1998), 39–43.

24. "Explorer I Finds Cosmic Ray Spurt," *New York Times*, Mar. 7, 1958.

25. George H. Ludwig, "Cosmic-Ray Instrumentation in the First U.S. Earth Satellite," *Review of Scientific Instruments*, 30 (1959), 223–29.

26. M. I. Glassner, "The Frontiers of Earth—and of Political Geography: The Sea, Antarctica and Outer Space," *Political Geography Quarterly*, 10 (1991), 422–37.

27. Edward L. Beach and J. Baylor Roberts, "*Triton* Follows Magellan's Wake," *National Geographic*, 118 (November 1960), 585–615; Edward L. Beach, *Around the World Submerged: The Voyage of the Triton* (New York, 1962).

28. Beach, *Around the World Submerged*, 5–6, 45, 52, 100–02, 114, 237, 252.

29. Ibid., 118, 158.

30. Ibid., 45, 55, 91, 116, 215, 217.

31. Siddiqi, *Challenge to Apollo*, 252.

32. Ibid., 253–54, 265; Hall and Shayler, *Rocket Men*, 123–24.

33. Gagarin, *Road to the Stars*, 65; Siddiqi, *Challenge to Apollo*, 244; M. Scott Carpenter, L. Gordon Cooper, Jr., John H. Glenn, Jr., Virgil I. Grissom, Walter M. Schirra, Jr., Alan B. Shepard, Jr., and Donald K. Slayton, *We Seven, by the Astronauts Themselves* (New York, 1962), 7–9.

34. Hall and Shayler, *Rocket Men*, 91–94, 97; *Gherman Titov, First Man to Spend a*

Day in Space: The Soviet Cosmonaut's Autobiography as Told to Pavel Barashev and Yuri Dokuchayev (New York, 1962), 71–72, 79, 89; Gherman Titov and Martin Caidin, *I Am Eagle!: Based on Interviews with Wilfred Burchett and Anthony Purdy* (New York, 1962), 81; Gagarin, *Road to the Stars*, 138; Carpenter, et al., *We Seven*, 25.

35. Siddiqi, *Challenge to Apollo*, 266; Hall and Shayler, *Rocket Men*, 127.

36. Siddiqi, *Challenge to Apollo*, 273–74; Gagarin, *Road to the Stars*, 99–100, 141–42.

37. Hall and Shayler, *Rocket Men*, 152.

38. Joseph L. Zygielbaum, trans., "The First Man in Space. Soviet Radio and Newspaper Reports on the Flight of the Spaceship, Vostok," Jet Propulsion Laboratory, Pasadena, CA, May 1, 1961, 1, 3, 5, 6, 7.

39. Ibid., 6, 7, 10, 11.

40. Ibid., 1, 17, 18; Siddiqi, *Challenge to Apollo*, 280, 283; Hall and Shayler, *Rocket Men*, 155; Gagarin, *Road to the Stars* 160.

41. V. Belyaov, et al., eds., "Yuriy Gagarin's Star Voyages: Documents . . . ," *Izvestiia TsK KPSS [Political Archives of the Soviet Union]*, 5 (1991), 110–11, 112, 115. Trans. by Gregory Afinogenov.

42. Zygielbaum, "First Man in Space," 15.

43. L. Lebedev, B. Lyk'yanov, and A. Romanov, *Sons of the Blue Planet* (1971), trans. Prema Pande (Washington, DC, 1973), 13; Frantz Fanon, *The Wretched of the Earth* (1961), trans. Richard Philcox, comm. Jean-Paul Sartre and Homi K. Bhabha (New York, 2004), 61.

44. Hall and Shayler, *Rocket Men*, 162–73.

45. *Gherman Titov*, 4, 92, 102.

46. Ibid., 87, 100, 110; Titov and Caidin, *I Am Eagle*, 171.

47. J. R. McNeill, "The Environment, Environmentalism, and International Society in the Long 1970s," in *The Shock of the Global: The 1970s in Perspective*, ed. Niall Ferguson, et al. (Cambridge, MA, 2010), 275.

48. Launius and McCurdy, *Robots in Space*, 64–65, 220–21.

49. Carpenter, et al., *We Seven*; Gagarin, *Road to the Stars*, 131.

50. Carpenter, et al., *We Seven*, 104, 110, 181.

51. Ibid., 27, 106, 243, 261–62, 290; Hall and Shayler, *Rocket Men*, 95, 99; Titov and Caidin, *I Am Eagle*, 162.

52. Carpenter, et al., *We Seven*, 198, 295, 324, 346; see also John M. Logsdon with Roger D. Launius, *Exploring the Unknown: Selected Documents in the History of the U.S. Civil Space Program*. Vol. VII: *Human Spaceflight: Projects Mercury, Gemini, and Apollo* (Washington, DC, 2008), 226, 231.

53. *Oxford English Dictionary*, s.v., "cyborg"; Launius and McCurdy, *Robots in Space*, 198.

54. Siddiqi, *Challenge to Apollo*, 455.

55. Carpenter, et al., *We Seven*, 26; Hall and Shayler, *Rocket Men*, xxx; Zygielbaum, "First Man in Space," 5, 7.

56. Siddiqi, *Challenge to Apollo*, 250, 251; Lebedev, Lyk'yanov, and Romanov, *Sons of the Blue Planet*, 4.

57. Carpenter, et al., *We Seven*, 153–55, 227, 251, 323.

58. Virgil "Gus" Grissom, *Gemini: A Personal Account of Man's Venture into Space* (New York, 1968), 113; Robert Poole, *Earthrise: How Man First Saw the Earth* (New Haven, CT, 2008), 27.

59. http://www.jfklibrary.org/Research/Ready-Reference/JFK-Speeches/Address -at-Rice-University-on-the-Nations-Space-Effort-September-12-1962.aspx.

60. Gagarin, *Road to the Stars*, 90; transcript of BBC interview with Elena Gagarin, Radio 4 Random Edition (Apr. 11, 2011); Carpenter, et al., *We Seven*, 51, 69.

61. "Major Titov's Flight—and the Future," *New Scientist*, 247 (Aug. 10, 1961), 320; Gagarin, *Road to the Stars*, 154; Titov and Caidin, *I Am Eagle*, 89.

62. Gagarin, *Road to the Stars*, 91; Titov and Caidin, *I Am Eagle*, 89.

63. Siddiqi, *Challenge to Apollo*, 263, 270; Carpenter, et al., *We Seven*, 78, 261, 263, 302.

64. Stuart Banner, *Who Owns the Sky?: The Struggle to Control Airspace from the Wright Brothers On* (Cambridge, MA, 2008), 268, 278–80.

65. Arthur H. Dean, "The Second Geneva Conference on the Law of the Sea: The Fight for Freedom of the Seas," *American Journal of International Law*, 54 (1960), 751–89; Banner, *Who Owns the Sky?*, 284, 285.

66. Dean, "Law of the Sea," 753, 762–70; Peter Redfield, "The Half-Life of Empire in Outer Space," *Social Studies of Science*, 32 (2002), 791–825.

67. Siddiqi, *Challenge to Apollo*, 580–89.

68. Ibid., 293–94.

69. Ibid., 369–73; Hall and Shayler, *Rocket Men*, 196–214.

70. Siddiqi, *Challenge to Apollo*, 523–24; Hall and Shayler, *Rocket Men*, 260–62.

71. Siddiqi, *Challenge to Apollo*, 725, 728–29.

72. Henry R. Hertzfeld and Ray A. Williamson, "The Social and Economic Impact of Earth Observing Satellites," *Societal Impact of Spaceflight*, ed. Steven J. Dick and Roger D. Launius (Washington, DC, 2007), 238–40; J. R. McNeill, "The Environment, Environmentalism, and International Society in the Long 1970s," in *The Shock of the Global*, 264; Poole, *Earthrise*, 64–65.

73. The event remains understudied, but see Wikipedia, s.v., "*Our World* (TV special)" and embedded links.

74. Marshall McLuhan, *Understanding Media: The Extensions of Man* (New York, 1964), 3, 343.

75. Narender K. Sehgal, "Soviet Rocket puts India in Space Club," *Nature*, 255 (May 8, 1975), 98–99; Narender K. Sehgal, "Indian Diary," ibid., 256 (July 31, 1975), 361.

76. John Clute and Peter Nicholls, eds., *The Encyclopedia of Science Fiction* (London, 1999), s.v., *Astounding, Orbit, Star Science Fiction Stories*.

77. Stanislaw Lem, *Solaris*, trans. Joanna Kilmartin and Steve Cox (New York, 1970), 72.

78. Arthur C. Clarke, "The Sentinel," in *The Collected Stories of Arthur C. Clarke* (New York, 2000), 301–08.

79. Dwight Steven-Boniecki, *Live TV from the Moon* (Burlington, ON, 2010), 34, 55–78, 107.

80. Denis Cosgrove, "Contested Global Visions: One-World, Whole-Earth, and

the Apollo Space Photographs," *Annals of the Association of American Geographers*, 84 (1994), 270–94; Denis Cosgrove, *Apollo's Eye: A Cartographic Genealogy of the Earth in the Western Imagination* (Baltimore, 2001); Poole, *Earthrise*; Andrew Chaikin, "Live from the Moon: The Societal Impact of Apollo," *Societal Impact of Spaceflight*, 53–66.

81. J. R. McNeill, "The Environment, Environmentalism, and International Society," 263–78; Lebedev, Lyk'yanov, and Romanov, *Sons of the Blue Planet*, 98.

82. R. Buckminster Fuller, *Operating Manual for Spaceship Earth* (Carbondale, IL, 1969), 46, 52–53, 77, 87; Peder Anker, *From Bauhaus to Ecohouse: A History of Ecological Design* (Baton Rouge, LA, 2010), 68–112.

83. "Million Milers of the Air," *Popular Mechanics*, 64 (August 1935), 210; press releases for July 11, 1967, Box 1, folder 22, and Box 383, folder 26, Accession I, Pan American World Airways Records, University of Miami.

11. Army and Navy Surplus

1. "Rounding Up Stars in 80 Ways," *Life*, Oct. 22, 1956, 87–92; "Around the World in 80 Days," *Variety*, Oct. 23, 1956; "Mammoth Show," *New York Times*, Oct. 18, 1956.

2. Thomas C. Renzi, *Jules Verne on Film: A Filmography of the Cinematic Adaptations of His Works, 1902 through 1997* (Jefferson, NC, 1998), 20; Orson Welles and Peter Bogdanovich, *This Is Orson Welles*, ed. Jonathan Rosenbaum (New York, 1992), 346, 396; Chuck Berg and Tom Erskine, eds., *The Encyclopedia of Orson Welles* (New York, 2003), 5–6; Simon Callow, *Orson Welles: Hello Americans* (London, 2006), 280–81.

3. Callow, *Hello Americans*, 284, 287, 292–98; Berg and Erskine, *Encyclopedia of Orson Welles*, 5–6; *Boston Herald,* Apr. 29, 1946; *Boston Evening American*, Apr. 29, 1946; *Philadelphia Daily News*, May 15, 1946; *Philadelphia Record*, May 15, 1946.

4. Callow, *Hello Americans*, 302, 303–06, 321–22; Berg and Erskine, *Encyclopedia of Orson Welles*, 5–6; Welles and Bogdanovich, *This Is Orson Welles*, 394–96; *The New Yorker*, June 8, 1946, 48, 50; Orson Welles, "Around the World in 80 Days," stage directions and lyrics for Fogg song in Act I, production script (c. 1945/46), Houghton Library, Harvard University.

5. Callow, *Hello Americans*, 290; Bogdanovich, *This Is Orson Welles*, 402; *Don't Tread on Me: The Selected Letters of S. J. Perelman*, ed. Prudence Crowther (New York, 1987), 178–79.

6. D. A. Farnie, *East and West of Suez: The Suez Canal in History, 1854–1956* (Oxford, 1969), 718–45; Wm. Roger Louis and Roger Owen, eds., *Suez 1956: The Crisis and Its Consequences* (Oxford, 1989).

7. John Darwin, *Britain and Decolonisation: Retreat from Empire in the Post-war World* (London, 1988); C. N. Hill, *A Vertical Empire: The History of the UK Rocket and Space Programme, 1950–1971* (London, 2001).

8. Yves Michaud, *Qu'est-ce que la globalisation?* (Paris, 2004), 111.

9. Clive Ponting, *A New Green History of the World: The Environment and the Collapse of Great Civilizations,* rev. ed. (New York, 2007), 332–33; J. R. McNeill, "The En-

vironment, Environmentalism, and International Society in the Long 1970s,"
in *The Shock of the Global: The 1970s in Perspective,* ed. Niall Ferguson, et al.,
(Cambridge, MA, 2010), 271.

10. [John] Lennox Cook, *The World Before Us* (London, 1955), [7], 14–15, 256.

11. Ibid., 9, 170, 134.

12. "H.R.H. The Prince Philip, Duke of Edinburgh Introduces to Members the
Narrative of His Round-the-World Tour," *National Geographic,* 112 (November
1957), 583–626; Alistair Boyd, *Royal Challenge Accepted: Around the World on Five
Pounds* (London, 1962; rpt. 2006), 19–21, 29, 45, 105–06.

13. Boyd, *Royal Challenge Accepted,* 102, 107, 111, 155, 167, 182, 192, 222.

14. Wendy Myers, *Seven League Boots: The Story of My Seven-Year Hitch-Hike round
the World* (London, 1969), 8, 21, 28, 48, 57, 66, 192.

15. Ibid., 22, 32, 43, 70–72, 212.

16. 'Adnān Ḥusnī Tallū, *Al-Quwah wa-al-iqtidār fī buḥūr al-asfār* (Damascus, Syria,
1993), 39, 42, 50–51, 131; translations by Dzavid Dzanic.

17. Anthony Sampson, *Empires of the Sky: The Politics, Contests and Cartels of World
Airlines* (New York, 1984), 87–89; John Stroud, "Airliner Evolution in the
Postwar Era," in *Modern Air Transport: Worldwide Air Transport from 1945 to the
Present,* ed. Philip Jarrett (London, 2000), 11–34.

18. John Gunn, *High Corridors: Qantas, 1954–1970* (St. Lucia, Australia, 1988),
109, 110; "World Flight Completed," *Sydney Morning Herald,* Jan. 9, 1958;
Qantas Empire Airways: A Gazette, vol. 24, no. 1 (January 1958), 1–3; ibid., no. 2
(February 1958), 27; Qantas timetables for 1958, the "Kangaroo Route" and
Southern Cross Route, Eastbound and Westbound, Qantas Heritage Collection, Qantas Domestic Terminal T3, Mascot, New South Wales; fares for 1958
courtesy of Des Sullivan, Manager, Qantas Heritage Collection.

19. Itinerary for first PAA "around-the-world" flight, 6/17–30/1947, Box 1, folder
22, Accession I, Pan American World Airways Records, University of Miami
(Pan Am Recs.); "Father Knickerbocker," Box 204, folder 18, Accession I, ibid.;
Pan Am application to Civil Aeronautics Board, Aug. 1, 1950, Box 10, folder 1,
Accession I, ibid.; memoranda re: Qantas, Box 900, folder 6, Accession II,
ibid.; Robert Gandt, *Skygods: The Fall of Pan Am* (New York, 1995), 179.

20. *Qantas Empire Airways: A Gazette,* vol. 24, no. 1 (January 1958), 3; ibid., vol. 24
no. 2 (February 1958), 18, 27.

21. Wilmot Hudson-Fysh, *Visit to Moscow* (Sydney, 1958), 1–2, 4–6; "Qantas Chief
Sees Lag in Tourist Drive," *Sydney Morning Herald,* Dec. 4, 1958.

22. "Jets to Give Globe a Tighter Girdle," *New York World-Telegram and Sun,* Mar.
6, 1959; Gunn, *High Corridors,* 153–55; R. E. G. Davies, *A History of the World's
Airlines* (London, 1964), 479; Wilmot Hudson-Fysh, *Airlanes* (October 1959);
Sampson, *Empires of the Sky,* 105–06; *Qantas Empire Airways: A Gazette,* 25, no. 9
(September 1959), 2–3, 6.

23. Wolfgang Langewiesche, "The New Jet Liners," *Harper's Magazine,* June 1958,
50; *Qantas Empire Airways: A Gazette,* vol. 24, no. 10 (October 1958), inside
cover; ibid., 25, no. 3 (March 1959), 20; ibid., no. 9 (September 1959), 5.

24. *Qantas Empire Airways: A Gazette,* vol. 25, no. 12 (December 1959), 10–11.

25. Gunn, *High Corridors*, 110; Andrew Brookes, *Flights to Disaster* (Shepperton, UK, 1996), 7.

26. Sampson, *Empires of the Sky*, 83; Davies, *World's Airlines*, 425–37; "Pan Am to Fly New World Service," *New York Times*, Oct. 9, 1959; Press releases for July 11, 1967, Box 1, folder 22, and Box 383, folder 26, Accession I, Pan Am Recs.; Simon Bennett, "Victim of History? The Impact of Pan Am's 'Imperialistic' Past on Its Capacity to Adjust to Deregulation," *Risk Management*, 4 (2002), 23–29.

27. Bill Cormack, *A History of Holidays, 1812–1990* (London, 1998), 83.

28. Beatrice Cobb, *On a Clipper Trip around the World* (Morganton, NC, 1948), 82; E[nrique] Aguilar, *114 Horas de vuelo alrededor del mundo* (Barcelona, 1957), 128–29; *Qantas Empire Airways: A Gazette*, 25, no. 9 (September 1959), 8, 10.

29. Chester L. Cooper, "Fly Now, Pay Later," *Foreign Policy*, 20 (1975), 165–69; Anant Gopal Sheorey, *This Great Small World* (Bombay, 1969), 48, 49, 184, 185.

30. Aguilar, *114 Horas de vuelo alrededor del mundo*, 129.

31. Hieronymus Fenyvessy, *My Dream Came True: A Journey round the World*, trans. Gregor Kirstein (Cologne & Detroit, [1958]), 9–15, 32, 47, 54, 93, 111; Anīs Manṣūr, *Ḥawla al-ʿālam fī 200 yawm (Around the World in 200 Days)* (Cairo, 1964), 12, 329–30; translations by Dzavid Dzanic.

32. Diosdado Macapagal, *An Asian Looks at South America* (Manila, 1966); *Billboard*, Sept. 12, 1960, 9 (Israel Phil.); *Jet*, Dec. 21, 1961, 62 (Quincy Jones); *Black Belt* (January 1969), 38–39 (Ito).

33. Arnold J. Toynbee, *East to West: A Journey round the World* (New York, 1958), 219, 220; Manṣūr, *Ḥawla al-ʿālam fī 200 yawm*, 19, translation by Dzavid Dzanic.

34. Aguilar, *114 Horas de vuelo alrededor del mundo*, 86, 199; Ira Wolfert, "What It's Like to Fly around the World," *Travel*, Nov. 1967, 220, 226; "Global Flight in Day," *San Francisco Chronicle*, Jan. 8, 1971; "A Once in a Lifetime Experience," *Fort Lauderdale News*, July/Aug. 1981.

35. S. J. Perelman, excerpts from *Westward Ha!* (1948) in *The Most of S. J. Perelman* (New York, 1980), 310, 312, 395; Perelman, *Eastward Ha!* (New York, 1977), 11–12, 91, 122–25.

36. Bruce Brown, dir., *The Endless Summer* (1966).

37. Harry Pidgeon, *Around the World Single-Handed: The Cruise of the "Islander"* (New York, 1932), 4–10, 190, 232; Alain Gerbault, *In Quest of the Sun: The Journal of the "Firecrest"* (London, 1937), 34–35, 38–39, 47, 57, 73, 135, 201, 211, 229, 263; Louis Bernicot, *The Voyage of Anahita Single-Handed round the World*, trans. Edward Allcard (London, 1953), 13–15, 38, 100, 145, 184. See also Donald Holm, *The Circumnavigators: Small Boat Voyagers of Modern Times* (London, 1975).

38. Holm, *Circumnavigators*, 179–84, 346–47; Jon Adams, *The Cruise of the Jest* (Los Gatos, CA, 2007).

39. Robin Lee Graham, "A Teen-Ager Sails the World Alone," *National Geographic*, 134 (October 1968), 449.

40. Ibid., 449, 450, 454, 455, 457; Robin Lee Graham with Derek L. T. Gill, *Dove* (New York, 1972), 16.

41. Graham, "Teen-Ager Sails the World," 457, 458, 482.

42. Francis Chichester, *Gipsy Moth Circles the World* (New York, 1967), 1; John Rowland, *Lone Adventurer: The Story of Sir Francis Chichester* (Worcester, UK, 1968), 35–125.

43. Chichester, *Gipsy Moth*, 2, 3; Francis Chichester, *Along the Clipper Way* (New York, 1966).

44. Chichester, *Gipsy Moth*, 7.

45. Ibid., xv, 19; Harold Evans, *My Paper Chase: True Stories of Vanished Times, an Autobiography* (London, 2009), 280–81.

46. Chichester, *Gipsy Moth*, 27, 41, 46, 47, 70, 72, 73, 81, 84, 88.

47. Ibid., 108, 110, 115, 116, 130, 137.

48. Ibid., 144–46; Gerbault, *In Quest of the Sun*, 11.

49. Chichester, *Gipsy Moth*, 170, 191.

50. Ibid., 199, 222, 223–24.

51. Ibid., 29, 70, 224, 226, 258.

52. Nigel Barley, *Not a Hazardous Sport* (London, 1988); Patrick Laviolette, *Extreme Landscapes of Leisure: Not a Hap-Hazardous Sport* (Farnham, UK, 2011).

53. Robin Knox-Johnston, *A World of My Own: The First Ever Non-Stop Solo round the World Voyage* (London, 2010), iv–v, 1.

54. Bernard Moitessier, *The Long Way*, trans. William Rodarmor (Dobbs Ferry, NY, 1995), 5, 7; Nicholas Tomalin and Ron Hall, *The Strange Last Voyage of Donald Crowhurst* (New York, 1995), 19–20; Knox-Johnston, *World of My Own*, 2.

55. Moitessier, *Long Way*, 60, 72, 74–75, 76, 77, 78–79, 89, 114, 117.

56. Ibid., 151, 163.

57. Knox-Johnston, *World of My Own*, 63, 122, 174.

58. Ibid., 142.

59. Ibid., 25; Moitessier, *Long Way*, 32.

60. Knox-Johnston, *World of My Own*, 70.

61. Ibid., 84, 85, 183, 201.

62. Ibid., 239; Moitessier, *Long Way*, 177, 181.

63. Knox-Johnston, *World of My Own*, 16; Moitessier, *Long Way*, 168; Tomalin and Hall, *Last Voyage of Donald Crowhurst*, 33–70.

64. Tomalin and Hall, *Last Voyage of Donald Crowhurst*, 71–205, 235–45.

65. Moitessier, *Long Way*, 72; Tomalin and Hall, *Last Voyage of Donald Crowhurst*, 206–34.

66. Graham, "Teen-Ager Sails the World," 467; Robin Lee Graham, "World-roaming Teen-ager Sails On," *National Geographic*, 135 (April 1969), 475–76, 484–88; Graham, *Dove*, 156.

67. Graham, *Dove*, 71, 92, 156; Robin Lee Graham, "Robin Sails Home," *National Geographic*, 138 (October 1970), 527, 531. (Graham stressed the tension with his father in his book, not in *National Geographic*.)

68. Graham, *Dove*, 140–41, 180.

69. Ibid., 155.

70. See http://www.sailspeedrecords.com and http://www.joshuaslocumsociety intl.org/solo/solotable.htm.

71. Philip José Farmer, *The Other Log of Phileas Fogg* (New York, 1973), cover, 82.

72. Gandt, *Skygods*, 8–9.

73. J. C. Hafele and Richard E. Keating, "Around-the-World Atomic Clocks: Predicted RelativisticTime Gains," *Science*, n.s., vol. 177, no. 4044 (July 14, 1972), 166–68; Hafele and Keating, "Around-the-World Atomic Clocks: Observed Relativistic Time Gains," ibid., 168–70.

74. "Crossing Aviation's Almost Forgotten Frontier," *Flying Magazine*, 78 (June 1966), 101–03; polar route brochure, October 1977, Box 1, folder 23, Accession I, Pan Am Recs.; Ivor Davis, "Around the World in 48 Hours," *Aircraft*, Dec. 1977, Box 307, folder 12, ibid.

75. Ian Goold, "The Modern Jet Airliner—the Trailblazers," in *Modern Air Transport*, 179, 182–85; Francis Spufford, *Backroom Boys: The Secret Return of the British Boffin* (London, 2003), 39–70.

76. "Concorde around the World. August 1987," pp. [2–3], [4], British Airways Archive (BAA), Waterside, Harmondsworth, UK; "Concorde World Air Cruise 1988," pp. 3–5, 19, BAA; "Concorde Cabin Crew Brief—Round the World Cruise, October, 1989," 1, 5, BAA; *British Airways News*, Oct. 24, 1986, 2, BAA; *BA News*, Oct. 31, 1986, 1; *BA News*, Apr. 28, 1989, 12.

77. http://www.fai.org/records.

78. Robert J. Millichap, "Airline Markets and Regulation," in *Modern Air Transport*, 35, 43–52; Mike Hirst, "The Modern Jet Airliner—the Trendsetters," ibid., 188–96; Bennett, "Victim of History?" 29–35; Sampson, *Empires of the Sky*, 110–14, 126–32.

79. Sampson, *Empires of the Sky*, 113–14.

80. "The Great World Cruise of 1955," Cunard Steamship Co., Folder 109, Kemble Maritime Ephemera, Huntington Library (KME); "Norwegian America Line Carefree Cruises, 1965–1966," Norske Amerikalinje, Folder 3, KME; "For a Perfect Vacation, How does this Sound?" Oct. 1957, Folder 2, American President Lines, KME; pamphlet for a 1961 tour by steamship, Donald L. Ferguson Cruises, Folder 1, KME; schedule for 1972–73, Folder 21, American President Lines, KME; "Around the World in 80 Days," 1962, Folder 84, Holland-Amerika, KME; "14th Holland America World Cruise," 1973, ibid., KME; Heddy Kraemer, *More Time Than Money: A Retired Couple Travel around the World for Twenty Months on Freight Ships* (Brooklyn, NY, 1963), 5–6, 9. Some shippers continued to use both canals; see the Svenska Amerika Linien advertisements for 1956, 1959, 1960, Folders 38 and 39, KME.

81. Bruce Watson, "World's Unlikeliest Bestseller," *Smithsonian*, Aug. 2005, 76–81.

82. David Kunst and Clinton Trowbridge, *The Man Who Walked around the World: A True Story* (New York, 1979), 11–17, 21, 23, 246.

83. Ibid., 47, 108–09, 164, 168, 178.

84. Ted Simon, *Jupiter's Travels: Four Years on One Motorbike* (London, 1979), 23, 153, 385.

85. Ibid., 14, 17, 22, 419.

86. Ibid., 71, 297.

87. Ibid., 22–23, 183, 404.

88. Ibid., 121, 178, 319.

89. Ibid., 17, 21, 24, 97, 111, 149–50, 182, 315.

90. Ibid., 272–73, 321.

91. Ranulph Fiennes, *To the Ends of the Earth: The Transglobe Expedition—the First Pole-to-Pole Circumnavigation of the Globe* (New York, 1983), maps between pp. 10 and 11, 15–16.

92. Ibid., 38, 61, 435–48.

93. Ibid., 24, 33, 124, 137, 138–39.

94. Nicholas Coleridge, *Around the World in 78 Days* (New York, 1985), 2–5.

95. Ibid., xii–xiii, 4, 82, 114, 124.

96. Jeffrey M. Pilcher, *Cantinflas and the Chaos of Mexican Modernity* (Wilmington, DE, 2001), 163–75; "Michael Todd's 80 Days Is Grand Entertainment," *The Times of India*, Sept. 29, 1960.

97. Farmer, *Other Log of Phileas Fogg*, 255, 260.

98. Michael Palin, *Around the World in 80 Days with Michael Palin* (London, 1989), 56.

99. Interview with Michael Palin, *Around the World in 80 Days* (London: BBC Video, 2007).

100. Palin, *Around the World in 80 Days*, 9.

101. Ibid., 46, 156, 176, 213.

102. Ibid., 256.

103. Ibid., 182, 185, 187, 236, 240.

104. "A Race Against Time," *The Times of India*, Aug. 16, 2001.

12: Magellan redivivus

1. António Lobo Antunes, *The Return of the Caravels*, trans. Gregory Babassa (New York, 2002), 87, 122, 168.

2. Michael Russell Rip and James M. Hasik, *The Precision Revolution: GPS and the Future of Aerial Warfare* (Annapolis, MD, 2002), 69–70; David E. Hoffman, *The Dead Hand: The Untold Story of the Cold War Arms Race and Its Dangerous Legacy* (New York, 2009); Andrew E. Kramer, "Seeking a Fix, by Russian Satellite," *New York Times*, Apr. 4, 2007.

3. Rigel Crockett, *Fair Wind and Plenty of It: A Modern-day Tall Ship Adventure* (Toronto, 2005), 383.

4. http://www.cunard.com/Destinations/World-Voyages-and-Exotic-Voyages/World-Voyage-Search. See also the private jet circumnavigations organized annually by National Geographic: http://www.nationalgeographicexpeditions.com/expeditions/around-the-world-jet-tour/detail.

5. Ellen MacArthur, *Taking On the World: A Sailor's Extraordinary Solo Race around the Globe* (Camden, ME, 2003), 312–13.

6. http://www.volvooceanrace.com/; Christopher Clarey, "In France, the Vendée Globe Race Gains Popularity," *New York Times*, Nov. 11, 2008.

7. http://www.voile.banquepopulaire.fr/Maxi-Trimaran-Banque-Populaire-V/Tentative-de-record-Trophee-Jules-Verne-2010-2011-2907.html.

8. Michael Palin, *Around the World in 80 Days with Michael Palin* (London, 1989), 73, 119.

9. Kay Cottee, *First Lady: A History-Making Solo Voyage around the World* (Melbourne, Australia, 1989), 58; Jesse Martin, *Lionheart: A Journey of the Human Spirit* (London, 2001), 86, 93–95; MacArthur, *Taking On the World*, 246.

10. Pete Goss, *Close to the Wind* (New York, 1998), 213, 217, 220–21.

11. Ibid., 235, 256, 260; http://www.petegoss.com/journey-to-date-vendee-globe.php.

12. Dan Bilefsky, "Dutch Court Blocks Girl, 13, from Sailing Solo around the World," *New York Times*, Aug. 29, 2009.

13. http://www.sailspeedrecords.com/round-the-world-non-stop.html; http://www .sailspeedrecords.com/historical-list-of-offshore-world-records.html; http:// www.sailspeedrecords.com/other-kinds-of-sailing-records.html.

14. Sophie Tedmanson, "Teenage round-world sailor Jessica Watson 'hasn't gone far enough,' " *The Times* (London), May 5, 2010; http://yachtpals.com/zac-sunder land-record-4178.

15. http://www.telegraph.co.uk/news/worldnews/europe/netherlands/9029662/ Teenage-sailor-Laura-Dekker-becomes-youngest-to-circumnavigate-the -globe.html.

16. "Knox-Johnston Quits the 'Ruined' Oceans," *Sunday Times*, Apr. 29, 2007.

17. http://www.cbs.com/shows/amazing_race; *The Amazing Race: The First Season*, episode 103, Paris (Hollywood, 2005).

18. Nicole Borie-Chalard and Yvon Borie, *Le nouveau Tour du monde de Jules Verne en 44 jours: 8 juillet–20 août 2000* ([Lissac], FR., 2000), 48, 49.

19. http://www.newmediastudies.com/art/mk.htm.

20. Kareem Abdulrahman, "Poets' Projects," review of Bakhtyar Ali, *Ghazalnus w Baghakani Khayal*, in the *Times Literary Supplement*, Dec. 5, 2008, 21.

21. Ted Simon, *Dreaming of Jupiter* (London, 2007), 10–11, 177; Ewan McGregor and Charley Boorman with Robert Uhlig, *Long Way Round: Chasing Shadows Across the World* (New York, 2004), 27, 238.

22. Simon, *Dreaming of Jupiter*, 8, 14, 18, 125, 263, 356, 377.

23. Ibid., 19, 128, 144, 178, 420, 424.

24. Elizabeth Pennisi, "Satellite Tracking Catches Sharks on the Move," *Science*, n.s., vol. 310, n. 5745 (Oct. 7, 2005), 32–33; John Roach, "Great White Breaks Distance, Speed Records for Sharks," http://news.nationalgeographic.com/ news/2005/10/1006_051006_shark_fastest.html.

25. http://www.circumnavigators.org/membership/.

26. http://www.bluemoment.com/seymour.html; Cottee, *First Lady*, 31, 79, 132, 135, 158, 213; MacArthur, *Taking On the World*, 240; Martin, *Lionheart*, 88, 103–04, 187.

27. http://www.sorcerer2expedition.org/version1/HTML/main.htm.

28. http://www.planetsolar.org/vision.html.

29. Ffyona Campbell, *The Whole Story: A Walk around the World* (London, 1996), 60, 272.

30. Colin Angus, *Beyond the Horizon: The Great Race to Finish the First Human-Powered Circumnavigation of the Planet* (Toronto, 2007), 3, 6, 11; http://www.nationalgeo

graphic.com/adventure/best-of-adventure-2007/achievements/colin-angus
-julie-wafaei.html. The Guinness authorities have since established guidelines
for human-propelled circumnavigations that did not apply to Angus-Wafaei;
see http://www.angusadventures.com/circumnavigations.html.

31. http://www.pbs.org/newshour/bb/transportation/may97/earhart_5-28a.html.

32. Carroll V. Glines, *Around the World in 175 Days: The First Round-the-World Flight*
    (Washington, DC, 2001), 172.

33. Bertrand Piccard, "Around at Last!," *National Geographic*, 196 (September
    1999), 32, 34, 35, 36, 37, 38, 39, 43.

34. "First in History; High Officials Greet the Plane as It Ends Hop at Fort
    Worth," *New York Times*, Mar. 3, 1949; Ross Perot, Jr., "Around the World by
    Helicopter," *Popular Mechanics*, July 1983, 73–75, 127–28; http://www.australian
    geographic.com.au/journal/on-this-day-dick-smiths-around-the-world-solo
    -flight.htm; http://www.polarfirst.com/index.php?option=com_frontpage&Ite
    mid=1; http://www.fai.org/records/; Jeana Yeager and Dick Rutan with Phil
    Patton, *Voyager* (New York, 1987).

35. Steve Fossett with Will Hasley, *Chasing the Wind: The Autobiography of Steve Fos-
    sett* (London, 2006), 163, 219, 228.

36. http://www.solarimpulse.com.

37. Richard Goldstein, "Walter M. Schirra Jr., an Original Astronaut, Dies at 84,"
    New York Times, May 4, 2007.

38. John M. Logsdon with Roger D. Launius, *Exploring the Unknown: Selected Docu-
    ments in the History of the U.S. Civil Space Program*. Vol. VII: *Human Spaceflight:
    Projects Mercury, Gemini, and Apollo* (Washington, DC, 2008), 752; Robert
    Zimmerman, *Leaving Earth: Space Stations, Rival Superpowers, and the Quest for
    Interplanetary Travel* (Washington, DC, 2003), 14–15; Howard E. McCurdy,
    The Space Station Decision: Incremental Politics and Technological Choice (Baltimore,
    1990).

39. Roger D. Launius and Howard E. McCurdy, *Robots in Space: Technology, Evolu-
    tion, and Interplanetary Travel* (Baltimore, 2008), 50–61.

40. Zimmerman, *Leaving Earth*, 48–80, 443; David Hitt, Owen Garriott, and Joe
    Kerwin, *Homesteading Space: The Skylab Story* (Lincoln, NE, 2008).

41. Michael Griffin, "Continuing the Voyage: The Spirit of Endeavour," speech
    to the Royal Society of London, Dec. 1, 2006, in *Leadership in Space: Selected
    Speeches of Michael Griffin* (Washington, DC, 2008), 41, 43.

42. http://www-pao.ksc.nasa.gov/shuttle/resources/orbiters/orbiters.htm.

43. Ibid.

44. Launius and McCurdy, *Robots in Space*, 133–34; Asif A. Siddiqi, *Challenge to
    Apollo: The Soviet Union and the Space Race, 1945–1974* (Washington, DC, 2000),
    766–68, 777–80; Zimmerman, *Leaving Earth*, 195, 373.

45. Zimmerman, *Leaving Earth*, 19, 96, 103, 146, 181, 282, 285, 439.

46. Rachel Nowak, "NASA Space Biology Program Shows Signs of Life," *Science*,
    n.s., vol. 268, no. 5210 (April 28, 1995), 497.

47. Launius and McCurdy, *Robots in Space*, 128–30.

48. See the ASE Web site at www.space-explorers.org.

49. Zimmerman, *Leaving Earth*, 172, 256; http://www.jsc.nasa.gov./Bios/htmlbios/al-saud.html.

50. Ibid., 292–96, 301–02, 319. See also Jeffrey Manber, *Selling Peace: Inside the Soviet Conspiracy That Transformed the U.S. Space Program* (Burlington, ON, 2009).

51. Information Office of the State Council of the People's Republic of China, *China's Space Travel* (Beijing, 2004), 83, 118, 157; http://www.spacex.com; "China's Shenzhou 9 spacecraft returns to Earth," *Guardian*, June 29, 2012.

52. Siddiqi, *Challenge to Apollo*, 777–81; Zimmerman, *Leaving Earth*, 35–46.

53. http://opensat.cc.

54. http://projectoscar.wordpress.com; http://www.amsat.org/amsat-new/About Amsat/amsat_history.php; http://www.cubesat.org/index.php/about-us/mission -statement.

55. Launius and McCurdy, *Robots in Space*, 85; Alistair Boyd, *Royal Challenge Accepted* (London, rprt. 2006), viii.

56. Haruki Murakami, *Sputnik Sweetheart* (1999), trans. Philip Gabriel (New York, 2001), 8, 63, 98, 117, 177, 179.

57. "Alan Bean's In-Flight Diary," in *Homesteading Space*, 474, 481, 488–89, 514. For an imaginative recent summary of the early Circumnavigator's Paradox, see Umberto Eco, *The Island of the Day Before*, trans. William Weaver (New York, 1994), 252–60.

58. Andrei Ujica, dir., *Out of the Present* (Chicago, 2001).

59. Rachel Donadio, "Interstellar Ramadan," *New York Times*, Dec. 9, 2007. Sultan Salman Abdulaziz al-Saud, who had gone into space in 1985 as a payload specialist on the shuttle *Discovery*, orbited during Ramadan, though this was not discussed at the time. See http://www.jsc.nasa.gov/Bios/htmlbios/al-saud .html.

60. http://science.nasa.gov/science-news/science-at-nasa/2002/08apr_atomic clock.

61. http://science.nasa.gov/missions/nimbus; Richard A. Kerr, "Has Stratospheric Ozone Started to Disappear?" *Science*, n.s., vol. 237, no. 4811 (July 10, 1987), 131–32; http://ozone.unep.org/new_site/en/index.php.

62. http://landsat.gsfc.nasa.gov; Siddiqi, *Challenge to Apollo*, 799–800.

63. Philip Taubman, "In Death of Spy Satellite Program, Lofty Plans and Unrealistic Bids," *New York Times*, Nov. 11, 2007; Kenneth Chang, "A NASA Satellite Will Track the Carbon Dioxide on Earth," *New York Times*, Feb. 22, 2009; http://oco.jpl.nasa.gov.

64. Zimmerman, *Leaving Earth*, 126; Ujica, *Out of the Present*.

65. Jonathan Lethem, *Chronic City* (New York, 2009), 68–72, 381.

66. Don Delillo, "Human Moments in World War III" (1983), in Delillo, *The Angel Esmeralda: Nine Stories* (New York, 2011), 25, 26, 44; Viktor Pelevin, *Omon Ra* (1992), trans. Andrew Bromfield (New York, 1994), 94; Lethem, *Chronic City*, 169.

67. William J. Broad, "Surveillance Is Suspected as Main Role of Spacecraft," *New York Times*, May 22, 2010; Kramer, "Seeking a Fix, by Russian Satellite."

68. J. E. S. Fawcett, *Outer Space: New Challenges to Law and Policy* (Oxford, 1984);

William J. Broad, "A Bailout That Could Be Practically Free (Yes, There's a Catch)," *New York Times*, Apr. 2, 2009; Marc Kaufman and Dafna Linzer, "China Criticized for Anti-Satellite Missile Test," *Washington Post*, Jan. 19, 2007.

69. "Spacemen are from Mars," *The Economist*, Sept. 29, 2007, 16.

70. http://hubblesite.org/the_telescope/hubble_essentials.

71. Dániel Apai, et al., "Lessons from a High-Impact Observatory: The Hubble Space Telescope's Science Productivity Between 1998 and 2008," *Publications of the Astronomical Society of the Pacific*, 122 (July 2010), 808–26.

72. http://hubblesite.org/newscenter/archive/releases/2005/04/image/a.

73. Fawcett, *Outer Space*, 32–36; Donald J. Kessler and Burton G. Cour-Palais, "Collision Frequency of Artificial Satellites: The Creation of a Debris Belt," *Journal of Geophysical Research,* 83, A6 (1978), 2637–46; Evan I. Schwartz, "The Looming Space Junk Crisis," *Wired*, June 2010, 174–80.

74. Michael Klesius, "Space Trash and Treasure," *National Geographic*, 211 (January 2007), 29; William J. Broad, "Debris Spews into Space in Collision of Satellites," *New York Times*, Feb. 11, 2009; Kenneth Chang, "Debris Gives Space Station Crew Members a 29,000-m.p.h. Close Call," *New York Times*, June 29, 2011.

75. http://www2.jpl.nasa.gov/magellan/; http://www.esa.int/esaSC/120382_index _0_m.html.

Illustration Credits

1. World map after Ptolemy, c. 1482. Courtesy of the John Carter Brown Library at Brown University.
2. Map of the "Moluccas" from Antonio Pigafetta, "Journal of Magellan's Voyage," c. 1525 (in French), MS 351. Courtesy of the Beinecke Rare Book and Manuscript Library, Yale University.
3. "Ambassador's Globe" (c. 1525, Nuremberg?). Courtesy of the Beinecke Rare Book and Manuscript Library, Yale University.
4. Sir Francis Drake, oil on panel, unknown artist, c. 1580, NPG 4032. © National Portrait Gallery, London.
5. *The Relation of a Wonderfull Voiage made by William Cornelison Schouten of Horne* (London, 1619), title page. By permission of the Huntington Library, San Marino, California.
6. Willem Janszoon Blaeu world map with portraits, c. 1618. Courtesy of the John Carter Brown Library at Brown University.
7. Herman Moll, *A Map of the World, Shewing the Course of Mr. Dampier's Voyage Round It: From 1679 to 1691* (London, 1697). Courtesy of the John Carter Brown Library at Brown University.
8. Terrestrial pocket globe (GLB0028), Nicolas Lane, London, gores produced in 1776; plates altered in 1779 to commemorate James Cook's death in Hawai'i. © National Maritime Museum, Greenwich, UK.
9. James Holman, *A Voyage round the World . . .*, 2 vols. (London, 1834–35), vol. I, frontis. Courtesy of Harvard College Library, Widener Library.
10. Léon Benett, frontispiece for Jules Verne, *Le Tour du monde en quatre-vingts jours* (Paris, 1873). Courtesy of Harvard College Library, Widener Library.
11. *Le Tour du monde en 80 jours d'après le roman de Jules Verne (Around the World in 80 Days after the Novel by Jules Verne)* (Paris: Société Française de Jeux et Jouets, c. 1915). Courtesy of Princeton University Library, Graphic Arts Collection.
12. Li Gui, *Huan you di qiu xin lu* (China, 1877), foldout map. Courtesy of Harvard-Yenching Library.
13. Cartoon from *New York World*, Jan. 26, 1890. Private Collection.
14. Paul S. Junkin, *A Cruise around the World . . .* (Creston, IA, 1910?), photograph facing p. 40. By permission of the Huntington Library, San Marino, California.
15. Hamburg-Amerika Line brochure, "Around the World—110 Days—SS *Victoria Luise*," 1912. By permission of the Huntington Library (Kemble Maritime Ephemera), San Marino, California.

16. International Mercantile Marine Co., "Dancing Round the World," 1920s. Courtesy of the Huntington Library (Kemble Maritime Ephemera), San Marino, California.

17. Autographed photograph of "F. K. Koenig Warthausen" with Tanim (later renamed Felix), c. 1930. Private Collection.

18. John Henry Mears, *Racing the Moon (and Winning)* . . . (New York, 1928), photograph facing p. 96. Private Collection.

19. "Laika—First Space Traveler," Soviet postcard, 1958. Private Collection.

20. Photograph of Explorer 1 held up by (*left to right*), William H. Pickering, James Van Allen, and Werner von Braun, 1958. Courtesy of NASA.

21. *Gherman Titov, First Man to Spend a Day in Space* (New York, 1962), frontispiece. Courtesy of Harvard College Library, Cabot Science Library.

22. Philatelic first cover, Qantas Inaugural Round-the-World Flight, Jan. 14, 1958. Private Collection.

Index

Page numbers in *italics* refer to maps.

About the Author

JOYCE E. CHAPLIN is the James Duncan Phillips Professor of Early American History at Harvard University. She has taught at five different universities on two continents and an island and in a maritime studies program on the Atlantic Ocean. She received her B.A. from Northwestern University and her M.A. and Ph.D. from the Johns Hopkins University and was awarded a Fulbright Grant for study in the United Kingdom. She is the author of four previous books of nonfiction, including *The First Scientific American: Benjamin Franklin and the Pursuit of Genius* (2006), a finalist for the *Los Angeles Times* Book Prize and winner of the Annibel Jenkins Biography Prize. She lives in Cambridge, Massachusetts.

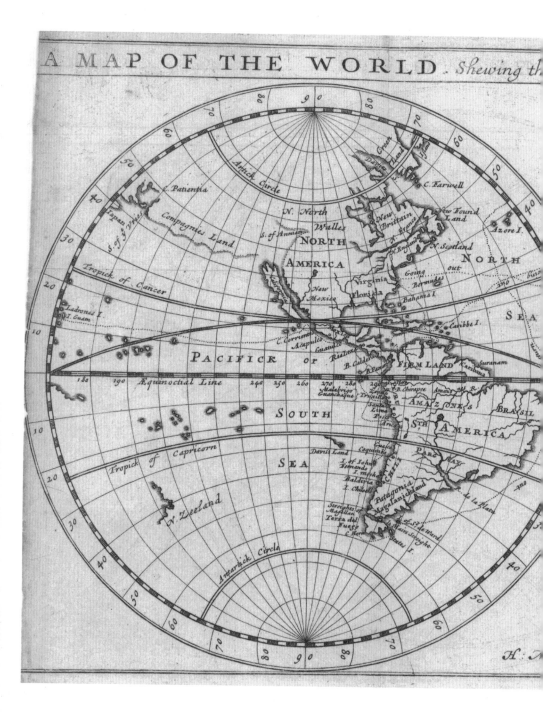

C. Patientia

Japan of y Dutes

Compagnies Land

S. of Annian

N. North Walles

NORTH AMERICA

Davis Land

Groin Land

C. Farwell

New Brittain

N. France

New Found Land

Azore I.

N. England

N. Scotland

NORTH

Going out

Virginia Florida

Bermudas

Bahama I.

SEA

Antick Circle

Tropick of Cancer

Ladrones I.
I. Guam

New Mexico

C. Corrientes

Acapulco

Guatimala

R. de Leon

Caribbe I.

B. Caldera

R. Peru

FIRM LAND

Carthagena

Suranam

PACIFICK or

Æquanoctial Line

180   190   240   250   260   270   280   290   300

Malabrigo
Guanchaque

B. Chirapoto

Trucillo

Amazon R.

Santa

Lima

Paita

AMAZONES

BRASIL

SOUTH

Sth AMERICA

Coquimbo

Davis Land

I. of John
Fernando

I. micha

Baldivia

I. Chiloe

Coquimbo

Chili

PARA GUAY

Tropick of Capricorn

SEA

Patagonia
Magalanickland

R. de la Plata

N. Zeeland

Streights of
Magellan

Terra del
Fuego

C. Her

C. S. de Wind

Maire Streight

Strate I.

Artartick Circle

H: N:

ML